T0171766

Lecture Notes in Networks and Systems

Volume 238

The series "Lecture Notes in Networks and Systems" publishes the latest developments in Networks and Systems—quickly, informally and with high quality. Original research reported in proceedings and post-proceedings represents the core of LNNS.

Volumes published in LNNS embrace all aspects and subfields of, as well as new challenges in, Networks and Systems.

The series contains proceedings and edited volumes in systems and networks, spanning the areas of Cyber-Physical Systems, Autonomous Systems, Sensor Networks, Control Systems, Energy Systems, Automotive Systems, Biological Systems, Vehicular Networking and Connected Vehicles, Aerospace Systems, Automation, Manufacturing, Smart Grids, Nonlinear Systems, Power Systems, Robotics, Social Systems, Economic Systems and other. Of particular value to both the contributors and the readership are the short publication timeframe and the world-wide distribution and exposure which enable both a wide and rapid dissemination of research output.

The series covers the theory, applications, and perspectives on the state of the art and future developments relevant to systems and networks, decision making, control, complex processes and related areas, as embedded in the fields of interdisciplinary and applied sciences, engineering, computer science, physics, economics, social, and life sciences, as well as the paradigms and methodologies behind them.

Indexed by SCOPUS, INSPEC, WTI Frankfurt eG, zbMATH, SCImago.

All books published in the series are submitted for consideration in Web of Science.

More information about this series at http://www.springer.com/series/15179

Priyadarsi Nanda · Vivek Kumar Verma ·
Sumit Srivastava · Rohit Kumar Gupta ·
Arka Prokash Mazumdar

Editors

Data Engineering for Smart Systems

Proceedings of SSIC 2021

 Springer

Editors
Priyadarsi Nanda
School of Electrical and Data Engineering
University of Technology Sydney
Sydney, NSW, Australia

Sumit Srivastava
Manipal University Jaipur
Jaipur, Rajasthan, India

Arka Prokash Mazumdar
Department of Computer Science
and Engineering
Malviya National Institute of Technology
Jaipur, Rajasthan, India

Vivek Kumar Verma
Manipal University Jaipur
Jaipur, Rajasthan, India

Rohit Kumar Gupta
Manipal University Jaipur
Jaipur, Rajasthan, India

ISSN 2367-3370   ISSN 2367-3389 (electronic)
Lecture Notes in Networks and Systems
ISBN 978-981-16-2640-1   ISBN 978-981-16-2641-8 (eBook)
https://doi.org/10.1007/978-981-16-2641-8

This Springer imprint is published by the registered company Springer Nature Singapore Pte Ltd.
The registered company address is: 152 Beach Road, #21-01/04 Gateway East, Singapore 189721,
Singapore

Preface

This book is one-pot solution for authors to showcase their work among the research communities, application warehouses and the public–private sectors. This series is an opportunity to gather research scientific works related to data engineering concept in the context of computational intelligence consisted of interaction between smart devices, smart environments and smart interactions, as well as information technology support for such areas. Data from mentioned areas also need to be stored (after their gathering) in intelligent database systems and to be processed (using smart and intelligent approach). The aim of this series is to make available a platform for the publication of books on all aspects of single- and multi-disciplinary research on these themes to make the latest results available in a readily accessible form. The innovations provide the latest software tools to be showcased in the book series through publications. The high-quality content with broad range of the topics pertaining to the book series will be peer-reviewed and get published on suitable recommendations.

This book will provide state of the art to research scholars, scientists, industry learners and postgraduates from the various domains of engineering and related fields, as this incorporates data science and the latest innovations in the field of engineering with their paradigms and methods that employ knowledge and intelligence in the research community. This book comprises its scope ranging from data science for smart systems which are having inbuilt capabilities of handling the research challenges and problems related to energy-aware sensors, smart city projects, wearable devices, smart healthcare solutions, smart e-learning initiatives and social implications of IoT. Further, it extends its coverage to the different computational aspects involved in various domains of engineering such as complex security solutions for data engineering, communication networks, data analytics, machine learning, integrating IoT data with external data sources and data science approaches for smart systems.

Secondarily, this book provides the technological solutions to non-engineering and sciences domain as it contains the fundamental innovation in the field of engineering which turns to be a real solution for their problems. Also, it includes the paradigms which support industries for the development of solutions in favor of society and make everyone's life easier.

Sydney, Australia   Priyadarsi Nanda
Jaipur, India   Vivek Kumar Verma
Jaipur, India   Sumit Srivastava
Jaipur, India   Rohit Kumar Gupta
Jaipur, India   Arka Prokash Mazumdar

Contents

Using Machine Learning, Image Processing and Neural Networks to Sense Bullying in K-12 Schools: Enhanced 1
Lalit Kumar, Palash Goyal, Karan Malik, Rishav Kumar, and Dhruv Shrivastav

Feature-Based Comparative Study of Machine Learning Algorithms for Credibility Analysis of Online Social Media Content 13
Utkarsh Sharma and Shishir Kumar

Smart Support System for Navigation of Visually Challenged Person Using IOT 27
Tuhin Utsab Paul and Aninda Ghosh

Identity-Based Video Summarization 37
Soummya Kulkarni, Darshana Bhagit, and Masooda Modak

Security Testing for Blockchain Enabled IoT System 45
A. B. Yugakiruthika and A. Malini

Two-Dimensional Software Reliability Model with Considering the Uncertainty in Operating Environment and Predictive Analysis 57
Ramgopal Dhaka, Bhoopendra Pachauri, and Anamika Jain

Object Recognition in a Cluttered Scene 71
Rashmee Shrestha, Mandeep Kaur, Nitin Rakesh, and Parma Nand

Breast Cancer Prediction on BreakHis Dataset Using Deep CNN and Transfer Learning Model 77
Pinky Agarwal, Anju Yadav, and Pratistha Mathur

A Comprehensive Tool Survey for Blockchain to IoT Applications 89
A. B. Yugakiruthika and A. Malini

Hybrid Ensemble for Fake News Detection: An Attempt 101
Lovedeep Singh

Detection of Abnormal Activity at College Entrance Through Video Surveillance 109
Lalit Damahe, Saurabh Diwe, Shailesh Kamble, Sandeep Kakde, and Praful Barekar

Audio Peripheral Volume Automation Based on the Surrounding Environment and Individual Human Listening Traits 123
Adit Doshi, Helly Patel, Rikin Patel, Brijesh Satasiya, Muskan Kapadia, and Nirali Nanavati

A Survey: Accretion in Linguistic Classification of Indian Languages 133
Dipjayaben Patel

The Positive Electronic Word of Mouth: A Research Based on the Relational Mediator Meta-Analytic Framework in Electronic Marketplace 147
Bui Thanh Khoa

A Review: Web Content Mining Techniques 159
Priyanka Shah and Hardik B. Pandit

GPS-Free Localization in Vehicular Networks Using Directional Antennas 173
Parveen, Sushil Kumar, and Rishipal Singh

An Improved Scheme in AODV Routing Protocol for Enhancement of QoS in MANET 183
Amit Kumar Bairwa and Sandeep Joshi

Knowledge Management Framework for Sustainability and Resilience in Next-Gen e-Governance 191
Iqbal Hasan and Sam Rizvi

Enriching WordNet with Subject Specific Out of Vocabulary Terms Using Existing Ontology 205
Kanika, Shampa Chakraverty, Pinaki Chakraborty, Aditya Aggarwal, Manan Madan, and Gaurav Gupta

Automatic Detection of Grape, Potato and Strawberry Leaf Diseases Using CNN and Image Processing 213
Md. Tariqul Islam and Abdur Nur Tusher

Sentiment Analysis on Global Warming Tweets Using Naïve Bayes and RNN 225
Deborah T. Joy, Vikas Thada, and Utpal Srivastava

Novel and Prevalent Techniques for Resolving Control Hazard 235
Amit Pandey, Abdella K. Mohammed, Rajesh Kumar, and Deepak Sinwar

Sound Classification Using Residual Convolutional Network 245
Mahesh Jangid and Kabir Nagpal

Dimensionality Reduction Using Variational Autoencoders 255
Parv Dahiya and Sahil Garg

Computational Intelligence for Image Caption Generation 263
Sahil Garg and Parv Dahiya

An Empirical Statistical Analysis of COVID-19 Curve Through Newspaper Mining 271
Shriya Verma, Sonam Garg, Tanishq Chamoli, and Ankit Gupta

Wearable Monopole Antenna with 8-Shaped EBG for Biomedical Imaging 281
Regidi Suneetha and P. V. Sridevi

Speech Signal Processing for Identification of Under-Resourced Languages 291
Shweta Sinha

A Unique ECG Authentication System for Health Monitoring 299
Kusum Lata Jain, Meenakshi Nawal, and Shivani Gupta

Assamese Dialect Identification From Vowel Acoustics 313
Priyankoo Sarmah and Leena Dihingia

A Review on Metaheuristic Techniques in Automated Cryptanalysis of Classical Substitution Cipher 323
Ashish Jain, Prakash C. Sharma, Nirmal K. Gupta,
and Santosh K. Vishwakarma

An Empirical Study of Different Techniques for the Improvement of Quality of Service in Cloud Computing 333
Chitra Sharma, Pradeep Kumar Tiwari, and Garima Agarwal

Contribution Analysis of Scope of SRGAN in the Medical Field 341
Moksh Kant, Sandeep Chaurasia, and Harish Sharma

Machination of Human Carpus 353
Sumit Bhardwaj, Bhuvidha Singh Tomar, Adarsh Ankur,
and Punit Gupta

Building Machine Learning Application Using Oracle Analytics Cloud 361
Tarun Jain, Mahek Agarwal, Ashish Kumar, Vivek Kumar Verma,
and Anju Yadav

**An Empirical Analysis of Heart Disease Prediction Using Data
Mining Techniques** 377
Ashish Kumar, Sivapuram Sai Sanjith, Rajkishan Cherukuru,
Vivek Kumar Verma, Tarun Jain, and Anju Yadav

**2D Image to Standard Triangle Language (STL) 3D Image
Conversion** 391
Manoj K. Sharma and Ashish Malik

**Improving Recommendation for Video Content Using
Hyperparameter Tuning in Sparse Data Environment** 401
Rohit Kumar Gupta, Vivek Kumar Verma, Ankit Mundra,
Rohan Kapoor, and Shekhar Mishra

**Importance and Uses of Telemedicine in Physiotherapeutic
Healthcare System: A Scoping Systemic Review** 411
Saurabh Kumar, Ankush Sharma, and Priyanka Rishi

**Feature Exratction of PTTS System and Its Evaluation
by Standard Statistical Method Mean Opinion Score** 423
Sunil Nimbhore, Suhas Mache, and Sidhharth Mache

**Computational Analysis of a Human–Robot Working Alliance
Trust in Robot-Based Therapy** 431
Azizi Ab Aziz and Wadhah A. Abdulhussain

**Novel Intrusion Prevention and Detection Model in Wireless
Sensor Network** 443
Neha Singh and Deepali Virmani

**Shadow Detection from Real Images and Removal Using Image
Processing** 451
Sumaya Akter Usrika and Abdus Sattar

**A Study on Pulmonary Image Screening for the Detection
of COVID-19 Using Convolutional Neural Networks** 461
Shreyas Thakur, Yash Kasliwal, Taikhum Kothambawala,
and Rahul Katarya

**Analyzing Effects of Temperature, Humidity, and Urban
Population in the Initial Outbreak of COVID19 Pandemic in India** 469
Amit Pandey, Tucha Kedir, Rajesh Kumar, and Deepak Sinwar

**FOFS: Firefly Optimization for Feature Selection to Predict
Fault-Prone Software Modules** 479
Somya Goyal

**Deceptive Reviews Detection in E-Commerce Websites Using
Machine Learning** 489
Sparsh Kotriwal, Jaya Krishna Raguru, Siddhartha Saxena,
and Devi Prasad Sharma

**A Study on Buying Attitude on Facebook in the Digital
Transformation Era: A Machine Learning Application** 497
Bui Thanh Khoa, Ho Nhat Anh, Nguyen Minh Ly,
and Nguyen Xuan Truong

**Performance Evaluation of Speaker Identification in Language
and Emotion Mismatch Conditions on Eastern and North Eastern
Low Resource Languages of India** 511
Joyanta Basu, Tapan Kumar Basu, and Swanirbhar Majumder

Autonomous Wheelchair for Physically Challenged 521
B. Kavyashree, B. S. Aishwarya, Mahima Manohar Varkhedi,
S. Niharika, R. Amulya, and A. P. Kavya

Watermarking of Digital Image Based on Complex Number Theory 531
Nadia Afrin Ritu, Ahsin Abid, Al Amin Biswas, and M. Imdadul Islam

**A Computer Vision Approach for Automated Cucumber Disease
Recognition** 543
Md. Abu Ishak Mahy, Salowa Binte Sohel, Joyanta Basak,
Md. Jueal Mia, and Sourov Mazumder

**An Automated Visa Prediction Technique for Higher Studies Using
Machine Learning in the Context of Bangladesh** 557
Asif Ahmmed, Tipu Sultan, Sk. Hasibul Islam Shad, Md. Jueal Mia,
and Sourov Mazumder

**COVID-19 Safe Guard: A Smart Mobile Application to Address
Corona Pandemic** 569
Soma Prathibha, K. L. Nirmal Raja, M. Shyamkumar, and M. Kirthiga

**A Novel Mathematical Model to Represent the Hypothalamic
Control on Water Balance** 581
Divya Jangid and Saurabh Mukherjee

**Importance of Deep Learning Models to Perform Segmentation
on Medical Imaging Modalities** 593
Preeti Sharma and Devershi Pallavi Bhatt

**A Multi-component-Based Zero Trust Model to Mitigate
the Threats in Internet of Medical Things** 605
Y. Bevish Jinila, S. Prayla Shyry, and A. Christy

A Modified Cuckoo Search for the *n*-Queens Problem 615
Ashish Jain, Manoj K. Bohra, Manoj K. Sharma,
and Venkatesh G. Shankar

A Survey on Diabetic Retinopathy Detection Using Deep Learning 621
Deepak Mane, Namrata Londhe, Namita Patil, Omkar Patil,
and Prashant Vidhate

A Survey on Alzheimer's Disease Detection and Classification 639
D. T. Mane, Mehul Patel, Madhavi Sawant, Karina Maiyani,
and Divya Patil

**An Approach for Graph Coloring Problem Using Grouping
of Vertices** 651
Prakash C. Sharma, Santosh Kumar Vishwakarma, Nirmal K. Gupta,
and Ashish Jain

**Role of PID Control Techniques in Process Control System:
A Review** 659
Vandana Dubey, Harsh Goud, and Prakash C. Sharma

**Investigating Cancer Survivability of Juvenile Lymphoma Cancer
Patients Using Hard Voting Ensemble Technique** 671
Amit Pandey, Rabira Galeta, and Tucha Kedir

Author Index 679

Editors and Contributors

About the Editors

Dr. Priyadarsi Nanda is Senior Lecturer at the University of Technology Sydney (UTS) with more than 27 years of experience specializing in research and development of cybersecurity, IoT security, Internet traffic engineering, wireless sensor network security, and many more related areas. His most significant work has been in the area of intrusion detection and prevention systems (IDS/IPS) using image processing techniques, Sybil attack detection in IoT-based applications, and intelligent firewall design. In cybersecurity research, he has published over 80 high-quality refereed research papers including transactions in computers, transactions in parallel processing and distributed systems (TPDS), future generations of computer systems (FGCS) as well as many ERA Tier A/A* conference articles. In 2017, his work in cybersecurity research has earned him and his team the prestigious Oman Research Council's National Award for best research. Dr. Nanda has successfully supervised 8 HDR at UTS (5 PhD + 3 Masters), and currently, supervising 8 Ph.D. students.

Vivek Kumar Verma is Assistant Professor (Senior Grade) in the Department of Information Technology, School of Computing & Information Technology, Manipal University Jaipur. His areas of expertise include data engineering, image processing, and natural language processing. He has published more than 35 research articles in peer-reviewed international journals, book series, and conference proceedings.

Sumit Srivastava is Professor of Information Technology with expertise in the domain of data analytics and image processing. He is also Senior Member of IEEE. He has published more than 100 research papers in peer-reviewed international journals, book series, and conference proceedings. He is having two edited volumes of Springer Nature Book series in his account. He has supervised more than 10 Ph.D. scholars in the areas of data science, machine learning, and natural language processing.

Rohit Kumar Gupta is Assistant Professor (Senior Grade) in the Department of Information Technology, School of Computer Science and IT, Manipal University Jaipur. He is currently pursuing a Ph.D. in the area of content distributed network from Malaviya National Institute of Technology, Jaipur. His research interests include cloud computing, network technologies, and IoT. He has published over 25 research articles in peer-reviewed international journals, book series, and conference proceedings.

Arka Prokash Mazumdar is Assistant Professor (Senior Grade) in the Department of Computer Science and Engineering, Malviya National Institute of Technology Jaipur, India. He has completed his Ph.D. from IIT Patna and M.Tech. from NIT Durgapur. His research interests include information-centric networks, cloud computing, and IoT. He has published over 50 research articles in peer-reviewed international journals, book series, and conference proceedings.

Contributors

Wadhah A. Abdulhussain Relational Machines Group, Human-Centred Computing Lab, School of Computing, Universiti Utara Malaysia, Sintok, Kedah, Malaysia

Ahsin Abid Department of CSE, Jahangirnagar University, Dhaka, Bangladesh

Md. Abu Ishak Mahy Department of CSE, Daffodil International University, Dhaka, Bangladesh

Garima Agarwal Manipal University Jaipur, Jaipur, India

Mahek Agarwal SCIT, Manipal University Jaipur, Jaipur, India

Pinky Agarwal Manipal University Jaipur, Jaipur, India

Aditya Aggarwal Netaji Subhas Institute of Technology, Delhi, India

Asif Ahmmed Department of CSE, Daffodil International University, Dhaka, Bangladesh

B. S. Aishwarya Vidyavardhaka College of Engineering, Mysuru, Karnataka, India

R. Amulya Vidyavardhaka College of Engineering, Mysuru, Karnataka, India

Ho Nhat Anh Industrial University of Ho Chi Minh City, Ho Chi Minh City, Vietnam

Adarsh Ankur Department of Electronics and Communication Engineering, Amity School of Engineering and Technology, Amity University, Noida, Uttar Pradesh, India

Azizi Ab Aziz Relational Machines Group, Human-Centred Computing Lab, School of Computing, Universiti Utara Malaysia, Sintok, Kedah, Malaysia

Amit Kumar Bairwa Manipal University Jaipur, Jaipur-Ajmer Express Highway, Dehmi Kalan, Jaipur, Rajasthan, India

Praful Barekar Computer Technology Department, Yeshwantrao Chavan College of Engineering, Wanadongri, Nagpur, India

Joyanta Basak Department of CSE, Daffodil International University, Dhaka, Bangladesh

Joyanta Basu CDAC Kolkata, Salt Lake, Sector-V, Kolkata, India

Tapan Kumar Basu Department of Electrical Engineering, IIT, Kharagpur, West Bengal, India

Y. Bevish Jinila Sathyabama Institute of Science and Technology, Chennai, India

Darshana Bhagit SIES Graduate School of Technology, Navi Mumbai, India

Sumit Bhardwaj Department of Electronics and Communication Engineering, Amity School of Engineering and Technology, Amity University, Noida, Uttar Pradesh, India

Devershi Pallavi Bhatt University, Manipal University, Jaipur, Rajasthan, India

Al Amin Biswas Department of CSE, Jahangirnagar University, Dhaka, Bangladesh

Manoj K. Bohra School of Computing and Information Technology, Manipal University Jaipur, Jaipur, India

Pinaki Chakraborty Netaji Subhas Institute of Technology, Delhi, India

Shampa Chakraverty Netaji Subhas Institute of Technology, Delhi, India

Tanishq Chamoli Chandigarh College Of Engineering And Technology (Degree Wing), Chandigarh, India

Sandeep Chaurasia Department of Computer Science and Engineering, Manipal University Jaipur, Jaipur, India

Rajkishan Cherukuru SCIT, Manipal University Jaipur, Jaipur, India

A. Christy Sathyabama Institute of Science and Technology, Chennai, India

Parv Dahiya Amity University Haryana, Gurugram, Haryana, India

Lalit Damahe Computer Technology Department, Yeshwantrao Chavan College of Engineering, Wanadongri, Nagpur, India

Ramgopal Dhaka Department of Mathematics & Statistics, Manipal University Jaipur, Jaipur, Rajasthan, India

Leena Dihingia Indian Institute of Technology Guwahati, Guwahati, India

Saurabh Diwe Computer Technology Department, Yeshwantrao Chavan College of Engineering, Wanadongri, Nagpur, India

Adit Doshi Department of Computer Engineering, SCET, Surat, India

Vandana Dubey School of Computing & Information Technology, Manipal University Jaipur, Jaipur, India

Rabira Galeta College of Informatics, Bule Hora University, Bule Hora, Ethiopia

Sahil Garg Amity University Haryana, Gurugram, Haryana, India

Sonam Garg Chandigarh College Of Engineering And Technology (Degree Wing), Chandigarh, India

Aninda Ghosh Distronix, Kolkata, India

Harsh Goud Department of Electronics and Communication Engineering, Indian Institute of Information Technology Nagpur, Nagpur, India

Palash Goyal Mount Carmel School, Delhi, India

Somya Goyal Department of Computer and Communication Engineering, SCIT Manipal University Jaipur, Jaipur, Rajasthan, India

Ankit Gupta Chandigarh College Of Engineering And Technology (Degree Wing), Chandigarh, India

Gaurav Gupta Netaji Subhas Institute of Technology, Delhi, India

Nirmal K. Gupta School of Computing and Information Technology, Manipal University Jaipur, Jaipur, India

Punit Gupta Department of Computer and Communication Engineering, Manipal University Jaipur, Jaipur, India

Rohit Kumar Gupta Manipal University Jaipur, Jaipur, Rajasthan, India

Shivani Gupta Vellore Institute of Technology, Chennai, TN, India

Iqbal Hasan Department of Computer Science, Jamia Millia Islamia University, New Delhi, India

Sk. Hasibul Islam Shad Department of CSE, Daffodil International University, Dhaka, Bangladesh

M. Imdadul Islam Department of CSE, Jahangirnagar University, Dhaka, Bangladesh

Anamika Jain Department of Mathematics & Statistics, Manipal University Jaipur, Jaipur, Rajasthan, India

Ashish Jain School of Computing and Information Technology, Manipal University Jaipur, Jaipur, India

Kusum Lata Jain Manipal University Jaipur, Jaipur, RJ, India

Tarun Jain SCIT, Manipal University Jaipur, Jaipur, India

Divya Jangid Faculty of Mathematics and Computing, Department of Computer Science, Banasthali Vidyapith, Banasthali, Tonk, Rajasthan, India

Mahesh Jangid Department of CSE, Manipal University Jaipur, Jaipur, Rajasthan, India

Sandeep Joshi Manipal University Jaipur, Jaipur-Ajmer Express Highway, Dehmi Kalan, Jaipur, Rajasthan, India

Deborah T. Joy Amity University Gurgaon, Gurugram, Haryana, India

Md. Jueal Mia Department of CSE, Daffodil International University, Dhaka, Bangladesh

Sandeep Kakde Electronics Engineering Department, Yeshwantrao Chavan College of Engineering, Wanadongri, Nagpur, India

Shailesh Kamble Computer Technology Department, Yeshwantrao Chavan College of Engineering, Wanadongri, Nagpur, India

Kanika Netaji Subhas Institute of Technology, Delhi, India

Moksh Kant Department of Computer Science and Engineering, Manipal University Jaipur, Jaipur, India

Muskan Kapadia Department of Computer Engineering, SCET, Surat, India

Rohan Kapoor Manipal University Jaipur, Jaipur, Rajasthan, India

Yash Kasliwal Department of Computer Science, Delhi Technological University, New Delhi, India

Rahul Katarya Department of Computer Science, Delhi Technological University, New Delhi, India

Mandeep Kaur Department of Computer Science and Engineering, Sharda University, Greater Noida, Uttar Pradesh, India

A. P. Kavya Vidyavardhaka College of Engineering, Mysuru, Karnataka, India

B. Kavyashree Vidyavardhaka College of Engineering, Mysuru, Karnataka, India

Tucha Kedir College of Informatics, BuleHora University, Bule Hora, Ethiopia

Bui Thanh Khoa Industrial University of Ho Chi Minh City, Ho Chi Minh City, Vietnam

M. Kirthiga Sri Sairam Engineering College, Chennai, Tamil Nadu, India

Taikhum Kothambawala Department of Computer Science, Delhi Technological University, New Delhi, India

Sparsh Kotriwal School of Computing and Information Technology, Manipal University Jaipur, Jaipur, India

Soummya Kulkarni SIES Graduate School of Technology, Navi Mumbai, India

Ashish Kumar SCIT, Manipal University Jaipur, Jaipur, India

Lalit Kumar Gazelle Information Technologies, Delhi, India

Rajesh Kumar College of Informatics, Bule Hora University, Bule Hora, Ethiopia

Rishav Kumar Mount Carmel School, Delhi, India

Saurabh Kumar Faculty of Physiotherapy, SGT University, Gurugram, Haryana, India

Shishir Kumar Department of Computer Science & Engineering, Jaypee University of Engineering and Technology, Guna, India

Sushil Kumar Jawaharlal Nehru University, New Delhi, India

Namrata Londhe JSPM'S Rajarshi Shahu College of Engineering, Pune, Maharashtra, India

Nguyen Minh Ly Industrial University of Ho Chi Minh City, Ho Chi Minh City, Vietnam

Sidhharth Mache Dr. Babasaheb Ambedkar Marathwada University, Aurangabad, MH, India

Suhas Mache R. B. Attal Arts, Science and Commerce College Georai, Beed, MH, India

Manan Madan Netaji Subhas Institute of Technology, Delhi, India

Karina Maiyani JSPM's Rajarshi Shahu College of Engineering, Pune, Maharashtra, India

Swanirbhar Majumder Department of Information Technology, Tripura University, Tripura, India

Ashish Malik Manipal University Jaipur, Jaipur, Rajasthan, India

Karan Malik Computer Science Engineering, USICT- Guru Gobind Singh Indraprastha University, Delhi, India

A. Malini Department of Computer Science, Thiagarajar College of Engineering, Madurai, India

D. T. Mane JSPM's Rajarshi Shahu College of Engineering, Pune, Maharashtra, India

Deepak Mane JSPM'S Rajarshi Shahu College of Engineering, Pune, Maharashtra, India

Pratistha Mathur Manipal University Jaipur, Jaipur, India

Sourov Mazumder Department of CSE, Daffodil International University, Dhaka, Bangladesh

Shekhar Mishra Manipal University Jaipur, Jaipur, Rajasthan, India

Masooda Modak SIES Graduate School of Technology, Navi Mumbai, India

Abdella K. Mohammed Faculty of Informatics, IOT, Hawassa University, Hawassa, Ethiopia

Saurabh Mukherjee Faculty of Mathematics and Computing, Department of Computer Science, Banasthali Vidyapith, Banasthali, Tonk, Rajasthan, India

Ankit Mundra Manipal University Jaipur, Jaipur, Rajasthan, India

Kabir Nagpal Department of CSE, Manipal University Jaipur, Jaipur, Rajasthan, India

Nirali Nanavati Department of Computer Engineering, SCET, Surat, India

Parma Nand Department of Computer Science and Engineering, Sharda University, Greater Noida, Uttar Pradesh, India

Meenakshi Nawal Poornima College of Engineering, Rajasthan Technical University, Jaipur, RJ, India

S. Niharika Vidyavardhaka College of Engineering, Mysuru, Karnataka, India

Sunil Nimbhore Dr. Babasaheb Ambedkar Marathwada University, Aurangabad, MH, India

K. L. Nirmal Raja Sri Sairam Engineering College, Chennai, Tamil Nadu, India

Bhoopendra Pachauri Department of Mathematics & Statistics, Manipal University Jaipur, Jaipur, Rajasthan, India

Amit Pandey College of Informatics, Bule Hora University, Bule Hora, Ethiopia

Hardik B. Pandit P.G. Department of Computer Science and Technology, Sardar Patel University, Anand, India

Parveen Guru Jambheshwar University of Science and Technology, Hisar, Haryana, India

Parveen Jawaharlal Nehru University, New Delhi, India

Dipjayaben Patel Vivekanand College For Advanced Computer And Information Science, Surat, Gujarat, India

Helly Patel Department of Computer Engineering, SCET, Surat, India

Mehul Patel JSPM's Rajarshi Shahu College of Engineering, Pune, Maharashtra, India

Rikin Patel Department of Computer Engineering, SCET, Surat, India

Divya Patil JSPM's Rajarshi Shahu College of Engineering, Pune, Maharashtra, India

Namita Patil JSPM'S Rajarshi Shahu College of Engineering, Pune, Maharashtra, India

Omkar Patil JSPM'S Rajarshi Shahu College of Engineering, Pune, Maharashtra, India

Tuhin Utsab Paul St. Xavier's University, Kolkata, India

Devi Prasad Sharma School of Computing and Information Technology, Manipal University Jaipur, Jaipur, India

Soma Prathibha Department of Information Technology, Sri Sairam Engineering College, Chennai, Tamil Nadu, India

S. Prayla Shyry Sathyabama Institute of Science and Technology, Chennai, India

Jaya Krishna Raguru School of Computing and Information Technology, Manipal University Jaipur, Jaipur, India

Nitin Rakesh Department of Computer Science and Engineering, Sharda University, Greater Noida, Uttar Pradesh, India

Priyanka Rishi Faculty of Physiotherapy, SGT University, Gurugram, Haryana, India

Nadia Afrin Ritu Department of CSE, Jahangirnagar University, Dhaka, Bangladesh

Sam Rizvi Department of Computer Science, Jamia Millia Islamia University, New Delhi, India

Sivapuram Sai Sanjith SCIT, Manipal University Jaipur, Jaipur, India

Priyankoo Sarmah Indian Institute of Technology Guwahati, Guwahati, India

Brijesh Satasiya Department of Computer Engineering, SCET, Surat, India

Abdus Sattar Daffodil International University, Dhaka, Shukrabad, Bangladesh

Madhavi Sawant JSPM's Rajarshi Shahu College of Engineering, Pune, Maharashtra, India

Siddhartha Saxena School of Computing and Information Technology, Manipal University Jaipur, Jaipur, India

Priyanka Shah Kadi Sarva Vishwavidyalaya University, Gandhinagar, India

Venkatesh G. Shankar School of Computing and Information Technology, Manipal University Jaipur, Jaipur, India

Ankush Sharma School of Physiotherapy, BUEST, Baddi, HP, India

Chitra Sharma Manipal University Jaipur, Jaipur, India

Harish Sharma Department of Computer Science and Engineering, Manipal University Jaipur, Jaipur, India

Manoj K. Sharma School of Computing and Information Technology, Manipal University Jaipur, Jaipur, India

Prakash C. Sharma School of Computing and Information Technology, Manipal University Jaipur, Jaipur, India

Preeti Sharma University, Manipal University, Jaipur, Rajasthan, India

Utkarsh Sharma Department of Computer Science & Engineering, Jaypee University of Engineering and Technology, Guna, India

Rashmee Shrestha Department of Computer Science and Engineering, Sharda University, Greater Noida, Uttar Pradesh, India

Dhruv Shrivastav Mount Carmel School, Delhi, India

M. Shyamkumar Sri Sairam Engineering College, Chennai, Tamil Nadu, India

Lovedeep Singh Punjab Engineering College, Chandigarh, India

Neha Singh University School of Information, Communication and Technology, Guru Gobind Singh Indraprastha University, Dwarka, Delhi, India

Rishipal Singh Guru Jambheshwar University of Science and Technology, Hisar, Haryana, India

Shweta Sinha Amity University, Gurugram, Haryana, India

Deepak Sinwar Department of Computer and Communication Engineering, Manipal University, Jaipur, India

Salowa Binte Sohel Department of CSE, Daffodil International University, Dhaka, Bangladesh

P. V. Sridevi Andhra University College of Engineering (A), Andhra University, Visakhapatnam, AP, India

Utpal Srivastava Banasthali University, Jaipur, Rajasthan, India

Tipu Sultan Department of CSE, Daffodil International University, Dhaka, Bangladesh

Regidi Suneetha Andhra University College of Engineering (A), Andhra University, Visakhapatnam, AP, India

Md. Tariqul Islam Khulna University of Engineering and Technology (KUET), Khulna, Bangladesh

Vikas Thada Amity University Gurgaon, Gurugram, Haryana, India

Shreyas Thakur Department of Computer Science, Delhi Technological University, New Delhi, India

Pradeep Kumar Tiwari Manipal University Jaipur, Jaipur, India

Bhuvidha Singh Tomar Department of Electronics and Communication Engineering, Amity School of Engineering and Technology, Amity University, Noida, Uttar Pradesh, India

Nguyen Xuan Truong Industrial University of Ho Chi Minh City, Ho Chi Minh City, Vietnam

Abdur Nur Tusher Daffodil International University (DIU), Dhaka, Bangladesh

Sumaya Akter Usrika Daffodil International University, Dhaka, Shukrabad, Bangladesh

Mahima Manohar Varkhedi Vidyavardhaka College of Engineering, Mysuru, Karnataka, India

Shriya Verma Chandigarh College Of Engineering And Technology (Degree Wing), Chandigarh, India

Vivek Kumar Verma SCIT, Manipal University Jaipur, Jaipur, India; Manipal University Jaipur, Jaipur, Rajasthan, India

Prashant Vidhate JSPM'S Rajarshi Shahu College of Engineering, Pune, Maharashtra, India

Deepali Virmani Department of Computer Science Engineering, Bhagwan Parshuram Institute of Technology, Rohini, Delhi, India

Santosh K. Vishwakarma School of Computing and Information Technology, Manipal University Jaipur, Jaipur, India

Santosh Kumar Vishwakarma School of Computing and Information Technology, Manipal University Jaipur, Jaipur, India

Anju Yadav SCIT, Manipal University Jaipur, Jaipur, India

A. B. Yugakiruthika Department of Computer Science, Thiagarajar College of Engineering, Madurai, India

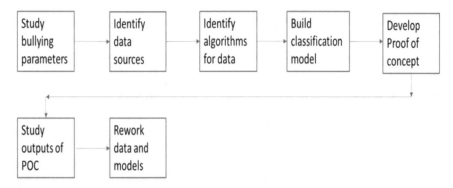

Fig. 1 Research methodology

Table 1 How to identify a victim of bullying

Victim parameter	Data source
Show deteriorating grades	Academic report
Regular absenteeism	Attendance records
Being picked on, pushed, punched, etc. (physically harassed)	CCTV images
Downfall of social skills	Registration for school events and extracurricular groups
Suffering from learning or mental disabilities	Counselor reports
Unwilling to go to school regularly	Attendance records
Random bruises, missing belongings or torn clothing	CCTV images
Prone to attacks of anxiety	School medical reports
Alone at lunch-breaks	CCTV images
Experiences regular nightmares	Medical room report
Starts bullying younger or weaker kids to vent out the frustration	CCTV images

are used to filter the data. In the proposed model, it is planned to integrate AI tools with school infrastructure and data like cameras with microphones, student portals containing community forums where students and teachers interact, attendance of students, their scorecards, etc. Data used for analysis and a complete infrastructural setup is discussed below in Tables 1 and 2.

Table 2 Bullying perpetrator parameters and data sources

Perpetrator parameters	Data source
Rude behavior with students as well as teachers	CCTV images, audio, school report
Low grades	School report
Lacks empathy or guilt	School report
Feeling of entitlement because of being good in school, sports or belonging to a prominent family	School report
Short-tempered and having emotional outbursts	CCTV images, audio, school report
Usually popular or among a big group	CCTV images, audio, school report
Regularly get into trouble with authority	School report

2.2 Integration with School Infrastructure

Recordings of playground, common-area (like corridor, locker rooms, etc.) and classes will be taken along with their audios via cameras that are installed. More detailed information associated with students will be extracted from student portals. This will further help in understanding and analyzing student behavior and personality. This raw information is then reworked into structured data, which will supplement the learning algorithm in predictions and analysis. Based on this, appropriate action can be taken by school authorities. This method can also be inverted and be used to discern the victims of harassment.

Data Analysis:

Photos uploaded by the students on community forums along with acquired video footage from CCTV are split into images (see Figs. 2 and 3) and ran through the algorithm to identify: drugs, number of faces, anxiety attacks, crying, isolation, fighting, torn clothes, bruises, sleeping, smoking, hard drinks, gore, explicit and adult content. The image classifier uses CNN and pre-trained network called Darknet-19 (trained

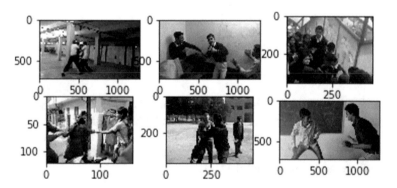

Fig. 2 Sample images (in case of bullying) from dataset

Fig. 3 Sample images (non-bullying) from dataset

on ImageNet dataset), along with LSTM which provides the capability to process sequence of image data with feedback connections. When a bullying scenario is detected by the CNN-LSTM architecture, faces of all those involved or present in the particular frame are detected using the popular object detection algorithm You Only Look Once (YOLO). These faces are input into another simple CNN, namely a Siamese network, which tries to identify the perpetrator by matching the input with the student database already present.

Audios which are mapped to text using Google Cloud Speech API alongside comments at student community are further used to determine the following features: tone, amplitude and pitch of voice, language used (explicit or not), uppercase text, text length and sentiments, threatening statements, trolling, unpleasant comments and distasteful words. Other attributes like low attendance, incompetent grades, enrolment in extracurricular activities, teacher–student interaction, frequency of councilor appointments and behavior report by staff will also be considered. Data for these inputs is converted from physical form into digital form first.

2.3 Techniques Involved

Convolution neural network is a type of neural network (NN) that provides the capability to convert pixels into well-structured data [5]. CNNs replicate the function of the frontal lobe of the human brain (cerebral cortex), which is responsible for processing audio visual stimulus in humans. To process images, CCTV footage is spliced into still images and then each frame is analyzed to extract the crucial and important features which can be further analyzed for more refined results.

Proposed network architecture for video classification is shown in Fig. 4. It has been already shown that adding LSTM (which extract global temporal features) after CNN, the local temporal features obtained from optical flow are also of great importance [10].

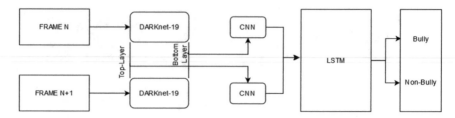

Fig. 4 Architecture of model

Optical flow is due to displacement of boundaries or movement of object across two consecutive frames. Optical flow is best used for action recognition, so its functionality is mimicked by taking two consecutive frames as input to training model. These consecutive frames are fetched to pre-trained CNN model (Darknet-19). Then, both of the frame values from the bottom layer of pre-trained model are fed as an input to additional CNN. Additional CNN network is now to supposed to learn from the local motion features including the invariant features by comparing the both frame. The top layer of the pre-trained network is also fetched to another additional CNN to learn from the comparison of high-level features of both the frames. Furthermore, the output of both the additional CNN is fully connected with LSTM cell, which further classifies the images as bully and non-bully.

When the image is classified as bully, the YOLO algorithm, which is an object detection algorithm, is used to detect all the faces present in the "bully" frame, using a vector of bounding boxes (see Fig. 5). It uses a convolutional neural network (CNN) to complete this task. The structure of the CNN used and a brief working is shown in Fig. 5. The input image is divided into grids or regions, and bounding boxes are predicted for each of these regions. The prediction vectors for each region contain six elements, namely [x, y, w, h, c, p], where x, y, w and h are the coordinates of the bounding boxes, c is the confidence and p is the probability of occurrence of faces, i.e., if the algorithm fails to detect any faces, p becomes equal to 0 and the remaining values are not defined.

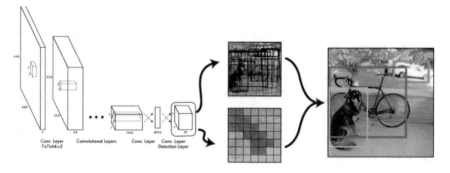

Fig. 5 YOLO algorithm

For unambiguous and noise-free detection, threshold values are set, in order to filter-out disturbances. Only regions that have their probability (p) value exceeding the threshold are considered. To further improve quality of predictions, Intersection over Union (IoU) and non-max suppression are performed. Non-max suppression helps in eliminating any bounding regions that might have detected an already detected feature or face. The final bounding boxes received after the complete procedure are sent as inputs into a Siamese network for facial recognition.

Since only one picture of each student will be present in the database, training a deep learning algorithm to map the received images to one stored in the dataset will not be feasible. Due to this, a Siamese network, another variation of the convolutional neural network that is a one-shot learning algorithm is used. It gives high performance in applications and scenarios where there is not a large amount of data belonging to each data class.

A Siamese network consists of two parallel neural networks, with the same weights and parameters. It consists of convolutional, pooling and dense layers just like a regular CNN, but what distinguishes it from it is that it is bereft of a softmax/sigmoid activation at the last layer. Thus, the encoding of an image is received as an output instead of a category. The contrastive loss or difference between encodings received from the two parallel networks is calculated to ascertain the similarity between the input images.

The faces inside the bounding boxes predicted by the YOLO algorithm are extracted and provided as input to one of the two parallel networks. The other network receives images from the database of students, and the difference of the received encodings in calculated. The image having the least difference is considered to be "identified" as one of those present in the bullying frame. To fasten the process and reduce computational needs, the encodings of all students in the database were generated in advance using the same Siamese network, and only difference calculation was done during the test phase. Immediate alerts, including the identity of those involved, are then sent to the authorities. Figure 6 displays the working of a Siamese network for facial recognition.

For audio analysis, we have mapped CCTV voice output with Google Cloud Speech API which uses CNN and provides real-time streaming of speech recognition and conversion from audio to text. This lies within our CCTV and Google Cloud Storage for sentiment analysis to be done on it.

For text analysis, mentioned words are mapped with sentiment analysis dictionary ("Liu and Hu opinion lexicon" containing 6800 positive and negative words [4]). To extract feature from the text, multinomial Naïve Bayes is used (refer Fig. 7). It helps in classification of bad words and rude comments from audio-to-text converted files as well as from student's community portal including explicit and vulgar remarks, text length as well as usage of upper-case letters in community forums, etc. All such attributes are then listed for feature extraction by the model.

Even data like medical records, attendance, grades and teachers remark from student portals is used and processed for feature extraction. Logistic regression modeling is used for this analysis.

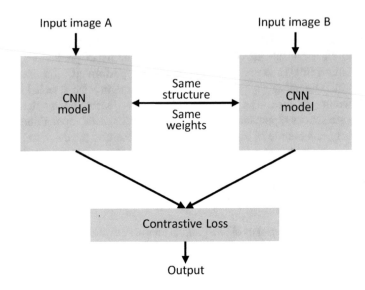

Fig. 6 Siamese network for facial recognition

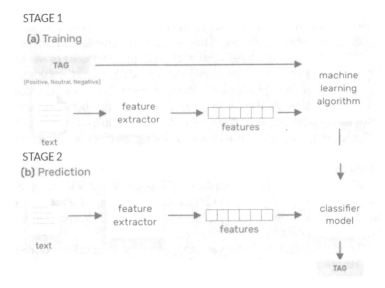

Fig. 7 Sentiment analysis

2.4 Implementation Infrastructure

Proposed solution can easily be integrated and implemented in schools, colleges, institutes and other places prone to bullying. A data pipeline can be built for fetching data at real time with predictive analytical capabilities. Infrastructure for such a solution is a one-time investment and will provide a great benefit to the coming future making schools bullying free, a dream most people never thought would become a reality.

Detailed information and diagrammatic representation of recommended architecture is provided below:

As shown in Fig. 8, data streams from CCTV are split as audio and video media streams. Audio is then sent to Google speech-to-text API which is used for sentiment analysis. In case of video, clips are sliced down into multiple frames based on time and frame rates. These images are analyzed to detect any signs of bullying or any other sort of unethical action via the use of defined CNN model.

From student portals, web scraping is implemented to obtain information about posts, tags, comments, open chats and for records which can be taken directly from the school database. All these attributes are fetched to be deployed into the model (on cloud premises), and results can be fetched remotely via Internet.

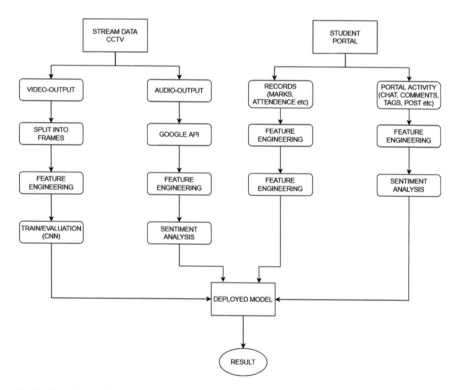

Fig. 8 Data flow/infrastructure architecture flow diagram

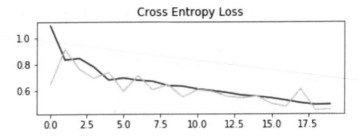

Fig. 9 Cross-entropy loss (train dataset is represented by blue curve, and test dataset is represented by orange curve)

3 Result

To analyze the accuracy in initial phases, labeled data was split into training and testing data in ratio of 7:3, respectively. At initial stages, with the gathered data our model is able to predict with accuracy of 87%.

Cross-entropy loss was used as the loss function to analyze the performance of the deployed model [11]. It is used to measure the performance of a classification model, with outputs between 0 and 1. As the predicted label diverges from the actual label, the penalty or the loss increases. Equation (1) represents cross-entropy loss for N number of classes.

$$C.E.\ Loss = -\sum_{1}^{N} y_{o,c} \log(p_{o,c}) \tag{1}$$

where N = Number of classes.

$y_{c,o}$ = binary indicator if class label c is correct for observation o.

$p_{o,c}$ = predicted probability that o lies in c.

The charts below depict cross-entropy loss (refer Fig. 9) which helps in measuring the performance of the classification model. The output represents a probability value between 0 and 1 [6] which shows our model's classification accuracy (refer Fig. 10).

4 Improvements

Future improvements can include the model being used as a measure to check for depressed victims so as to avoid suicide and self-harm tendencies and help in countering it.

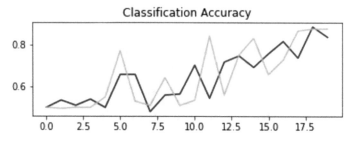

Fig. 10 Classification accuracy (train dataset is represented by blue curve, and test dataset is represented by orange curve)

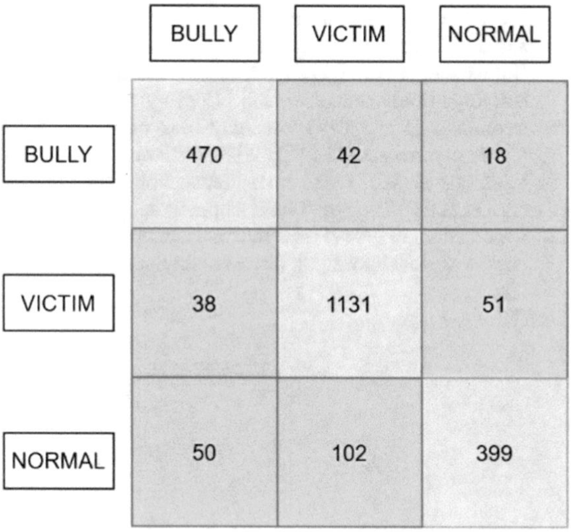

Fig. 11 Confusion matrix plotted on labeled data having sample space of 2000 to check accuracy of model

	BULLY	VICTIM	NORMAL
BULLY	470	42	18
VICTIM	38	1131	51
NORMAL	50	102	399

5 Conclusion

The proof-of-concept solution developed by the authors was successful in identifying the bullying situation (refer Fig. 11) and hence the bully and the victim. The output was more accurate for audio-visual inputs. For the unstructured data, the accuracy of prediction will have to be improved by interlinking of the outputs with the audio-visual parameters. Overall, it was a successful proof of concept.

Acknowledgments We would like to thank Mr. Kirit, the CEO of Gazelle Information Technologies PVT LTD, for his expert advice and a supply of required resources for the implementation of this project.

References

1. National Academies of Sciences, Engineering, and Medicine (2016) Preventing bullying through science, policy, and practice. The National Academies Press, Washington, DC. https://doi.org/10.17226/23482
2. Poeter D (2011) Study: a quarter of parents say their child involved in cyberbullying. pcmag.com. http://www.pcmag.com/article2/0,2817,2388540,00.asp
3. StopBullying.gov (2020) What is bullying. https://www.stopbullying.gov/bullying/what-is-bullying. Accessed 28 Jan 2020
4. Liu B (2020) Opinion mining, sentiment analysis, opinion extraction. Cs.uic.edu. https://www.cs.uic.edu/~liub/FBS/sentiment-analysis.html#lexicon. Accessed 16 June 2020
5. WildML (2020) Understanding convolutional neural networks for NLP. http://www.wildml.com/2015/11/understanding-convolutional-neural-networks-for-nlp/. Accessed 28 Jan 2020
6. Brownlee J (2020) How to classify photos. Machine learning mastery. https://machinelearningmastery.com/how-to-develop-a-convolutional-neural-network-to-classify-photos-of-dogs-and-cats/. Accessed 28 Jan 2020
7. Kumar L et al (2020) Using machine learning, image processing & neural networks to sense bullying in K-12 schools. Asian J Converg Technol
8. Redmon J et al (2015) You only look once: unified, real-time object detection. Cornell University. Accessed 24 Sept 2020. (arXiv.org)
9. Koch G et al (2015) Siamese neural networks for one shot image recognition
10. Ng J et al (2015) Beyond short snippets: deep networks for video classification
11. Readthedocs.io (2017) Loss functions: cross entropy. https://ml-cheatsheet.readthedocs.io/en/latest/loss_functions.html. Accessed 25 Sept 2020

Feature-Based Comparative Study of Machine Learning Algorithms for Credibility Analysis of Online Social Media Content

Utkarsh Sharma and Shishir Kumar

Abstract As the use of social media is growing these days in context of information sharing, the necessity for reality check is also the need of the hour. As more of the users rely on the news spread on these social media platforms, more are the chances of spreading rumours through these contents. To avoid the escalation of hoaxes, the content must be checked and properly labelled as fact or fake by human perception or by using widely used classification algorithms. In this paper, we are performing a feature-based survey of four popular machine learning-based algorithms for the classification of content on social media as credible or fake. The platform which we will consider for data collection will be Twitter API, which is freely distributed by Twitter. We propose the feature selection by using ant colony optimization (ACO) algorithm. The algorithms we consider for comparison are decision tree, Naïve Bayes, random forest and SVM. All these algorithms are tested on some predefined features of tweets which are provided by the Twitter API, and also, some additional features are also extracted by feature engineering on data. We provide our findings on the basis of accuracy, precision and recall measures calculated for all of the algorithms. We also applied cross-validation with tenfold on these algorithms to get the appropriate measure of accuracy.

Keywords Machine learning · OSN · Decision tree · Random forest · Naïve Bayes · SVM · ACO

1 Introduction

Online social networks (OSN) are a stage to which people can join in and that permits them to interconnect with one another and make them able to share their own views, encouraging the making of a virtual informal community with mutual agreement and social relations among individuals that generally share same interests, exercises or genuine associations. OSN is also affecting human social life in a big way [1].

U. Sharma (✉) · S. Kumar
Department of Computer Science & Engineering, Jaypee University of Engineering and Technology, Guna, India

© The Author(s), under exclusive license to Springer Nature Singapore Pte Ltd. 2022
P. Nanda et al. (eds.), *Data Engineering for Smart Systems*, Lecture Notes in Networks and Systems 238, https://doi.org/10.1007/978-981-16-2641-8_2

Table 1 Number of active
users worldwide for major
OSN

OSN platform	No. of active users (World Wide)
Facebook	2.45 billion
Twitter	330 million
LinkedIn	575 million
Foursquare	55 million
Google +	440 million

OSN nowadays attracted millions of people to express and to also to gain insight from others views [2]. According to statia [3], the number of users of only Facebook is 2.45 (billions) worldwide, and only India has 269 million users of Facebook and has an estimate of rising this number up to 445 million till 2023. Table 1 shows the number of active users worldwide for major online social networking sites.

Just like Facebook, now there are several of online social media platforms are available such as Twitter, WhatsApp, Instagram, etc. As from the statistics gathered, it is much obvious that Facebook tops the chart in the number of active users exists worldwide but still when it comes to assess human behaviour and credibility of facts, researchers tend to prefer microblogging sites rather than Facebook. A microblogging site just like any blogging site allows a user to share or express his/her views in form of text or audio/visual content but just with the restriction of character count of 140 [4]. Microblogs are hotspots for examination with respect to forecast of patterns or early alerts just as for hearing data about the open point of view. And the most popular microblogging site available as of now is Twitter.

The number of users of Twitter worldwide is 330 million and increasing. Twitter allows a user to share the content with a label or tag which is represented by a Hashtag (#), all the users can post or search the contents related to a particular hashtag. It allows us to analyse or classify the responses of the users on Twitter on the basis of different hashtags. The suitability of Twitter for data analysis also comes from the fact that it allows us to extract the tweets runtime free of cost just with few limitations on the number of tweets to be accessed.

But as there are lot of benefits with the microblogging sites, there exists some severe drawbacks also. People use these platforms to spread rumours and false news and mitigate hoax among community, which is a major shortfall of online social network, and it seems for now it cannot be controlled.

2   Related Work/Existing Technologies

The machine learning algorithms are mainly classified under two categories, which are (a) supervised learning and (b) unsupervised learning. The supervised learning category is of those problems in which we have a set of training labels for classification and we categorize our data only to those labels. On the other hand, the unsupervised type of algorithms works for the problems in which we do not have any kind of labels for classification rather we group our data based on the properties, data with similar properties is grouped in same cluster and so on.

Already a number of machine learning algorithms have been used by the researchers for the classification of fake content on the social media. Almost all of the supervised category of machine learning algorithms has been tested for this task of categorization as clearly, we have labels in our dataset to distinguish the contents as 'fake' or 'real'. In this paper for our consideration, we will be describing the performance of four major machine learning algorithms namely:–Decision tree, SVM, KNN and random forest. Following is the brief introduction of these algorithms:

2.1 Decision Tree Induction

The class labelled training tuples-based learning of decision trees is known as decision tree induction which is the most widely used algorithm for classification. The basic structure of a decision tree algorithm [5] is a recursively defined partitioning mechanism based on numerical value of the nodes. The node having the higher index values will be split in to further nodes. The overall representation of this method looks like a flowchart structure in which the root node is the target node and all of the leaf nodes are the class labels to be identified. The transition from the root node to the subsequent internal nodes is based on the outcome of the decision to be taken by testing some attribute. This process continues until we reach from the root node to the leaf nodes based on our decisions and those decisions becomes basis of our classification model. The problem with the decision tree algorithm is of overfitting and underfitting, which can later be overcome by using the random forest approach. A variation of decision tree algorithm is used by Djukova and Peskov [6] which is termed as complete decision tree. Decision tree algorithm is widely used by statisticians for classification of fake news such as Singh et al. [7] and Adewole et al. [8].

The main feature of decision tree method is the selection of attribute which is a heuristic measure for selecting a criterion for splitting the data. The most common measures for attribute selection are: (i) Gini index and (ii) information gain. The formulas for calculating Gini index and information gain are represented in Eqs. 1 and 2, respectively.

$$Gini(D) = 1 - \sum_{i=1}^{m} p_i^2 \tag{1}$$

where is the probability that a tuple in belongs to class D and is C_i estimated by $|C_i, D|/|D|$. The sum is computed over m classes.

$$Info(D) = - \sum_{i=1}^{m} p_i log_2(p_i) \tag{2}$$

2.2 Support Vector Machine (SVM)

The support vector machine is a technique for classification which will work for both linear and nonlinear data. It is also used for regression problems also [5]. The main principle of SVM is that it transforms the input points into a plane of higher dimension by some nonlinear or linear mapping. Then it tries to search a hyperplane to separate the data or to classify the points into different categories. By an efficient nonlinear mapping, it is always possible to find a hyperplane which separates the data. The procedure it uses to find the hyperplane is based on support vectors which are the training tuples and margins which are related to the support vectors. There may be a scenario that we have more than one hyperplane in our data plane separating our points, then we need to select the most appropriate hyperplane according to best separation or the distance between the points and the plane. There comes the use of margin feature, we select the hyperplane with the highest marginal difference for our categorization approach of the future datasets. One of the drawbacks of SVM approach is that it will ignore the outliers or we can say that it does not perform well when the dataset has a high amount of noise.

In practical, SVM uses a kernel trick which is termed as SVM Kernel which transforms the data points of lower dimensions to higher dimension. It is done because there are some points which cannot be made separable in the same plane so for them these points need to be transformed by some polynomial function to plot them on to new plane. Basically, three types of kernel are used with the SVM algorithm namely—linear kernel, polynomial kernel and radial basis function kernel. Hadeer ahmed et al. [9] used linear SVM algorithm with TF-IDF as extraction of feature technique for classification of fake news articles. There are several applications of SVM in field of classification such as heart disease classification [10], analysis and identification of traffic [11], classification of mechanical faults [12].

2.3 Random Forest

This is basically an ensemble kind of classification algorithm in which a strategy of divide and conquer works. The basis of this algorithm is same as that of decision tree algorithm but here, the number of trees is more than one. So, as the name suggests, it is a forest comprising of various trees. Each of the tree of the forest will classify the data points in to some categories and out of all the classification approaches a best one will be selected by a voting mechanism. Apart from decision tree, some other basic algorithms of classification can be used with multiple variations, and later, the best of all those variations can be selected as the final outcome.

Due to its multiple iterations, it is considered as the most accurate and robust algorithm of all; also, the problem of overfitting is minimized in random forest. It is mostly used for the feature extraction process of machine learning models. It also outperforms other classification algorithms in terms of time consumption [13].

Due to its characteristics, it is used in various feature extraction problems such as suggested by Kaur et al. [14] for text recognition of Gurmukhi scripts.

These are the four basic steps of this algorithm:

1. The dataset is searched for the initial random samples.
2. Out of each sample selected construct a decision tree or a classification strategy and calculate the result.
3. Perform a voting among the outcomes.
4. The result with the highest voting percentage will be selected as the final output.

2.4 K-Nearest Neighbour (KNN)

This algorithm uses the similarity of features to predict the labels for the new data points; basically, it is a non-parametric classification algorithm as it does not assume anything about the underlying data used. The basic principle of this algorithm is that it will first learn about the structure of the data presented for training, and then on the basis of proximity of the likeliness of the neighbours, it will choose the appropriate class. Suppose point 'X' has three neighbours (that 3 is the K in the name of the algorithm), (as shown in Fig. 1) out of which two belongs to class 'A' and one is of class 'B', then based on the maximum closeness to the class 'A', it will categorize 'X' as a point of class 'A'. Due to its slow pace of learning, it is also considered as the lazy learner algorithm. The performance of this algorithm is mostly dependent on the number of 'K' we select. That also becomes its disadvantage because every time we have to assume the value of this parameter. Apart from these drawbacks, one significant advantage of this algorithm is that it is easy to implement and robust to noise and outliers. It also outperforms SVM algorithm as feature selection according to Li et al. [15].

Fig. 1 KNN algorithm example

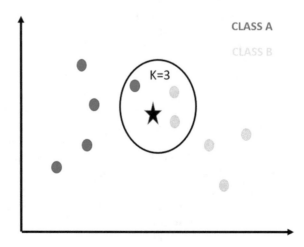

3 Ant Colony Optimization

The ant colony optimization algorithm (ACO) is a probabilistic strategy for taking care of computational issues which can be diminished to discovering great ways through graphs. Ant colony optimization algorithm (ACO) was presented by Dorigo M. [16] as a novel nature-enlivened metaheuristic for the arrangement of hard combinatorial optimization (CO) issues. ACO has a place with the class of metaheuristic, which are inexact calculations used to get sufficient answers for hard CO issues in a sensible measure of calculation time. The motivating wellspring of ACO is the scavenging conduct of genuine ants. While scanning for nourishment, ants at first investigate the region encompassing their home in an arbitrary way. When a subterranean insect finds a nourishment source, it assesses the amount and the nature of the sustenance and conveys some of it back to the home. Amid the arrival trip, the subterranean insect stores a substance pheromone trail on the ground. The amount of pheromone saved, which may rely upon the amount and nature of the sustenance, will control different ants to the nourishment source.

Algorithm 1 The Basic ACO Algorithm [17]

- Set all initial parameters and initialize value of pheromone
- while (! End constraints) do
- Perform Ant traversal
- Search locally for the solution (optional)
- Update local or global Pheromone values
- end

4 Feature Selection Using ACO

As referenced before given a list of n features, the issue of feature selection is to identify subset of minimal size m less than n, while holding a reasonably high exactness in speaking to the initial n features.

For feature selection task using the ACO algorithm, some perspectives should be kept in mind [18]. The principal issue is tended by how the graph is to be visualized containing ants and the feature set, i.e. the proper mapping of feature selection problem to ACO-compatible problem. The fundamental principle of ACO is to find a shortest path from a given cost graph. We model the features as the nodes of the graph and the edges connecting to the nodes will be the selection of the next appropriate feature. The final solution will be a traversal of ant with passing through a minimum number of nodes in the graph and that satisfies the constraints for stopping the traversal. The major benefit through any ACO algorithm is the heuristic-based best probabilistic solution.

Any feature selection based on ACO algorithm basically works on three main rules which are pheromone evaporation rule, transitional probability rule and the pheromone depositional rule.

The probability p(i, j) that ant k will include the feature set (i, j) into its path is given by the following Eq. 3.

$$p_{i,j}^k = \begin{cases} \dfrac{\tau_{i,j}^\alpha \cdot \mu_{i,j}^\beta}{\sum_{e_{iv} \in E(s^p)} \tau_{iv}^\alpha \cdot \mu_{iv}^\beta} & \text{if } e_{ij} \in E(s^p) \\ 0 & otherwise \end{cases} \tag{3}$$

where $E(s^p)$ is the set of the features which could be added to the solution. Whereas τ and μ are the variables representing the effect of pheromone trail and the heuristic information, respectively. α and β are the coefficients which control weather the probability is dependent on the heuristic value or the pheromone value.

After all ants complete their path, the pheromone value at each node will be updated so that all ants in the next iteration should not converge to the same path in the first iteration. For this the pheromone $\tau_{i,j}$ is updated by the rule in Eq. 4. The evaporation rate is represented by ρ and the amount of pheromone laid by the ant k on the feature set (i, j) is represented by $\Delta\tau_{i,j}^k$, the number of ants will be n:

$$\tau_{i,j} = (1 - \rho) \cdot \tau_{i,j} + \sum_{k=1}^{n} \Delta\tau_{i,j}^k \tag{4}$$

The amount of pheromone each ant k deposit on the edge ei, j is given by Nemati et al. [19] as:

$$\Delta\tau_{i,j}^k = \phi \cdot \gamma\left(s^k(t)\right) + \psi \cdot \frac{\left(n - \left|s^k(t)\right|\right)}{n} \quad if \ i \in S^k(t) \tag{5}$$

In Eq. 5, $\gamma\left(s^k(t)\right)$ is the measurement of performance of the classifier, whereas ϕ and ψ are the parameters controlling the effect of performance of classifier and length of the feature subset $S^k(t)$. The value of ϕ ranges between 0 and 1 and for ψ it is $1 - \phi$. Giving preference to performance of classifier more than the length of feature subset, we set the value of $\phi = 0.7$ and $\psi = 0.3$.

5   System Architecture for Analysis

Already a lot of research work is done for the determination of truthiness of news content available online, and moreover, all of those approaches follow a same flow of action. We will look at the literature available on the prospect of credibility analysis. Kumar et al. [20] proposed an approach which talks about the use of shallow classifiers based on particle swarm optimization (PSO) for rumour veracity detection of tweets available for a particular hashtag. In their model they collected the data from the Twitter API, which is the widely used source of collecting the data for analysis. Then they extracted the features from the collected tweets based on various criterion namely content-based features like bag of words, part of speech, etc. and pragmatic features like presence of emoticons and anxiety related words, sentiments etc. Also, they considered some network-based features which are given by the API itself such as no. of followers, no. of retweets. Out of all these features, they extracted some features by using the PSO-based classifier and compared the performance with all of the features and of the extracted features and their results shown a significant improvement in the accuracy.

Zubiaga and Ji [21] proposed an epistemology for information verification of tweets. They basically targeted the tweets about natural disaster and specifically about hurricanes. The study suggests that apart from the features available on Twitter, the image-based features can add up a subsequent amount of accuracy in to the results. In their research they basically collected the tweets containing the images of hurricane related disasters. They provided a reputation score based on features of the image and other text-based features and author related info.

Shu et al. [22] presented a tool for study of facts available online. They built that tool for data collection from some fact checking sites such as Politifact and Buzzfeed, and then based on the topic of fake and true news, they searched tweets related to those categories. They used social engagement of tweets as a basic feature, and for calculating the temporal engagement, they used an RNN and LSTM-based algorithm. Further, they used SVM, LR and GNB for feature-based detection.

On the basis of our study, we propose a model (shown in Fig. 2) for credibility analysis of online social media content (mostly tweets) using ACO-based feature

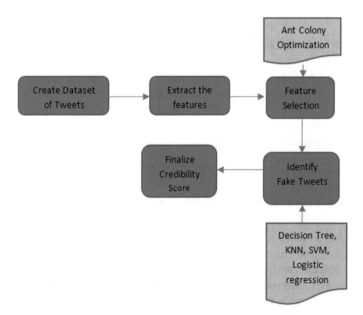

Fig. 2 System architecture

extractor, and for classification of tweets, we used decision tree, KNN, SVM and also random forest algorithm.

The model for this credibility analysis will be comprising of the following components:

- Data collection for classification (Twitter API).
- Feature extraction
- Feature selection (using ACO).
- Apply machine learning algorithm (decision tree, KNN, random forest, SVM)
- Calculate accuracy

For collection of data, we made use of some fact checking website like politifact, where they provide the facts taken from various social media platforms and the assessment of those facts by their experts as true or fake. We crawled some news topics which were fake and on the basis of those topics we searched the tweets related to that particular hashtag (#). We categorized the tweets based on false news related topics as fake and others as real manually.

In the feature extraction process, we make use of both API provided features and some extracted features from the tweet. A total of 18 features are extracted for the classification task; the features we used are listed as:

Features Provided by API

1. Retweet count,
2. Favourite count,
3. User statuses count,
4. User verified,
5. Statuses followers count,
6. Friends followers count,
7. User has URL.

Features Extracted from Text.

1. No of question marks
2. No of exclamation marks
3. No of hashtags
4. No of mentions
5. No of URLs
6. No of colon marks
7. No of words
8. Polarity
9. Length of text
10. Syntactic features
11. Diminishing statements

After applying ACO as feature extractor, it gives us these five features (shown in Table 2) as the most promising features for classification.

Feature	Percentage use
Friends followers count	27.05
Length	19.57
Statuses followers count	18.12
User statuses count	16.95
No. of words	12.94

Table 2 Features selected by ACO algorithm

Then by using these five features, we perform the classification using the machine learning algorithms. The performance of the algorithms is compared on the basis of accuracy measure. The accuracy of the algorithms is measured by the help of confusion matrix. The formula for accuracy as mentioned by Chouhan et al. [23] is shown in Eq. 6:

$$Accuracy = (TP + TN)/(TP + TN + FP + FN) \qquad (6)$$

where TP = True Positive, TN = True negative, FP = False Positive and FN = False Negative.

6 Results

We calculated the tweets related to the events of natural disaster, specifically hurricanes with the hashtag (#) Hurricane. For performing the algorithms on the dataset, we used the R language and R-studio IDE. The accuracy measure of the algorithms is compared before feature selection and after feature selection. A significant improvement is seen in all of the algorithms after feature selection with largest in decision tree around 21% and the smallest in the KNN around 8%. The results with KNN vary with a great effect if we change the number of neighbours. For our computation, we selected the number of neighbours as 10 (Approx. the square root of the number of samples taken in the test set). A wordcloud for both of the categories of tweets (fake and real) is also shown in Fig. 3.

7 Conclusion and Future Scope

The research work suggests that the impact of feature selection is important in the case of classification problems, as also suggested by our results (around 21%). The performance result for the SVM is found to be best among all of the algorithms for the above-mentioned dataset Fig. 4. The performance of all of the algorithms do tends to decrease as we try to test them on some other dataset collected for some

Fig. 3 Wordcloud for the tweets with real news (Top) and for the tweets of fake news (Bottom)

other issues (or hashtag). So, the further work will be in the direction of making the result more consistent and not effected by the topics selected or change of the dataset. Also, the result of other feature selection algorithms is needed to be compared for the dataset we collected and to observe which feature selection algorithm performs better. Also, the classification of fake tweets can be done on the basis of fuzzy logic [24] as we cannot state some script as totally fake.

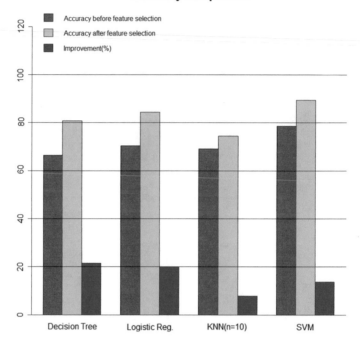

Fig. 4 Accuracy comparison for before and after feature selection

References

1. Luceri L, Braun T, Giordano S (2019) Analyzing and inferring human real-life behavior through online social networks with social influence deep learning. Appl Netw Sci 4:34. https://doi.org/10.1007/s41109-019-0134-3
2. Ferrara E (2014) Measurement and analysis of online social networks systems. In: Alhajj R, Rokne J (eds) Encyclopedia of social network analysis and mining. Springer, New York, NY
3. https://www.statista.com/statistics/268136/top-15-countries-based-on-number-of-facebook-users/
4. Alhajj R, Rokne J (eds) (2014) Encyclopedia of social network analysis and mining. https://doi.org/10.1007/978-1-4614-6170-8
5. Han J, Kamber M (2006) Data mining: concepts and techniques, 2nd edn. Morgan Kaufmann Publishers
6. Djukova EV, Peskov NV (2007) A classification algorithm based on the complete decision tree. Pattern Recognit Image Anal 17:363–367. https://doi.org/10.1134/S1054661807030030
7. Singh M, Bansal D, Sofat S (2016) Behavioral analysis and classification of spammers distributing pornographic content in social media. Soc Netw Anal Min 6:41. https://doi.org/10.1007/s13278-016-0350-0
8. Adewole KS, Han T, Wu W et al (2018) Twitter spam account detection based on clustering and classification methods. J Supercomput. https://doi.org/10.1007/s11227-018-2641-x
9. Ahmed H, Traore I, Saad S (2017) Detection of online fake news using n-gram analysis and machine learning techniques. In: Traore I, Woungang I, Awad A (eds) Intelligent, secure, and dependable systems in distributed and cloud environments. ISDDC 2017. Lecture notes in computer science, vol 10618. Springer, Cham

10. Vijayashree J, Sultana HP (2018) A machine learning framework for feature selection in heart disease classification using improved particle swarm optimization with support vector machine classifier. Program Comput Soft 44:388–397. https://doi.org/10.1134/S0361768818060129
11. Zhu Y, Zheng Y (2020) Traffic identification and traffic analysis based on support vector machine. Neural Comput Appl 32:1903–1911. https://doi.org/10.1007/s00521-019-04493-2
12. Zhi-qiang J, Hang-guang F, Ling-jun L (2005) Support vector machine for mechanical faults classification. J Zheijang Univ-Sci A 6:433–439. https://doi.org/10.1631/jzus.2005.A0433
13. Dong Y, Zhang Y, Yue J et al (2016) Comparison of random forest, random ferns and support vector machine for eye state classification. Multimed Tools Appl 75:11763–11783. https://doi.org/10.1007/s11042-015-2635-0
14. Kaur RP, Kumar M, Jindal MK (2020) Newspaper text recognition of Gurumukhi script using random forest classifier. Multimed Tools Appl 79:7435–7448
15. Li S, Harner EJ, Adjeroh DA (2011) Random KNN feature selection-a fast and stable alternative to Random Forests. BMC Bioinform 12:450. https://doi.org/10.1186/1471-2105-12-450
16. Dorigo M, Stützle T (2004) Ant Colony Optimization Theory, in Ant Colony Optimization, MIT Press, pp. 121–152.
17. Dorigo M, Birattari M, Stutzle T (2006) Ant colony optimization. IEEE Comput Intell Mag 1(4):28–39. https://doi.org/10.1109/MCI.2006.329691
18. Sohafi-Bonab J, Aghdam MH (2010) Erratum: feature selection using ant colony optimization for text-independent speaker verification system. In: Cai Z, Hu C, Kang Z, Liu Y (eds) Advances in computation and intelligence. ISICA 2010. Lecture notes in computer science, vol 6382. Springer, Berlin, Heidelberg
19. Nemati S, Basiri ME, Ghasem-Aghaee N, Hosseinzadeh Aghdam M (2009) A novel ACO–GA hybrid algorithm for feature selection in protein function prediction. Expert Syst Appl 36(10):12086–12094. ISSN 0957–4174, https://doi.org/10.1016/j.eswa.2009.04.023
20. Kumar A, Sangwan SR, Nayyar A (2019) Rumour veracity detection on twitter using particle swarm optimized shallow classifiers. Multimed Tools Appl 78:24083–24101. https://doi.org/10.1007/s11042-019-7398-6
21. Zubiaga A, Ji H (2014) Tweet, but verify: epistemic study of information verification on Twitter. Soc Netw Anal Min 4:163. https://doi.org/10.1007/s13278-014-0163-y
22. Shu K, Mahudeswaran D, Liu H (2019) FakeNewsTracker: a tool for fake news collection, detection, and visualization. Comput Math Organ Theory 25:60–71. https://doi.org/10.1007/s10588-018-09280-3
23. Chouhan SS, Kaul A, Singh UP (2019) Image segmentation using computational intelligence techniques: review. Arch Comput Methods Eng 26:533–596. https://doi.org/10.1007/s11831-018-9257-4
24. Madani Y, Erritali M, Bengourram J et al (2019) A multilingual fuzzy approach for classifying Twitter data using fuzzy logic and semantic similarity. Neural Comput Appl. https://doi.org/10.1007/s00521-019-04357-9

Smart Support System for Navigation of Visually Challenged Person Using IOT

Tuhin Utsab Paul and Aninda Ghosh

Abstract The paper discussed about an IOT-based smart navigation support system that will assist in the movement of visually challenged person within room or outdoor environment. The device proposed in the paper uses camera to capture real-time video, and then, it processes those for object detection of obstacles or human figures and transmits the knowledge gained by object detection as speech signals to the person. Based on those speech instructions, the person may navigate in indoor or outdoor environment. The device is an image-based tracking system for precise locomotive assistance and tracking. The device is capable of real-time video acquisition, object detection including human, doors, stairs and obstructions and converting the knowledge to speech and transmitting via earphone to the user. Using this device, visually challenged persons will be much more independent in case of movement.

Keywords Smart navigation system · Voice modulated · Visually challenged people · IOT · Mapping

1 Introduction

The growth of IOT had enabled the establishment of connection between the electronic medical equipment with the IOT grid. This paper discussed about a device that uses an image-based tracking system within the system for precise locomotive assistance for the visually challenged people. This IOT-based device, named as Vision-X, can be used for guiding any visually challenged person to navigate within a building or in outdoor. Vision-X connects a camera to a processing unit that analyzes the video in a real-time scenario and does object tracking and detection and converts the knowledge into audio signals that are conveyed to the person via an earphone. The user's locomotion data is sent to the server for generating a digital map of the places visited for any future requirement. The main objective of the device is to help

T. U. Paul (✉)
St. Xavier's University, Kolkata, India

A. Ghosh
Distronix, Kolkata, India

© The Author(s), under exclusive license to Springer Nature Singapore Pte Ltd. 2022   27
P. Nanda et al. (eds.), *Data Engineering for Smart Systems*, Lecture Notes in Networks and Systems 238, https://doi.org/10.1007/978-981-16-2641-8_3

visually challenged person to navigate independently and safely. The device uses various algorithms for video processing and object detection and audio generation. The processing unit is an ARM processor. Two cameras are used as input device for accurate analysis of depth in video. This device is portable and can be carried in a small pouch or worn as a belt and is highly power efficient. This runs on rechargeable lithium-ion batteries, and a single charge is sufficient for running the device for nearly 24 h.

2 Literature Review

Volodymyr Ivanchenko [1] in 2008 described clear path guidance guiding blind and visually impaired wheelchair users along a clear path that uses computer vision to sense the presence of obstacles or other terrain features and warn the user accordingly. Since multiple terrain features can be distributed anywhere on the ground, and their locations relative to a moving wheelchair are continually changing, it is challenging to communicate this wealth of spatial information in a way that is rapidly comprehensible to the user. Elchinger [2] in 2008 proposed Utilizing QR Code. This paper proposed a barcode-based system to help the visually impaired and blind people identifying objects in the introduced environment. The system is based on the idea of utilizing QR codes (two-dimensional barcode) affixed to an object and scanned using a camera phone equipped with QR reader software. The reader decodes the barcode to a URL and directs the phone's browser to fetch an audio file from the web that contains a verbal description of the object. Their proposed system is expected to be useful in real-time interaction with different environments and to further illustrate the potential of new idea. Bighamy [3] in 2010 introduced VizWiz: LocateIt. This method enables blind people to take a picture and ask for assistance in finding a specific object. The request is first forwarded to remote workers who outline the object, enabling efficient and accurate automatic computer vision to guide users interactively from their existing cell phones. A two-stage algorithm is presented that uses this information to conduct users to the appropriate object interactively from their phone. They produced an app in iOS and app is live in iTunes, currently, iPhone users can use it, and now they are trying to make the app for android mobiles. Jing Su in September 2010 [2] introduced Timbremap. A sonification interface enabling visually impaired users to explore complex indoor layouts using off-the-shelf touch screen mobile devices. This is achieved using audio feedback to guide the user's finger on the device's touch interface to convey geometry. The user study evaluation shows Timbremap is effective in conveying nontrivial geometry and enabling visually impaired users to explore indoor layouts. Roberto Manduchi in 2010 explored blind guidance using mobile computer vision. Here, they focused on the usability of a wayfinding and localization system for persons with visual impairment. This system uses special color markers, placed at key locations in the environment, which can be detected by a regular camera phone. Three blind participants tested the system in various indoor locations and under different system settings. Quantitative performance results are reported for the different system settings, and the reduced field of view setting is the most challenging one. The role of frame rate and of marker size did not result fully clear in the experiments. Bradski and Kaehler [1] in 2011 explored pedestrian crossing. A mobile cloud collaborative approach is provided here for context-aware outdoor navigation, where the computational power of resources made available by cloud computing providers for real-time image processing. The system architecture also has the advantages of being extensible and having minimal infrastructural reliance, thus allowing for wide usability. They have developed an outdoor navigation application with integrated support for pedestrian crossing guidance and report experiment results, which suggest that the proposed approach is promising for real-time crossing guidance for blind. Boris Schauertey in 2012 proposed to find lost things. Here, they proposed a computer vision system that helps blind people to find lost objects. To this end, they combine color and SIFT-based object detection with signification to guide the hand of the user toward potential target object locations. This way, it is able to guide the user's attention and effectively reduce the space in the environment that needs to be explored. They verified the suitability of the proposed system in a user study. They experimentally demonstrated that the system makes it easier for visually impaired users to find misplaced items, especially if the target object is located at an unexpected location. Kanungo et al. [4] in June 2013 invented object recognition in mobile phone. In this work, they described main features of software modules developed for Android smartphones that are dedicated for the blind users. The main module can recognize and match scanned objects to a database of objects, e.g., food or medicine containers. The two other modules are capable of detecting major colors and locate direction of the maximum brightness regions in the captured scenes. The paper was concluded with a short summary of the tests of the software aiding activities of daily living of a blind user.

3 Product Modeling

The main objective of designing the device was to make it economical and portable so that it can be used by common man and can be easily carried along. The device is segregated into various blocks as follows (Fig. 1):

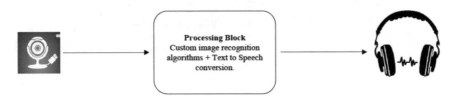

Fig. 1 Basic outline of the devise

Block	Hardware used
Input block	2 USB Camera more than 10 MP
Processing block	Raspberry Pi B + 1.2 GHz with 1 GB RAM
Power block	Lithium-ion power bank of 5 V
Output block	Bluetooth/wired earphone

The device is constructed by connecting the various hardware devices as mentioned above. The figures below show the various components of the device and the connections between them (Figs. 2, 3, and 4).

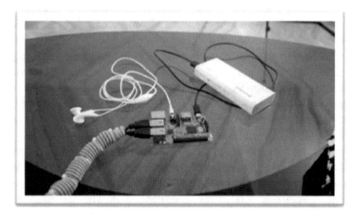

Fig. 2 Components—Raspberry Pi 3 B+ , Earphone, Power bank

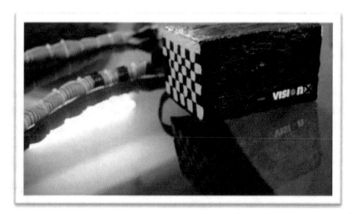

Fig. 3 Eyepiece

Fig. 4 Raspberry Pi 3 B+ processor board

4 Algorithms

The device executes various algorithms at different stages to arrive at the required output. Various algorithms such as ground and obstruction detection, human detection, k-means algorithm [5], door detection, speech generation are executed in sequence to achieve the desired result. The process flow diagram is given above (Fig. 5).

To start using the device, first, it needs to power on the device and let the OS boot up for the first time, application starts after the boot up process is complete, and for the first time, software needs to be calibrated according to height of the user so as to identify possible interested regions within a captured frame, e.g., identifying amount of ground area to be present for subsequent movement. Once calibrated the hardware works independently and starts taking snaps of the environment continuously and processes the snaps and gets to a conclusion whether the path in front is safe or not, if not then finding the direction where a most probable safe path can exist depending on the clearance in front of user. The device doesn't classify all the objects in front but does classify some special cases which seem to play important role in navigation of a visually impaired person, e.g., doors, staircases, human and ground. The feedback provided to the user is based on speech signals. According to previous medical research, for any visually impaired person, the ears are the next powerful sense organs and it is quite reliable to use voice feedback. But keeping ears busy all the time with the speech output will not be a great advantage in hand rather it will interfere with the natural navigation system of human body. Whenever a major change in environment is observed, this information is given to user in the form of speech signals. Major change in environment includes the appearing of a staircase or door, which doesn't allow a natural flow of walk, thus enabling the user to know about the obstacles beforehand.

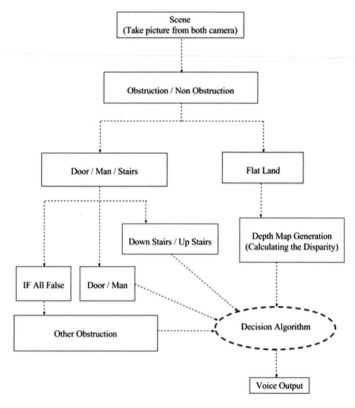

Fig. 5 Process flow diagram

The flow of the task that is performed is as follows:

I.   Acquire image from the camera.
II.   Send the digital image to the device via USB cable or Bluetooth.
III.   Followed by object segmentation.
IV.   After the classes of objects are segregated out, the next task is to extract the features out of them based on some predefined criteria.
V.   The features so extracted, form the base for classification of the particular object into door, staircase, human or open ground.
VI.   Once the classification is done, a judgment algorithm is run so as to decide whether a major change has been observed in the environment or not. If observed, user is made aware of the change in the form of a suitable phrase in a language he has selected while installation of the system.
VII.   The previous steps are followed repeatedly until the device is switched of or the battery has been drained out.

Various state-of-the-art algorithms such as k-means, canny edge detection, ROI detection, contour detection algorithms are used on the video in real time for object detection and identification [6].

Having 2 cameras within the system introduced stereo image disparity. Having the 2 stereo images the disparity is calculated based on the angle of the camera mount and the ROI using HOG features within the simultaneous frames. The depth accuracy achieved was around 75% + which is significant for an embedded device.

The various methodologies used for human face detection, stair detection and door detection are:

Human face detection–Haar cascade feature detection is used. The complete face detection cascade has 38 stages with over 6000 features. Nevertheless, the cascade structure results in fast average detection times. On a difficult dataset, containing 507 faces and 75 million subwindows, faces are detected using an average of 10 feature evaluations per subwindow. In comparison, this system is about 15 times faster than the implementation of the detection system constructed by Rowley et al.

Staircase detection—The Hough Transform is an algorithm presented by Paul Hough in 1962 for the detection of features of a particular shape like lines or circles in digitalized images. In its classical form, it was restricted to features that can be specified in a parametric form. However, a generalized version of the algorithm exists which can also be applied to features with no simple analytic form, but it is very complex in terms of computation time. Although Hough Transform is a standard algorithm for a line or circle detection it has weak points, especially its computational complexity. However, a faster version called Fast Hough Transform has been developed.

Door detection—The method is based on the canny edge detection algorithm and the Hough Transform and uses fuzzy logic to determine the probability of the existence of doors that fit a set of predefined rules. The implementation of the algorithm is composed of a few core steps: region of interest (ROI) preprocessing, door line extraction and door classification [7–9]. A heuristic method is used to extract parallel lines from the line database. The parallel lines are used to find the four corners of the door. The heuristic method employs three metrics to determine the likelihood of line segments being part of a door. One metric is assigned to vertical line pairs, the second metric is assigned to the top horizontal line and the third metric is assigned to the bottom horizontal line. High metric values indicate more confidence that the line combination constitutes a door.

Detailed discussions of these algorithms are beyond the scope of this paper.

5 Output

The outputs are shown for various cases such as door, obstructions, human, stairs. (Fig. 6).

Door detection algorithm separates the door from the background and finds the optimum distance from the person wearing the device. A secondary camera is there to find the clear portion (green patch) of floor for effective locomotion (Fig. 7).

Human detection algorithm finds the face of the person interacting, and it provides an audio feedback of the person's history (unknown or already registered) (Fig. 8).

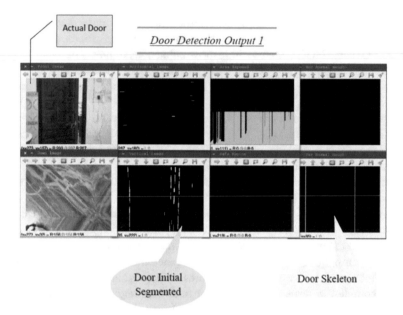

Fig. 6 Door detection 1

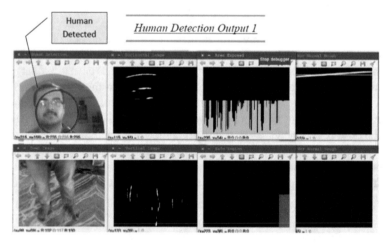

Fig. 7 Human detection 1

Staircase detection algorithm finds the sharp edges of the stairs and use projections into horizontal plane to find the optimum distance and response time to reach that stair.

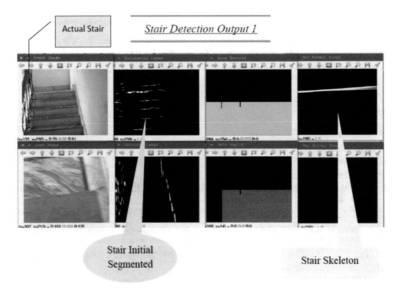

Fig. 8 Stair detection 1

The accuracy received in a test run of over 300 times are as follows:
True Positive–85%
True Negative–79%
False Positive–10%
False Negative–15%

6 Conclusion

Our device, vision-X, is an amalgamation of video processing, object detection and speech generation. This device is a buildup of various devices connected together to function in unison and generate the desired output. Cameras take real-time video, raspberry pi processor processes the video to detect human, objects, stairs and doors and finally, the findings are converted to speech and transmitted by the microphone to the user. A number of state-of-the-art algorithms work to detect various objects precisely. Multiple trial runs have shown that the device is successful in more than 98% cases in indoor setup consisting of human, doors, stairs and obstructions. Even in outdoor setup, trials show a success rate of approximately 95% cases. The main risk in using the device in outdoor setting is the time lapse in detecting moving objects like car. This is the challenge that needs to be overcome. But, in current scenario, this device can greatly help the visually challenged persons in movement in indoor. They can move independently and confidently with the use of this device. This device is portable and of low cost. Future research can be directed to real-time detection of fast-moving object so that this device can be used in outdoor conditions also. Moreover, the movement of the person can be real-time uploaded vis GPS system so that his/her movements can be monitored remotely and his/her safety can be ensured.

References

1. Bradski G, Kaehler A (2008) Learning OpenCV: computer vision with the OpenCV library.
2. Elchinger GM (1981) Mobility cane for the blind incorporating ultrasonic obstacle sensing apparatus. Patent no.: US 4280204 A
3. Cheeseman P, Self M, Kelly J, Stutz J (1988) Bayesian classification seventh national conference on artificial intelligence
4. Kanungo T, Mount DM, Netanyahu NS, Piatko CD, Silverman R, Wu AY (2002) An efficient k means clustering algorithm: analysis and implementation. IEEE Trans Pattern Anal Mach Intell 24(7). (Almaden Res. Center, San Jose, CA, USA)
5. Ray S, Turi RH (1999) Determination of number of clusters in k-means clustering and application in colour image segmentation
6. Acharya T, Ray AK (2005) Image processing: principles and applications
7. Campbell et al (2014) Mobility device and method for guiding the visually impaired, united states patent. Patent no.: US 8,825,389 B1
8. Cover T, Hart P (1967) Nearest neighbor pattern classification. IEEE Trans Inf Theory 13(1)
9. Heitz G, Gould S, Saxena A, Koller D (2008) Cascaded classification models: combining models for holistic scene understanding. In: Advances in neural information processing systems, vol 21

Identity-Based Video Summarization

Soummya Kulkarni, Darshana Bhagit, and Masooda Modak

Abstract The process of automatic video summarization involves identification of frames automatically that contains essential information, which would not have been possible if it was not for machine learning. There are lots of algorithms that can be used for summarization of videos based on face recognition and speaker verification, for instance: face recognition could be performed by caffe-based model containing facial landmarks, principal component analysis, discrete cosine transform, 3D acceptance methods, Gabor wavelets method, etc. This work has been centered on a caffe-based model containing facial landmarks for face recognition in an efficient manner. There are a lot of factors affecting face recognition like accuracy, time limitations, process speed, and availability. With all these in mind, caffe-based model was selected for being accurate and having fast computation time.

Keywords Video summarization · Face recognition · Speaker verification · Facial embeddings · Caffe model

1 Introduction

Due to the increasing volume of video content on the Web, and the human effort taken to process it, new technologies need to be researched to develop efficient indexing and search techniques to manage effectively and efficiently the huge amount of video data. Among the current upcoming technologies based on video processing, video summarization is one of them. Extracting a certain set of frames based on the features processed from the actor's face as well as voice provided by the user as an input is combined into a video. This approach of summarizing a video helps reduce the effect the original video has on storage as well as saves the time of a user to understand the

S. Kulkarni · D. Bhagit (✉) · M. Modak
SIES Graduate School of Technology, Navi Mumbai, India
e-mail: masooda.modak@siesgst.ac.in

© The Author(s), under exclusive license to Springer Nature Singapore Pte Ltd. 2022
P. Nanda et al. (eds.), *Data Engineering for Smart Systems*, Lecture Notes in Networks and Systems 238, https://doi.org/10.1007/978-981-16-2641-8_4

video. The proposed approach contains four phases which are preparing the dataset, face recognition, speaker recognition, and feature extraction and summarization on unseen video.

Face recognition implemented in this work is based on FaceNet [12], which computes a 128-d embedding that quantifies the face itself. This has been made possible because of two parameters viz. input data to the network and triplet loss function. FaceNet uses these two parameters to extract a compact 128-D facial embedding and train the output based on LMNN [11]. The triplet loss function works on the basis of separating the positive pair of thumbnails from the negative pair of thumbnails using a distance margin. With the help of this process, the neural network is able to quantify faces by creating distinctive facial embeddings.

However, speaker recognition consists of two sections, identifying the person speaking which is performed during the training of the speaker models and verifying the voice input of the person speaking which is being performed during validation. Verification of a speaker is done based on MFCC [3] and GMM [1] along with Python's 'python_speech_features' library [6] which resulted in an accuracy of 95.29% on a custom made dataset.

2   Proposed System

Video summarization technique in this work, takes into account the identity of a person based on their face and voice features. Now the target frames based on face recognition on combining with the speaker verification section ensure that the person is identified correctly. Therefore, by specifically targeting the frames which contain both the target person's face and voice, ensures that the video primarily contains only the target person, thereby making the video a proper summary of the input raw video. The architecture of the entire system is as shown in Fig. 1.

The proposed technology is composed of three sections: facial recognition, speaker recognition, and target frame extraction. The face recognition part con-

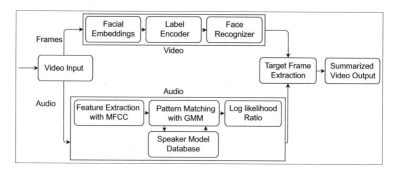

Fig. 1 Proposed system architecture

Fig. 2 Illustration of 68-D facial landmarks

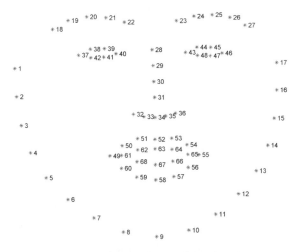

Fig. 3 Face identification

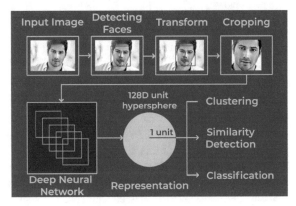

sists of face detection and face identification. Face detection is implemented using FaceNet [8] which computes a 128-d embedding that helps in identifying the face. Face identification is implemented using a support vector machine [9] to identify faces.

This work is centered around face detection framework [4] which uses OpenCV's pretrained caffe-based deep learning model containing facial landmarks as shown in Fig. 2 which could be used to map facial features extracted from frames to detect faces. These annotations are extracted with the help of Dlib's facial landmarks model [2] which was trained on the iBUG 300-W dataset which contains 68 point landmarks. Then these facial features are passed on to support vector machine (SVM) to identify faces depending on the dataset provided to train the network. Dlib library itself maps the facial features of the trained dataset with the identified face from the frames to predict the person being identified. Figure 3 explains this process in a much better way.

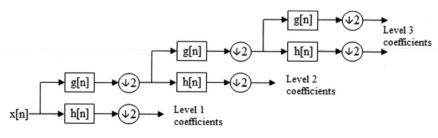

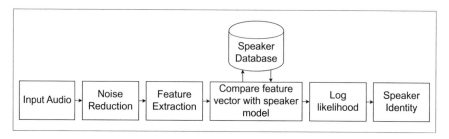

Fig. 4 Discrete wavelet transform

Fig. 5 Speaker identification

Before using the audio for the speaker recognition process, the data has to be preprocessed to make it clean and noise-free. The denoising process consists of decomposing the original signal using discrete wavelet transform (DWT) as shown in Fig. 4, then thresholding the detail coefficients based on ambient noise level and reconstructing the signal. The speaker verification part uses the Mel frequency cepstral coefficient (MFCC) estimation combined with the GMM approach to verify that the voice belongs to the detected person. The segment is selected if the proportion in which the target person's face and voice appears is greater than a threshold that is identified based on experimental results. The selected segments are combined to make the output audio on which pattern matching is performed from the trained set of utterances for all the speakers and segments based on verified results are extracted and combined to form the summarized video. The structure of speaker identification is mentioned clearly in Fig. 5.

3 Methodology

Individual frames are extracted from the input video, and face detection is performed to obtain ROI using 'res10_300 × 300_ssd_iter_140000'—a caffe-based DL face detector model [7, 10] which quantifies faces with a threshold factor of 0.5. Then to extract 128-D facial features from these features, 'openface_nn4.small2.v1.t7'—a torch deep learning model [5] is used. Then, scikit-learn's support vector machines (SVM) is used to identify faces based on these features and store the indexes of the corresponding frames. Based on these indexes, audio from the main input is extracted using Python's 'speech_recognition'. Features from this audio is extracted using Mel frequency cepstrum coefficients (MFCC) with 'Python_speech_features' library which returns a feature vector. These feature vectors are compared with the stored scikit-learn's Gaussian mixture model (GMM) to verify the identity.

4 Experimental Setup

4.1 Process

Anaconda is selected as the backbone of this work with Python 3.7.7. Using OpenCV's inbuilt methods, the video was taken as an input and separated into frames. These individual frames are passed on to embedded, detector, recognizer, and LabelEncoder models for performing face recognition which returned a list of indexes of frames corresponding to the respective actor's faces. A smoothing algorithm was implemented on this generated list to remove stutters in the video and alter the list. Based on this list, audio from the source video is extracted and trimmed into various segments. These segments are used as input for speaker verification with the help of pretrained GMM speaker models and identified results are segregated based on the required subject and merged with the respective frames during that specific time-frame. After the completion of face recognition and speaker verification of every frame as well as audio segments, the final result is combined to form a summarized video.

4.2 Deployment

This work has been locally hosted with the help of flask framework acting as a server where front end is built with HTML, CSS, and JavaScript. The homepage takes video files as well as target subject as input, sends it to the video processor which performs the face as well as speaker recognition and sends the summarized video as output to result page.

5 Results

5.1 Testing of the Model

The flask server has been deployed using the command shown in Fig. 6, then the port in which the application is hosted is accessed with a browser which opens the landing page as shown in Fig. 7. This page takes a video as an input as well as the name of

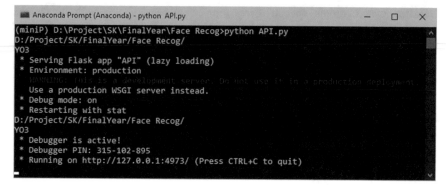

Fig. 6 Command to deploy model to flask

Fig. 7 Landing page

the subject present in the video. The application then processes for a while before redirecting into the output page as shown in Fig. 8 which displays the summarized video.

5.2 Comparative Analysis

There are existing video summarization techniques using various methods like frame difference, histogram difference, the sum of conditional variance, etc., but the amount of compression, as well as condensation of summary in our case, is better for the pro-

Fig. 8 Output page

Table 1 Comparative analysis of summarization techniques

Summarized methods	DCR	SS	CR (%)
Existing methods			
Frame difference	−2.685	0.7286	74
Histogram difference	−2.812	0.7377	74.1
Sum of conditional variance	−4.751	0.8261	83.3
Our method			
Face recognition and speaker verification	−9.418	0.9040	79.9

Where,
Data compression ratio (DCR) = 1–(Uncompressed Size/Compressed Size)
Space saving (SS) = 1–(Compressed Size/Uncompressed Size)
Condensed ratio (CR) = [1–(Output Frames/Input Frames)] × 100

vided input video. Experiments with the above-mentioned methods was performed based on the parameters: Data compression ratio (DCR), space saving, and condensed ratio (CR) (Table 1).

6 Conclusion

This work of identity-based video summarization is more accurate and well defined as well as provides the feasibility of identity verification for target frame extraction reliably. Face recognition and speaker verification for person-based video summarization has opened up a whole new world of possibilities providing a lot of space for improvement in all the three technologies. If these technologies are improved further and integrated with faster techniques for summarization, then that may even lead to the implementation of much faster and reliable video summarization based on the information being provided till that particular time.

References

1. Campbell WM, Sturim DE, Reynolds DA (2006) Support vector machines using GMM super-vectors for speaker verification. IEEE Signal Process Lett 13(5):308–311
2. Dlib C++ Library, http://blog.dlib.net/
3. Hussain H, Salleh S, Ting C, Ariff A, Kamarulafizam I, Suraya R (2011) Speaker verification using Gaussian mixture model (GMM). In: 5th Kuala Lumpur international conference on biomedical engineering 2011. Springer, pp 560–564
4. Khan M, Chakraborty S, Astya R, Khepra S (2019) Face detection and recognition using opencv. In: 2019 international conference on computing, communication, and intelligent systems (ICCCIS). IEEE, pp 116–119
5. Model and Accuracies-OpenFace https://cmusatyalab.github.io/openface/models-and-accuracies/
6. Python-Speech-Features. https://python-speech-features.readthedocs.io/en/latest/
7. Qiao S, Ma J (2018) A face recognition system based on convolution neural network. In: 2018 Chinese automation congress (CAC). IEEE, pp 1923–1927
8. Schroff F, Kalenichenko D, Philbin J (2015) Facenet: a unified embedding for face recognition and clustering. In: Proceedings of the IEEE conference on computer vision and pattern recognition, pp 815–823
9. Sun X, Zhang Q, Wang Z (2009) Face recognition based on NMF and SVM. In: 2009 Second international symposium on electronic commerce and security, vol 1. IEEE, pp 616–619
10. Vaishali SS (2019) Real-time object detection system using caffe model. Int Res J Eng Technol 6(5):5727–5732
11. Weinberger KQ, Saul LK (2009) Distance metric learning for large margin nearest neighbor classification. J Mach Learn Res 10(2)
12. William I, Rachmawanto EH, Santoso HA, Sari CA et al (2019) Face recognition using facenet (survey, performance test, and comparison). In: 2019 Fourth international conference on informatics and computing (ICIC). IEEE, pp 1–6

Security Testing for Blockchain Enabled IoT System

A. B. Yugakiruthika and A. Malini

Abstract The IoT and blockchain were the biggest technology that was growing in day to day life. The testing as well as validating is the one that deals with all the errors. The testing and validating are heterogeneous, and it has its own challenges. The emphasis on formal confirmation procedures has implied that less consideration has been given toward testing systems. The scope of this work relies on security testing for blockchain-enabled IoT system. The blockchain is used to store the database and provides security from unauthorized user. Right now, we talk about plan, testing strategies, furthermore, instruments to permit security testing at scale of blockchain and IoT applications. A framework influences Ethereum for authentication, reputation, identity of IoT gadgets. Gadgets are enlisted in a keen agreement by means of a web interface and send cryptographically marked messages to a stage that validates them by blockchain. Then by using security testing, it determines the data and the resources are protected and also it will be preserved from vulnerabilities and checks if it provides better platform for authorized users.

Keywords Internet of Things · Blockchain · Security testing · Smart contract · Digital identity

1 Introduction

IoT is confronting personality, security and interoperability issue. Current frameworks depend on client–server model that will before long be unacceptable because of the fast increment in the quantity of devices associated with the Internet. Blockchain is shared, dispersed and decentralized record that permits advancement of decentralized applications. This proposition analyzes the idea of its utilization for enlistment and the executives of IoT gadgets. A framework that comprises of a smart contract, web interface, device and a better platform has been created. Frameworks clients, substances, register gadgets inside a smart contract with their control data by means of

A. B. Yugakiruthika · A. Malini (✉)
Department of Computer Science, Thiagarajar College of Engineering, Madurai, India
e-mail: amcse@tce.edu

© The Author(s), under exclusive license to Springer Nature Singapore Pte Ltd. 2022   45
P. Nanda et al. (eds.), *Data Engineering for Smart Systems*, Lecture Notes in Networks and Systems 238, https://doi.org/10.1007/978-981-16-2641-8_5

a web interface. Gadgets create a sign messages by utilizing private key which is sent to the stage alongside control data and related verification. The received messages are approved by utilizing blockchain, which provides validation, respectability and non-repudiation.

The challenges related to the selection of blockchain are the need to distinguish significant use cases that would provide profit from the reconciliation of blockchain innovation. The Internet of Things (IoT) has taken quite some time that has been related to security shortcomings and difficulties and specialists then the associations have started investigating the utilization of blockchain to making sure about the IoT. Associations like IOTA and the Trusted IoT Alliance have started to concentrate on IoT security through the use of blockchain.

The distributed edge IoT gadgets gather furthermore to communicate information. The IoT frameworks depend upon this information to offer propelled types of assistance, mechanization highlights and customized encounters to end clients. IoT frameworks are dynamic and conveyed. They contain incorporate gadgets, versatile applications, entryways, cloud administrations, investigation and AI forms, organizing framework, web administrations, stockpiling frameworks, haze layers and clients. These frameworks will read information and then record it. This paper addressed two different technologies.

1.1 Blockchain

In the framework, the ledger is characterized as blockchain that maintains the transaction records. In a decentralized framework on a shared system, the blockchain is applied. The full copies of records are stored in each node of the blockchain. On the endorsement of each transaction, these record will be updated in the ledger.

1.2 Internet of Things

A quick developing arrangement of advancements that help the change of business and strategic. The IoT has arrived at different degrees of development across divisions, for example, shopper, transportation, vitality, medicinal services, manufacturing, retail and money related.

The IoT is the systems administration of physical gadgets such as associated vehicles, savvy structures, mechanical control frameworks, automation and applies autonomy frameworks and different things inserted with hardware, programming, sensors, actuators and system availability that empower these items to trade information. This paper describes an overview of blockchain technology and blockchain that enable to be used to secure the IoT.

The objective of this paper is to build a smart contract via web interface by using blockchain for registration and control of IoT devices. Then the proposed security testing enables the tester or developer to identify the vulnerabilities that produce the large impacts on the performance and development of smart contract through web interface. This paper proposes a methodology to provide a secured environment with reduced energy consumption and increased performance.

2 Literature Review

Upon surveying a number of papers published in the past decade, we understood the challenges in blockchain enabled IoT, the importance of testing and the approaches taken toward semantic interoperability and in general, thus identifying room for improvement on the subject. In wearable system, there is numerous number of states which lead to problem. To avoid, they used semantic similarity clustering algorithm and combinatorial test based on FSM. The FSM design is based on black box testing and then convert that to regular expression. The combinatorial testing is used to debug the errors [1]. They proposed a BlendCAC for effective access, service and information to transfer in large amount. In this paper, they transfer large number of resources. The resources can be accessed by unauthorized user. To secure the resources, they use blockchain [2]. The IoT is leveraged from big data and cloud computing. To protect the data, they use blockchain technique. They also explain the importance of blockchain in this paper [3].

They proposed fog computing that combines one or more physical devices with processing and sensing capability. The aim of this paper is to secure the authentication using blockchain. They used Ethereum software for blockchain to reduce the fraud control [4]. They proposed a third-party auditor to check the integrity of data based on user demand. They used cloud to save encrypted block of data [5]. They proposed bitcoin transaction. The error will occur when transaction of cryptocurrency is high. To check, the satisfaction of all test suite is done by property-based testing. They build new blockchain system containing the test suite that is satisfied [6].

A useful tool which proposes an algorithm is called optimal node deployment. In this paper, they measure the performance and test the relationship between the communication and blockchain model by using this algorithm [7]. Proposes a comparative testing based on read or write data and process or record transaction of blockchain. The bottleneck problem can be easily identified. They used Ethereum software for blockchain to reduce the fraud control [8]. The repeatable testing occurred for IoT application. The present paper provides an efficient analysis of different testing techniques for IoT system. In addition to that, we were using blockchain that provides privacy and reliability concerns in the IoT [9].

For IoT gadgets, a decentralized to control instrument methodology is proposed and it will be appropriate for many situations. The control instrument depends on the idea of blockchain [10]. To detect information, IoT was utilized by blockchain. The practical application is to manage the expense of the device owner and that provides an absolute, perpetual log and enables simple admission to their contractions [11].

3 Methodology

The Umair Khalid et al. proposed authentication scheme and access control based on fog computing and blockchain technology. Similarly, the authentication scheme provides a mechanism of proof of work algorithm [11]. It is a consensus algorithm and is used to ensure the transaction and create new chain of blocks. This algorithm takes huge amount of energy. In order to reduce, this paper proposes a proof-of-stake algorithm. The proof-of-stake will validate or mine a block of transactions based on the coins they have. The objective of this algorithm is to reduce the energy consumption while verifying each blocks.

4 An Overview of Blockchain Technology

Generally, in blockchain, the transaction is grouped and then converted into blocks. Each block will contain the hash value of the previous blocks. The data in each block will be digitally signed to confirm the integrity of data that was recorded during transactions. The three components of blockchain are:

4.1 The Network of Autonomous Nodes

Free hubs will produce, register and exchange distributed records independently. To exchange the records, the approval from neither local authority nor third party is very important. To maintain the consistency of records, all hub in the blockchain will work together. Each and every hub will run a modified instrument called an agreement [12]. The most important procedure in hub is accord. It provides an effective method to refresh the blockchain due to lot of exchanges occurred. Most of the hubs have approved similar exchanges by establishing the guarantee.

The main objective is to keep the records inadequate number of framework with peer address and state of art [13]. Agreement components protect against malicious and that can also degenerate the trustworthiness of records by (a) altering exchanges (b) semantically unpermitted exchanges (for example, "double-spending" and moving not-claimed resources in a cryptographic money setting) or (c) obstructing the acknowledgment and booking of right exchange demands.

This approach picked during the advancement of the blockchain administration makes preparations for explicit assaults. These assaults are not absolutely specialized in nature. With bitcoin's "Evidence of Work," for instance, there is a monetary disincentive to dealing with 51% of the hash power inside the system [14]. Increasing 51% of the mining hash rate would conceivably permit an aggressor to twofold spend coins or adjust an ongoing history of exchanges. What's more that noxious gathering could basically utilize their hash power toward the way toward digging to create gains for themselves [15]. Likewise, increasing 51% hash rate and spreading vindictive exchanges would quickly demolish trust in the digital money.

4.2   A Ledger

A ledger is the refile for recording and adding up to financial exchanges estimated regarding a fiscal unit of record by account type, with charges and credits in discrete segments and a starting money-related parity and consummation money-related parity for each record. Every exchange is investigated and recorded. Except for the overall record, every record comprises of comparative individual records [16]. For instance, the entirety of the individual sellers you work with is remembered for your buys record. To finish a record-to-record exchange, you survey the sorts of records accessible, select the proper records and enter the exchange utilizing a twofold section bookkeeping framework.

4.3   The Distributed Database

A ledger is worked after some time as a new exchanges are included, and it is accessible and imitated across hubs in the framework [17]. Each hub on the system has its own duplicate of the database and can get to the historical backdrop of any. The blockchain has a specific size of digital money that will drive prerequisites for capacity that limit inside IoT and other gadgets that have the records.

5   Smart Contracts

A smart contract is a keen agreement inserted in a code. There are a variety of laws in the code in which the agreement consents to comply with one another. The understanding is automatically maintained until the predefined rules are met. Keen agreements give systems to effectively overseeing tokenized resources and access rights between at least two gatherings. One can consider it like a cryptographic box that opens worth or get to it. Keen agreements offer an open and obvious solution in a few lines of code to install management guidelines and market logic, which can be reviewed and enforced through the dominant component agreement. From elements within and outside the blockchain, a shrewd agreement can be summoned. Of these components, the infuse details from the savvy contract that is applicable to the savvy contract.

Whenever executed effectively, brilliant agreements could give exchange security better than customary agreement law, in this way diminishing coordination expenses

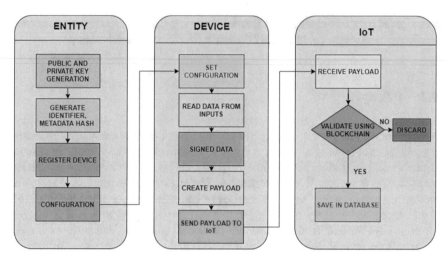

Fig. 1 Architecture

of reviewing and authorization of such understandings [18]. They can follow the exhibition of the understanding progressively. Keen agreements diminish the exchange expenses of understandings by significant degrees; explicitly, they lessen the expenses of (I) agreeing, (II) formalization and (III) implementation.

6 Blockchain-Enabled IoT

Blockchain is used in IoT to enhance privacy and security [19]. Blockchain at its center is a cryptographically made sure about, disseminated record that takes into account the protected exchange of information between parties. Conventional IoT frameworks are reliant on a concentrated design.

Data are sent from the gadget to the cloud, where the information is prepared to utilize investigation and afterward sent back to the IoT gadgets. With billions of gadgets set to join IoT arranges in the coming years, this kind of brought together framework has constrained adaptability, uncovered billions of powerless focuses which bargain organize security and will turn out to be staggeringly costly and slow if outsiders need to continually check and verify every single smaller scale exchange between gadgets [17].

7 Workflow Method

The working architecture of this system was explained in Fig. 1. It consists of entity device and IoT platform. In entity, the device will be registered. Then, the public key and private key will be generated for registered device. Then, the registered device will get and set the configuration. Then, it reads data from the input and after that, the digital signature will be allocated to the data, and it creates the payload in the device. Then, the IoT platform receives the payload and validate by using the blockchain. If the data was valid, it will be saved in the database otherwise it will not.

8 Security Testing

This testing is a kind of programming testing that plans to reveal vulnerabilities of the framework and confirm that its information and assets are shielded from potential interlopers. The importance behind Security Testing is to distinguish every single imaginable escape clause and shortcomings of the product framework which may bring about lost data [20]. The primary objective of Security Testing is to recognize the dangers in the framework and measure its likely weaknesses, so the dangers can be experienced and the framework doesn't quit working or cannot be misused. It likewise helps in distinguishing all conceivable security dangers in the framework and causes engineers to fix the issues through coding. Then, security testing is applied to this system to check the vulnerabilities and remove it by using this testing.

9 Proposed Approach

This paper proposed four different concepts. They are device identity, message authentication, firmware hashing and device reputation.

9.1 Device Identity

Once the registered device is not revealing its private properties by using Merkle tree, it will take the public key or its representation as user id. A Merkle tree sums up all the exchanges in a square by creating a computerized unique mark of the whole arrangement of exchanges, in this way empowering a client to check whether an exchange is remembered for a square.

Merkle trees are made by over and over hashing sets of hubs until there is just one hash left. They are developed from the base up, from hashes of individual exchanges [21]. Each leaf hub is a hash of conditional information, and each non-leaf hub is

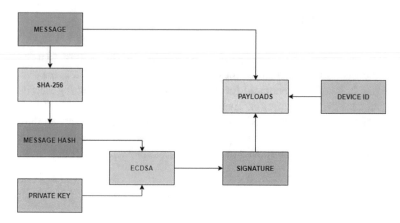

Fig. 2 Generating the signature

a hash of its past hashes. Merkle trees are paired and subsequently require a much number of leaf hubs. In the event that the quantity of exchanges is odd, the last hash will be copied once to make a much number of leaf hubs [22].

9.2 Message Authentication

Each and every message was signed as well as validated by using blockchain on receiver's end. A digital signature is the detail of an electronic archive that is utilized to recognize the individual sending information. DS makes it conceivable to learn the non-twisting status of data in a report that was marked and to check whether the mark has a place with the key certificate. There are two prospects to validate first sign-then-encode and then scramble then sign [23]. The process of digital signature was explained in Fig. 2. The beneficiary of accepting the scrambled information and mark on it that is first checks the mark utilizing sender's open key. Subsequent to guaranteeing the legitimacy of the mark, he at that point recovers the information through unscrambling utilizing his private key. Further, the collected data are validated and was explained in Fig. 3.

9.3 Firmware Hashing

This will confirm that the device has valid firmware. So as to confirm the uprightness and realness of the encoded information, other cryptographic calculations are required. Firmware minimizing is a potential assault situation if more than one firmware picture has been scrambled utilizing the equivalent legitimate key [24].

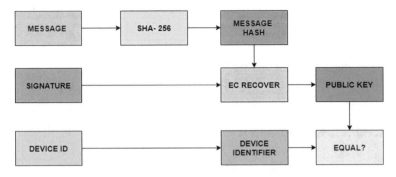

Fig. 3 Validation

9.4 Device Reputation

Based on Web of rules, gadgets can frame a system of trust. The more marks obtained from other legitimate gadgets. Device notoriety is best at forestalling extortion when misrepresentation experts cooperate as a feature of a friend consortium that empowers them to share worldwide extortion insight by means of an information base [25]. Misrepresentation experts can utilize this information to settle ongoing choices dependent on the exercises both inside their own industry and across others—which represent the most serious hazard to them.

10 Blockchain Technology for IoT Security

Blockchain can help secure IoT gadgets. IoT gadgets can be arranged either to utilize open blockchain administrations or to speak with private blockchain hubs in the cover over a protected API. The IoT framework allows IoT gadgets to find one another safely and with the help of administration key the encode will be exchanged from one machine to another by fusing the blockchain into the security structure and approve the trustworthiness and genuineness of programming picture refreshes, just as strategy refreshes [26].

In view of the potential designs point by point in this report, an IoT gadget will speak with a blockchain exchange hub by means of an API, permitting even obliged gadgets to partake in the blockchain management [27, 28]. The following is an utilization case for IoT disclosure that bolsters the enlistment of an IoT gadget into an exchange hub. The IoT gadget should initially be provisioned with qualifications that can be utilized to demonstrate approval so as to be added to an exchange hub. This qualification provisioning must be done in a made sure about condition that shields against dangers of a specific IoT gadget biological system [29]. The following is used to secure the IoT by using the blockchain technology:

1. Scalable IoT discovery
2. Trusted communication
3. Message authentication
4. IoT configuration and updates
5. Secure firmware image distribution and update.

11 Result and Discussion

The IoT is generally expected to work in a secured environment with a minimum time delay. Therefore, the IoT devices can communicate with each other and exchange data in a secured environment. Through web interface devices will be registered in a smart contract and send the signed messages by cryptographic algorithm and it will be validated by using blockchain. It uses proof-of-stake algorithm in order to reduce the energy consumption and increase the performance. Finally, the security testing will be done to remove the vulnerabilities and protect from unauthorized users.

12 Conclusion

Testing is a novel approach for IoT as well as blockchain which tests the overall performance, usability, conformance, security of a device. But, this paper focuses on security testing based on blockchain-enabled IoT system that would broaden the performance and scalability of the device encouraging small businesses to push their devices into the market. This paper describes a prototype to create a service that exclusively tests the security for IoT devices by using the blockchain platform in particular.

References

1. Li C, Zhang LJ (2017) A blockchain based new secure multi-layer network model for internet of things. In: 2017 IEEE international congress on internet of things (ICIOT), June 2017, pp 33–41
2. Aichernig BK, Tappler M (2016) Symbolic input-output conformance checking for model-based mutation testing. Electr Notes Theor Comput Sci 320:3–19
3. Sompolinsky Y, Zohar A (2013) Secured high-rate transaction processing in Bitcoin. https://eprint.iacr.org/2013/881.pdf
4. Jing Q, Vasilakos AV, Wan J, Lu J, Qiu D (2014) Security of the internet of things: perspectives and challenges. Wirel Netw 20(8):2481–2501
5. Dorri A, Kanhere SS, Jurdak R, Gauravaram P (2017) Blockchain for IoT security and privacy: the case study of a smart home. In: IEEE Percom workshop on security privacy and trust in the internet of thing

6. Borgohain T, Kumar U, Sanyal S (2015) Survey of security and privacy issues of internet of things. Department of Instrumentation Engineering Assam Engineering College Cornel University Library, pp 1–7
7. Ranger S (2018) What is the IoT? Everything you need to know about the internet of things right now. http://www.zdnet.com/article/what-is-the-internet-of-things-everything-you-need-to-know-about-the-iot-right-now/
8. Khalid U, Asim M, Baker T, Hung PCK, Tariq MA, Rafferty L (2020) A decentralized lightweight blockchain-based authentication mechanism for IoT systems
9. Hang L, Kim D-H (2019) Design and implementation of an integrated IoT blockchain platform for sensing data Integrity
10. Novo O (2018) Blockchain meets IoT: an architecture for scalable access management in IoT
11. Christidis K, Devetsikiotis M (2016) Blockchains and smart contracts for the internet of things. IEEE Access 4:2292–2303
12. Dorri A, Kanhere SS, Jurdak R (2017) Towards an optimized blockchain for IoT. In: Proceedings of the second international conference on internet-of-things design and implementation, pp 173–178
13. Buterin V (2014) A next-generation smart contract and decentralized application platform. https://github.com/ethereum/wiki/wiki/White-Paper. (Dec 2014)
14. Narula N, Vasquez W, Virza M (2018) zkledger: privacy-preserving auditing for distributed ledgers. In: 15th USENIX symposium on networked systems design and implementation, NSDI 2018, Renton, WA, USA, 9–11 April 2018, pp 65–80
15. Alharby M, van Moorsel A (2017) Blockchain-based smart contracts: a systematic mapping study. arXiv:1710.06372
16. Shao QF, Jin CQ, Zhang Z, Qian WN (2017) Blockchain: architecture and research progress. Chin J Comput 40(157):1–21
17. Natoli C, Gramoli V (2016) The blockchain anomaly. In: IEEE 15th international symposium on network computing and applications (NCA), pp 310–317. (Oct 2016)
18. Lewenberg Y, Sompolinsky Y, Zohar A (2015) Inclusive block chain protocols. In: Financial cryptography and data security (FC)
19. Baker T, Asim M, Tawfik H, Aldawsari B, Buyya R (2017) An energy-aware service composition algorithm for multiple cloudbased IoT applications. J Netw Comput Appl 89:96–108
20. Abbas N, Asim M, Tariq N, Baker T, Abbas S (2019) A mechanism for securing IoT-enabled applications at the fog layer. J Sens Actuator Netw 8:16
21. Jan, M.A., Nanda, P., He, X., Tan, Z., Liu, R.P.: A robust authentication scheme for observing resources in the internet of things environment. In: 2014 IEEE 13th International Conference on Trust. Security and Privacy in Computing and Communications, pp. 205–211 (2014).
22. Padma M, KasiViswanath N, Swat T (2019) Blockchain for IoT application challenges and issues
23. Panarello A, Tapas N, Merlino G, Longo F, Puliafito A (2018) Blockchain and IoT integration: a systematic survey. Sensors 18:2575
24. Jefferies N, Mitchell C, Walker M (1996) A proposed architecture for trusted third party services. In: Cryptography: policy and algorithms, Springer pp 98–104
25. Shafagh H, Hithnawi A, Duquennoy S (2017) Towards blockchain-based auditable storage and sharing of IoT data. In: Proceedings of the 9th ACM cloud computing security workshop (CCSW 2017), Dallas, TX, USA, 3 Nov 2017, pp 45–50
26. Moinet A, Darties B, Baril J-L (2017) Blockchain based trust & authentication for decentralized sensor networks. arXiv:1706.01730
27. Ahmad A, Bouquet F, Fourneret E, Le Gall F, Legeard B (2016) Model-based testing as a service for IoT platforms. Springer International Publishing, Cham, pp 727–742
28. Khan MI, Lawal IA (2020) Sec-IoT: a framework for secured decentralised IoT using blockchain-based technology
29. Larimer D (2014) Delegated proof-of-stake (DPOS)

Two-Dimensional Software Reliability Model with Considering the Uncertainty in Operating Environment and Predictive Analysis

Ramgopal Dhaka, Bhoopendra Pachauri, and Anamika Jain

Abstract In the current scenario, software plays a vital role in our daily life, e.g., research, medical, military, engineering, industries, home appliances, etc. To develop high reliable software system, many researchers are working on software reliability modeling. In the literature, a few researchers have included uncertainty in the operating environment and most of them are in one dimension in which the reliability enhancement depends on testing time. In this article, two-dimensional SRGM has proposed with uncertainty in the operating environment. model Cobb–Douglas production function has been used to convert one-dimensional to two-dimensional model in which the combined effect of the testing effort and used resources is considered. Further, the predictive analysis has been done for better understanding of prediction. The numerical results have been compared with existing SRGMs to validate the proposed model using five statistical comparison criteria.

Keywords Software reliability · SRGMs · Predictive analysis · Cobb–Douglas production function

1 Introduction

The pace of technological development and introduction of new technologies are growing at a breakneck speed. Software system is an integral part of these latest technologies. As the importance and integration of software system increases in society and industry, any error in the software can cause considerable operational and economic damages. For successful operation of any software, it is important to minimize the possibility of software failure with high reliability. The reliability of a software can be defined, "the probability that the system will perform without failure under certain conditions for a specific period of time". Software failure prediction is a research area that is of utmost importance to the software companies and developers as their main focus is to develop a reliable software. Many SRGMs have been developed but most of them fall short for today's complex software. The present

R. Dhaka · B. Pachauri (✉) · A. Jain
Department of Mathematics & Statistics, Manipal University Jaipur, Jaipur, Rajasthan, India

© The Author(s), under exclusive license to Springer Nature Singapore Pte Ltd. 2022
P. Nanda et al. (eds.), *Data Engineering for Smart Systems*, Lecture Notes in Networks and Systems 238, https://doi.org/10.1007/978-981-16-2641-8_6

SRGMs can be categorized as one-dimensional and two-dimensional models. One-dimensional SRGMs consider only testing time. Whereas two-dimensional SRGMs use the effect of testing time as well as testing effort. The Cobb–Douglas production function is used in a two-dimensional model. The Cobb–Douglas function is branded in economics as the production/utility function. This function was presented by Knut Wicksell (1851–1926) and tried in statistics by Cobb and Douglas [1]. Two-dimensional SRGMs were developed by Kapur et al. [2] in a perfect debugging environment. Pachauri et al. [3] extended a two-dimensional SRGM in an imperfect debugging.

In the published literature, most of the SRGMs have been studied under the common assumptions that when the software failure occurs, the caused fault is immediately removed, and no error is introduced because of fault removal process. The testing and the operating environment of the software development are assumed to be the same. Based on these assumptions, the NHPP-based SRGM was developed by Goel and Okumoto [4] in 1979. This model is also called the NHPP exponential model. Then S-shaped SRGM was given by Yamada et al. [5]. The inflection S-shaped SRGM dependent on the failure rate of each detectable fault was discussed in [6]. Pachauri et al. [7] considered an inflection S-shaped curve as the fault reduction factor. Pradhan et al. [8] introduced a SRGM with fault dependency and various time lags, to predict and quantify the reliability. Huang et al. [9] proposed that in the fault removal process the correction time should not be ignored.

Recently, many SRGMs have been discussing considering the uncertainty in the operating environment [10–16]. Song et al. [15] introduced a SRGM with Weibull fault detection factor. Lee et al. [16] discussed an SRGM considering the syntax error and analyzed release time policy with sensitivity analysis. Song et al. [10] introduced an SRGM using three-parameter fault detection. Pachauri et al. [17] studied an SRGM under the fuzzy paradigm with optimal release time. The cost-reliability has been calculated. There are some other software reliability models [2, 7, 18–22] that do not involve uncertainty in the operating environment.

To provide invaluable advance warning of failure, we use predictive analysis techniques. The aim of predictive analysis is to make prediction about future outcome based on historical data. Song et al. [11] have done predictive analysis in their SRGM.

Motivated from [2, 3, 11], SRGM is introduced. The model is two dimensional with the uncertainty in the operating environment. To understand the predictive capability of this model, the predictive analysis has been done for better understanding of prediction. The calculation of the cumulative number of faults for the SRGM with mathematical derivation is shown in Sect. 2. To validate the model, comparative study through numerical examples is discussed in Sect. 3. Finally, the conclusion is given in Sect. 4.

2 Software Reliability Modeling

The brief discussion about the general NHPP SRGM and mathematical modeling of the proposed model is given in this section. The basic idea about NHPP is discussed in the next subsection.

2.1 A General NHPP SRGM

In the NHPP model, the number of failures practiced up to time t is denoted by $N(t)(t \geq 0)$ [21] and it follows the memory less property. In the one-dimensional NHPP-based SRGM, the chance of exactly n faults/failures occurs in between the given time interval (0, t) is defined as,

$$\Pr\{N(t) = n\} = \frac{(m(t))^n e^{-m(t)}}{n!}, \quad n = 0, 1, 2, 3 \ldots \tag{1}$$

where m(t) is a mean value function (MVF) of NHPP. The rate of change in MVF shows the failure intensity and is denoted as,

$$\lambda(t) = \frac{d\,m(t)}{dt}. \tag{2}$$

Recently, the uncertainty in the operating environment has been considered by [12]. In this model, the rate of change in MVF with uncertainty in the operating environment is given as:

$$\frac{d\,m(t)}{d(t)} = \eta[b(t)][N - m(t)], \tag{3}$$

where η is a random variable denotes the uncertainty of in the operating environments with the p.d.f. g, $b(t)$ is the fault detection rate and N the total number of faults [12]. After solving Eq. (3) with initial condition $m(0) = 0$ we get:

$$m(t) = \int_{\eta} N\left(1 - e^{-\eta \int_0^t b(x)dx}\right) dg(\eta), \tag{4}$$

where η is a PDF with parameters $\alpha \geq 0$ and $\beta \geq 0$. η follows the gamma distribution. Then the MVF $m(t)$ after applying the random variable η, is given as:

$$m(t) = N\left(1 - \frac{\beta}{\beta + \int_0^t b(s)ds}\right)^{\alpha}.$$

(5)

Using this basic concept, the proposed model is given in next subsection.

2.2 Proposed Model

Here, two-dimensional NHPP-based SRGM has been proposed with uncertainty in the operating environment by using following assumptions [3, 15]:

a. The software failure phenomena follow NHPP.
b. The detection rate of fault is proportionate to the residual faults in the system.
c. The effect of uncertainty factor is represented by-product of detection rate of fault $b(t)$ and a r.v. η.
d. A software failure can occur during operation, caused by unresolved faults in the system.
e. To represents the effect of the used resources and testing effort, Cobb–Douglas production function is used.

Song et al. [11] recently discussed NHPP SRGM in which uncertainty in the operating environment has been considered and the MVF and detection rate function is given as,

$$m(t) = N\left(1 - \frac{\beta}{\beta + \ln\left(\frac{a+e^{bt}}{1+a}\right)}\right)^{\alpha},$$

(6)

and

$$b(t) = \frac{b}{1 + ae^{-bt}}, \quad a, b > 0,$$

(7)

where $b(t)$ is the function of fault detection rate, b is the detection rate of fault, and a is the inflection factor.

Cobb–Douglas production function is mathematically defined as [1–3],

$$T = s^{\gamma}u^{1-\gamma}, 0 \leq \gamma \leq 1,$$

(8)

where $\mathcal{T}$ denotes the combined effect factor of testing (used) resource and testing time. s is the testing time, u considers used testing resources, and γ is the output elasticities. Here, after using this function in Eq. (6), the new MVF is,

$$m(s, u) = N \left(1 - \frac{\beta}{\beta + \ln\left(\frac{a + e^{b(s^{\gamma} u^{1-\gamma})}}{1+a} \right)} \right)^{\alpha}.$$

(9)

In the next section, parameter estimation with numerical result is discussed.

3 Numerical Result and Discussion

To validate the performance of the new model, we used two historical data sets [21, 22] as shown in Table 1. The parameters values have been obtained using the curve fitting tool in MATLAB and are given in Tables 2 and 3. The performance of the model has been attained in terms of mean square error (MSE), root mean square error (RMSE), sum of square error (SSE), Adjusted R-square $(Adj\ R^2)$, and R-square (R^2). The SSE for predicted value (Pre SSE) is SSE for some part of data in which the full data set is not considered.

For DS1, the estimated parameters values, MSE, SSE, RMSE, R^2, and $(Adj\ R^2)$ criteria values are shown in Table 2. The MSE, SSE, and RMSE values of proposed model are 0.556, 8.835, and 0.7454. The values of R-square 0.9948 and Adjusted R-square 0.9931 for new model.

For DS2, the estimated parameters values, MSE, SSE, RMSE, R^2, and $(Adj\ R^2)$ criteria values are shown in Table 3. The MSE, SSE, and RMSE values of the proposed model are 3.3219, 53.15, and 1.8226. The values of R-square 0.9969 and Adjusted R-square 0.9959 for new model.

Calculated values of MSE, SSE, and RMSE for new model are less than the other existing models. The values of Adjusted R-square and R-square for new model are greater than the other existing models. Based on these results, it can be said that the new model gives improved results and considerably better goodness of fit.

The graphical representation of the MVF with time is shown in Figs. 1 and 2. The proposed model curve is best fitting to actual data with comparison of another

Table 1 Summary of data sets

Data set	Time (t)	Total testing hours	Total fault	Description	References
Data set-1 (DS1)	21 (weeks)	7476 (hours)	26	Telecommunication system test data	[21]
Data set-2 (DS2)	22 (days)	93 CPU hours	86	The pattern of discovery of error	[22]

Table 2 Comparative study with existing models for DS1

No	Model	Estimated value	MSE	SSE	RMSE	R^2	Adj R^2
1	GO [4]	$\hat{a} = 3923854.73,$ $\hat{b} = 3.2 \times 10^{-7}$	3.8672	73.477	1.9665	0.9582	0.9535
2	Y-DS [5]	$\hat{a} = 39.82198,$ $\hat{b} = 0.1104$	1.4938	28.382	1.2222	0.9838	0.9820
3	K- SRGM 3 [18]	$\hat{A} = 24.989,$ $\hat{p} = 0.1385$ $\hat{b} = 0.1385,$ $\hat{\alpha} = 0.1012$	1.2295	20.902	1.1088	0.9881	0.9851
4	P-Vtub [12]	$\hat{a} = 1.0985,$ $\hat{\alpha} = 1.5176,$ $\hat{b} = 1.2978,$ $\hat{\beta} = 11.3848,$ $\hat{N} = 25.7412$	0.7178	11.485	0.8472	0.9935	0.9913
5	S-3PFD [10]	$\hat{a} = 0.038,$ $\hat{\beta} = 0.002,$ $\hat{b} = 0.292,$ $\hat{N} = 26.889,$ $\hat{c} = 1488.598,$	0.7590	12.144	0.8712	0.9931	0.9908
6	SONG-P [11]	$\hat{a} = 108232.819,$ $\hat{\alpha} = 0.2176,$ $\hat{b} = 1.0047,$ $\hat{\beta} = 155.5011,$ $\hat{N} = 47.7965$	0.5864	9.382	0.7658	0.9947	0.9929
7	Proposed (new) Model	$\hat{a} = 2254,$ $\hat{\alpha} = 0.6508,$ $\hat{\beta} = 1.296,$ $\hat{b} = 0.07461,$ $\hat{N} = 27.65,$ $\hat{\gamma} = 0.3863$	**0.556**	**8.335**	**0.7454**	**0.9948**	**0.9931**

models. The graphical representation of the relative error is shown in Figs. 3 and 4. The new model curve is closer to zero with comparison of another models. The relative error curve confirms the ability to provide better accuracy.

Table 3 Comparative study with existing models for DS2

No	Model	Estimated value	MSE	SSE	RMSE	R^2	Adj R^2
1	GO [4]	$\hat{a} = 153,$ $\hat{b} = 0.0414$	25.190	503.818	5.019	0.9706	0.9691
2	Y-DS [5]	$\hat{a} = 94.25,$ $\hat{b} = 0.1929$	7.645	152.883	2.765	0.9911	0.9906
3	K- SRGM 3 [18]	$\hat{A} = 10.97,$ $\hat{p} = 1.152,$ $\hat{b} = 1.026,$ $\hat{\alpha} = 0.8902$	9.865	177.557	3.141	0.9896	0.9879
4	P-Vtub [12]	$\hat{a} = 15.78,$ $\hat{\alpha} = 0.0367,$ $\hat{\beta} = 7.08,$ $\hat{b} = 1.291,$ $\hat{N} = 60.13$	292.20	4967.6	17.094	0.7099	0.6417
5	S-3PFD [10]	$\hat{a} = 1.117,$ $\hat{c} = 136.9,$ $\hat{\beta} = 0.2511,$ $\hat{N} = 89.31$ $\hat{b} = 0.2656$	6.602	112.231	2.5694	0.9934	0.9919
6	SONG-P [11]	$\hat{a} = 58.4,$ $\hat{\alpha} = 0.8222,$ $\hat{\beta} = 0.2918,$ $\hat{N} = 92.21$ $\hat{b} = 0.3154$	6.506	110.605	2.551	0.9935	0.9920
7	Proposed (new) Model	$\hat{a} = 6.067,$ $\hat{\alpha} = 7.3E - 07$ $\hat{\beta} = 1.216,$ $\hat{b} = 0.0733$ $\hat{N} = 99.39,$ $\hat{\gamma} = 0.7092$	**3.3219**	**53.15**	**1.8226**	**0.9969**	**0.9959**

3.1 Predictive Analysis

In this paper, we are using the DS2 to predictive analysis. The 75% of the data is used to estimate the parameters, and the remaining 25% is used to predict the model performance. For comparison, same criteria are used for all the models. We obtained PreSSE (Predictive SSE) for comparison.

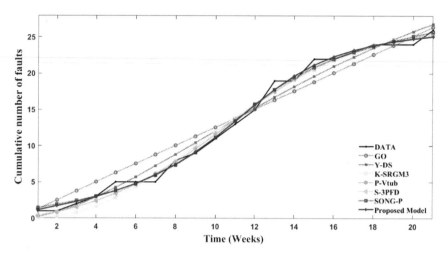

Fig. 1 MVF of various models for DS1

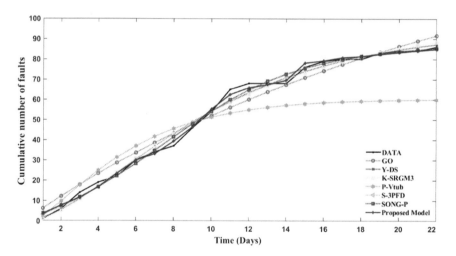

Fig. 2 MVF of various models for DS2

The estimated parameters values MSE, RMSE, SSE, R^2, and $\left(Adj \ R^2\right)$ criteria values are shown in Table 4. The MSE, SSE, and RMSE values of new model are 5.656, 62.219, and 2.3783, respectively, which are less than the values of other comparison models. The values of R-square 0.9946 and Adjusted R-square 0.9921 for new model, which are bigger than the values of other comparison models. The PreSSE values of new model are 2.5856. The PreSSE values of new model are less than the values of other comparison models.

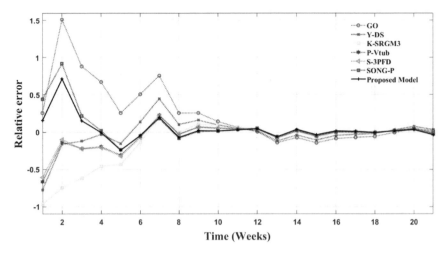

Fig. 3 RE curve for all models for DS1·

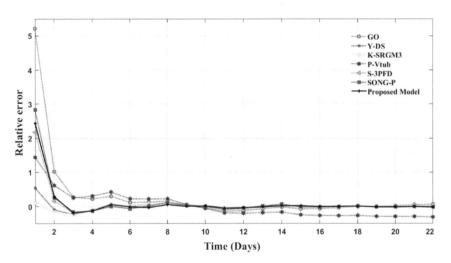

Fig. 4 RE curve for all models for DS2

Based on these results, we can say that the new model gives better results and considerably better goodness of fit. The value of the PreSSE shown that the new model has considerably predictability. Figure 5 shows that the goodness of fit and prediction of the mean value function for various models for DS2.

Table 4 Comparative study and prediction with existing models for DS2

No	Model	Estimated value	MSE	SSE	RMSE	R^2	Adj R^2	PreSSE
1	GO [4]	$\hat{a} = 762.2,$ $\hat{b} = 0.006952$	17.32	259.8	4.162	0.9773	0.9758	1314.1
2	Y-DS [5]	$\hat{a} = 100,$ $\hat{b} = 0.1807$	8.731	131.4	2.955	0.9885	0.9878	81.526
3	K-SRGM-3 [18]	$\hat{A} = 2.583,$ $\hat{p} = 0.5484,$ $\hat{b} = 2.947,$ $\hat{\alpha} = 1.034$	107.6	1398	10.373	0.8778	0.8496	8649.7
4	P-Vtub [12]	$\hat{a} = 1.419,$ $\hat{\alpha} = 0.7894,$ $\hat{\beta} = 6.275,$ $\hat{b} = 0.7635,$ $\hat{N} = 90.58$	9.327	111.9	3.054	0.9902	0.9870	11.37

(continued)

Table 4 (continued)

No	Model	Estimated value	MSE	SSE	RMSE	R^2	Adj R^2	PreSSE
5	S-3PFD [10]	$\hat{a} = 2.692,$ $\hat{c} = 60.48,$ $\hat{\beta} = 1.218,$ $\hat{b} = 0.2539,$ $\hat{N} = 90.24$	10.30	123.58	3.2091	0.9892	0.9856	43.85
6	SONG-P [11]	$\hat{a} = 46.95,$ $\hat{\alpha} = 0.824,$ $\hat{\beta} = 0.3572,$ $\hat{b} = 0.3125,$ $\hat{N} = 93.98$	8.999	108	2.9999	0.9906	0.9874	3.635
7	Proposed (new) Model	$\hat{a} = 2.647,$ $\hat{\alpha} = 0.000549,$ $\hat{\beta} = 0.5675,$ $\hat{b} = 0.04153,$ $\hat{N} = 103.3,$ $\hat{\gamma} = 0.9634$	**5.656**	**62.219**	**2.3783**	**0.9946**	**0.9921**	**2.5856**

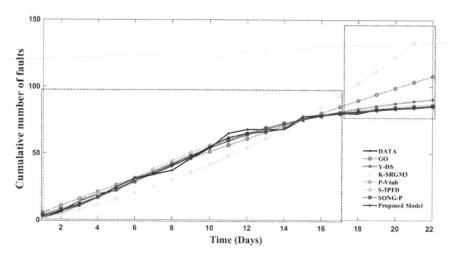

Fig. 5 Prediction of MVF for various models for DS2

4 Conclusions

The software development and testing are done in a predefined environment that may change accordingly. Therefore, we are considering the uncertainty in operating environments. Cobb–Douglas production function has been used to model the effect of testing effort and used resources. In this article, we discussed a two-dimensional SRGM by considering the uncertainty in software operating environments and did predictive analysis. Predicted the values for model and compared proposed model with several existing SRGMs based on six criteria (MSE, SSE, RMSE, R-square, Adjusted R-square, and PreSSE). The result has shown that proposed model has significantly goodness of fit and predict the value better than the other existing models. In the future, the proposed model may be improved using soft-computing techniques or adding some new reliability factors like fault reduction factor, change point, etc.

References

1. Cobb CW, Douglas PH (1928) A theory of production. Am Econ Rev 18:139–165
2. Kapur PK, Aggarwal AG, Kaur G (2012) Two-dimensional multi-release software reliability modeling and optimal release planning. IEEE Trans Reliab 61:1–11
3. Pachauri B, Kumar A, Raja S (2019) Imperfect software reliability growth model using delay in fault correction. In: Deep K, Jain M, Salhi S (eds) Performance prediction and analytics of fuzzy, reliability and queuing models. Asset analytics (Performance and Safety Management). Springer, Singapore. https://doi.org/10.1007/978-981-13-0857-4_8
4. Goel AL, Okumoto K (1979) Time-dependent error-detection rate model for software reliability and other performance measures. IEEE Trans Reliab 28:206–211

5. Yamada S, Ohba M, Osaki S (1983) S-shaped reliability growth modeling for software fault detection. IEEE Trans Reliab 32:475–484
6. Ohba M (1984) Inflexion S-shaped software reliability growth models. In: Osaki S, Hatoyama Y (eds) Stochastic models in reliability theory. Springer, Berlin, Germany, pp 144–162
7. Pachauri B, Kumar A, Dhar J (2015) Incorporating inflection S-shaped fault reduction factor to enhance software reliability growth. Appl Math Model 39(5):1463–1469
8. Pradhan V, Kumar A, Dhar J, Gupta M (2018) A software reliability model incorporating fault dependency considering time delay. Int J Pure Appl Math 22:1527–1535
9. Huang CY, Lin CT (2006) Software reliability analysis by considering fault dependency and debugging time lag. IEEE Trans Reliab 55(3):436–450
10. Song KY, Chang IH, Pham H (2017) A Three-parameter fault-detection software reliability model with the uncertainty of operating environments. J Syst Sci Syst Eng 26:121–132
11. Song KY, Chang IH, Pham H (2019) NHPP software reliability model with inflection factor of the fault detection rate considering the uncertainty of software operating environments and predictive analysis. Symmetry 11:521. https://doi.org/10.3390/sym11040521
12. Pham H (2014) A new software reliability model with Vtub-Shaped fault detection rate and the uncertainty of operating environments. Optimization 63:1481–1490
13. Chang IH, Pham H, Lee SW, Song KY (2014) A testing-coverage software reliability model with the uncertainty of operation environments. Int J Syst Sci Oper Logist 1:220–227
14. Song KY, Chang IH, Pham H (2018) Optimal release time and sensitivity analysis using a new NHPP software reliability model with probability of fault removal subject to operating environments. Appl Sci 8:714. https://doi.org/10.3390/app8050714
15. Song KY, Chang IH, Pham H (2017) A software reliability model with a Weibull fault detection rate function subject to operating environments. Appl Sci 7:983. https://doi.org/10.3390/app7100983
16. Lee HH, Chang IH, Pham H, Song KY (2018) A software reliability model considering the syntax error in uncertainty environments, optimal release time and sensitivity analysis. Appl Sci 8:1483. https://doi.org/10.3390/app8091483
17. Pachauri B, Kumar A, Dhar J (2013) Modeling optimal release policy under fuzzy paradigm in imperfect debugging environment. Inf Soft Tech 55(11):1974–1980
18. Kapur PK, Pham H, Anand S, Yadav K (2011) A unified approach for developing software reliability growth models in the presence of imperfect debugging and error generation. IEEE Trans Reliab 60:331–340
19. Roy P, Mahapatra GS, Dey KN (2014) An NHPP software reliability growth model with imperfect debugging and error generation. Int J Reliab Qual Saf Eng 21(2):1–32
20. Pachauri B, Kumar A, Dhar J (2014) Software reliability growth modeling with dynamic faults and release time optimization using ga and maut. Appl Math Comp 242:500–509
21. Pham H (2006) System software reliability. Springer, London, UK
22. Tohma Y, Jacoby R, Murata Y, Yamamoto M (1989) Hyper-Geometric distribution model to estimate the number of residual software fault. In: Proceedings of the thirteenth annual international computer software & applications conference, Orlando, FL, USA, pp 610–617. https://doi.org/10.1109/CMPSAC.1989.65155

Object Recognition in a Cluttered Scene

Rashmee Shrestha, Mandeep Kaur, Nitin Rakesh, and Parma Nand

Abstract Object Recognition is an important and challenging computer vision task, which plays a major role in various applications such as robotics, medical imaging, recognition and verification systems, information retrieval systems and so on. Within the past few years, many single object tracking systems have come into picture. However, in the presence of many objects, object recognition task becomes difficult, especially when parts of the object are fully or partially occluded. In this paper, an object recognition system based on deep learning techniques is proposed. RetinaNet Model has been used for object detection and identification. RetinaNet model has demonstrated to work well with both small scale as well as dense objects.

Keywords CNN · Computer vision · Object recognition · Deep learning · Neural network · RetinaNet

1 Introduction

Object recognition is a process in which specific objects are identified in an image or video sequence. The accuracy of object recognition has considerably improved in recent years due to the use of neural network in the object recognition process [1–3]. The use of RetinaNet model for this project has also proven to work effectively on a significant number of objects [4, 5].

RetinaNet Architecture: The architecture consists of four major components as shown in Fig. 1 [6–11]:

1. Bottom-up Pathway—Here the feature maps at different scales are calculated, regardless of the size of the input image.
2. Top-down pathway and Lateral connections—Spatially denser feature maps are unsampled from higher pyramid levels using top-down pathway. The top-down

R. Shrestha · M. Kaur (✉) · N. Rakesh · P. Nand
Department of Computer Science and Engineering, Sharda University, Greater Noida, Uttar Pradesh, India
e-mail: mandeep.kaur@sharda.ac.in

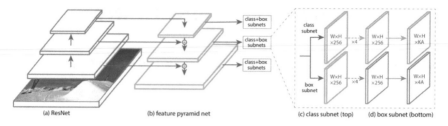

Fig. 1 RetinaNet model architecture [11]

and bottom-up layers having the same spatial size are then merged together using lateral connections.

3. Classification subnetwork—Here the probability of an object being present at each spatial location for each anchor box and object class is predicted [11].

4. Regression subnetwork—Here the offset for the bounding boxes from the anchor objects for each ground-truth object is reverted [11].

2 Methodology

In this section, we have mainly focused on the workflow of the system along with the methods used in our system. Training Data: Our model was initially trained with COCO (Common Objects in Context) dataset. COCO is a large scale dataset designed specifically for the purpose of object detection, segmentation and captioning generation [10]. There are originally 91 object categories in COCO, out of which 80 object categories are labelled and segmented images.

Pre-processing: Here Image Generator Function which was included in the deep learning package of keras was used in order to achieve data augmentation. The images were also normalized in order to reduce the impact of uneven illumination and noise on feature extraction to obtain better features.

Feature Extraction based on Deep Neural Network: More powerful and distinct features can be extracted using deep neural network. The recognition systems based on these features also tend to learn faster and have a higher recognition rate. RetinaNet was used to extract image features and this network was pre-trained with COCO dataset.

Model Learning/Recognition: RetinaNet Model (A type of Convolutional Neural Network) was trained based on the features extracted from the above dataset. The model worked effectively and gave good recognition results on a significant number of objects (Fig. 2).

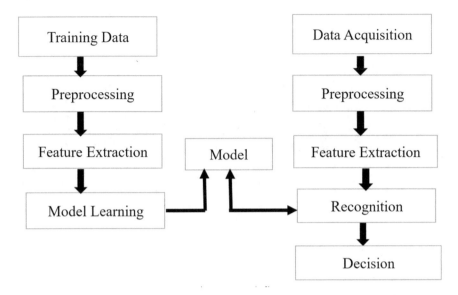

Fig. 2 Workflow

3 Test Results

The following section depicts the images before and after detection along with the console results obtained for each image. The model can predict around 80 possible objects so far. It also prints the name and the confidence percentage of each object detected in the image. The before and after detection for samples 1, 2 & 3 are shown in Figs. 3, 4 and 5.

• **Sample 1**

Fig. 3 Before and after detection for sample 1

• **Sample 2**

Before Detection   **After Detection**

Fig. 4 Before and after detection for sample 2

• **Sample 3**

Before Detection   **After Detection**

Fig. 5 Before and after detection for sample 3

Console result for above image (Fig. 3):

Person: 76.331 Motorcycle: 71.071

Person: 83.830 Dog: 94.580

Person: 96.081 Car: 55.414

Person: 96.669 Person: 86.197

Console result for above image (Fig. 4):

Cup : 79.316 Person: 99.344

Cup: 66.110 Person: 92.911

Cup: 83.249 Person: 86.394

Cup: 92.639 Person: 98.521

Laptop: 83.086 Person : 86.497

Console result for above image (Fig. 5):

Person: 96.867

Horse: 98.641

Likewise, the model has been tested with 30 random images in order to verify the recognition results. It has proven to work well with both small scale as well as dense objects and given good recognition results.

4 Quantitative Analysis

The following section depicts the images before and after detection along with the console results obtained for each image. The model can predict around 80 possible objects so far. It also prints the name and the confidence percentage of each object detected in the image. The before and after detection for samples 1, 2 & 3 are shown in Figs. 3, 4 and 5. Mean Average Precision (mAP) is used as an evaluation metric. The average precision for the object categories are mentioned below in Table 1.

Table 1 Average precision for classes

Class	Average precision
Motorcycle	0.741
Bird	0.655
Bottle	0.292
Airplane	0.747
Cat	0.910
Person	0.592
Boat	0.564
Dining table	0.554
Chair	0.370
Train	0.910
Couch	0.730
Bicycle	0.656
Horse	0.746
Bus	0.756
Cow	0.642
Car	0.654
TV	0.653
Potted plant	0.360
Sheep	0.643
Dog	0.828

5 Conclusion

An object recognition model based on deep learning is used and implemented. The model is capable of identifying a number of different objects based on the user's needs. It also a gave better result in terms of illumination, occlusion and time consumption. The object recognition in the sample scenes took about 5–10 s. So far, RetinaNet was able to detect high number of objects in a given scene and gave good recognition results.

References

1. Dorner J, Kozák Š (2015) Object recognition by effective methods and means of computer vision. IEEE
2. Jiang B, Li X, Yin L, Yue W, Wang S (2019) Object recognition in remote sensing images using combined deep features. IEEE
3. Meera MK, Shajee Mohan BS (2016) Object recognition in images. IEEE.
4. Mr Sudharshan DP, Ms Raj S (2018) Object recognition in images using convolutional neural network. IEEE
5. Yan L, Wang Y, Song T (2017) An incremental intelligent object recognition system based on deep learning. IEEE
6. Singh J, Shubham J (2015) Object detection by colour threshold method. Int J Adv Res Comput Commun Eng (IJARCCE)
7. Tiwari M, Dr Singhai R (2017) A review of detection and tracking of object from image and video sequences. Int J Comput Intell Res
8. Khurana P, Sharma A (2016) A survey on object recognition and segmentation techniques. IEEE
9. Rasool Reddy K, Hari Priya K, Neelima N (2015) Object detection and tracking–a survey. IEEE
10. Internet source. www.github.com
11. Internet source. https://developers.arcgis.com/

Breast Cancer Prediction on BreakHis Dataset Using Deep CNN and Transfer Learning Model

Pinky Agarwal, Anju Yadav, and Pratistha Mathur

Abstract Breast cancer (BC) is a type of disease where cells grow uncontrollably and attack on normal cells. Death rate caused by BC can be reduced by early recognition of the disease. In the recent years, convolution neural network and its various architectures that support transfer learning showed significant improvement in the classification of malignant and non-malignant tumors. In this study, we have performed performance analysis on deep CNN and four popular CNN-based architectures: VGG16, VGG19, MobileNet, and ResNet 50 on publically available BreakHis dataset for breast cancer classification on histopathological images. Among classifiers, VGG16 has best performed with highest 94.67% accuracy, 92.60% precision, 85.21% f1-score, and 80.52% recall value.

Keywords Deep learning · Breast tumor classification · Histopathological breast cancer images · CNN · Transfer learning

1 Introduction

The rate of BC in women is in quite large amount of that witnessed in men. It is second prominent reason of death after lung cancer (LC) in females worldwide. As per the statements of World Health Organization (WHO), there are total 2.09 cases and 627,000 mortalities worldwide and it affects 21 million women each year [1]. For the examination of breast cancer, various imaging techniques are being used like magnetic resonance imaging (MRI), mammography, ultrasound, etc., [2]. Among all, whole slide images (histopathological) are denoted as gold standard for cancer identification.

Early identification and detection of BC become very essential to increase the rate of survival. Patients with the high risk are recommended annual MRIs. If MRI

P. Agarwal · A. Yadav · P. Mathur (✉)
Manipal University Jaipur, Jaipur, India
e-mail: pratishtha.mathur@jaipur.manipal.edu

A. Yadav
e-mail: anju.yadav@jaipur.manipal.edu

© The Author(s), under exclusive license to Springer Nature Singapore Pte Ltd. 2022
P. Nanda et al. (eds.), *Data Engineering for Smart Systems*, Lecture Notes in Networks and Systems 238, https://doi.org/10.1007/978-981-16-2641-8_8

shows any symptoms of the disease then biopsy is done. In the process of biopsy, tissue samples are taken out and send for the examination in the pathology lab. Here, these tissues are kept on glass slides and marked with hematoxylin (H) and eosin (E) stains to highlight cell structure. Hematoxylin is a purple dye which gives blue color to nuclei, whereas eosin (E) gives pinkish color to cytoplasm. Pathologist manually examines these slides with the help of microscope to get the symptoms of the disease. This process is known as Histopathology. A pathologist examines various features like color, shape, size of nuclei, and proportion of the cytoplasm in an image. The manual examination of these slides with microscope is widely accepted clinical standard and the accuracy of the results highly rests on the practice, intelligence, workload and frame of the mind of pathologists.

But as the number of cancer patients is increasing every year, manual analysis and examination of the tissues are very painstaking and time-consuming task. According to [3], concordance rate among the pathologist for the dense tissue is 73 and 77% for less-dense tissue.

Therefore, we perceive a need to design an automated computer-assisted detection and classification system that can relieve the workload of pathologists and also avoid misdiagnosis of the cancer patients. In the recent years, artificial intelligence (AI) in the way of natural language processing, computer vision (CV), image processing, machine learning (ML), and deep learning (DL) has made it possible to identify, classify and quantify structures and patterns in medical images [4]. DL algorithms can take out valuable information from massive image dataset and can help to avoid subjective variances caused by manual analysis.

The paper is contributing to the research by evaluating and comparing the performances of various transfer learning (TL) models and deep CNN on microscopic biopsy dataset. We have evaluated classifiers on various performance matrices and the classifier with the highest f1-score is considered best for our unbalanced dataset. This work can further help researchers and pathologist to select the classifier with best accuracy and f1-score. We have organized the paper as follows: Sect. 2 describes the details of research work done by many researchers and description of the work is classified ML, DL, and TL algorithms. Section 3 describes the related concepts. Section 4 presents data preparation techniques. Sections 5 and 6 describe setup used for experiment and result and Sect. 7 concludes the work.

2 Related Work

Since last decade, many authors and researchers are putting significant efforts for breast cancer detection and classification from various imaging modalities using ML and DL techniques, and undoubtedly, many researchers have attained remarkable achievements. This classification study can be classified in two categories: ML and DL.

ML techniques manually extract the various features like local phase quantization (LPQ), local binary patterning (LBP), gray-level co-occurrence matrix (GLCM), scale invariant feature extraction (SIFT), and threshold adjacency statistics (TAS) [5]. ML techniques need more efforts and focused on small dataset and are low performing. While DL models can fetch more abstract features and patterns inevitably from large image dataset.

Zhang et al. [6] used series of two classifiers and ensemble them. The top classifier is a set of support vector machines (SVM) and classifier next to it is multilayer perceptron (MLP). Rejected samples from the first classifier went to second classifier (MLP). He used microscopic biopsy images and achieved 99.25% accuracy. Spanhol et al. [5] used dataset of 7909 microscopic biopsy image samples of 82 patients. They extracted various features like LBP, GLCM, SIFT, and TAS. They achieved 80–85% accuracy. Spanhol et. al. [7] used variation of AlexNet convolution neural network to classify microscopic biopsy breast cancer images and improved the accuracy from 4–6%. Jain et al. [8] utilized five mainstream ML calculations K nearest-neighbor (KNN), logistic regression (LR), random forest (RF), support vector machine (SVM), decision tree (DT) BreakHis dataset [9].

Above stated proposals were based on the binary classification. For multiclass classification, Araújo (2017) et al. [9] offered a CNN model that can classify the images in four types of classes: benign class, normal tissue, invasive carcinoma, and in situ carcinoma, and in two different classes: carcinoma and non-carcinoma. He achieved the accuracy of 77.8% for four classes and 83.3% for two classes. Han et al. [10] proposed a deep learning framework for multiclass classification and scored the accuracy of 93.2% on histopathological images of BreakHis. Nawaz et al. [11] designed DL-based CNN for multiclass classification on BreakHis histopathological image dataset and attained the accuracy of 95.4%. In 2019, Dabeer et al. [12] suggested a model based on CNN to classify histopathology images and they achieved 99.8% accuracy for true class. For their experiment purpose, they used two open dataset. In Li et al. [13] used clustering algorithms and CNN to extract both cell level and tissue level patches from histology images and got 98% accuracy. The results obtained were better than traditional machine learning algorithms.

3 Related Theories

In this section, we present detailed explanation of the concepts used in this study.

3.1 Convolution Neural Network (CNN)

CNN is known as popular neural network and is getting attention of programmers because of it high performance ability in object detection and recognition [14]. CNN models need very less preprocessing. While traditional ML algorithms used enough labor to extract the features, CNN models are trained to learn these features themselves. We can perform feature extraction as well as classification using CNN model. CNN architecture is composed of many layers. Convolutional layer performs convolution on an input image and applies activation function (RELU is most preferred)

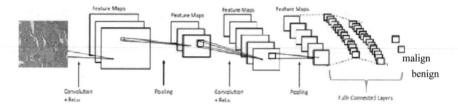

Fig. 1 Architecture of a CNN model

on matrix. After that, pooling layer reduces the dimensionality, and then, output is flattened to become the input for fully connected layer to categorize the images. There are many learning parameters that can be adjusted to enhance the efficiency and accuracy of the model like optimizers, learning rate, batch size, etc., in CNN architecture model. Figure 1 shows complete architecture of a CNN model.

3.2 Transfer Learning

Transfer learning (TL), as the name suggests, is an idea of transferring the information (features, weights) gathered from one model to another related model [15]. In TL, a model is pre-trained on large dataset using deep CNN approach. To make this model applicable, its output layer is trained (fine tuning) on smaller related dataset. There are two widely used transfer learning approaches: (1) pre-trained model as a feature extractor, (2) fine tuning of output layer of pre-trained model.TL does not need random initialization of weights hence making learning faster and easier for many deep learning applications. Fine tuning is carried out by retaining some of the desired pre-trained layers and retraining of the output layer as per need.

4 Data Preparation and Preprocessing Techniques

There are various data processing techniques like data augmentation and normal-ization, etc. To improve the efficiency of the model, in the following section, we describe them briefly.

4.1 Dataset

The BreakHis dataset was first publically by the researcher Spanhol et al. [5] in year 2016. The dataset consists of sample images (7909) from 82 BC patients.

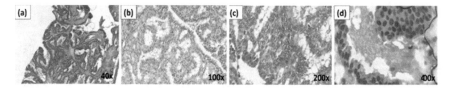

Fig. 2 Breast malignant tumor slides at various magnification levels: **a** 40×, **b** 100×, **c** 200× and **d** 400×

These are RGB sample images with 700 × 460 size. The dataset is collected at four dissimilar magnification levels, namely 40×, 100×, 200×, and 400×. The dataset is classified into benign and malign categories by experienced pathologists and further classified into eight subcategories (four for each). For benign class, subcategories include: fibroadenoma (F), adenosis (A), phyllodes tumor (PT), and tubular adenoma (TA), and malign subclasses include: lobular carcinoma (LC), ductal carcinoma (DC), mucinous carcinoma (MC), and papillary carcinoma (PC). Figure 2 shows microscopic biopsy images at different magnification factors. A benign tumor does not match with any of the malignancy criteria like disruption of any basement membranes, mitosis, metastasize and cellular atypia, etc. Basically, benign is non-cancerous tumor, whereas malign is cancerous tumor that can spread into another distant cell region and can be the cause of the death of a person. The samples were obtained by excisional biopsy or surgical open biopsy (SOB) method.

4.2 Data Augmentation

From the various studies, it is proved that small and limited data size may over-fit the model [16]. Data augmentation helps to increase training sample size by various transformation techniques like sheer, zoom, rotation, horizontal flip, and vertical flip and hence improves models performance. Table 1 shows different augmentation techniques used by us in this work.

Table 1 Data augmentation parameters and values

	Parameter	Values
1	Rotation	40
2	Width	0.2
3	Height	0.2
4	Shear	0.2
5	Zoom	0.2

4.3 Normalization

Image normalization is an important step to acquire the similar range of values for the sample images. After normalization images are feed into CNN model. It helps in fast convergence of the model thereby improving performance of the overall model. There are various methods to normalize datasets like logistic, Z-score, min–max, lognormal, tanh, etc. We have normalized Images using min–max normalization method to give them a range of [0, 1]. Min–Max Normalization formula is given below (see Eq. 1). Here, x denotes original value of x and z is the normalized value of x. Min(x) and max(x) are the lowest and extreme values of observations.

$$z = \frac{X - \min(x)}{[\max(x) - \min(x)]} \tag{1}$$

5 Proposed Method

In this study, we have done performance analysis on four popular convolution-based architectures VGG16, VGG19 [17], ResNet50 [18], MobileNet [19], and one deep CNN architecture to classify histopathological breast images. Flowchart and CNN layer architecture of for the proposed CNN are shown in Fig. 3. The flowchart mainly consists of eight steps.

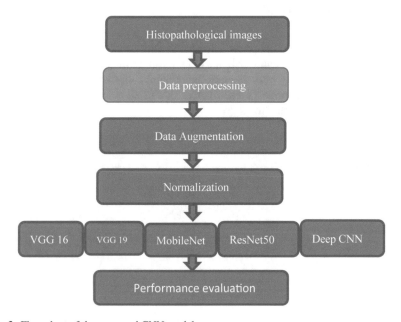

Fig. 3 Flow chart of the proposed CNN model

Table 2 Deep CNN architecture

Layer features	L1	L2	L3	L4	L5	L6
Layer type	Convo	Pool	Convo	Pool	Convo	Pool
Channel	32	–	32	–	64	–
Filter size	3×3	–	3×3	–	3×3	–
Pool size	–	2×2	–	2×2	–	2×2
Activation	Relu	–	Relu	–	Relu	–
No. parameters	896	0	9248	0	18,496	0

Deep CNN architecture is shown in Table 2. Input image is resized to $254 \times 254 \times 3$ and then we used 32 kernels of size 3×3. Output of conv1 is given to max-pooling layer1 to decrease the size of input image. Output image size of max pool1 is $127 \times 127 \times 3$. After this, we have 2 more convolutional layer and each layer is followed by max-pool layer. Image size after third max-pool layer is reduced to $34 \times 34 \times 60$ and flattened for fully connected layer (FC). In FC1, there are 64 neurons and we used rectified linear activation function (RELU) at all the layers except FC2. There are 2 neurons and softmax activation function is used at FC2 for binary classification. We used Adam optimizer with learning (LR) rate 0.001.

Histopathological images are visually complex therefore by using transfer learning approach, we can use extract more complex features from the source dataset. Transfer learning approach has many more advantages like: fast convergence of the network by avoiding random initialization of weights and reduces computation complexities.

5.1 Experimental Setup

For transfer learning approach, Images are resized to 224×224 pixels according to the default input size of these models. Adam optimizer is used with learning rate 0.001 for all the models. Batch size is agreed to 32 with 100 epochs in each model. RELU is used to train fully connected layer and dropout rate is set to 0.5 to avoid over-fitting problem. To fine tune transfer learning models, top layer is modified to classify the images in binary classes: benign and malign rather than 1000 classes as proposed in ImageNet competition. ImageNet weights are used for weight initialization in all the pre-trained models thereby avoiding random weight initialization. Detailed architecture explanation of VGG16, VGG19, MobileNet, and ResNet50 can be seen in [17–19]. Windows10 operating system (OS) with 32 GB RAM and Intel(R) Core(TM) i78700K 3.7 GHz processor is used. Also, Keras package with Tensor flow as backend for training and testing of models is used.

5.2 Evaluation Criteria

We have used accuracy, F1-score, precision, and recall as performance matrices to evaluate our models. For the given value of TN_{MM} = True Positive (TP), FP_{BM} = False Positive (FP), TN_{MM} = True Negative(TN), and FN_{MB} = False Negative(FN). We can mathematically compute recall, precision, F1-score, and Accuracy as follows:

$$Accuracy = \frac{TN_{MM} + TP_{BB}}{TP_{BB} + FP_{BM} + TN_{MM} + FN_{MB}} * 100$$

$$Precision = \frac{TP_{BB}}{FP_{BM} + TP_{BB}} * 100$$

$$Recall = \frac{TP_{BB}}{TP_{BB} + FN_{MB}} * 100$$

$$F1score = 2 * \frac{Recall * Precision}{Recall + Precision}$$

6 Results and Discussion

The proposed model is applied to deep CNN, VGG16, VGG 19, MobileNet, and ResNet50 for the classification of BC images. For the performance analysis of various classifiers, we have calculated learning curves (LC) (see Fig. 4) and ROC curve (see

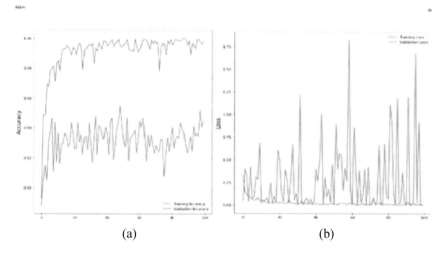

(a) (b)

Fig. 4 VGG16 training and validation accuracy curve **a** training and **b** validation loss curve

Fig. 5). We have also calculated precision, recall, and f1-score along with accuracy as performance matrices as our dataset is unbalanced (see Table3). From Table 3, it is clearly visible that the performance of VGG16, VGG19, and deep CNN is quite

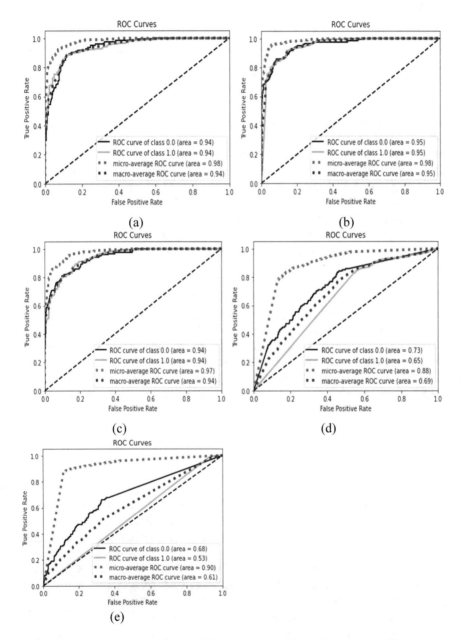

Fig. 5 ROC curves for **a** VGG16, **b** VGG19, **c** Deep CNN, **d** MobileNet, **e** ResNet50

Table 3 Performance analysis of deep CNN, VGG16, VGG19, MobileNet, and ResNet50 in terms of precision, recall, F1-score, and accuracy

Classifier	Precision (%)	Recall (%)	f1-score (%)	Accuracy (%)
VGG 16	92.60	80.52	85.21	94.67
VGG19	89.01	71.37	76.78	93.21
Deep CNN	92.48	58.61	61.89	90.45
MobileNet	61.40	62.44	61.87	84.62
ResNet50	81.42	51.18	49.13	88.38

significant, whereas MobileNet and ResNet50 have shown poor performance on same dataset.VGG16 turned out as the best classifier with highest 94.67% accuracy, 92.60% precision, 80.52% recall, and 85.21% f1-score. VGG 19 is second best classifier with 76.78% f1-score, 93.21% accuracy, 71.37% recall, and 89.01% precision. Worst accuracy 84.62% and f1-score 49.13% is scored by MobileNet and ResNet50, respectively.

Figure 4 presents learning curve of VGG16. Accuracy learning curves (see Fig. 4a) show variations in training and validation accuracy in contrast to the number of epochs, whereas loss learning curve (see Fig. 4b) shows fluctuations in training and validation loss against number of epochs.

Receiver operating characteristic curve (ROC) can measure separation capability of a classifier between two classes. For higher value of area under the curve (AUC) model performs better. Figure 5 shows ROC curve to analyze the performance of classifiers in two classes (benign and malign). VGG16 covers the maximum 95% area for both class 0 and class 1. Again, for macro average ROC curve, maximum 98% area is covered by VGG16 and this is the main reason for the best performance of this model. ResNet50 has provided lowest value of AUC for both class 0 (0.68) and class 1(0.53) when compared with other classifiers and this cut down the performance of this classifier in significant amount.

7   Conclusion

In this paper, we have performed comparative analysis of various transfer learning models and one deep CNN on publically available BreakHis dataset. CNN model and other transfer learning models have shown state of our performance on histopathological images. With 94.67% accuracy VGG16 scored highest accuracy and can help pathologist to get an idea about the type of tumor, whereas REsNet50 and MobileNet have failed to extract the features properly and showed worst performance for BreakHis histopathological images.

For future work, the accuracy is not 100% therefore there is a room for improvement. We can assemble two or more transfer learning models and can extract the features from them and can use any classifier at the top of these models to get better results.

References

1. Breast Cancer. http://www.who.int/cancer/prevention/diagnosis-screening/breast-cancer/en/
2. Wang L (2017) Early diagnosis of breast cancer. Sensors17(1572):1–20
3. Elmore JG, Longton GM, Carney PA, Geller BM, Onega T, Tosteson AN, O'Malley FP (2015) Diagnostic concordance among pathologists interpreting breast biopsy specimens. JAMA 313(11):1122–1132
4. Obulesu O, Mahendra M, ThrilokReddy M (2018). Machine learning techniques and tools: a survey. In: 2018 international conference on inventive research in computing applications (ICIRCA). IEEE, pp 605–611. (July 2018)
5. Spanhol FA, Oliveira LS, Petitjean C, Heutte L (2016) A dataset for breast cancer histopathological image classification. IEEE Trans Biomed Eng 63:1455–1462. https://doi.org/10.1109/TBME.2015.2496264
6. Zhang Y, Zhang B, Coenen F, Lu W (2013) Breast cancer diagnosis from biopsy images with highly reliable random subspace classifier ensembles. Mach Vision Appl 24:1405–1420. https://doi.org/10.1007/s00138-012-0459-8
7. Spanhol FA, Oliveira LS, Petitjean C, Heutte L (eds) (2016) Breast cancer histopathological image classification using convolutional neural networks. In: 2016 international joint conference on neural networks (IJCNN). IEEE, Vancouver, BC
8. Jain T, Verma VK, Agarwal M, Yadav A, Jain A (2020) Supervised machine learning approach for the prediction of breast cancer. In: 2020 international conference on system, computation, automation and networking (ICSCAN). IEEE, pp 1–6. http://web.inf.ufpr.br/vri/breast-cancer-database. (July 2020)
9. Araújo T, Aresta G, Castro E, Rouco J, Aguiar P, Eloy C et al (2017) Classification of breast cancer histology images using convolutional neural networks. PLoS ONE 12:e0177544. https://doi.org/10.1371/journal.pone.0177544
10. Han Z, Wei B, Zheng Y, Yin Y, Li K, Li S (2017) Breast cancer multi-classification from histopathological images with structured deep learning model. Sci Rep 7:4172. https://doi.org/10.1038/s41598-017-04075-z
11. Nawaz M, Sewissy AA, Soliman THA (2018) Multi-class breast cancer classification using deep learning convolutional neural network. Int J Adv Comput Sci Appl 9:316–322. https://doi.org/10.14569/IJACSA.2018.090645
12. Dabeer S, Khan MM, Islam S (2019) Cancer diagnosis in histopathological image: CNN based approach. Inf Med Unlocked 16:100231
13. Li Y, Wu J, Wu Q (2019) Classification of breast cancer histology images using multi-size and discriminative patches based on deep learning. IEEE Access 7:21400–21408
14. Albawi S, Mohammed TA, Al-Zawi S (2017) Understanding of a convolutional neural network. In: 2017 international conference on engineering and technology (ICET). IEEE, pp 1–6. (Aug 2017)
15. Torrey L, Shavlik J (2010) Transfer learning. In: Handbook of research on machine learning applications and trends: algorithms, methods, and techniques. IGI Global, pp 242–264
16. Van Dyk DA, Meng XL (2001) The art of data augmentation. J Comput Graph Stat 10(1):1–50
17. Simonyan K, Zisserman A (2014) Very deep convolutional networks for large-scale image recognition. Computing research repository (CoRR), abs/1409.1556. [Ref list]

18. He K, Zhang X, Ren S, Sun J (2016) Deep residual learning for image recognition. In: Proceedings of the IEEE conference on computer vision and pattern recognition, pp 770–778.
19. Howard AG, Zhu M, Chen B, Kalenichenko D, Wang W, Weyand T, Andreetto M, Adam H (2017) Mobilenets: efficient convolutional neural networks for mobile vision applications. arXiv:1704.04861

A Comprehensive Tool Survey for Blockchain to IoT Applications

A. B. Yugakiruthika and A. Malini

Abstract The system generally has lot of errors. The IoT was still a growing area. To deal with all the errors, the testing as well as validating plays a major role. The testing and validating are easy to use. The system is insufficient for the solution of testing. In this paper, we provide an overview on test methods, tools, and methodologies for both the blockchain as well as Internet of Things, its software and its devices. We all are still lagging behind the Software Engineering community in the past decades.

Keywords Internet of Things · Software testing · Blockchain

1 Introduction

The Internet Of Things is generally transferred data over network without human-to-human interaction and human-to-computer interactions. It provides a unique identifier. The major concept of IoT is fascinating as well as exciting. It provides a secure ecosystem and block of IoT architecture. The IoT still come through security, monitoring, sensor, real-time information [1].

Blockchain is generally a database that contains growing set of information and records. The entire chain no master to hold it. The nodes that are participating will have the copy of the chain.it is generally a cryptocurrency system contain set of rules and policies. The policies are applicable to emission of token, and the rule is applicable to peer management and network communication [2]. The problem in security of data made limitations in large technology of IoT. However, the blockchain is used to identify and access the control of data in IoT [3].

The blockchain-enabled IoT is still missing to settle the privacy as well as the reliability concerns in the IoT. It is generally used to track the billions and millions of connected devices, the transaction, and the communication that happened between the devices. It will eliminate the occurrence of single point failure. The aim of the

A. B. Yugakiruthika (✉) · A. Malini
Thiagarajar College of Engineering, Madurai, India

A. Malini
e-mail: amcse@tce.edu

© The Author(s), under exclusive license to Springer Nature Singapore Pte Ltd. 2022
P. Nanda et al. (eds.), *Data Engineering for Smart Systems*, Lecture Notes in Networks and Systems 238, https://doi.org/10.1007/978-981-16-2641-8_9

algorithm used by blockchain is to make the user data more private. The blockchain contains infinite history of records from an IoT devices.

The blockchain such as Bitcoin and Ethereum that suffers from design issues that prevent from direct application to IoT. The IoT has also less power and storage capacity which must be tested and to ensure that the constraints are met correctly [4]. The IoT has composition layer such as cloud, fog, and edge. There will be two types on testing the IoT as well as the blockchain: (1) the implementation of test contains single client and more than one smart contracts (2) the focus will be fully based on theories and implementation of future work. We have been given less attention toward testing methodologies [4].

The blockchain is divided in to two categories. The one is turing and other one is non-turing. The turing complete gives the theoretical representation of any computation that needs to be completed. However, non-turing complete limits the action that is available in that specific language. This makes the code that moves into the blockchain ledger and distributively executed. The non-turing will easily analyze and easily predict the runtime [5].

The level of testing blockchain and IoT is increasing gradually. The key benefit of using the blockchain as well as IoT is the trust will be built, it reduces the cost and it accelerate its transaction [5]. This paper is a way to identifying the research questions and the issues on testing as well as validation of IoT, Sect. 2 briefly explain the knowledge about testing, methodologies, and tools. It highlights the base concepts and shortcomings. An overview for testing IoT systems and blockchain is given. A summary of the research is given on Sect. 3. In Sect. 4, the solutions that are available for testing IoT-enabled blockchain solutions are given. In Sect. 6, the final remarks are given.

2 Background

Testing is generally a process of identifying the defects and errors. It will identify the failure causes between the actual result and expected result. The four main reasons for software testing are the quality of product, it saves money, it provides security, and it provides satisfaction to the customer.

IoT is a collection of hardware, software, and objects connected to Internet that can sense and interact with the environment and surroundings through internet [6]. In this paper, we discussed about the types of testing and different testing tools for blockchain as well as the IoT. For testing, the methods and techniques are similar.

2.1 Levels of Testing and Its Methods

There is more number of testing depends on the scope and objective of test. The different testing is defined below as follows [5]:

Unit Testing: Testing of individual equipment or programming units or gatherings of related units [5]. It comprises on confining each piece of the framework and shows that individual parts fit its necessities and functionalities.

Integration Testing: Programming and additionally equipment segments are joined and tried to check the connection among them and how did they perform together [5, 7].

System Testing: Testing a total, coordinated framework to check the framework's consistency and conduct inside the predefined prerequisites [5].

Acceptance Testing: Formal testing is directed to decide if a framework fulfills its acknowledgment standards and to empower a client, a customer, or other approved substance to decide if to acknowledge the framework [5]. To test the system under test (SUT), there are different methods used namely, white-box testing [5, 7], gray-box testing [5, 6], and black-box testing [5, 6].

White-box Testing: The SUT is on the whole noticeable and referred to, and, all things considered, this data can be utilized to make test scenarios. Furthermore, white-box testing isn't limited to failure recognition, but at the same time, it can also identify errors.

Black-box Testing: The SUT inside content is covered up, and just information about the framework's or module's data sources and yields is known, being nearer to true utilize circumstances.

Gray-box Testing: A blend of the two past strategies is utilized. Data about the internals of the SUT is used, in any case; however, tests are led under sensible conditions, where just failures are recognized.

2.2 Testing the Internet of Things

There are different types of test across all the system components and architecture of IoT. The types of test are recommended to ensure the scalability, security, performance of IoT, and its applications. By extricating fundamental testing ideas and different configurations from these examination works, we plan an adaptable IoT testing structure that can test an enormous number of IoT gadgets. Hence, the types of testing are:

Edge Testing: By testing the more low-level parts of IoT systems, as small scale controllers (for example Arduino) and programmable rationale controllers (PLC). Testing approaches like implanted framework testing can be commonly used to perform tests on the edge layer, declaring the edge gadgets against their determination [8].

Fog Testing: Tests with respect to the center point layer on IoT systems, ordinarily made out of gateways. Delicate product testing approaches can be flawlessly applied since the gadgets that have a place with this layer have, regularly, an agreeable measure of figuring force and memory, running full working frameworks (for example Linux). Moreover, since this is the availability empowering agent layer, associating the restriction gadgets and the Internet fundamentally, it should cover network testing [6] and security testing [8].

Cloud Testing: Cloud testing tends to the need of test the remarkable quality worries of the cloud foundation, for example, enormous versatility and dynamic design. This field has open difficulties and issues of its own, and they are broadly investigated in the literature[8].

Interoperability Testing: In a perfect world, the convention definition is the highest quality level of innovation determination—it indicates precisely how gadgets arrange and speak with one another. Actually, determinations are dependent upon translation, and understandings differ generally. Regardless of whether your framework meets the convention determination precisely, you can in any case lose business and market notoriety by neglecting to manage another framework that doesn't. Quality Logic's interoperability trying takes care of that issue [8].

Security Testing: The purpose of security testing is to determine the information's are protected from the unauthorized user and uncover all the threats from the system [8]. The objective is to verify the application is free from viruses and malicious programs.

Network impact Testing: This testing, which represents Immediate Post-Concussion Assessment and Cognitive Test, is a typical testing convention in the territory of mind injury. It is a modernized device that tests an individual's intellectual (thinking and memory) capacities [8].

Performance and real-time Testing: This testing covers analysis, load testing, real-time stream, and time bound under velocity and variety.

End user application Testing: This testing includes functional and non-functional of an IoT application includes the experience of user and usability testing [9].

2.3 Testing the Blockchain

To ensure trust, analyzers must guarantee that all the parts of a blockchain are working consummately and that all applications are cooperating with it in a confided in way. A portion of the center tests that ought to be run incorporate utilitarian, execution, API, hub testing, and other specific tests. So, here is the thing that they involve more or less:

Functional Testing: This is an all-encompassing procedure that assesses crafted by different useful pieces of the blockchain (for example smart contracts).

API Testing: In the blockchain environment, it tests the interacting address between the application. It checks to verify that the formatted requests and replies are handled properly [10].

Performance Testing: It distinguishes execution bottlenecks, recommends the strategies for adjusting the framework and audits if the application is prepared for propelling.

Node Testing: Every single heterogeneous hub on the system must be tried freely to guarantee smooth collaboration [11]. In blockchain improvement, which regularly follows Agile practices, the move left way to deal with testing is picking up prominence. Doing a progression of tests as from the get go in the advancement lifecycle as conceivable permits limiting the quantity of deformities that could be found in the application's lifecycle later on when the effect on a business can be unfavorable.

3 Research Challenges

In previous section, we have presented the different types of testing and methodologies that have been developed and studied across hardware as well as the software. The IoT solution is based upon the hardware and software. The challenges of blockchain in IoT are:

Scalability: It identifying with the size of blockchain record that may prompt centralization as it's developed after some time and required some record the board which is throwing a shadow over the eventual fate of the blockchain innovation.

Power and time: The encryption process is required in blockchain enabled IoT that provides a way of biological systems that are exceptionally assorted and involved gadgets that have totally different processing capacities, and not every one of them will be equipped for running a similar encryption at ideal speed calculation [12].

Storage: In nodes, the ledger has to be stored by themselves. In order to store transactions and Device ID, the central server is eliminated by the blockchain. The time passes when the size of the ledger is increased [13]. The big hurdle is storage. The smart devices have very low storage capacity, and also, it has wide range of capabilities.

3.1 IoT Testing Tools

Wireshark: Wireshark analyzes the real-time network traffic and security. It is an open-source tool and provides free access. The network issue is solved by troubleshooting in Wireshark.

Shodan: The Internet-connected devices are identified by Shodan using filters. It is also called as search engine in which the client receives metadata send by the server.

Tcpdump: Wireshark yet it doesn't have GUI as that of Tcpdump does comparable employments. The TCP/IP was sent and got different bundles over the system by order line utility.

Thingful: For Internet of Things, the Thingful is a web search tool. By means of the Internet, it permits and secures interoperability between a huge number of items This IoT testing instrument likewise to control how information is utilized and engages to take progressively conclusive and important choices.

SOASTA cloud test: SOASTA is generally used to test the website of cloud. The web application is tested automatically by creating customized test. The predefined test is also used by the customer.

3.2 Blockchain Testing Tools

Ethereum Tester: It is a testing library accessible as a Github repo and also an open source. Its arrangement is truly simple with a sensible API support for different Testing prerequisites [14].

Populus: To build a smart contract the Populus is developed with python framework. The py.test is created in the system. In DApp series, the Populus development is more simple compare to other framework.

Truffle: It acquires great testing highlights, for example, mechanized agreement testing. The structure holds abilities past simply testing usefulness inside the blockchain application. It is also called as Ethereum engineers.

Embark: On creating decentralized applications (dApps), a sudden spike in demand for different frameworks or hubs is created. It has incorporations with Ethereum blockchain, IPFS, and a decentralized correspondence stages, for example, Whisper and Orbit.

3.3 Blockchain and IoT Testing Tools

Ganache Truffle: Ganache is an ethereum customer which one can use for Ethereum improvement. Ganache is a piece of Truffle environment. You can utilize ganache for the DAPP, and once it is created and tried on the ganache you can convey your DAPP on ethereum customer like geth or equality.

Ethereum: Ethereum is an open-source, based on blockchain and decentralized programming stage utilized for its own digital currency, ether. It empowers smart contracts and distributed applications to be constructed and run with no vacation, misrepresentation, control, or impedance from an outsider.

4 IoT and Blockchain Integration

In IoT, the privacy and reliability are missing. The blockchain is used to solve the missing link. The foundation element of IoT solution is blockchain. The blockchain is decentralized and has trustful capabilities. The smart devices history of immutable records is stored in blockchain enabled IoT. Without the need of centralized authority, this feature enables the smart device autonomous function [15]. The significant demonstration is performed for cloud and IoT processing. For devices to run on a more resilient ecosystem was created. The single point failure was eliminated by decentralized approach.

The way in which the blockchain technology was distinguished to fathom adaptability, security, and unwavering quality issues identified with the IoT. Blockchain goes about as a computerized correspondence layer between IoT sensors just as a vault for the information they create and transfer [16]. For instance, IoT gadgets in transportation compartments can follow area as well as screen temperature, vibration and whether a bundle has been messed with.

Blockchain systems have developed as a promising advancement in light of their capacity to certify the honesty of information shared among constituents in multiparty process joint effort. IoT has risen as a strategy for overcoming any issues between assets and their related business forms. Incorporating IoT and blockchain underpins trusted multiparty forms that connect physical world things to business process registering conditions [17].

Beyond cryptographic forms of money and brilliant agreements, blockchain advancements can be applied in various regions where IoT applications are such as detecting, information stockpiling, personality the boar, time-stamping administrations, keen living applications, keen transportation frameworks, wearable's, flexible chain the board, portable swarm detecting, digital law, and security in mission critical situations [18].

5 Opportunities

In this area, we will depict the possibilities to utilize the innovation of blockchain technology in the IoT.

5.1 Privacy

Blockchain exchanges utilizing the advanced character created by open key cryptography and the hash calculation. IoT applications with classified data can utilize this instrument to hide the true personality of the system [19].

5.2 Exchange of Data and Estimation of Cash

Trade of money related and PC information: Utilization of keen city sensors identified with swarms for offering digital services to city inhabitants. Cash can be fundamental for pulling in individuals from the network to shrewd urban communities and applications to utilize propelled assets.

5.3 Enlistment of Records for Records and Reviews

IoT information applications are shipped utilizing a framework having a place with various associations. Gracefully, chain checking centers around following and observing assets all over the graceful chain [20].

5.4 Smart Understanding

The agreement is a wise computerized framework incorporated with the framework, which is executed if the authoritative states of the agreement are met. An intercession contract for brilliant, independent exchanges between the gatherings to trade resources or questionable work with individuals from a blockbuster arrange. For instance, IoT applications can utilize insightful sensor information transport with those that have a place with different parts and the information that the sensors are supposed to sell [21].

6 Comparison of Testing Solutions

An outline comparison on the accessible tools for testing IoT arrangements is given on Table 1. Testing abilities of each arrangement are broke down by the perception of various factors.

For testing certain ancient rarities, there is an absence of tools. The objective of security testing is to measure its vulnerabilities and risk in the framework so that it does not quit working or misused. Through coding, these issues are fixed with the help of the designers. In the framework, the all conceivable security hazards were distinguished [22].

It is perceptible that most of the arrangements have been introduced in the writing, be that as it may, the greater parts of them are simply scholarly and there is no entrance to its source code or the product bundle [23]. Relatively, the arrangements accessible

Table 1 Comparison of the software tools on testing both the IoT and blockchain

Tool name	Open source	Browser extension	Operating system	Programming language
Ethereum	Yes	Chrome, Firefox, Opera, Metamask	Windows	Python, Java
Populus	Yes	Chrome, Firefox	Windows	Python
Truffle	Yes	Chrome, Firefox	Windows	Python, Java
Embark	Yes	Chrome, Firefox	Windows	Python, Java
Ganache Truffle	Yes	Chrome, Firefox	Windows	Python, Java
Wireshark	Yes	Chrome, Firefox	Linux, macOS, BSD, Unix, Solaris	C,C++
Shodan	Yes	Chrome, Firefox	Windows, Linux	Java
Tcpdump	Yes	Chrome, Firefox	Windows	C
Thingful	No	Chrome, Firefox	Windows	Java
SOASTA cloud	Yes	Chrome, Firefox	Windows	XML, Java, Python

to be utilized are scant and a large portion of them are shut source, diminishing the chance of broadening the device functionalities, or improving it by the methods for augmentations or module.

7 Result and Discussion

By the analysis, the tools are covering each and every test levels starting from unit testing until acceptance testing. Generally, the tools enable to test all levels but in some cases, it provides only partial support for testing the functionalities. The gaps may appear in the solution based on the supporting languages and the working platform. Example the Ethereum tool focuses on the transaction and device profile. The Gnache truffle provides better platform to work with blockchain-integrated IoT applications. Shodan is best to test the IoT, and it will be helpful to find the device that was connected to Internet. Blockchain and IoT help the user to take decision while building and deploying.

It is recognizable that most of the arrangements have been introduced in the literature, maybe the greatest parts of them are academic, and there is no admittance to its source code or the product bundle. Relatively, the arrangements accessible to be utilized are scant and a large portion of them are shut source, lessening the chance of expanding the apparatus functionalities or improving it by the methods for augmentations or modules. Here it can likewise be noticed that a few devices are as it were accessible on far off test runners which can lessen the capacity to test explicit necessities of specific arrangements and raise security concerns.

8 Conclusion

This paper survey's motivation is to push the peruse to comprehend an alternate part of blockchain innovation, which prominently affected cryptographic cash applications. Flexibly chain and IoT observing applications, these are promising fundamental structures of blocking advancements. In this paper, we have examined the structure of the Internet-of-Things applications. At that point, we offered the choice of applying blockchain IoT. Besides, we have introduced a few research difficulties and opportunities of blockchain in IoT. Inside this, we consider that there is a lot of old-known difficulties that are presently affecting straightforwardly the IoT frameworks and that further work must be sought after on the advancement of testing arrangements, mechanization methodology for testing, and constant reconciliation highlights.

References

1. Fan K, Bao Z, Liu M, Vasilakos AV, Shi W (2020) Dredas: decentralized, reliable and efficient remote outsourced data auditing scheme with blockchain smart contract for industrial IoT
2. Chen J-C, Lee N-Y, Chi C, Chen Y-H (2018) Blockchain and smart contract for digital certificate. In: IEEE international conference on applied system innovation
3. Jeon JH, Kim KH, Kim JH (2018) Block chain based data security enhanced IoT server platform
4. Lee CH, Kim K-H (2018) Implementation of IoT system using blockchain with authentication and data protection
5. Liao C-F, Bao S-W, Cheng C-J, Chen K (2017) On design issues and architectural styles for blockchain-driven IoT services
6. Coetzee L, Eksteen J (2011) The internet of things-promise for the future? An introduction. In: 2011 IST-Africa conference proceedings, May 2011, pp 1–9
7. Macrinici D, Cartofeanu C, Gao S (2018) Smart contract applications within blockchain technology: a systematic mapping study
8. Li S, Xu LD, Zhao S (2015) The internet of things: a survey. Inf Syst Front 17(2):243–259
9. Christidis K, DevetsikIoTis M (2016) Blockchains and smart contracts for the internet of things. IEEE Access 4:2292–2303
10. Whitmore A, Agarwal A, Da Xu L (2015) The internet of things—a survey of topics and trends. Inf Syst Front 17(2), pp 261–274. (20)
11. Zhang ZK, Cho MCY, Wang CW, Hsu CW, Chen CK, Shieh S (2014) IoT security: ongoing challenges and research opportunities. In: 2014 IEEE 7th international conference on service-oriented computing and applications, Nov 2014, pp 230–234
12. Khan MI, Lawal IA (2020) Sec-IoT: a framework for secured decentralized IoT using blockchain-based technology
13. Xu T, Wendt JB, Potkonjak M (2014) Security of IoT systems: design challenges and opportunities. In: Proceedings of the 2014 IEEE/ACM international conference on computer-aided design. IEEE Press, 2014, pp 417–423
14. Ghazi AN, Petersen K, Borstler J (2015) Heterogeneous systems testing techniques: an exploratory survey. Springer International Publishing, Cham, pp 67–85
15. Jiang ZM, Hassan AE (2015) A survey on load testing of large-scale software systems. IEEE Trans Softw Eng 41(11):1091–1118
16. Kshetri N (2017) Can blockchain strengthen the internet of things? IT Prof 19(4):68–72
17. Blockchain Disruptive Use Cases (2016). https://everisnext.com/2016/05/31/blockchain-disruptive-use-cases/. Accessed 1 Feb 2018
18. Bocek T, Rodrigues BB, Strasser T, Stiller B (2017) Blockchains everywhere-a use-case of blockchains in the pharma supply-chain. In: Proceedings of the IFIP/IEEE symposium on integrated network and service management (IM), Lisbon, Portugal, 8–12 May 2017
19. Banafa A (2017) IoT and blockchain convergence: benefits and challenges-IEEE Internet of things. https://IoT.ieee.org/newsletter/january-2017/IoT-and-blockchain-convergence-benefits-and-challenges.html. Accessed 31 Aug 2017.
20. Mohanta BK, Panda SS, Jena D (2018) An overview of smart contract and use cases in blockchain technology
21. Li Z, Kang J, Yu R et al (2017) Consortium blockchain for secure energy trading in industrial internet of things. IEEE Trans Ind Inf (99):1
22. Zhang Y, Kasahara S, Shen Y, Jiang X, Wan J (2018) Smart contract-based access control for the internet of things. (Yuanyu Zhang, Member, IEEE, Shoji Kasahara, Member, IEEE, Yulong Shen, Member, IEEE, Xiaohong Jiang,Senior Member, IEEE, and Jianxiong Wan)
23. Mohan N, Kangasharju J (2016) Edge-fog cloud: a distributed cloud for internet of things computations. In: 2016 cloudification of the internet of things (CIoT), Nov 2016, pp 1–6

Hybrid Ensemble for Fake News Detection: An Attempt

Lovedeep Singh

Abstract Fake news detection has been a challenging problem in the field of machine learning. Researchers have approached it via several techniques using old statistical classification models and modern deep learning. Today, with the growing amount of data, developments in the field of NLP and ML, and an increase in the computation power at disposal, there are infinite permutations and combinations to approach this problem from a different perspective. In this paper, we try different methods to tackle fake news, and try to build, and propose the possibilities of a hybrid ensemble combining the classical machine learning techniques with the modern deep learning approaches.

Keywords Fake news detection · NLP · Ensemble techniques

1 Introduction

Fake news spreads through a variety of mediums in many forms. It can spread via word of mouth in the form of audio or the form of images via social media platforms like Facebook, WhatsApp, and Twitter. One of the primitive forms of fake news has been text. This paper discusses possibilities to tackle this category and is an attempt toward a hybrid ensemble for fake news detection with the form of fake news being textual.

2 Related Work

Ensemble techniques [1] appear promising in the field of machine learning. The ideal goal of an ensemble technique is to incorporate the positive qualities of different models, while avoiding the bias present in a particular model. People have tried various approaches to classify text using machine learning methods on top of a

L. Singh (✉)
Punjab Engineering College, Chandigarh, India

© The Author(s), under exclusive license to Springer Nature Singapore Pte Ltd. 2022
P. Nanda et al. (eds.), *Data Engineering for Smart Systems*, Lecture Notes in Networks and Systems 238, https://doi.org/10.1007/978-981-16-2641-8_10

preprocessed data, preprocessing done via various NLP techniques [2]. Researchers have utilized classification models such as SVM, KNN, logistic regression etc. as well as experimented with deep learning techniques [3]. They have also tried to combine algorithms sequentially or parallelly [4] to improve the accuracy of the models.

3 Proposed Method

3.1 Dataset

There are some good datasets available for the detection of fake news. We use the LIAR [5] dataset in the training and testing of our model. The dataset classifies the news into six categories, three bending toward truth, three bending toward false. We make this categorization broader by combining three positive categories as true and the three negative categories as false with the hope that with more data in each category, the models will identify more differences and perform better. The dataset had in actual 12 columns for each label. We use only selected columns and introduce some new entries to the columns.

3.2 Preprocessing

While it makes more sense to use all the columns present in the dataset, we do not use them. Using all columns in building a machine learning model may improve the performance of the model, but will also introduce a bias in the model for a specific type of dataset. While we cannot totally sway away from the bias introduced by the dataset's political nature, we can avoid creating dependency on a large number of unique features that are not available in many practical scenarios. In practical life, whenever we encounter a piece of news, and we are not sure whether the news is genuine or fake, it is because we do not have enough parameters to make a sound conclusion We want our machine to find those elementary features in case they exist, which can help us say with some certainty whether the news is fake or genuine.

3.3 Removal of Columns

The dataset initially contained ['ID', 'label', 'statement', 'subject(s)', 'speaker', 'speaker's job title', 'state info', 'party affiliation', 'barely true counts', 'false counts', 'half true counts', 'mostly true counts', 'pants on fire', 'venue/location']. We keep only two columns label and statement. After removing other columns, we are left with ['label', 'statement'].

3.4 Addition of New Columns

We add some new features that are not specific to any domain or structure of the dataset. The only requirement is that the data present is in textual form. The intent is to try and figure out the intricate differences between fake and true news independent of the domain. We want to answer questions that are common in all domains. These are the features that are captured in the linguistic aspect of the language being used to convey the information rather than in the domain knowledge available to an expert. There are a whole lot of linguistic features available, or one can think of. We limit ourselves to the following features, namely Readability[1] (ease of understanding of the text), CountPunc (total count of the punctuations), SentimentScore[2] (reflects the tone, mood of the speaker displayed in the text), CountWord (total count of the words).

3.5 Word Embeddings

To deal with text in an ML problem, one way is to follow the regular ML routine to ask the right questions and cleverly figure out the hidden features, which are the essential differentiators, and train the model on those features. This model is likely to be more efficient in terms of complexity, but requires good experience and knowledge in the underlying problem statement domain. Another approach is to try including all the features and let it be on the machine to figure out the differentiating ones. Although a more relaxed approach, this approach involves higher complexity. In NLP, we have made progress to represent full text in the form of vectors through word embeddings. Two of the most common approaches are TFIDF and Doc2Vec. It is but obvious that while transformation, we lose some of the features in the text as we are moving from a higher dimension to a lower-dimensional space. Nevertheless, if we are still able to retain the differentiating features, we will be able to train an ML model with a good level of accuracy. We have used both TFIDF and Doc2Vec in our experiments.

3.6 The ML Models

There are many models available in the classification domain of ML. We have used SVM, KNN, logistic regression, and random forest in our experiments.

---

[1] https://pypi.org/project/textstat/.

[2] https://pypi.org/project/vaderSentiment/.

3.7 The ANN

Artificial neural networks (ANNs) have been the pioneers of deep learning and are the way to solve complicated, ill-defined problems of the modern deep learning. We have also used ANN in our experiments.

3.8 The Hybrid Ensemble

The underlying motivation is the same as in any ensemble technique. We intent to use the goodness of the classical ML models with the capabilities of the modern deep learning pioneer—the ANN.

3.9 The Architecture

We trained each model (from SVM, KNN, logistic regression, and random forest) using different combinations of the additional features we introduced, and also by using TFIDF and Doc2Vec individually. We also trained the ANN using all the features we introduced earlier; we did not try any combination of features as ANN will automatically adjust weights to best use the given features. Finally, coming to the hybrid ensembles, we have four versions of these. In all of the hybrid models, we use only 60% of training data to train the classical models; then, we get their predictions on 40% of the remaining training data. Finally, we use this 40% of the remaining training data to train our hybrid model, an ANN. We experiment with each of these scenarios to see which one yields the best possible model and whether we can achieve any success with such hybrid ensembles. In Hybrid V1, we provide prediction vectors of the classical models (SVM, KNN, logistic regression, random forest) built using all the linguistic features along with all the linguistic features to ANN. This provides ANN with the opinion of other four statistical models along with the features to form its own opinion. In Hybrid V2, we only provide prediction vectors of classical models built using all linguistic features to ANN. In Hybrid V3 and V4, we feed prediction vectors of classical models trained on TFIDF and Doc2Vec, respectively, to ANN. Figure 1 depicts the hybrid architecture used. It is to be noted that the rightmost connection is ON only for Hybrid model V1 and is OFF for the other three Hybrid models (V2, V3, and V4).

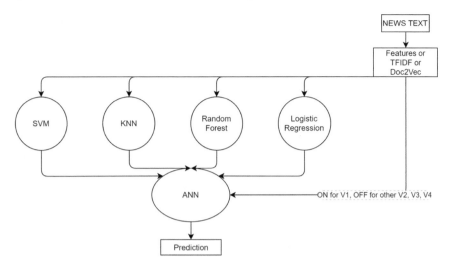

Fig. 1 The hybrid architecture

Table 1 SVM

Features	Test (%)	Validation (%)
Readability	56.35	52.02
CountPunct	56.27	51.48
SentimentScore	56.35	52.02
CountWord	56.35	53.89
All Features	56.43	53.19
TFIDF	61.72	61.29
Doc2Vec	56.35	52.02

4 Experiments

We conducted experiments using each of the features, including the complete set for each classical model. We also used TFIDF and Doc2Vec individually on each classical model. The results have been captured in Tables 1, 2, 3, and 4. Finally, we experimented with ANN. The results of the experiment have been summarized in Table 5.

5 Discussion

We can see that the accuracy was best in TFIDF with the model as SVM. TFIDF performed comparatively better in all the models. Even in the hybrid models, the best performing model is hybrid ensemble V3, which uses the prediction vectors

Table 2 KNN

Features	Test (%)	Validation (%)
Readability	51.78	50.39
CountPunct	44.59	47.51
SentimentScore	53.20	52.02
CountWord	49.88	52.88
All Features	57.06	52.10
TFIDF	57.06	55.61
Doc2Vec	54.14	49.14

Table 3 Logistic regression

Features	Test (%)	Validation (%)
Readability	56.35	52.02
CountPunct	56.35	52.02
SentimentScore	56.35	52.02
CountWord	56.35	52.18
All Features	56.35	53.27
TFIDF	61.64	59.50
Doc2Vec	56.35	52.02

Table 4 Random forest

Features	Test (%)	Validation (%)
Readability	55.41	51.40
CountPunct	56.20	51.01
SentimentScore	55.85	50.62
CountWord	56.35	53.97
All Features	53.04	53.82
TFIDF	61.56	61.29
Doc2Vec	53.67	52.02

Table 5 ANN

Features	Test (%)	Validation (%)
All Features	55.85	54.05
TFIDF	57.77	57.17
Doc2Vec	56.35	52.02
Hybrid V1	56.35	53.74
Hybrid V2	56.59	53.89
Hybrid V3	61.64	60.51
Hybrid V4	56.35	52.02

of classification models trained using TFIDF. One possible explanation for this is the higher dimension of data in TFIDF. We always lose some information in the data whenever we perform some processing to convert it form natural language to a machine understandable form, which is mathematical. It appears that TFIDF is able to retain some essential information which helps in separating fake news from the true news. This higher space, higher complexity, higher accuracy trend have been promising and is likely to work even in future. If we come up with more elementary representations of textual data close to the natural language and train on a humongous corpus of data, we are likely to surpass the accuracy offered by TFIDF. All other variations in the feature vectors and models showed an ambivalent trend and did not promise much hope. Doc2Vec did not perform well since the average number of words in a statement was less, and the total number of statements also could not compensate for yielding essential relations between the words during the three-layered shallow neural net training and failed to yield a meaningful Doc2Vec. Coming to the hybrid ensemble models, these results depict that we are unlikely to achieve success with such architecture models owing to several reasons. First, the hybrid models were based on ANNs, but we did not have a considerable amount of quality data to train ANN well. Second, it may be that the hybrid models are bound by the best performing model underneath and cannot perform better than it. This is just a hypothesis. Today, we have ensemble models that perform better than the underneath building models. Third, we experimented with a limited number of linguistic features, we may explore more features that may help us in improving the models, but it is going to be hard to determine which features are essential and why they are so. We could always use other NLP techniques such as N -grams and experiment with different deep learning models such as CNN and RNN variants like LSTM and get different results. Fourth, even if we succeed in determining the differentiating features, our models will be successful only in a particular type of textual forms of fake news that play with those features. Whenever our model encounters something different, it will be tricked, and all predictions might be totally wrong. Fifth, even if we progress significantly in the linguistic feature department, we will still be left to deal with the expert knowledge department. We will have fake news which will be entirely equivalent to the true version in a grammatical sense, but will be flawed due to the message being conveyed. Such news demands expert domain knowledge and cannot be dealt with by relying on pure linguistic features. To build a truly capable fake news detection model, we will have to incorporate domain knowledge in some form or the other. All code and data used in this paper are available at GitHub[3] for further experimentation.

---

[3] https://github.com/singh-l/hybrrid_FN_dat_.

6 Conclusion

Fake news detection continues to be an interesting and complex domain of research. We discussed it is difficult to develop a truly robust machine learning model due to various factors. Although, there can be numerous ways to approach this problem such as linguistic approach, complex social networks (pattern of news generation, news flow pattern, source nodes, etc.), vector spaces (Word2Vec, Doc2Vec, TFIDF, etc.). It appears as if the expert domain knowledge still is likely not replaceable by such static machine learning models. Dynamic algorithms which in effect use dynamic data as their ground truth could be a sound way forward in this area.

References

1. Huang F, Xie G, Xiao R (2009) Research on ensemble learning. In: 2009 international conference on artificial intelligence and computational intelligence, Shanghai, 2009, pp 249–252. https://doi.org/10.1109/AICI.2009.235
2. Falessi D, Cantone G, Canfora, G (2010) A comprehensive characterization of NLP techniques for identifying equivalent requirements. In: ESEM 2010-Proceedings of the 2010 ACM-IEEE international symposium on empirical software engineering and measurement. https://doi.org/10.1145/1852786.1852810
3. Thota A, Tilak P, Ahluwalia S, Lohia N (2018) Fake news detection: a deep learning approach, SMU data science review 1(3), Article 10. https://scholar.smu.edu/datasciencereview/vol1/iss3/10
4. Mahabub A (2020) A robust technique of fake news detection using Ensemble Voting Classifier and comparison with other classifiers. SN Appl Sci 2:525. https://doi.org/10.1007/s42452-020-2326-y
5. Wang WY (2017) "Liar, Liar Pants on Fire": a new benchmark dataset for fake news detection. In: Proceedings of the 55th annual meeting of the association for computational linguistics (Short Papers), Vancouver, Canada, July 30-Aug 4 2017. c 2017 Association for computational linguistics, pp 422–426. https://doi.org/10.18653/v1/P17-2067

Detection of Abnormal Activity at College Entrance Through Video Surveillance

Lalit Damahe, Saurabh Diwe, Shailesh Kamble, Sandeep Kakde, and Praful Barekar

Abstract Most of the colleges are having the facility to park vehicles for students and staff members in distinct locations. As the location of parking inside campus has been assigned by the college authority in the restricted areas, but some abnormal activity is found related to vehicle parking. So the main objective of this paper is to detect students who park their vehicles inside college or prohibited area. The sample video frames are used and the extracted image from the video frame can be further utilized to extract information of color, logo and number plate recognition. The vehicle number then add to database for the identification of student vehicle for further punishment. Different algorithms, i.e., CNN, ANN, RNN are tested on sample dataset for blue color, logo and number plate detection and CNN found satisfactory result in comparison ANN and RNN.

Keywords CNN · ANN · RNN · Extraction · Segmentation · Recognition

1 Introduction

At present scenario, uses of vehicle is drastically increased throughout city and also in our college premises. All the vehicles have unique vehicle identification number for identification purpose. The ID is a license number which refers as a legal license. Each vehicle has its own number plate which is placed at back and front of the vehicle as shown in Fig. 1. There are some number plate identification techniques are available to obtain proper recognition. Such as getting the location of number plate in vehicle and then select that image. The most presiding and fundamental step is to track the exact location of number plate from the image panel which is obtained by the system. The limit of a license plate is acknowledged by shape

L. Damahe (✉) · S. Diwe · S. Kamble · P. Barekar
Computer Technology Department, Yeshwantrao Chavan College of Engineering, Wanadongri, Nagpur, India

S. Kakde
Electronics Engineering Department, Yeshwantrao Chavan College of Engineering, Wanadongri, Nagpur, India

© The Author(s), under exclusive license to Springer Nature Singapore Pte Ltd. 2022
P. Nanda et al. (eds.), *Data Engineering for Smart Systems*, Lecture Notes in Networks and Systems 238, https://doi.org/10.1007/978-981-16-2641-8_11

Fig. 1 Captured Image
(Sample)

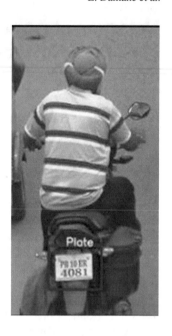

analysis and color analysis method. The spots which are unwanted are removed by parsing. Detection of abnormal activity on YCCE entrance gate through video surveillance is a practice through which the college authority would be able to trace the student who is bringing his or her vehicle up in staff parking rather than parking the vehicle in student parking. Here the abnormal activity means, if a student tries to bring his/her vehicle inside the college, so through video surveillance it will detect. This will reduce accidents inside college and as students park their vehicles at faculty parking, so faculties don't get space for parking their vehicles, so this problem will be solved. But the core challenge in computer vision is to create a proper machine learning model which is capable of identifying and localizing multiple objects. Due to advancement of technology for applications, it is easier to develop as compared to past years. Tensor flow is open source framework which helps to construct, train and deploy object detection models.

2   Previous Work

The article presented by R. R. Sontakke et al., is providing a overview of different methods are available for logo matching [1] and recognition. Mainly, edge detectors and shape detectors are discussed for the purpose of logo matching and also mentioned the difficulties in logo matching as different images contains different logos with different size and shapes. The technique is proposed by Jain et al. is based on automatic license plate recognition [2] using OpenCV. This presented method is alternative of ALPR System which uses Python and OpenCV. The neural network is used to recognize the character. The technique proposed by Sharma et al. [3] is based on character recognition using neural network and which is developed for isolated handwritten English character (A to Z). This method proves the satisfying results for character recognition. The technique proposed by Hashmi et al. [4] is based on the automatic no. plate detection using deep learning for solving complex task. The license plate has been recognized using template matching technique which generated a list of possible plates in an image by matching various features. The technique presented by Sharma et al. [5] is based on signature and Logo detection using deep CNN. It deals with signature and logo detection from a repository of scanned documents, which can be used for document retrieval using Signature or logo information. The technique, presented by Tafti et al. [6], is based on logo detection using pose clustering and momentums. A sample of iron logo is selected because it has not any regular geometric form and has special complexity. For the vehicle logo detection, the article is presented by Thubsaeng et al. [7] based on images of front and back views of vehicle. The proposed method is a two-stage scheme which combines convolutional neural network (CNN) and pyramid of histogram of gradient (PHOG) features. The technique proposed by Atikuzzaman et al. [8] is based on HAAR feature-based classifier to detect license plate and convolution neural network for recognizing class letters. Gupta et al. [9] presented a technique based on CNN for detecting no. plate of multiple languages. This paper suggests a new strategy to detect no. plate and comprehend the nation, language and layout of no. plates. YOLOv2 sensor with ResNet is proposed for detection and modified Convolutional neural network architecture is suggested to classify. The technique proposed [10–17] is based on the detecting contours.

3 Proposed Methodology

3.1 Input Video

To capture a video, create a video capture object. So, first sample video of a student who violating the rule is captured. Then, in the python code need to insert the path of input video.

Syntax:

cv2.VideoCapture(0): First camera or webcam.

cv2.VideoCapture(1): Second camera or webcam.

cv2.Video Capture("file name.mp4"): Video file as a video input for processing.

3.2 Processing

Now the video which is given as input is processed further to detect the blue color. While detecting the blue color of the college uniform one problem was faced i.e. the color of sky is also blue. The background object has to be removed otherwise it was detecting the whole blue objects which was present in the background. Because, our target was to detect the student uniform and not the background objects.

$$\text{Wavelength of Blue} \sim 490 - 450nm \sim 610$$

So, blue color was detected by keeping the wavelength of the color blue in our mind.

3.3 Logo Detection

First, sample logo in the database will be inserted. Then, with help of the tensor flow library, we will detect the logo on the shirt. Nowadays, due to advancement of technology applications are easier to develop as compared to past years. TensorFlow is open source framework which is easy to construct, train and deploy object detection models. So using this library, logo on student's shirt is being detected.

3.4 Number Plate Detection

In this concept, optical character recognition [OCR] is used to recognize the characters on the number plate of vehicle. As an input license plate recognition takes the image of vehicle and outputs the character written on license plate.

The flowchart as shown in Fig. 2, depicts how character segmentation is performed. Firstly, the input image is processed and converted from RGB to GRAY

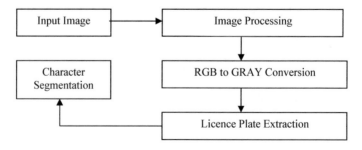

Fig. 2 Flowchart of character segmentation

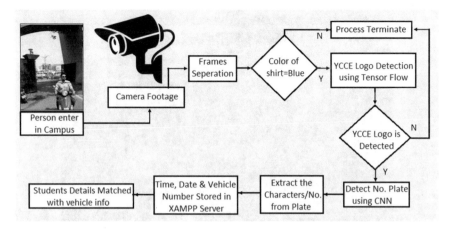

Fig. 3 Flowchart of proposed System

which makes it more suitable for further processing and perform various functions on it for maximizing results. Hence, the noise removal technique proves to be very beneficial. Noise removal is the technique where the excess part of the image is removed which is of noise and on which no processing can be done. After the Noise Removal is done, the license plate is extracted, and after extraction further processing is done. The characters are recognized from the number plate and for proper recognition of character, character segmentation is done, where the characters are extracted from the number plate in readable form and then further processing is performed.

The following flowchart as shown in Fig. 3, depicts how the module exactly works and how further processing is done. When the person enters the campus, his/her uniform is checked, if its blue color then the system will further check whether the logo is there on the uniform or not. If there is a logo, it will be detected using Tensorflow. After detecting the logo, it will particularly see whether its YCCE logo or not, after confirming the college logo, it will detect the number plate on the vehicle using CNN algorithm. Then, it will extract the characters from the number plate and make it into readable form and the characters will be stored in database. Time, date and number plate are stored in database using XAMPP hosting and hence student can be identified using this information that is stored in database.

4 Results and Discussion

4.1 Color Detection and Number Plate Extraction

As Fig. 4 indicates that color is detected in below image using OpenCv and number plate is extracted using convolution neural Network (CNN). As shown in

Fig. 4 Output of number plate

Fig. 4 output is displayed that is number plate is detected and displayed in green color, which is added to database.

4.2 Database

When new registrations are done in the database then the user table updated with new entry. The database has the entries for registration, the basic information is number plate, and it is unique as per time and date. Sample Database entries are presented in Fig. 5.

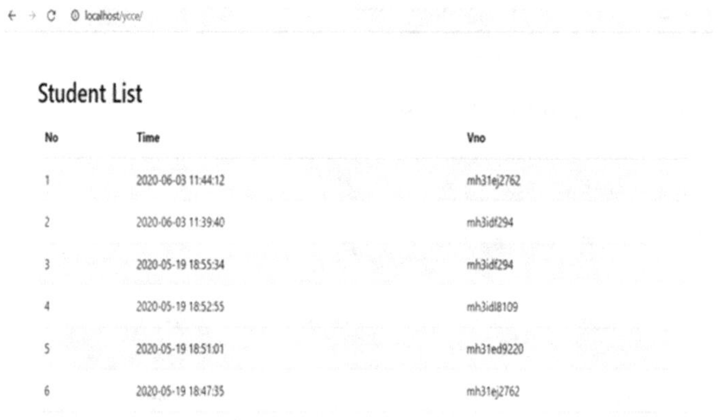

← → C ① localhost/ycce/

Student List

No	Time	Vno
1	2020-06-03 11:44:12	mh31ej2762
2	2020-06-03 11:39:40	mh3idf294
3	2020-05-19 18:55:34	mh3idf294
4	2020-05-19 18:52:55	mh3idl8109
5	2020-05-19 18:51:01	mh31ed9220
6	2020-05-19 18:47:35	mh31ej2762

Fig. 5 Output of database entry

Table 1 Comparison of CNN, ANN and RNN

Sample video clips	CNN			ANN			RNN		
	Blue color detection	Logo detection	No. plate detection and storing in database	Blue color detection	Logo detection	No. plate detection and storing in database	Blue color detection	Logo detection	No. plate detection and storing in database
Sample 1	Yes	Yes	Yes	Yes	Yes	No	Yes	Yes	No
Sample 2	Yes	Yes	Yes	Yes	No	No	Yes	No	No
Sample 3	Yes	Yes	No	Yes	No	No	Yes	No	No
Sample 4	Yes	Yes	Yes	Yes	Yes	No	Yes	No	No
Sample 5	Yes	No	No	Yes	No	No	Yes	No	No
Sample 6	Yes	Yes	Yes	Yes	Yes	No	Yes	No	No

Output of the various algorithms is presented in Table 1, for blue color, logo and number plate detection. So if we compare all algorithms namely ANN, RNN, CNN with the actual output table with reference to the parameters mentioned CNN proves to be good with other algorithm. It gives the good result and it matches with the actual outcome. As presented in Table 2, it is the actual output which is expected. The summary of True positive and False Positive is presented in Table 3.

Figures 6, 7 and 8 presented the comparison of accuracy for CNN, ANN, RNN with actual output in terms of all the parameter namely Blue color detection, Logo detection and Number plate detection. It can be clearly observed that all the algorithms work perfectly when compared with actual output of Blue color detection. To detect the logo, algorithms like ANN and RNN are not that accurate as presented in Fig. 7. It gives unpredictable results when compared with actual result. It works for some sample but not for other samples. Similarly, Fig. 8 is presented for the comparison of number plate detection. Algorithms like ANN and RNN are performing not so good context with actual or expected results. Figure 9 shows F1 score comparison of proposed approach.

Among three figures, there is one algorithm namely CNN which shows the good results when compared with actual or expected results in terms of all the parameters. This is the prime reason why we have used CNN algorithm in our system. Similarly proof of samples is provided of working of various algorithms in terms of all those parameters. The main reason why RNN algorithm fails because training an RNN is a very difficult task. It cannot process very long sequence and here since the input is video so there is long sequence of functions. So CNN proves to be the good algorithm in such kind of situations.

5 Conclusion

Image processing is an integral part in high-end applications like object detection or classification includes number plate recognition. Sometimes scanning number plate goes unsuccessful by using shape analysis. But successful development of character recognition is achieved, using template matching algorithm. Detecting number plate characters during day time work precisely but it gets counterproductive in case of night time. Proposed method detect blue color of shirt, matching logo and detected number plate and extracted data stored in database. Instead of analyzing large video file of camera steaming available in the database captured by video-camera, the proposed system automatically separate out video segments in which the peoples are breaking rules or policies. Hence tracing of the person is possible in minimum amount of time. The Proposed scheme successfully detects the abnormal activity using CNN algorithm and it outperforms ANN and RNN. The system can be further extended for recognizing vehicle number and person driving that vehicle.

Table 2 Actual output

Sample videos	Blue color detection	Logo detection	No. plate detection and storing in database
Sample 1	Yes	Yes	Yes
Sample 2	Yes	Yes	Yes
Sample 3	Yes	Yes	Yes
Sample 4	Yes	Yes	Yes
Sample 5	Yes	Yes	Yes
Sample 6	Yes	Yes	Yes

Table 3 Summary of comparison

Samples	Actual	Blue Color			Logo			No. plate			No. of TP			No. of FP		
		CNN	ANN	RNN	CNN	ANN	RNN	CNN	ANN	RNN	CNN	ANN	RNN	CNN	ANN	RNN
Sample 1	Yes (1)	1	1	1	1	1	1	1	0	0	3	2	2		1	1
Sample 2	Yes (1)	1	1	1	1	0	0	0	0	0		1	1		2	2
Sample 3	Yes (1)	1	1	1	1	0	0	0	0	0	2	1	1	1	2	2
Sample 4	Yes (1)	1	1	1	1	1	0	1	0	0	3	2	1		1	2
Sample 5	Yes (1)	1	1	1	0	0	0	0	0	0	1	1	1	2	2	2
Sample 6	Yes (1)	1	1	1	1	1	0	1	1	0	3	2	1		1	2
											15	9	7	3	9	11

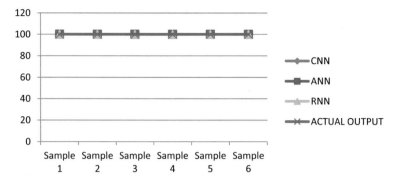

Fig. 6 Comparison of blue color

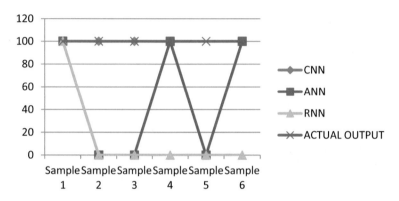

Fig. 7 Comparison of logo detection

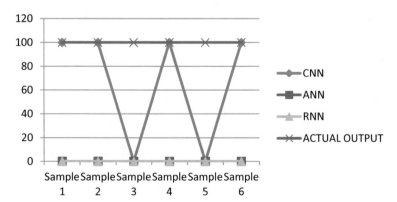

Fig. 8 Comparison of No. plate detection

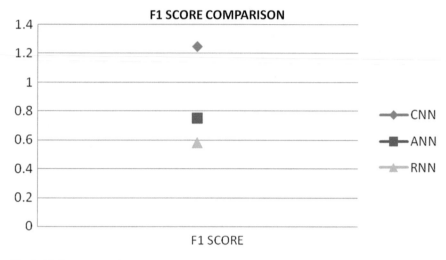

Fig. 9 F1 Score comparison

References

1. Sontakke RR, Damahe LB (2016) Logo matching and recognition: a concise review. World conference on futuristic trends in research and innovation for social welfare (Startup Conclave), Coimbatore, pp 1–6. https://doi.org/10.1109/STARTUP.2016.7583958)
2. Jain P, Chopra N, Gupta V (2014) Automatic license plate recognition using openCV. Int J Comput Appl Technol Res 3(12):756–761
3. Sharma A, Chaudhary DR (2013) Character recognition using neural network. Int J Comput Appl Technol Res 4(4):662–667
4. Hashmi S, Kumar K, Khandelwal S, Lochan D, Mittal S (2019) Real time license plate recognition from video streams using deep learning. Int J Inform Retr Res 9:65–87. https://doi.org/10.4018/IJIRR.2019010105
5. Sharma N, Mandal R, Sharma R, Pal U, Blumenstein M (2018) Signature and logo detection using deep CNN for document image Retrieval. 416–422. https://doi.org/10.1109/ICFHR-2018.2018.00079
6. Tafti M, Rajaee S, Sadeghi H, Tabatabaeifar S (2011) Logo detection using pose clustering and momentums. Int J Adv Sci Eng Inform Techno 1. https://doi.org/10.18517/ijaseit.1.4.79
7. Thubsaeng W, Kawewong A, Patanukhom K (2014) Vehicle logo detection using convolutional neural network and pyramid of histogram of oriented gradients, 34–39. https://doi.org/10.1109/JCSSE.2014.6841838
8. Atikuzzaman Md, Asaduzzaman Md, Islam Md (2019) Vehicle number plate detection and categorization using CNNs. https://doi.org/10.13140/RG.2.2.30125.64483
9. Gupta J, Saini V, Garg K (2020) Multilanguage number plate detection using convolutional neural networks. Int J Eng Trends Technol 68(8):18–23
10. Hatwar RB, Kamble SD, Thakur NV, Kakde S (2018) A review on moving object detection and tracking methods in video. Int J Pure Appl Mathem 118(16):511–526
11. Naaz H, Rathkanthiwar S, Kakde S (2016) Implementation of hybrid algorithm for image compression and decompression. Int J Eng Res 5(5):398–403. https://doi.org/10.17950/ijer/v5s5/514
12. Nirmalkar N, Kamble S, Kakde S (2015) A review of image forgery techniques and their detection. In: 2015 international conference on innovations in information, embedded and communication systems (ICIIECS), pp 1–5. IEEE

13. Channe R, Ambatkar S, Kakde S, Kamble S (2019) A brief review on: implementation of lossless color image compression. In: 2019 international conference on communication and signal processing (ICCSP), pp 0131–0134. IEEE. https://doi.org/10.1109/ICCSP.2019.8698027

14. Mahesh P, Kakde S, Yadav PM (2020) Implementation of MRI images reconstruction using generative adversarial network. In: Internet of things and big data applications, pp 161–169. Springer, Cham. https://doi.org/10.1007/978-3-030-39119-5_12

15. Awaghate A, Thakare R, Kakde S (2019) A brief review on: implementation of digital watermarking for color image using DWT method. In: 2019 international conference on communication and signal processing (ICCSP), pp 0161–0164. IEEE

16. Damahe LB, Thakur NV (2019) Review on image representation compression and retrieval approaches. In: Technological innovations in knowledge management and secision support, pp 203–231. IGI Global. https://doi.org/10.4018/978-1-5225-6164-4.ch009

17. Kashilani D, Damahe LB, Thakur NV (2018) An overview of image recognition and Retrieval of Clothing Items. 2018 international conference on research in intelligent and computing in engineering (RICE), San Salvador, pp 1–6. https://doi.org/10.1109/RICE.2018.8509041

Audio Peripheral Volume Automation Based on the Surrounding Environment and Individual Human Listening Traits

Adit Doshi, Helly Patel, Rikin Patel, Brijesh Satasiya, Muskan Kapadia, and Nirali Nanavati

Abstract Usage of audio peripheral is very common in nowadays. There are numerous works in literature that suggests that prolonged use of audio peripheral at a higher volume than the recommended level can induce partial hearing loss. The facts also suggest that over 60% of hearing loss is due to preventable causes. This paper embarks on automating the volume control of the audio peripheral using a supervised learning approach. Our paper gives a brief about the factors that human psychology for volume control depends on. Our work suggests that individual human habit (psychology) and surrounding type of environment forms the distinctive feature for volume control of audio peripheral. Modeling human psychology boils down to modeling individual human habit, forming the basis of our approach. We are using data points of the volume levels recorded in different noise dynamics and environments to capture the individual human habit and modeling using clustering algorithms. Further, we are using convolution neural networks trained on spectrogram representation of ESC 50 data for environment classification. After the integration of the models, the system adjusts the peripheral volume based on individual human listening traits and by understanding the surrounding type of environment. We are able to get 96% accuracy in environment classification and 92% accuracy in the listening traits classification model.

Keywords Peripheral volume automation · Environment classification · Noise-induced hearing loss · Listening traits

1 Introduction

Hearing is a mechanical sense which is one of the traditional five senses. It turns physical movement into the electrical signals that make up the language of the brain. Partial or total inability to hear is called hearing loss. A one-time exposure to extremely loud sounds or listening to loud sounds for a long time can cause hearing loss. Loud noise can damage cells and membranes in the cochlea. Listening to blaring noise for a long

A. Doshi (✉) · H. Patel · R. Patel · B. Satasiya · M. Kapadia · N. Nanavati
Department of Computer Engineering, SCET, Surat, India

© The Author(s), under exclusive license to Springer Nature Singapore Pte Ltd. 2022   123
P. Nanda et al. (eds.), *Data Engineering for Smart Systems*, Lecture Notes in Networks and Systems 238, https://doi.org/10.1007/978-981-16-2641-8_12

Table 1 Advisable duration at different sound levels referred from [2]

Sound level (dBA)	Max. duration of exposure	Environments
85 dBA	8 h	City traffic
95 dBA	1 h	Audio device average use
105 dBA	4 min	Audio device at max volume
110 dBA	100 s	Shouting into ear
150 dBA	10 s	Firecrackers

time can overwork hair cells in the ear, which can cause these cells to die. Medical experts and agencies have published studies on the advisable amount of time a human ear can expose at different noise levels [1] (Table 1).

Human psychology plays an important role in the habit formation and retention phase in the majority of humans [3]. In the following paper, we are targeting the formation of the habit of using audio peripherals at recommended volumes. As discussed above, one of the major causes of partial hearing loss is exposure to audio peripheral at high volumes for a prolonged time. We can reduce this time by using an intelligent system to control the peripheral volume. The system will make sure to reduce the volume as and when the surrounding noise is relatively low. So the long term reward of the system is that the user should be comfortable at hearing lower volumes.

Our study suggests that two important parameters for peripheral volume control are modeling individual human preference (psychology) and classifying the surrounding type of environment. So for modeling individual human preference, we have trained the clustering algorithm on a small manually collected dataset. The dataset consists of around 1500 data points collected in different noise dynamics. And for surrounding environment classification, we are using a convolution neural network trained on ESC 50 dataset [4]. The ESC-50 dataset is a labeled collection of 2000 environmental audio recordings suitable for benchmarking methods of environmental sound classification.

The rest of the paper is organized as follows. Section 2 gives the related work in this domain. Section 3 gives the main idea of our proposed method in which we describe the convolution neural network being used to classify surrounding environments, the use of the clustering model to understand the individual preferences. Section 4 shows the experimental setup and findings. In Sect. 5, we conclude this paper.

2 Related Work

Mentioned below is brief discussion of components like environment classification, noise cancellation systems, and dynamic volume equalizer covered in this paper.

In [5] gives an idea about the process to identify various types of activity, modes of communication and involvement of people in varied situations. The tools mainly used for modeling and training are recordings of microphones primarily gleaned from PDAs and consumer devices. In [6] highlights how discriminative spectro-temporal patterns are learnt by using the ability of deep convolutional neural networks (CNNs) which in turns makes them a good fit to using the ability of deep convolutional neural networks (CNNs). This study led to two major outcomes: first being CNN architecture for environmental sound classification and second being usage of audio data augmentation for overcoming the problem of data scarcity.

In [7] undertakes a study of urban high-school students in Malaysia to inspect the correlation between listening habits and underlying hearing risks of personal listening device usage. For this study, 177 samples between 13–16 years age group were analyzed to study their listening habits like duration, volume and symptoms of hearing loss. It also involved establishing their listening levels by asking them to set the volume to normal. It was concluded that mean measured listening level and listening duration for all subjects were 72.2 dBA and 1.2 h/day, respectively. The study also showed that listening level submitted by them were highly similar to measured levels ($P < 0.001$) (p-value is the probability of obtaining results at least as extreme as the observed results of a statistical hypothesis test). It also concluded that ones who listened to high volumes were also the ones who preferred to listen for longer time period ($P = 0.012$). The probability of males, i.e., ($P = 0.008$) listening at high volume turned out to be more than those of females. In [8] asserts that utterly damaging disability that is caused due to being exposed to noise type like recreational and occupational can be guaranteed prevented. The second most known sensorineural hearing deficit is noise-induced hearing loss after presbycusis (age-related hearing loss).

In [9] suggests a noise cancellation system that reduces the effects of ambient noise when a noise cancellation signal is added to a wanted signal. The downside of this system is that listener also becomes oblivious to the noise in his surroundings which at times is essential. In [10] covers the approach of volume control based on metadata inputs. So the system is working on technical components of audio like bass, treble, etc., and then adjusting the volume range. Lastly, in [11] highlights the broader and wider aspect about life by linking and producing a relation relation between color and human emotions thereby endorsing the evolved behavior. They have come up with K-means clustering algorithm to cluster individual human emotions.

3 Proposed Methodology

In this section, we describe our proposed overall methodology which consists of a convolution neural net for environment sounds classification and a fully connected network for classification of individual human listening preference.

3.1 Environment Sound Classification

In this section, we describe our proposed overall methodology which consists of a convolution neural net for environment sounds classification and a fully connected network for classification of individual human listening preference.

In the proposed system, we are incorporating environment sound classification to understand the importance of the surrounding environment. From the studies [12], we were able to know about environmental classification using deep neural networks. In the proposed system, we are also considering types of environment because every environmental noise has its own importance. For example, traffic noise is much more important compared to the noise in the park or cafe. So the system will reduce the volume level to a greater extent in the environment where the surrounding noise is important.

We have modeled a convolution neural network for environmental sound classification. Convolution nets generally work on image data. Hence, we are using the spectrogram methodology to represent audio data as image data. The model is trained on an ESC 50 dataset by Harvard [4]. The ESC-50 dataset is a labeled collection of 2000 environmental audio recordings suitable for benchmarking methods of environmental sound classification. The dataset consists of 5-s-long recordings organized into 50 semantical classes (with 40 examples per class) loosely arranged into 5 major categories as shown in Table 2.

A spectrogram is a visual representation of the spectrum of frequencies of a signal as it varies with time. The following Figs. 1 and 2 are an example of the spectrogram representation of audio data and the summary of the convolution neural network.

Table 2 Types of the environment in the dataset [4]

Animals	Natural soundscapes & water sounds	Human, non-speech sounds	Interior/domestic sounds	Exterior/urban noises
Dog	Rain	C<ying baby	Door knock	Helicopter
Rooster	Sea waves	Sneezing	click	Chainsaw
Pig	Crackling lire	Clapping	Keyboard typing	Siren
Cow	Crickets	Breathing	Door, wood creaks	Car horn
Frog	Chirping birds	Coughing	Can opening	
Cat	Water drops	Footsteps	Washing machine	Train
Hen	Wind	Laughing	Vacuum deaner	Church bells
Insects (flying)	Pouring water	Brushing teeth	!ng^{TM}	Airplane
Sheep	Toilet flush	Snoring	Clock tick	Fireworks
Crow	Thunderstorm	Drinking, sipping	Glass breaking	Hand saw

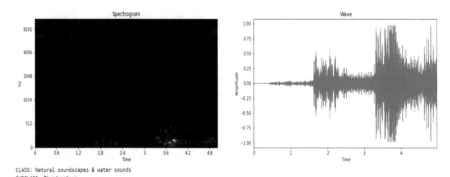

Fig. 1 Spectrogram representation of the audio

```
Layer (type)                    Output Shape            Param #
=================================================================
conv2d_1 (Conv2D)               (None, 126, 214, 64)    640

max_pooling2d_1 (MaxPooling2 (None, 63, 107, 64)        0

conv2d_2 (Conv2D)               (None, 61, 105, 128)    73856

max_pooling2d_2 (MaxPooling2 (None, 30, 52, 128)        0

conv2d_3 (Conv2D)               (None, 28, 50, 256)     295168

max_pooling2d_3 (MaxPooling2 (None, 14, 25, 256)        0

conv2d_4 (Conv2D)               (None, 12, 23, 256)     590080

max_pooling2d_4 (MaxPooling2 (None, 6, 11, 256)         0

flatten_1 (Flatten)             (None, 16896)           0

dense_1 (Dense)                 (None, 256)             4325632

dropout_1 (Dropout)             (None, 256)             0

dense_2 (Dense)                 (None, 50)              12850
=================================================================
Total params: 5,298,226
Trainable params: 5,298,226
Non-trainable params: 0
```

Fig. 2 Summary for environment classification model

3.2 Understanding Individual Human Listening Traits Using a Classification Algorithm

In the proposed system, we are trying to understand individual listening traits. The system will try to reduce the average peripheral volume over a gradual period of time. The system is designed user-specific so the user will be comfortable by the volume automation.

The cluster analysis was done on manually collected data. In total, 50 personal listening device users (18–22 years old) were interviewed to elicit their listening habits (e.g., listening duration, volume setting). Their listening levels were also determined by asking them to set their usual listening volume. Artificial noise environments were created just to measure listening levels at different noise dynamics. So from 36 different combinations of parameters in total 1800 data points were recorded. Using the dendrogram method, we were able to identify 4 different clusters to represent the collected information. The clustering was done using the hierarchical clustering algorithm. After clustering, the recorded data was stored into 4 clusters. A simple classification network was trained on that. In the system, there is a module where the user is advised to take a trait classification test. The result of the test is fed to the classification model and it outputs the cluster of the individual user. The following Fig. 3 is the dendrogram representation of the data.

Fig. 3 Dendrogram representation of the data

3.3   Integrated Working of the Components

In the proposed system, firstly the system will try to classify the individual user in one of four clusters. Once the cluster is assigned, periodically the system will be fed the spectrogram of the surrounding environment to classify the environment based on the importance of the environmental noise. Finally, the surrounding noise level (dB decibels) will be fed to a simple regression model to predict the volume level. The output of the regression model will be adjusted based on the output of the environment classification model and the cluster of the user.

4   Experimental Results and Analysis

4.1   Dataset and Experimental Setup

For the experiments we are using two datasets:

- ESC 50: The ESC-50 dataset is a labeled collection of 2000 environmental audio recordings suitable for benchmarking methods of environmental sound classification [4]. The dataset consists of 5-second-long recordings organized into 50 semantical classes (with 40 examples per class). Clips in this dataset have been manually extracted from public field recordings gathered by the Freesound.org project. The dataset has been pre-arranged into 5 folds for comparable cross-validation, making sure that fragments from the same original source file are contained in a single fold.
- Manually collected human listening traits dataset: In total, 50 personal listening device users (18–22 years old) were interviewed to elicit their listening habits (e.g., listening duration, volume setting). Their listening levels were also determined by asking them to set their usual listening volume. Artificial noise environments were created just to measure listening levels at different noise dynamics. So from 36 different combinations of parameters in total 1800 data points were recorded.

4.2   Result and Analysis

For environment classification, we are using ESC 50 for training the CNN model. As ESC 50 contains 2000 audio recordings the model was not able to train properly. So we added white noise to the dataset and augmented the data. And finally, CNN was trained on 4000 audio recordings (Tables 3 and 4).

For training, we are using 70% of the data, 20% for validation, and the rest 10% for testing the CNN model. The following are the results of the test data and a few examples of the predicted class by the model (Fig. 4).

Table 3 Data summary of the environment classification model

Audio recordings	4000
No of categories	50
The average length of audio	5

Table 4 Results of the environment classification model

Accuracy	0.9200
f1 score	0.9218
Precision score	0.9252
Recall score	0.9221

For individual human trait classification, we are using manually collected data. We ran hierarchical clustering on the collected data. After identifying four clusters, we modeled a simple multi-class classification network. To handle the class imbalance problem in classification we did resampling of the less dense classes. Out of 1800 data points, 60% used for training, 20% for validation, and rest 20% for testing. The following is the result of the classification network on test data (Table 5).

5　Conclusion and Future Work

We proposed an approach to address noise-induced hearing loss by using convolution neural nets and classification models. The generated results from the environment classification model were able to distinguish different types of environments and from the classification model was able to retain the listening traits of individuals. This integrated approach can be used to develop a more efficient volume automation system by collecting generalized data of the listening traits of the individuals.

The manually collected dataset can be used to understand listening traits based on the surrounding environment, noise levels, type of peripheral, and type of audio. Peripheral volume automation not only helps in reducing noise-induced hearing loss but also provides an ease of use to the users.

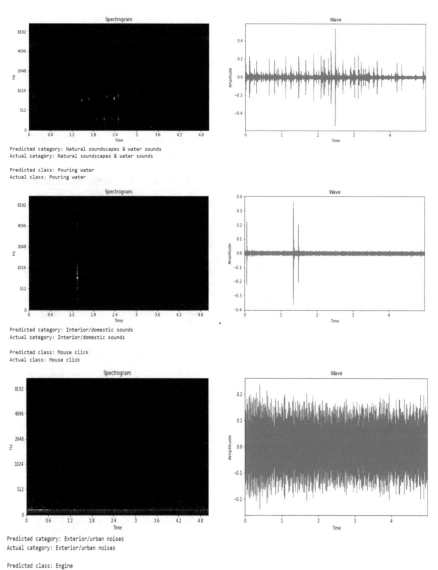

Fig. 4 Sample results from the environment classification model

Table 5 Results of listening traits classification model

Accuracy	0.9655
f1 score	0.9671
Precision score	0.9654
Recall score	0.9705

References

1. Health Care for Adults: Priorities for Improving Access and Affordability. Washington (DC): National Academies Press (US); 2016 Sep 6. 2, Hearing Loss: Extent, Impact, and Research Needs. (online) https://www.ncbi.nlm.nih.gov/books/NBK385309/
2. Centers for Disease Control and Prevention (online) https://www.cdc.gov/nceh/hearing_loss/what_noises_cause_hearing_loss.html
3. Gardner B, Lally P, Wardle J (2012) British J General Pract 62(605):664–666
4. Piczak KJ (2015) ESC: dataset for environmental sound classification. Harvard Dataverse V2
5. Ma L, Milner B, Smith D (2006) Acoustic environment classification. ACM Trans Speech Lang Process 3(2):1–22
6. Salamon J, Bello JP (March 2017) Deep convolutional neural networks and data augmentation for environmental sound classification. IEEE Signal Process Lett 24(3):279–283
7. Sulaiman AH, Seluakumaran K, Husain R (2013) Hearing risk associated with the usage of personal listening devices among urban high school students in Malaysia, Public Health 127(8)
8. Rabinowitz PM (2000) Noise-induced hearing loss. Am Fam Physician 61(9):2749–2760
9. Sibbald A, Alcock RD, Cirrus Logic Inc (2014) Noise cancellation system with gain control based on noise level. United States, US8737633B2
10. Ridmillers J, Gregory S, Carl N, Reden J (2017) Optimisation of volume and dynamic range through various playback devices. Russia, RU2631139C2
11. Kanagaraj, P. Anjana, S. Bavatarani, and D. Kumar, "A Study on Human Behavior-based Color Psychology using K-means Clustering," 2020 International Conference on Inventive Computation Technologies (ICICT), Coimbatore, India, 2020.
12. Piczak KJ (2015) Environmental sound classification with convolutional neural networks. 2015 IEEE 25th international workshop on machine learning for signal processing (MLSP), Boston, MA

A Survey: Accretion in Linguistic Classification of Indian Languages

Dipjayaben Patel⬤

Abstract The data on the internet is flourishing by leaps and bounds every second in structured and unstructured format. These data might be documents, Tweets, Articles, Images, Blogs, Reviews, Videos or audios in variety of languages and to classify such data or to analyze emotions from this data is a great demand of business nowadays and is a stimulating task. Machine learning and deep learning are spreading its wings in this field for automatic classification of such data and documents. This paper delves into contribution of the researchers in Indian Languages for information retrieval and classification with machine learning.

Keywords Machine learning · Indian language · Classifier · Poem · Feature extraction · POS

1 Introduction

Linguistic is predominant requisition for society to function efficiently. India is a multi-lingual country with hundreds of different languages being spoken in the country; however, there are 22 languages which are recognized by the constitution of India as official languages for communication.

Internet penetration in India has touched down to 50% of total population and still increasing by ten folds every year, this has resulted into regional languages actively used by the internet users as most devices nowadays are having local language options to operate the device which helps the person to browse, upload and edit in local lingos. India has a diverse and culturally enriched history and there is a lot of information nowadays which is uploaded in regional languages ranging from history, culture, religion, economics and academics which can benefit the entire world.

D. Patel (✉)
Vivekanand College For Advanced Computer And Information Science, Surat, Gujarat, India
e-mail: dipjaya_p_patel@yahoo.co.in

© The Author(s), under exclusive license to Springer Nature Singapore Pte Ltd. 2022
P. Nanda et al. (eds.), *Data Engineering for Smart Systems*, Lecture Notes in Networks and Systems 238, https://doi.org/10.1007/978-981-16-2641-8_13

Internet is a Pandora's Box which is yet to be explored to its full potential. There is a plethora of untapped useful information which can be helpful to users, but it is not identified by browsers due to language retrieval barriers of search engines.

Optimum utilization of multi lingual data extraction can only done by means of evolved symantec analysis in which regional languages inputs are to be introduced and implemented in matrix so that we can procure desired information by funneling available in the net and funneled into respective searches by user. Multi-language information retrieval has become the need of the hour to tap important information which we were unable to access till now. NLP has influential role to retrieve and classify such data, tweets and documents. Ample numbers of researchers are working to fetch and understand the data and emotions from these documents to determine the Psychology of human mind. Though huge amount of work has been done, due to diversity in Indian languages most of the languages are untouched or less work is done in these particular languages. Henceforth, researchers have a wide scope to explore in such areas.

To study and understand the multilingual literature, classification of documents is an initial stage. In the era of AI the machine learning algorithms are paving the path for classification. Machine learning algorithms are either supervised or unsupervised for labeled and unlabeled classes. Classification of the documents is a multistage process which is defined in following diagram (Fig. 1).

2   Literature Review

This review throws the light on the research work performed by various researchers in the field of classification with machine learning algorithms for various Indic Languages. Pal and Patel [1] and Kaur and Saini [2] have systematically studied the various classification models for Indic Languages. The study includes the 9 research papers in Hindi, 3 in Gujarati, 2 in English, 3 in Punjabi and 3 in other languages. However, apart from this lot of the work has been done in above-mentioned languages as well other languages like Bangla, Telugu, Urdu, Kannada etc.

2.1   Hindi Text

Pal and Patel [3] have classified Hindi poems on 4 criteria called Karuna, Shanta, Shringar and Veera. They have worked with a dataset of 180 poetries and implemented with Random forest and SGDC Classifier. The models trained and tested with 7 keyword features and 8 keyword features where both the models have better result with 7 keywords compare to 8 keywords which is 7.27% better with RF and 10% better with SGDC classifier. Hyper parameter tuning is applied for better precision and recall. Parameters used by SGDC Classifier were Alpha, Random State and Shuffle. Value of the loss parameter is set to "hinge". While RF used the

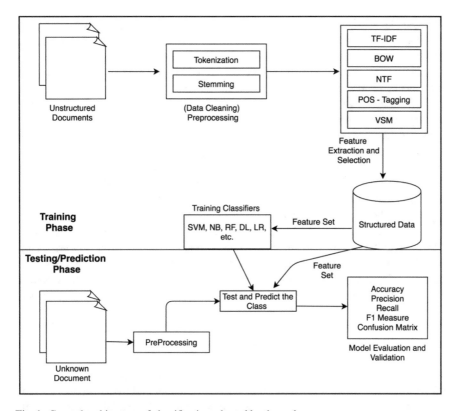

Fig. 1 General architecture of classification adopted by the authors

n_estimator, bootstrap and random state parameters. Performance tuning has been skillfully administered with total 70 combinations out of which 20 combinations comprise of SGD and the other 50 combinations of RF. During this research there were two obstacles which posed challenge to the researchers were the overlapping features found in two classes "Shringar" and "Shanta" which they sorted out by applying fuzzy logic and determination of "Unk" (unknown) words during POS tagging in which some of which are important words for feature extension could not be omitted. K-fold cross validator was used to validate the result and RF has better accuracy of 58.69%.

The model is developed by Pal and Patel [4], who has created 9 genres of the Hindi poems according to 'RAS' which are categorized into Shringar, Hasya, Adbuta, Shanta, Raudra, Veera, Karuna, Bhayanaka and Vibhasta. Data set has been created by scrapping the web to collect 55 poems and 10,531 words in which meaningful words are extracted by tokenization and stop word removal is implemented in preprocessing phase. Concept of TF and POS tagging has been used for feature extraction. To categorize the poems in 9 classes, the multi-class classifiers SVM and Naïve Bayes are used. The problem of overlapping has also been found in "Vibhasta" and

"Bhayanaka" class which has been solved by TF count of particular document. The resulted classes have concluded that the "Shringararas" class is dominant with the highest number of poetry in that class. The model provides better performance with SVM compared to Naïve Bayes. However, the only limitation of this research is, the numbers of poetry count of 55 which needs to be enhanced.

Bafna and Saini [5] have proposed this model to categorize Hindi poetry by using the supervised algorithms on the dataset of 450 poems. Dataset has been created by web scraping to fetch poetries of three genres which are "BalGeet", "UpdeshGeet" and "bhajans". TF-IDF is used to assign the weights to each token to generate vector space model. RF and Naïve Bayes classifiers are employed for classification which in turn is validated through confusion matrix where it is determined that performance of RF is better than Naïve Bayes.

Pal and Patel [6] have classified Hindi Poems into 3 classes Namely "Shringar", "Karuna" and "Veera" using 122 documents from the internet. For the feature extraction manual list of 266 stop words have been created without use of stemmer. Only unigram model was focused for feature extracted and in total 1238 features were extracted for experiment. They had trained and tested 5 ML algorithms Named SVM, KNN, Naïve Bayes, Decision Tree and Random Forest. RF algorithm was tried with 10 and 5 Decision trees which provided 40–56% of accuracy. KNN with 4 neighbors have outperformed accuracy of 52% and increase and decrease in these number of neighbors have negative impact on accuracy. Multinomial NB have total 64% of accuracy which is 8% more compared to Gaussian NB.DT with Entropy have an accuracy of 64% which is better than GiniIndex. Lastly, SVM with 'rbf' outperformed compared to SVM with 'linear' parameter and total accuracy is 52%.

Pal and Patel [7] have implemented classifiers to classify Hindi poetries into Three Classes "Shirangar", "Veera" and "Karuna". They had worked with dataset of 154 poems Hindi Poems which were manually extracted from the internet and even created the vocabulary list of 266 stop words for preprocessing. During the preprocessing phase K files have been created for K iterations.Concept of Frequency count of the word and Vector space model is used to extract the significant features. The data set is divided in the ratio of 80:20 for training and testing. They attempted to implement the model with the 5 algorithms namely SVM, RF, NaïveBayes, DTT and KNN in order to discover best classifier. Hyperplane with 'linear' is used in SVM classifier for categorization. For the implementation of DTT the "Gini Index" and "Entropy" has been implied to pick the root and nodes. Naive Bayes is developed with Probability. By trial and Error with all these algorithm it is observed that RF, NB and SVM outperform with more than 60% in certain folds while KNN and DTT is producing consistently low accuracy lesser than 60%.The authors have even suggested for the better performance with the use of "Chhand" and "Alankar" properties.

Harikrishna and Rao [8, 9] have developed the model to classify the 300 Hindi and 150 Telugu children stories, respectively, into three classes namely "Fable", "Folk-Tale" and "Legend". The data had been collected from the Blogs and Storybooks. Vector space model (VSM)" had constructed using individual term weight, TF, TF-IDF and POS density for the feature extraction. To convert the words into root words the lemmatization, POS tagging like NN, NNP, PRP, VM etc. with shallow parser have been used. Three classifiers NB, KNN, SVM were used on three parts of the stories introduction, main and climax. Finally, they analyzed that SVM has the highest accuracy and KNN had the minimum accuracy. Three measures precision, recall and F-measure had been improvized by the addition of the POS density and for the purpose of evaluation tenfold cross validation was used.

Harikrishna and Rao [10] have proposed the classification model for emotions from the sentences of the children stories. They had processed with 780 Hindi emotional sentences from 90 stories 30 of each category. POS tagging and emotion specific features were used. They have classified the emotions in 5 categories namely Happy, Sad, Anger, Fear and Neutral using three classifiers NB, KNN and SVM. Performance analysis has been performed using Precision, Recall, F-measures and Accuracy. Among three classifiers the SVM proved to be best with accuracy of 71.1% if they include only POS and ESF. However, it had been improvized by adding the genre information and lead to accuracy of 73.7% and F1-Measure as 0.73.

Bafna and Saini [11] have proposed the model to classify 565 Hindi poems with the Lazzy Machine Learning Algorithm named KNN and Linear regression. They had worked with the poems of four types "BAL GEET", "UPDESH GEET", "BHA-JAN" and "DESH BHAKTI GEET" which they have retrieved from the internet. To construct a term matrix TF and TF-IDF with the 50% threshold has been used. To train the model the labeled data had been employed, while to predict the class of the poetry the testing dataset was applied with the training: Testing ratio of 80:20. To evaluate the model accuracy, MSE and confusion matrix were used. KNN algorithm outperform over the regression for K = 8 on the other side residual error generated was 0.4 which is not significant.

Papacharissi et al. [12] have worked on analysis and classification of sentiments from the Hotel Review Dataset. Initially, the dataset consists of English data then they have converted it into Urdu using Google translate API and in turn these data again converted into Urdu-Roman using the Roman-Urdu Transliteration Tool. TF, TF-IDF and Normalized TF-IDF was employed to build Vector Space Model to extract the features. Dimensionality had been reduced by the removal of redundant attributes. Binary classifiers SVM, KNN, DT, NB, Ensemble Classifier, Ridge Classifier and Gradiant Classifier were used to classify the data set into two classes "Positive" and "Negative". The classifiers were trained by 1200 reviews and tested using 400 reviews and among all these classifiers SVM performed well with the accuracy of 97%. The performance has been analyzed by Confusion Matrix and from this matrix accuracy, sensitivity, precision and F1-Score had been calculated Bafna and Saini [11].

2.2   Gujarati Text

Naïve Bayes algorithm was used by the Researchers Rakholia and Saini [13] to classify Gujarati Documents. They have worked with total six predefined categories which comprises of Sports, Health, Entertainment, Business, Astrology and Spiritual. Also to experiment with the algorithm, 280 documents have been collected from Gujarati news web portals for each category. They have removed the stop words and created a list of stemmed words using manual observation technique which does not use any corpus using a JAVA language. TF-IDF has been used as feature selection and NB Model is validated through k-fold Cross Validation technique with and without feature selection. 10-fold Cross Validation technique has proved superior compared to the8-fold, sixfold, fourfold and twofold in case of without feature selection whereas the average accuracy of the model is determined as 88.96% with feature selection. Lastly, they mentioned that performance will increased by decreasing the number of influencing terms.

Rakholia and Saini [14] have constructed a corpus for a Gujarati language from the website of Gujarati news Paper and article. They have collected 126 documents by scrapping of the web with each document having more than 260 tokens. After the cleaning process, out of 32,540 tokens they kept 1125 unique tokens based on the lexical classes such as nouns, adverbs, adjectives etc. These lexical classes were considered as main classes that in turn consist of subclasses. Subclasses have most common and frequently used terms used in Gujarati language. Nouns have been classified on the base of "gender" and "number"(singular, plural) whereas the Two sub parts of the Adjectives are considered as" Consistent" and "Inconsistent". Gujarati vocabulary does not have a particular rule to classify the verb as "Transitive" or "Intransitive" instead, To identify pronoun they have used first, second and third person as well as number(singular or plural). Each stop word is assigned to the class label manually for each lexical category. They considered Term Frequency of each token for its significance in Gujarati literature and determined 190 stop words which are frequently used in all Gujarati documents.

Rakholia and Saini [15] have developed the Information Retrieval model by the application of Cosine Similarity based Vector Space Model particularly for Gujarati language. The documents were from seven different categories like Sports, Education, Health, Astronomy, Science, Politics and Entertainment.They had experimented with 1360 text and pdf files.Document Term Matrix has been formed to calculate the occurrence of each word in each document. To retrieve the information and for Gujarati language Vector Space Model and Cosine Similarity is used. Once the information was retrieved the Term Frequency(TF), Normalized Term Frequency(NTF) were used to find the influence of each term in the document. To remove the problem of more influence given to less important term had been removed by the Inverse Document Frequency (IDF). To evaluate the performance of the system the Recall and Precision values were calculated and the resultant values were 0.78 and 0.86, respectively.

2.3 English Text

Pal and Patel [16] have classified the movie reviews of English language from 2000 documents into two classes as positive and negative depends on the comments. Dataset was prepared by web scrapping, cleaning and stemming. For the feature extraction number of occurrence of the word in document had been considered. The classifier they applied was SVM, KNN and NB. Prominent part of the model is that they tried with different ratio of data set splitting for testing and training to determine best suitable model with best ratio of training and testing data. Ratios they tried with all these three classifiers are 50:50,60:40,70:30,80:20. Two main observations of the model are performance gets improvized consistently by increasing the size of training dataset for all algorithms and among all the algorithm the SVM has best accuracy about 80% and used 'rbf' as a parameter value for all the possible splitting of dataset.

Sreeja and Mahalakshmi [17] have suggested classification model to organize English poems on the base of Navrasa. 350 poetries of the leading Indian poets in the era of 1850–1950 were included in their dataset. The Corpus had been created manually by tagging the emotions in the poem and the algorithm to search and tag the emotions. The classifier applied were Naïve Bayes, Probabilistic Corpus Based Method (PCB) and Vector Space Model where NB outperformed.

2.4 Punjabi Text

Saini and Kaur [18] have developed a model named "Kavi" where they have categorized the Punjabi poetries in 9 Genres based on Navrasa. The dataset consist of 948 poetries with Gurumukhi script that has been retrieved from the internet. To tag the poems as per the emotions available in that particular poem 3 annotators were used. Feature extraction was performed as per the Linguistic, poetic and statistical features. For linguistic feature extraction, lexical features and syntactic feature of POS has been used. Sound rhyming pattern and title of the poem was used to extract poetic features and TF, TF-IDF were used as statistical measures to extract the features. Two classifiers NB and SVM were used to classify the poetries. For the performance analysis accuracy, precision, recall, F-measures and tenfold cross validation were used. Finally, they concluded that the accuracy of SVM is more compared to NB which is 70.02% and the title of the poem is most influential factor to achieve the accuracy.

Kaur and Saini [19] have made hard efforts to classify 240 Punjabi Poetries using 10 different machine learning algorithms to find most appropriate algorithm. They have classified the poems into 4 classes "Relation and Romantic", "Nature and Festival", "Linguistic and patriotic" and "Philosophy and Spiritual". The Bag of World and Vector Space Model were created using TF for feature extraction. The classifiers they tried to train and test were SVM, NB, KNN, DT, Hyperpipes, VFI, C4.5,

ADABoost, Bagging and PART in Weka Tool. After implementing all these models they determined that KNN, HP, NB and SVM model performed well for Punjabi Poetry categorization with accuracy of 52.92%, 50.63%, 52.75%, and 58.79%.

Kaur and Saini [20] have classified the 2034 Punjabi Poetries into four classes "Relation and Romantic", "Nature and Festival", "Linguistic and patriotic" and "Philosophy and Spiritual" keeping orthographic and phonemic poetic features in center. Feature Extraction has been performed by TF and TF-IDF. To train the model NB, KNN, SVM and hyper pipe algorithms were used where SVM have best performance of 71.98% accuracy. However, they observed that addition of these poetic features have no major significant effect on the performance.

2.5 Other Languages

Saini and Bafna [21] have worked with Sanskrit document to find the similarity among these documents.The dataset of 760 documents to implement the model has been formed by extracting 340 "Subhashits" and 420 "Stories" from the web. Total 1, 11,123 tokens has been found in tokenization phase and after clearing 76,836 tokens left for the experiment. To derive significant terms the Document Term Matrix was formed by TF (Term Frequency) and Document Synet matrix was formed using group frequency with threshold value of 75%. Document similarity had been calculated with the application of "Cosine Measure" on Document Synet Matrix where threshold value applied was 65%.the model was generally worked with Adjective, Adverb, Verb and Noun to identify unique term in Lemmatization Phase. For the model evaluation F1-score, Matthews Correlation Coefficient, accuracy and precision were used and finally derived that F1-Score, precision and accuracy were increased by 0.2 and Mathews Correlation Coefficient was declined by 0.1.

Gohil and Patel [22] have classified the multilingual political tweets in eight emotional classes. Dataset consist of 12,655 Tweets regarding general election 2019 in India from Twitter in three languages English, Hindi and Gujarati. From these by considering "Majority Votes" and "All Votes" two dataset were formed. Supervised learning and Hybrid Approach have been used for classification. Primary features generated by TF-IDF has been used with supervised learning, while hybrid approach worked with both primary and Secondary features where again TF-IDF was used for primary feature and SenticNet was used for secondary features. For the purpose of classification multilabel classifiers Logistic Regression, Multinomial Naïve Bayes and SVM were employed. To evaluate the classifiers the F-measure was used and they concluded that Hybrid approach with CS-SN outperformed compared to Machine learning Algorithm for English Language while the same algorithm have deprived result with Hindi and Gujarati Language.

Determination of document similarity of Marathi Documents has been performed using semantic–based dimension reduction method by Bafna and Saini [23].To generate this data set, total of 1206 documents were retrieved from the web which were categorized into 713 Verses and 493 Proses out of which 1,256,721 tokens were retrieved during tokenization and 56,345 tokens were left after the stop word removal process, so finally after lemmatization the left out words were 35,167. Algorithm works in two phases, in first phase the significant terms has been derived with morphological information such as noun, verb, adjective, voice and gender of the verb with the frequency count and the threshold value of 60% and Document Term Matrix (DTM) which has been generated using cosine measure. Synset matrix (DSMM) has been constructed for quality assessment to reduce the dimension of matrix where few words get more significance which were ignored in first phase. In this second phase the document similarity is calculated with values between 0 and 1.Out of total Documents 922 they found as similar and 87 as non-similar. To validate the model, precision, recall, accuracy, F1-Score and error rate are used where F1-score is higher and error rate is 0.1% (Table 1).

3 Conclusion

Indian languages have great range of literature with different scripts and grammatical formats which needs an attention of researchers to utilize it in right manner. Classification of such Literature is gaining popularity in contemporary era and a lot of work has been done for some Indic languages like English, Hindi, Punjabi, Marathi but much of the work yet to be done and is challenging due to morphological diversity [24]. Many of the models are working with small size of dataset and the result can be improvized by training the model by huge dataset. Survey of classification has been performed only with Machine Learning Algorithms, but lot of work has been done with Deep learning models such as ANN, CNN and RNN with better accuracy. Majority of the authors have used NB, SVM, RF, KNN and DT classifiers and it is observed that SVM is providing best results for poetry classification in any of the language. Some languages such as Gujarati and Sanskrit at the beginning stage of experiment where researchers can contribute to the community.

Table 1 Summary of the classification

Sr No	Author	Dataset and size	Language	Feature extraction	Classifiers used	Class	Performance analysis	Comment (Best classifier)
1	Pal and Patel [3]	180 Poems	Hindi	TF-IDF,POS Tagger	SGD, RF	4	Accuracy, K-Fold	RF (58.69%)
2	Pal and Patel [4]	55 Poems	Hindi	TF-IDF, POS Tagger	SVM,NB	9	Accuracy	SVM
3	Bafna and Saini [5]	450 Poems	Hindi	TF,VSM	RF,NB	3	Accuracy, confusion Matrix	RF
4	Rakholia and Saini [13]	280 Text Documents	Gujarati	TF-IDF	NB	6	Accuracy, K fold	NB (88.96%)
5	Gohil and Patel [22]	9056 ElectionTweet	English	TF-IDF, Cosine Measure	LR, NB, SVM, Hybrid Approach	8	F Measures	Hybrid Approach
6	Pal and Patel [6]	122 Poems	Hindi	TF, POS tagger	SVM,KNN, Naïve Bayes, DTT,RF	3	Accuracy	NB, RF
7	Pal and Patel [16]	2000 Movie Review	English	TF	SVM,KNN,NB	2	Accuracy	SVM
8	Pal and Patel [7]	154 Poems	Hindi	TF,VSM	SVM,RF, Naïve Bayes,DTT and KNN	3	Accuracy, K-fold	RF, NB, SVM
9	Harikrishna and Rao [8, 9]	300 Stories	Hindi	TF,TF-IDF and POS	NB, KNN, SVM	3	Precision, recall, F-measure	SVM
10	Harikrishna and Rao [10]	780 Stories	Hindi	POS,ESF	NB, KNN and SVM	5	Precision, Recall, F-measures, Accuracy	SVM
11	Bafna [11]	565 Poems	Hindi	TF, TF-IDF	KN, LR	4	Accuracy, MSE, confusion matrix	KNN

(continued)

Table 1 (continued)

Sr No	Author	Dataset and size	Language	Feature extraction	Classifiers used	Class	Performance analysis	Comment (Best classifier)
12	Saini and Kaur [18]	948 Poems	Punjabi	TF, TF-IDF	NB, SVM	9	Accuracy, precision, recall, F-measures10-fold	SVM
13	Papacharissi et al. [12]	1200 Hotel Reviews	Urdu Hindi	TF, TF-IDF and NTF-IDF	SVM, KNN, DT, NB, Ensemble Classifier, Ridge and Gradient	2	Accuracy, Sensitivity, Precision, F1-Score	SVM
14	Kaur and Saini [19]	240 Poetries	Punjabi	TF, Bag Of Word, VSM	Adaboost, Bagging, C4.5, Hyper pipes, KNN, NB, PART, SVM, Voting Feature Interval, Zero	4	Accuracy	KNN, HP, NB, SVM
15	Sreeja and Mahalakshmi [17]	350 Poetries	English	Manually tag own algorithm for frequency count	NB, VSM, Probability based	9	10 –Fold	NB
16	Kaur and Saini [20]	2034 Poetries	Punjabi	TF,TF-IDF	NB, SVM, KNN, Hyper pipes	4	Accuracy	SVM

References

1. Pal K, Patel BV (2017) A study of current state of work done for classification in Indian languages. Int J Sci Res Sci Technol 3(7):403–407. Available http://ijsrst.com/paper/1494.pdf
2. Kaur J, Saini J (2015) A study of text classification natural language processing algorithms for Indian languages. VNSGU J Sci Technol 4(1):162–167
3. Pal K, Patel BV (2020) Emotion classification with reduced feature set sgdclassifier, random forest and performance tuning, vol. 1235 CCIS. Springer Singapore
4. Pal K, Patel BV (2020) Model for classification of Poems in Hindi language based on ras. Smart Innov Syst Technol 141:655–661. https://doi.org/10.1007/978-981-13-8406-6_62
5. Bafna P, Saini JR (2020) Hindi poetry classification using eager supervised machine learning algorithms. 2020 international conference on emerging smart computing and informatics, ESCI 2020, pp 175–178. https://doi.org/10.1109/ESCI48226.2020.9167632
6. Pal K, Patel BV (2020) Automatic multiclass document classification of hindi poems using machine learning techniques. 2020 International conference on emerging technology INCET, 1–5. https://doi.org/10.1109/INCET49848.2020.9154001
7. Pal K, Patel BV (2020) Data classification with k-fold cross validation and holdout accuracy estimation methods with 5 different machine learning techniques. Proceedings 4th international conference on computing methodologies and communication ICCMC 2020, Iccmc, pp 83–87. https://doi.org/10.1109/ICCMC48092.2020.ICCMC-00016
8. Harikrishna DM, Rao KS (2015) Children story classification based on structure of the story. 2015 international conference on advances in computing, communications and informatics, ICACCI 2015, pp 1485–1490. https://doi.org/10.1109/ICACCI.2015.7275822
9. Harikrishna DM, Rao KS (2016) Classification of children stories in Hindi using keywords and POS density. IEEE international conference on computers communications and control IC4 2015. https://doi.org/10.1109/IC4.2015.7375666
10. Harikrishna DM, Rao KS (2016) Emotion-specific features for classifying emotions in story text. 2016 22nd national conference on communications NCC 2016, 2016. https://doi.org/10.1109/NCC.2016.7561205
11. Bafna P, Saini JR (2020) Hindi verse class predictor using concept learning algorithms. IEEE int.ernational Conference innovative mechanical industrial application, no. Icimia, pp 318–322. https://doi.org/10.1017/9781108552332.003.
12. Papacharissi Z et al (2013) Sentiment analysis of roman Urdu/Hindi using supervised methods. Ain Shams Eng J. 2(3):1093–1113
13. Rakholia RM, Saini JR (2017) Classification of Gujarati documents using Naïve Bayes classifier. Indian J Sci Technol 10,(5):1–9. https://doi.org/10.17485/ijst/2017/v10i5/103233
14. Rakholia RM, Saini JR (2016) Lexical classes based stop words categorization for Gujarati language. Proceedings - 2016 international conference on computing, communication and automation (Fall), ICACCA 2016. https://doi.org/10.1109/ICACCAF.2016.7749005
15. Rakholia RM, Saini JR (2017) Information retrieval for Gujarati language using cosine similarity based vector space model. Adv Intell Syst Comput 516:1–9. https://doi.org/10.1007/978-981-10-3156-4_1
16. Pal K, Patel BV (2020) Multi-class document classification: effective and systematized method to categorize documents. Int J Sci Res Sci Eng Technol 7(1):118–123. https://doi.org/10.32628/ijsrset207117
17. Sreeja P, Mahalakshmi GS (2016) Emotion recognition from poems by maximum posterior probability. Int J Comput Sci Inf Secur 14(CIC2016):36–43
18. Saini JR, Kaur J (2020) Kāvi: An annotated corpus of Punjabi poetry with emotion detection based on 'Navrasa.' Procedia Comput Sci 167(2019):1220–1229. https://doi.org/10.1016/j.procs.2020.03.436
19. Kaur J, Saini JR (2017) Punjabi poetry classification: the test of 10 machine learning algorithms. ACM international conference proceeding Series, vol Part F1283, pp 1–5. https://doi.org/10.1145/3055635.3056589

20. Kaur J, Saini JR (2018) Automatic classification of Punjabi poetries using poetic features. Int J Comput Intell Stud 7(2):124. https://doi.org/10.1504/ijcistudies.2018.10016073
21. Saini JR, Bafna PB (2020) Measuring the similarity between the sanskrit documents using the context of the corpus. Int J Adv Comput Sci Appl 11(5):140–145. https://doi.org/10.14569/IJACSA.2020.0110521
22. Gohil L, Patel D (2019) Multilabel classification for emotion analysis of multilingual tweets. Int J Innov Technol Explor Eng 9(1):4453–4457. https://doi.org/10.35940/ijitee.A5320.119119
23. Bafna PB, Saini JR (2020) Marathi document: similarity measurement using semantics-based dimension reduction technique. Int J Adv Comput Sci Appl 11(4):138–143. https://doi.org/10.14569/IJACSA.2020.0110419
24. Pal K, College S (2020) Computer linguistics for processing human language for artificial intelligence. Hard Appl 8:152–159

The Positive Electronic Word of Mouth: A Research Based on the Relational Mediator Meta-Analytic Framework in Electronic Marketplace

Bui Thanh Khoa

Abstract The electronic marketplace (e-marketplace), which gradually is becoming a popular concept in online business, is one of important distribution channels for businesses. However, businesses need to choose a good e-marketplace to place their sales booth to generate revenue. This study explored the positive electronic word of mouth (Positive e-WOM) about the e-marketplace brand and its influencing factors based on the relational mediator meta-analytic framework (RMMAF). Qualitative research was performed through in-depth interviews with seven experts related to the e-commerce industry, and quantitative research was done with the participation of 694 respondents through a self-administrated questionnaire. The research results confirmed the RMMAF result, with three relationship antecedents, including habituation, e-marketplace reputation, and social media communication. The two mediators that makeup Positive e-WOM were customer trust and commitment. Some managerial implications were proposed for the e-marketplace businesses to create the positive electronic word of mouth.

Keywords Electronic marketplace · Habituation · E-Marketplace reputation · Social media communication · Positive electronic word of mouth

1 Introduction

Social distance and isolation are gradually becoming familiar terms from February 2020, when the Covid-19 epidemic broke out worldwide. Covid-19 has slowed, even stopped the economic growth of many countries worldwide [1, 2]. However, electronic commerce (e-commerce) has helped the national economy to cope with this pandemic partly. E-commerce has fully played its advantages to overcome the disadvantages during the Covid-19 pandemic, such as purchasing and selling goods through the Internet. E-commerce not only helps limit the contact but also helps consumers have more choices, and more convenience in shopping activity [2, 3].

B. T. Khoa (✉)
Industrial University of Ho Chi Minh City, Ho Chi Minh City, Vietnam
e-mail: buithanhkhoa@iuh.edu.vn

© The Author(s), under exclusive license to Springer Nature Singapore Pte Ltd. 2022
P. Nanda et al. (eds.), *Data Engineering for Smart Systems*, Lecture Notes in Networks and Systems 238, https://doi.org/10.1007/978-981-16-2641-8_14

The electronic marketplace model (e-marketplace) appeared to provide both sellers and buyers with easy and secure location based on inheriting the strengths and limiting the weaknesses of the e-commerce sites. Moreover, a pioneer in improving the e-marketplace must mention Amazon [4]. Demand for online purchases increased, leading to a boom in the number of sellers participating in this channel. However, as a matter of course, a crowded "market" is difficult to manage. The mix of counterfeit, fake is also more, and even these goods are more blatant and open than traditional trade. According to US user behaviour research results by Northwestern University's Spiegel Research Center, online reviews can increase purchase rates by 380%, lucrative prey, leading to a problem [5]. Many e-marketplaces were established, and the transition from the B2C transaction method to an e-marketplace has resulted in fierce competition. Because of such fierce competition; companies, that operate under the e-marketplace model, need to develop specific strategies for inspiring customer loyalty, such as acquisitions or word of mouth [6]. Besides, electronic word-of-mouth (e-WOM) studies have been popularized in e-commerce [7] or social commerce [8] in recent years. Research on consumer e-WOM in the context of e-marketplace is still limited. Therefore, e-WOM studies in the context of e-marketplace will positively contribute to theories related to online consumer behaviour. It is possible to propose theoretical models, relationships between research concepts for further studies.

2 Literature Review

Simply put, Marketplace is a trading floor, where sellers and buyers gather to find each other easily [9]. In essence, the e-marketplace is like the traditional marketplace, where it allows sellers to rent a suitable location to promote activities such as displaying, introducing, buying, and selling products [10]. In other words, the e-marketplace is the "virtual market," where sellers and buyers access the same website to buy and sell goods.

The strong growth of the internet has provided a better means for consumers to collect information and advice related to consumer behaviour by e-WOM [11, 12]. e-WOM allows users to receive, share, and select information effectively, to overcome obstacles in space and time [13]. It can be said that e-WOM gives consumers the power to influence other buyers through opinions about the products or services used [14, 15]. With consumers increasingly exposed to an increasing number of marketing messages and shrinking marketing budgets, marketers cannot overlook the advantage of social media's influence in general and e-WOM in particular [11].

2.1 The Antecedents of the Relationship

Habit is defined as what an individual normally does when they prefer that behaviour, which leads to the continuation of the same behaviour [16]. The habit has a big impact on how often a customer visits a website. Research also shows that habits will increase the intention to buy the next time through a customer's specific website if the customer has previously purchased on this website [17].

The reputation is related merely to the website's properties, and it is all attributes of the organization. More specifically, Casaló et al. [18] emphasized that a website is just one means of communication between customers and the organization. Besides, reputation also demonstrates the industry's popularity, known by experts and supported by customers.

By taking advantage of Web 2.0 technologies, companies use social network sites to promote and relay their brands [19]. Social media is changing traditional marketing communication [20]. Internet users are gradually shaping brand communication that was previously controlled and administered by marketers [21]. Many brands such as Starbucks, Zara, and Orange seek to connect with customers and enhance their brand communication using social media channels [22].

2.2 Customer-Focuses Relational Mediator

Trust is defined as a set of trust in specific transactions primarily with integrity (trustee honesty and keeping promises), benevolence (trustee caring and motivated for the trustor's benefit), competence (the trustee's ability to execute the trustor's requests), and predictability (the trustee's behaviour is best) [23].

Lacoeuilhe [24] defined brand commitment as a psychological concept that expresses an enduring emotional response and cannot give away. Thomson et al. [25] emphasized that this is a strong psycho-emotional relationship. This emotional chain is reflected in the interdependence and "friendship" between the customer and the brand [26].

2.3 Research Hypotheses

Palmatier et al. [27] identified and gave 18 suitable structures and appeared in the synthesis model. Although the relationship is bi-directional, and both sides often share the benefits of a strong relationship, some of the premise and outcomes have different effects from a measurement standpoint. In this theoretical framework, sellers will try to make a relationship marketing effort to strengthen the relationship with their customers, and the relational mediation element to create customer perception

Table 1 Summary of constructs in the theoretical model

The antecedents of the relationship	Customer-Focuses Relational Mediator	Customer-focused outcome
– Customer-focused antecedents: Buying habituation from the e-marketplace website (Buyer's dependence) – Seller-focused antecedents: the e-marketplace reputation (Seller expertise) – Dyadic antecedents: Social media communication between e-marketplace và customer (Communication)	– Trust – Commitment	Positive electronic word of mouth

of their relationship with the seller. The factors in the research model are presented in Table 1.

Accordingly, the study proposed the following hypotheses:

H1: Shopping habituation positively affects the trust of customers for e-marketplace

H2: Shopping habituation positively affects the commitment of customers to the e-marketplace

H3: The reputation of e-marketplace positively affects the trust of customers for e-marketplace

H4: The reputation of e-marketplace positively affects the commitment of customers to the e-marketplace

H5: Social media communication positively affects the trust of customers for e-marketplace

H6: Social media communication positively affects the commitment of customers to the e-marketplace

H7: Customers' trust positively affects the Positive e-WOM for the e-marketplace

H8: Customer commitment positively affects the Positive e-WOM for the e-marketplace.

3   Research Method

Qualitative research is carried out to build the theoretical basis as well as the scale for the research. Through desk research, research will learn the theory of buying behaviour, e-commerce through textbooks, articles, and news or updated data related to the topic. The study conducted in-depth interviews with 15 experts, including 05 university lecturers in e-commerce, business administration; 05 managing companies that manage the e-marketplace sites; and 05 customers regularly buying products

Table 2 Description of respondent information

	n	Percentage		n	Percentage
Occupation			Age group		
Pupil/Student	123	17.72	16–22	136	19.60
Lecturer	149	21.47	23–35	235	33.86
Office worker	342	49.28	36–45	232	33.43
House-wife	80	11.53	>45	91	13.11

on e-marketplace. Through in-depth interviews, the research has drafted a research model and built a scale for quantitative research.

The study surveyed 700 respondents in Ho Chi Minh City, Hanoi City, using a self-administered questionnaire to collect quantitative data. Respondents who regularly shop on major electronic marketplace sites such as Shopee, Amazon, and Lazada, and the respondents have a certain understanding of the e-marketplace which they have purchased, so they have the basic knowledge to answer questions. The method of sampling in quantitative data collection is purposive sampling method. After screening, there were 694 valid responses to the analysis. Collected data was processed using SPSS 23 and SmartPLS 3.7. Respondent information was presented in Table 2.

The study used a 5-point Likert scale, with 1: strongly disagree, and 5: strongly agree. The habitation (HABIT) was measured with four items [23]. The reputation of the e-marketplace (REPU) was measured with four items [28]. The research used four items to measure social media communication (SMC), combined from firm-created and user-generated social media communication [29]. Trust (TRU) included three items [30]. The commitment scale (COMM) represented three aspects: emotion, passion, and connection [25]. Finally, Positive electronic Word Of Mouth (PeWOM) was measured by three items [31].

4 Result

First, the study evaluated the scale of the research constructs. First, the study assessed the scale's reliability by Cronbach's Alpha coefficient (CA), with a CA threshold greater than or equal to 0.7 [32]. Next, the study confirmed the convergent validity through composite reliability (CR), average variance extracted (AVE), and outer loading. If CR is greater than or equal to 0.7, AVE is greater than or equal to 0.5, and Outer loading is greater than or equal to 0.708, the construct scales will converge [33]. Finally, the constructs' discriminant validity was assessed through the heterotrait-monotrait ratio (HTMT), with HTMT is less than 0.85, the two constructs will be discriminant [34]. The results in Table 3 have shown that all measurement constructs have reliability and validity; therefore, these scales used for further research.

Table 3 Results the reliability and validity of the scale of constructs

	CA	CR	AVE	Outer loading	HTMT				
					COMM	HABIT	REPU	SMC	TRU
COMM	0.917	0.947	0.857	[0.901–0.952]					
HABIT	0.921	0.944	0.809	[0.852–0.960]	0.556				
REPU	0.879	0.926	0.806	[0.885–0.919]	0.558	0.490			
SMC	0.932	0.952	0.832	[0.878–0.947]	0.601	0.508	0.528		
TRU	0.943	0.963	0.898	[0.944–0.953]	0.703	0.760	0.580	0.643	
PeWOM	0.864	0.917	0.787	[0.808–0.945]	0.824	0.733	0.782	0.693	0.814

It is suggested that the variance inflation factor (VIF) value should be less than 2 to present no multicollinearity between the independent constructs. According to Table 4, all VIF coefficients were less than 2; there was no multicollinearity amongst the studied constructs.

Besides, Table 5 also presented the coefficients R^2, f^2, and Q^2. R^2 is considered predictable when it is greater than 50%; however, in behavioural science, Hair et al. [33]consider that R^2 is greater than 20% is good. According to Table 5, R^2_{COMM} was 0.428, R^2_{TRU} was 0.617, and R^2_{PeWOM} was 0.762. Besides, effect size f^2 is also proposed to be considered, with effect values of the small, medium, large; corresponding, the value of f^2 is 0.02, 0.15, 0.35 [35]. The results showed that (1) HABIT, REPU, and SMC had small effect size for COMM, with $f^2_{HABIT\ ->\ COMM}$ was 0.082,

Table 4 VIF coefficients

	Commitment	Trust	Positive electronic Word of Mouth
Commitment			1.748
Habituation	1.394	1.394	
Reputation	1.407	1.407	
Social media Communication	1.455	1.455	
Trust			1.748

Table 5 R^2, f^2, and Q^2 values

	R^2	f^2						Q^2
		COMM	HABIT	REPU	SMC	TRU	PeWOM	
COMM	0.428						0.512	0.361
HABIT		0.082				0.464		
REPU		0.068				0.054		
SMC		0.127				0.152		
TRU	0.617						0.598	0.547
PeWOM	0.762							0.592

Table 6 Structural model path coefficients

	Beta	t value	P Values	Hypotheses	Result
HABIT - > TRU	0.498	8.634	0.000	H1	Supported
HABIT - > COMM	0.255	5.164	0.000	H2	Supported
REPU - > TRU	0.170	4.104	0.000	H3	Supported
REPU - > COMM	0.233	4.886	0.000	H4	Supported
SMC - > TRU	0.288	6.617	0.000	H5	Supported
SMC - > COMM	0.325	5.794	0.000	H6	Supported
TRU- > PeWOM	0.499	12.185	0.000	H7	Supported
COMM - > PeWOM	0.461	12.538	0.000	H8	Supported

$f^2_{REPU \to COMM}$ was 0.068, $f^2_{SMC \to COMM}$ was 0.127; (2) HABIT, REPU, and SMC, respectively, had a large, the small, and medium effect size for the TRU; (3) TRU and COMM had a large effect size with PeWOM, with $Q^2_{TRU \to PeWOM}$ was 0.598, $f^2_{COMM \to PeWOM}$ was 0.512. The study also evaluated Predictive Relevance Q^2, with threshold $Q^2 > 0$ [32]. As shown in Table 5, Q^2_{COMM} was 0.361, and Q^2_{TRU} was 0.547, so COMM and TRU indicate the path models predictive relevance for the PeWOM.

Table 6 showed the structural model path coefficients. The research evaluated the level of inter-construct effects in research models and testing research hypotheses. With a 99% confidence level, all the model relationships were covariant, and all the hypotheses were accepted.

5 Discussion

First, shopping habits have a positive impacts on trust (Beta = 0.498, sig. = 0.000) and customer commitment with the e-marketplace (Beta = 0.255, sig. = 0.000). Buying online brings many risks to customers because they cannot see the product directly, so to reduce risk, the purchase will depend on the customer's habits [17]. Consumers often visit the e-marketplace they have purchased, driven by past positive experiences rather than comparative reviews of perceived costs and benefits [23].

Secondly, the reputation of an e-marketplace site also positively affects trust (Beta = 0.17, sig. = 0.000) and customer commitment with the e-marketplace (Beta = 0.233, sig. = 0.000). Besides, the reputable e-marketplace in the e-commerce sector, like eBay or Amazon, is typical for the conclusion that reputation brings high profit and establishes customer loyalty [36, 37].

Thirdly, customers often have greater trust in e-marketplace sites if they have more frequent communication with the site (Beta = 0.288, sig. = 0.000). Besides, the greater the interaction between the e-market site administrator and the buyer will create a greater connection with the consumer (Beta = 0.325, sig. = 0.000). In the context of social media brand communication, Bruhn et al. [38], Khoa et al. [39]

noticed that the quality of peer parent in brand communities (i.e., Facebook brand fan page) positively impacts functional, experiential, and symbolic brand community benefits, consequently levering brand loyalty.

Last but not least, customers would talk well about the e-marketplace site and encourage their friends and relatives to buy on the e-marketplace site when they have trust (Beta = 0.499, sig. = 0.000) and have a connection (Beta = 0.461, sig. = 0.000). Corbitt et al. [40] concluded a positive relationship between trust and loyalty through acquisitions and firms' speaking well. Purnasari and Yuliando [41], Khoa [42] also showed that trust and cohesion positively affect the positive e-WOM for the food and beverage businesses and fashion designers.

6 Conclusion

Theoretically, these research results confirmed the RMMAF result, with three relationship antecedents, including habituation, e-marketplace reputation, and social media communication. The two mediators that makeup e-WOM were customer trust and commitment. Besides, this research result comes from the research object of the e-marketplace, different from B2C websites owned directly by businesses.

In practice, the e-marketplace businesses need to build trust for their customers through clear and strict rules for the sellers on their sites. It is also necessary to take appropriate action if consumers' interests are infringed by counterfeiting or of incorrect quality. Moreover, the owners of the e-marketplace need to have regular surveys to assess consumer satisfaction with sellers. Increasing promotions for loyal customers is necessary to increase commitment with the e-marketplace. The reputation of a marketplace can be judged by the seller and the buyer directly on the website. Finally, enterprises that own e-marketplace need to have a hotline to respond to customer complaints. Besides, building a fan page to provide information and receive feedback from customers is equally important.

Despite these efforts, this research still has some limitations. First, the study was conducted in Vietnam, a country where e-commerce was in the first phase of a boom. Therefore, it is necessary to research more developed countries on science and technology to compare factors affecting the positive e-WOM. Another limitation is the incomplete application of elements in the RMMAF model. Therefore, further studies can exploit this limitation to create a more general relational marketing model in the e-marketplace context.

References

1. Supadiyanto S (2020) (Opportunities) death of newspaper industry in digital age and Covid-19 Pandemic. J Messenger 12(2):192–207. https://doi.org/10.26623/themessenger.v12i2.2244
2. Khoa BT (2020) The role of mobile skillfulness and user innovation toward electronic wallet acceptance in the digital transformation Era. In: 2020 international conference on information technology systems and innovation (ICITSI), Bandung—Padang, Indonesia, pp 30–37: IEEE. https://doi.org/10.1109/icitsi50517.2020.9264967
3. Khoa BT, Huynh LT, Nguyen MH (2020) The relationship between perceived value and peer engagement in sharing economy: a case study of ridesharing services. J Syst Manag Sci 10(4):149–172. https://doi.org/10.33168/JSMS.2020.0210
4. Schneider RR (1995) Government and the economy on the amazon frontier. World Bank Publisher, Washington, D.C
5. Tom C (2020) How online reviews influence sales: evidence of the power of online reviews to shape customer behavior," Northwestern University's spiegel research center, Evanston, IL 602082019, vol 2020 Available: https://spiegel.medill.northwestern.edu/_pdf/Spiegel_Online%20Review_eBook_Jun2017_FINAL.pdf
6. Khoa BT, Nguyen HM (2020) Electronic loyalty in social commerce: scale development and validation. Gadjah Mada Int J Bus 22(3):275–299. https://doi.org/10.22146/gamaijb.50683
7. Yoo CW, Sanders GL, Moon J (2013) Exploring the effect of e-WOM participation on e-Loyalty in e-commerce. Decis Support Syst 55(3):669–678
8. Erkan I, Evans C (2016) The influence of eWOM in social media on consumers' purchase intentions: an extended approach to information adoption. Comput Hum Behav 61:47–55
9. Lee K-W, Tsai M-T, Lanting MCL (2011) From marketplace to marketspace: investigating the consumer switch to online banking. Electron Commer Res Appl 10(1):115–125
10. Pinem Y, Hidayanto AN, Shihab MR, Munajat Q (2018) Does quality disconfirmation in tourism e-marketplace lead to negative tourist emotions and behaviors?. In: 2018 international conference on information technology systems and innovation (ICITSI), Bandung—Padang, Indonesia, pp 248–253: IEEE. https://doi.org/10.1109/icitsi.2018.8695924
11. Hennig-Thurau T, Gwinner KP, Walsh G, Gremler DD (2004) Electronic word-of-mouth via consumer-opinion platforms: what motivates consumers to articulate themselves on the internet? J Interact Mark 18(1):38–52
12. Erkan I, Evans C (2018) Social media or shopping websites? The influence of eWOM on consumers' online purchase intentions. J Mark Commun 24(6):617–632
13. Cheung R (2014) The influence of electronic word-of-mouth on information adoption in online customer communities. Global Economic Review 43(1):42–57
14. Cheung MY, Luo C, Sia CL, Chen H (2014) Credibility of electronic word-of-mouth: informational and normative determinants of on-line consumer recommendations. Int J Electron Commerce 13(4):9–38. https://doi.org/10.2753/jec1086-4415130402
15. Chu S-C, Kim Y (2011) Determinants of consumer engagement in electronic word-of-mouth (eWOM) in social networking sites. Int J Advert 30(1):47–75
16. Gefen D (2003) TAM or just plain habit: a look at experienced online shoppers. J Organ End User Comput 15(3):1–13
17. Khoa BT (2020) Electronic loyalty in the relationship between consumer habits, groupon website reputation, and online trust: A case of the groupon transaction. J Theor Appl Inf Technol 98(24): 3947–3960
18. Casaló L, Flavián C, Guinalíu M (2008) The role of perceived usability, reputation, satisfaction and consumer familiarity on the website loyalty formation process. Comput Hum Behav 24(2):325–345
19. Kaplan AM, Haenlein M (2012) The britney spears universe: social media and viral marketing at its best. Bus Horiz 55(1):27–31
20. Khoa BT, Ly NM, Uyen VTT, Oanh NTT, Long BT (2021) The impact of social media marketing on the travel intention of Z travelers In: 2021 IEEE International IOT, electronics

and mechatronics conference (IEMTRONICS), Toronto, ON, Canada, pp 1–6: IEEE. https://doi.org/10.1109/IEMTRONICS52119.2021.9422610

21. Vova C (2019) Chatbots: A new way to communicate with your customers. Available: https://chatbotslife.com/chatbots-a-new-way-to-communicate-with-your-customers-e1ed3596c04f
22. Khoa BT (2020) The antecedents of relationship marketing and customer loyalty: a case of the designed fashion product. J Asian Finance Econ Bus 7(2):195–204. https://doi.org/10.13106/jafeb.2020.vol7.no2.195
23. Lin HH, Wang YS (2006) An examination of the determinants of customer loyalty in mobile commerce contexts. Inf Manag 43(3):271–282
24. Lacoeuilhe J (2000) L'attachement a la marque: Proposition d'une echelle de mesure. Recherche et Appl Mark 15(4):61–77. https://doi.org/10.1177/076737010001500404
25. Thomson M, MacInnis DJ, Whan Park C (2005) The ties that bind: measuring the strength of consumers' emotional attachments to brands. J Consumer Psychol 15(1):77–91. https://doi.org/10.1207/s15327663jcp1501_10
26. Cristau C (2001) "Définition, mesure et modélisation de l'attachement à une marque avec deux composantes: la dépendance et l'amitié vis-à-vis d'une marque," Thèse de doctorat en Sciences de gestion, Aix-Marseille 3. Marseille, France
27. Palmatier RW, Dant RP, Grewal D, Evans KR (2006) Factors influencing the effectiveness of relationship marketing: a meta-analysis. J Mark 70(4):136–153. https://doi.org/10.1509/jmkg.70.4.136
28. Yee BY, Faziharudean T (2010) Factors affecting customer loyalty of using Internet banking in Malaysia. J Electron Banking Syst 2010(2010):1–21. https://doi.org/10.5171/2010.592297
29. Schivinski B, Dabrowski D (2015) The impact of brand communication on brand equity through Facebook. J Res Interact Mark 9(1):31–53. https://doi.org/10.1108/jrim-02-2014-0007
30. Nguyen HM, Khoa BT (2019) The relationship between the perceived mental benefits, online trust, and personal information disclosure in online shopping. J Asian Finance Econ Bus 6(4):261–270. https://doi.org/10.13106/jafeb.2019.vol6.no4.261
31. Hung KH, Li SY (2007) The influence of eWOM on virtual consumer communities: social capital, consumer learning, and behavioral outcomes. J Advert Res 47(4):485–495. https://doi.org/10.2501/s002184990707050x
32. Nunnally JC, Bernstein I (1994) The assessment of reliability. Psychometric Theory 3(1):248–292
33. Hair JF, Hult GTM, Ringle C, Sarstedt M (2016) A primer on partial least squares structural equation modeling (PLS-SEM). Sage Publications, London
34. Hair JF, Sarstedt M, Ringle CM, Mena JA (2012) An assessment of the use of partial least squares structural equation modeling in marketing research. J Acad Mark Sci 40(3):414–433
35. Cohen J (2013) Statistical power analysis for the behavioral sciences. Elsevier Science, Burlington
36. Resnick P, Zeckhauser R (2002) Trust among strangers in Internet transactions: Empirical analysis of eBay's reputation system. Econ Internet E-Commer 11(2):23–25
37. Nguyen MH, Khoa BT (2019) Customer electronic loyalty towards online business: the role of online trust, perceived mental benefits and hedonic value. J Distrib Sci 17(12):81–93. https://doi.org/10.15722/jds.17.12.201912.81
38. Bruhn M, Schnebelen S, Schäfer D (2014) Antecedents and consequences of the quality of e-customer-to-customer interactions in B2B brand communities. Ind Mark Manage 43(1):164–176
39. Khoa BT, Nguyen TD, Nguyen VT-T (2020) Factors affecting customer relationship and the repurchase intention of designed fashion products. J Distrib Sci 18(2):198–204
40. Corbitt BJ, Thanasankit T, Yi H (2003) Trust and e-commerce: a study of consumer perceptions. Electron Commer Res Appl 2(3):203–215

41. Purnasari H, Yuliando H (2015) How relationship quality on customer commitment influences positive e-WOM. Agric Agric Sci Procedia 3:149–153. https://doi.org/10.1016/j.aaspro.2015.01.029
42. Khoa BT (2020) The impact of the personal data disclosure's trade-off on the trust and attitude loyalty in mobile banking services. J Promotion Manag 27(4):585–608. https://doi.org/10.1080/10496491.2020.1838028

A Review: Web Content Mining Techniques

Priyanka Shah⬤ and Hardik B. Pandit⬤

Abstract World Wide Web provides a powerful platform that stores and retrieves mass information. It becomes a time-consuming and uncomfortable task to search the information due to its unstructured and heterogeneous nature of data on the World Wide Web. Web mining is one of the popular techniques of data mining that is used to discover and extract useful information from web documents and its services. Web usage mining, web structure, and web content are three different categories of web data mining. Each of these categories has various methods, tools, and approaches to excerpt data from volume of information over the web. This review paper states various issues, while encountering information from the web and also states various problems occurred while finding appropriate information from the web. This paper also introduces different techniques and approaches of web content mining for different types of data. This paper also states various applications of web content mining.

Keywords Web mining · Web content mining · Web structure mining · Web usage mining

1 Introduction

The World Wide Web extends an incredible amount of data or information for mining research. Presently, a day's information over the web is tremendous and expanding habitually step by step and millions of web pages are modified every day [1]. Different languages are utilized to create web pages, and they give information and data in different mediums like content, pictures, photographs, sound, music, video, and so forth [2]. There is a requirement for checking what is published on different pages and controlling the quality of the web content. The World Wide Web is significant

P. Shah (✉)
Kadi Sarva Vishwavidyalaya University, Gandhinagar 382021, India

H. B. Pandit
P.G. Department of Computer Science and Technology, Sardar Patel University, Vallabh Vidyanagar, Anand 388120, India

for finding user's interest that is helpful for business planned and current situation of the world. While interaction with the web, the obstacles disincentive by the users is.

1.1 Finding Appropriate Information

Data on the web is sizeable, dynamic, and heterogeneous. Information acquisition by the users from the web, the following problems are confronted:

(a) **To find relevant information**: To extract relevant information becomes strenuous on the web because the search engines are based on exact keyword matching with restricted syntax and vocabulary.

(b) **To summarize the information**: Because of the heterogeneous data present over the web, getting a summary of web content pages is a challenging task [3].

(c) **To personalize the information**: Web personalization changes a website as demonstrated by the user's necessities through the examination of the user's navigational models. This gives an understanding to the user on the best way to adjust his website that is thusly helpful for his business too.

(d) **To categorize the information**: Labeling of web page to a predefined categorization plays for obtaining better results of information management and retrieval tasks. This issue deals with assigning a web page to at least one predefined category labels.

1.2 Creation of New Knowledge from the Web

From a large set of accessible contents on the web, a user has to create essential knowledge and information.

1.3 Personalizing Data

While the user interacts with the web, he/she prefers different visualization and content. This issue is involved with visualization and type of information.

1.4 Analyzing Individual User Preferences

In order to recognize the needs of the users, includes personalization of individual user, website design and management, customizing user information, etc. The web becomes noisy if it contains various kinds of information. Web mining techniques can be used to solve those issues.

2   Literature Review

This section presents literature appraisal of assortment of research papers as shown in Table 1.

Table 1 Literature review

Author/date	Focus	Method	Findings
Hamid Mughal [4]	Researcher focused on different web mining techniques, tools, approaches, and algorithms to devise information from the voluminous web	Summarization table of web mining techniques, tools, and algorithms comparative analysis of web usage mining techniques among its methodology, data gathering, data stores, advantages, and its algorithms	• Decision tree is an incredible distribution and analytical-based methodology which comprise of the root node, branches, and leaf nodes • A simple, impressive algorithm for classification is known as Naive Bayes classifier • SVM is a notable and basic AI and classification algorithm that can be utilized for precise and confining data sets • Neural organization is another web content mining approach which utilizes a back propagation algorithm that comprises of numerous layers, for example, input layer, some shrouded layers, and afterward output layer, each feeds the following layer till last layer (output)

(continued)

Table 1 (continued)

Author/date	Focus	Method	Findings
Bharanipriya and Prasad [5]	Researcher focused on various web mining techniques, various tasks, six steps, and various tools for Web content mining	Comparative analysis of web content mining tools among their tasks like records the data, extract structured data, extract unstructured data, and user friendly	• All the tools mechanize the business chore and aptly reclaim the web data • Screen-scrapper needs preceding knowledge of proxy server and some specifics of HTML and HTTP, whereas other tools do not involve any such knowledge, and it needs an Internet connection to run • Automation-any where 5.5 recording of activities given in different devices • Mozenda tool will not grant to mount without an Internet connection, and this is not the case with other tools
Johnson and Kumar Gupta [6]	Researcher focused on different web content mining techniques and patterns for extracting information from different types of data	Comparative analysis of different web content mining tools based on its tasks like records the data, extract structured data, extract unstructured data, and user friendly	• Topic tracking is functional in anticipates web content associated to a user's interest • To endorse a topic exactness consideration, summarization is taken • Customer support in areas of large- and small-scale industries handles on categorization • Mining facilitates frequently on clustering and information visualization • Future enhancement of web content mining includes generating the prediction model of user requirements and the semantic web

(continued)

Table 1 (continued)

Author/date	Focus	Method	Findings
Shoaib and Maurya [7]	Researcher discussed web mining, its grouping, and algorithms coupled with it	Comparative analysis of various algorithms of web mining algorithms like PageRank, SimRank, TF-IDF, k-nearest neighbor, Page Gather, and CDL4 based on various parameters like its working, input constraints, intricacy and their pros and cons, relevance, their technique, and regression analysis	• SimRank is used in various patterns to provide better results than Pag eRank • TF-IDF is used to trace the contents for similar words as in the query. It fails to recognize plurals and synonyms form of words • K-NN wields training data to discover objects and find the neighboring identical object • In contrast to CDL4, which is not extensively used due to complex rules Page Gather effortlessly access web server logs
Ananthi [8]	The researcher focus in on web content mining methods utilized for mining and use of web content mining. The analyst discussed about emerging techniques used for the extraction of data from web-based shopping destinations	Information extraction from online shopping system helps to improve the product specification and its features. Researcher discussed tree structure, hidden Markov model parses, 2D conditional random field, DOM tree, Ontology method, and structural semantic entropy	• Opinion mining is part of web content mining that extracts reviews of a customer about the product • Web mining is also used for keeping the personalized data of the user
Singh et al. [9]	Researcher focused on significance of web content mining, successful tools, and algorithms	Researcher discussed different algorithms like decision tree, k-nearest neighbor: Naive Bayes, support vector machine, neural network, cluster hierarchy construction algorithm (CHCA)	It is difficult to extract relevant information from voluminous web data and dynamic nature

3   Web Mining

The process of discovering and extracting useful information from large and huge data available on the net is termed as web mining. This complete procedure is classifies into five subparts [5].

3.1   Resource Finding

The main objective of this first task is to collect the data from the web such as e-papers, tweets, user comments, etc.

3.2   Selection of Information and Pre-processing

In this second task, we will select related information and clean irrelevant data from the actual data collected in resource finding by using pre-processing.

3.3   Generalization

This task is used to discover general patterns by applying data mining and machine learning techniques.

3.4   Analysis

In this task, we will interpret and verify the patterns which are discovered from the generalization.

3.5   Visualization

This task will choose sequence of the discovered knowledge (web pages) has to be presented in visual elements like charts, graph, and map using visualization tools.

4 Web Mining Techniques

Web mining consists of three techniques.

(1) Web Content Mining(WCM): To extract various kinds of sequential data from web.
(2) Web Usage Mining(WUM): To extort usage behavior of user on web.
(3) Web Structure Mining(WSM): To extract structure summary of web.

5 Web Content Mining (Wcm)

Technique of excerpt and identify the user-definite data like text, image, video, sound, or records such as lists and tables that are available on the web defines the web content mining [10]. Discovering queries from billions of web archives is a complex and tedious task. Web content mining extricates questioned information by performing different mining procedures and definite down the exploration.

6 Web Content Mining Techniques

There are different techniques to gibe data for the content mining approach as shown in Fig. 1. The following are four procedures depicted which utilize

(1) Unstructured Data
(2) Structured Data
(3) Semi-Structured Data
(4) Multimedia Data.

6.1 Unstructured Data

The lion share of the web content data is in unstructured text or hyper text form. Different techniques, objectives, and the methodology of web content mining for unstructured data are shown in Table 2.

6.2 Structured Data

Content mining for organized data is an imperative procedure. Utilizing this strategy, it is in simple to extricate data contrast with unstructured information. Coming up next are a few strategies, objectives, and procedure utilized for structured data mining (Table 3).

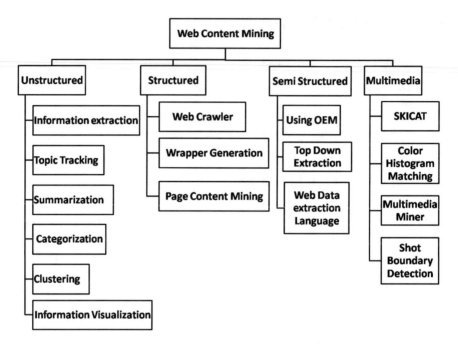

Fig. 1 Web content techniques [4]

6.3 For Semi-structured Data

Semi-structured is a type of organized information, manuscript in semi-**structured** information is syntactic [12]. Procedures, goals, and approaches that extricate semi-organized information are shown in Table 4.

6.4 Multimedia Data

This mining process is utilized for extorting multimedia data sets changed over informational index types into computerized media [14]. Table 5 describes the technology, objective, and methodology that are utilized for media data mining.

7 Applications of Web Content Mining

Web mining plays a significant role in the following areas to improve all services of website [16].

Table 2 Web content mining techniques for unstructured data

Technique	Objective	Methodology
Information extraction [4]	To transform unstructured text into a more structured form	• The methodology is used to extract data in order from the web • In the first step, the information is mined from extracted data, and then the missed out information is found using different types of rules • It follows the keywords and expressions and afterward discovers the association of keywords inside the content
Topic tracking [4]	To arrange the temporally ordered stories of news streams according to the events	• A few constructive samples are monitored by a stream of reports and acquire topic described • It monitors a stream of reports and discovers a similar subject described by a couple of positive examples • It studies the user profile by checking the documents viewed by the user • It predicts other documents related to an individual user's interests • The main drawback of this technique may diverge upon the interest when the exploration is on
Summarization [4]	To span in reduction of the document by maintaining the imperative topics and acquire the abstract	This technique is divided into two methods: 1. The excerpt method preferred a subset of phrases, sentences, and words to form the summary from the original text 2. The hypothetical method builds an inner linguistic representation and then uses the natural language generation technique to generate the summary • Outline may enclose such terminologies that do not exist in the creative document • Approach also provides information to user whether to read the topic or not

(continued)

Table 2 (continued)

Technique	Objective	Methodology
Categorization [4]	Data mining tools to consign the documents into a predefined position identify main themes	• This technique commutates figure in archive and ranks the document according to the topics • Web content mining tools give first or higher rank to those documents which have majority content on a particular topic • Categorization can be used in business and industries to provide customer support [9]
Clustering [4]	To group similar documents and create a cluster of it	This technique makes segment the arrangement of website pages into groups dependent on similarity and afterward relegate the labels to the association
Information visualization [4]	To discover related subjects from an exceptionally huge database available on the web	• Through visualization, similar documents can be initiated • In this, extraction and key term ordering strategy are utilized to fabricate a graphical representation • Large textual information is depicted in the form of maps where browsing facility is permitted • An easy visualization is availed by the user • User can easily interact without difficulty by chart by zooming, scaling, and making sub-maps

- The main application of web mining is to understand the purchase pattern and requirement of a customer from the web.
- Web content mining approaches are implemented in web opinion mining for extracting and analyzing opinions and reviews of customers about products or services [17].
- Web content mining is used to understand the web social communities and its social activities like interaction, sharing and recommending resources, their posts, comments, opinions, and so on.
- Using web content mining techniques, digital library services provide automated citation indexing.
  In the recommendation system, web content mining techniques are used to make recommendations with the help of its attributes.

Table 3 Web content mining techniques for structured data

Technique	Objective	Methodology
Web crawler [11]	The computer program that automatically bisects the web's hyperlink arrangement and loads each associated to a neighborhood capacity	• Web crawler is segregated into two types: Topic Crawler and Universal crawlers • Main undertaking of general crawler is to download all pages independent of their substance • Topic crawler downloads just pages of assertive points
Wrapper generation [11]	A computer program that coerce content of an express data source and decipher into a relational structure	• Web pages are rated by conventional search engines • In accordance with the interrogation, website pages are restored by utilizing estimation of page position • Output types based on the inclination of sources • The coverings will likewise give an assortment of Meta data
Page content mining [11]	It classifies these web pages by com paring page content rank	• The traditional search engine gives rank to different structured web pages • This technique is a structured data extraction technique that works on these web pages

Table 4 Web content mining techniques for semi-structured data

Technique	Objective	Methodology
Using OEM [13]	It encourages to discrete the data configuration on the net even more precisely	• In this method, significant in sequence are isolated from semi-organized information and are entrenched in a valuable information and segregated in the object exchange model (OEM)
Top down extraction [13]	It extricates objects from a bunch of rich web sources and modify into less unpredictable articles, until nuclear objects are extricated	• This technique is performed by traversing the structure of example object in preorder form • It also visits all its components and concatenates them in new resultant object • From resultant object, each new object is recognized and extracted in its component objects
Web data extraction language [13]	It changes over web information to organized information and conveys to end clients. It stores information as tables	• From the web data, a set of positive pages are given and generate extraction patterns • From single page with multiple data records, this technique will generate extraction patterns

Table 5 Web content mining techniques for multimedia data

Technique	Objective	Methodology
SKICAT [15]	The technique is an effective enormous information analysis producing the digital file of sky object	• It utilizes AI strategy to change these items over to human serviceable. It incorporates technique for picture handling, information arrangement which assists with characterizing exceptionally enormous characterization set
Color histogram matching [15]	It is the process of transformation of an image so that its histogram matches a specified histogram	• The methodology comprises of color histogram comparison and smoothing • Equalization attempts to discover correlation between color factor
Multimedia miner [15]	To extract images and videos from web, to store image features in a database, to coordinate the inquiry with picture and video, to mine picture data and to pursue the patterns in picture	• This technique contains four steps: 1. Data segmentation is carried out object-based representation from image, video, and audio; features are extracted, and additional information are added 2. Case Definition: From data segmentation, various patterns are extracted and cases are defined 3. Knowledge Representation: From various cases, knowledge is extracted depending on the pattern 4. Information Modeling: From knowledge representation, a information model is formed
Shot boundary detection [15]	It is a capability in which cardinally the precincts are differentiated between the video frames	• A basic camera shot unit is an un-busted progression of frames • Detecting the shot boundaries, frame distance comparison, escalate by log formula to constrict and develop the difference

- Different techniques are widely used in E-services include e-banking, search engines, blog analysis, e-learning, etc.

8 Conclusion

The paper elaborates web content mining techniques with the focus on web content mining that is performed using various approaches. These applications are listed showing its applicability in diversified domain. The techniques studied were based on four categories of data-structured data, unstructured data, semi-structured data, and multimedia data. The objectives and methodology of the said technique placed on exacting category of data are mentioned in the paper.

References

1. Gaikwad M, Naganath S, Pralhad S (2015) Web mining-types, applications, challenges and tools. Int J Adv Res Comput Eng Technol 4(5):2013–2015
2. NS, Shukla MKRK, Sharma P (2020) Web usage mining-a study of Web data pattern detecting methodologies and its applications in data mining. In: 2nd international conference data, engineering application, pp 1–6. https://doi.org/10.1109/idea49133.2020.9170690
3. Xia Xie WC, Fu Y, Jin H, Zhao Y (2020) A novel text mining approach for scholar information extraction from web content in Chinese, future generation computer systems. In: Future generation computer systems, vol 111, pp 859–872. https://doi.org/10.1016/j.future.2019.08.033
4. Hamid Mughal MJ (2018) Data mining: web data mining techniques, tools and algorithms: an overview. Int J Adv Comput Sci Appl 9(6):208–215. https://doi.org/10.14569/ijacsa.2018.090630
5. Bharanipriya V, Prasad VK (2011) Web content mining tools : a comparative study 4(1):211–215
6. Johnson F, Kumar Gupta S (2012) Web content mining techniques: a survey. Int J Comput Appl 47(11):44–50. https://doi.org/10.5120/7236-0266
7. Shoaib M, Maurya AK (2018) Comparative study of different web mining algorithms to discover knowledge on the web comparative study of different web mining algorithms to discover knowledge on the web
8. Ananthi J (2014) A survey web content mining methods and applications for information extraction from online shopping sites 5(3):4091–4094
9. Singh RK, Abdul APJ, Uit K (2017) A study on web content mining. 6(1):2015–2018. https://doi.org/10.18535/ijecs/v6i1.29
10. Vijiyarani S, Suganya ME (2015) Research issues in web mining. Int J Comput Technol 2(3):55–64. https://doi.org/10.5121/ijcax.2015.2305
11. Satish NR (2017) A study on applications, approaches and issues of web content mining. Int J Trend Res Develop 4(6):41–43
12. Tiwari KMD (2020) Social media data mining techniques: a survey. In: information and communication technology for sustainable development. Advances in intelligent systems and computing, Springer, vol 933, pp 978–981. https://doi.org/10.1007/978-981-13-7166-0_18
13. Mary XL, Silambarasan G (2017) Web content mining : tool, technique & concepts. 7(5):11656–11660

14. AD, Mahmood SSS, Ghani A (2019) Reputation-based approach toward web content credibility analysis. IEEE Access 7. https://doi.org/10.1109/access.2019.2943747
15. Kamde PM (2011) A survey on web multimedia mining. Int J Multimed Appl 3(3)
16. kumar TS (2012) A study: web data mining challenges and application for information extraction. IOSR J Comput Eng 7(3):24–29. https://doi.org/10.9790/0661-0732429
17. Ibukun N, Afolabi T (Covenant University, Ota, Nigeria), Makinde OS (Covenant University, Ota, Nigeria), Oladipupo OO (Covenant University, Ota (2019) Semantic web mining for content-based online shopping recommender systems. Int J Intell Inf Technol 15(4). https://doi.org/10.4018/ijiit.2019100103

GPS-Free Localization in Vehicular Networks Using Directional Antennas

Parveen, Sushil Kumar, and Rishipal Singh

Abstract Localization of vehicles is crucially important for network level operations and application level performance in Internet of Vehicles (IoV). Nowadays, vehicle positioning is being achieved using vehicle mounted global positioning system (GPS). But, it is highly expensive and not apt for wireless sensor networks with indoor applications. We propose GPS-free localization technique using vehicle-to-infrastructure (V2I) using directional antennas of road side units (RSUs). It is highly suitable for cost optimization of IoV-based applications both outdoor and indoor applications (absence of GPS signal). In this paper, we estimate Angle of Arrival (AOA) using the beacon messages released by RSUs. The proposed approach does not require any assumption of availability of prior knowledge of vehicle. The proposed technique for localization of vehicle is computed experimentally using ns-2 simulations. The comparison is made with state-of-art methods for localization using GPS and without GPS. The experimental results show that presented localization strategy exhibits more accuracy than the state-of-the-art methods.

Keywords Localization · Internet of vehicle (IoV) · Global positioning system (GPS) · Roadside unit (RSU) · Directional antenna

1 Introduction

Localization is an open research problem to backing the location-based applications where the GPS solutions are either not cost effective or technically infeasible such as indoor applications [1]. Vehicular localization is a computationally intractable problem and prone to positioning errors [2]. Nowadays, vehicular ad hoc networks (VANETs) have become prominent way to communicate the vehicles through wireless infrastructure components pertain to location awareness [3]. Recent literature

Parveen (✉) · R. Singh
Guru Jambheshwar University of Science and Technology, Hisar 125001, Haryana, India

Parveen · S. Kumar
Jawaharlal Nehru University, New Delhi 110001, India

© The Author(s), under exclusive license to Springer Nature Singapore Pte Ltd. 2022   173
P. Nanda et al. (eds.), *Data Engineering for Smart Systems*, Lecture Notes in Networks and Systems 238, https://doi.org/10.1007/978-981-16-2641-8_16

advocates variety of vehicular localization strategies to obtain accurate positions of moving vehicles based on either GPS or GPS-free localization methods [4–9].

1.1   GPS-Assisted and GPS-Free Localization

GPS-assisted localization involves vehicle-to-vehicle (V2V) communications for positioning in a cooperative way in vehicular networks. It is achieved to share position and range information for accurate positioning of vehicle. A minimum of three position-known neighboring vehicles is desirable for localization of the vehicle. The direct short-range communication (DSPC) is utilized for information exchange which causes message overhead [4, 5, 6]. GPS-free methods utilize RSUs to determine position. GPS-free methods involve techniques uses received signal strength and time of arrival time [7]. Current localization methods using RSU need the knowledge about the position of vehicle in advance or to equip the vehicle with the special set-up like RFIDreader [8, 9].

1.2   Motivation

In this paper, we propose a GPS-free positioning technique for vehicle localization utilizing vehicle-to-infrastructure (V2I) communications through road side units(RSUs)in a VANET using directional antennas. The vehicle is localized with its current position using the beacon messages received from RSUs. Only RSUs are installed with directional antennas and communication takes place over V2I. We present a novel localization method to overcome the drawback of above-mentioned literatures. The novelty of the presented algorithm is as follows: (1) GPS-free localization lower down the cost of the system, and it is also suitable when GPS range is not available. (2) Presented algorithm is free from the dedicated short-range communication (DSRC) among vehicles; as a result, it reduces the message overhead. (3) GPS-free algorithm does not need historical knowledge of vehicles exact position. (4) Presented algorithm also removes the requirement of specific hardware to be mounted on vehicles, which opens the door for variety of applications.

1.3   Problem Definition

The application of directional antennas for GPS-free localization of vehicles in VANET is novel. The objective is to obtain valuable positional and oriental information from RSU to compute accurate position of the moving vehicle. RSUs are installed with directional antenna with no specific hardware requirement imposed on the vehicles which keeps the localization cost optimal. The objective is to supplement

GPS-based localization by proposing a GPS-free localization method. The position of vehicle is computed using trigonometry-based calculation over a pair of straight lines informed by RSUs.

1.4 Paper Organization

The rest of the paper is organized as follows. Section 2 describes the proposed system model and the localization method. Section 3 presents the experimental results obtained through simulations. The work is concluded in Sect. 4 with special remarks on future scope of proposed work.

2 Proposed Model

This section describes the proposed system model and the GPS-free localization algorithm based on RSUs mounted with directional antenna.

2.1 System Model

The proposed system for localization of vehicles assumes that vehicles are moving on road, and RSUs are installed along the roadside mounted with directional antennas (see Fig. 1). The proposed system, moving vehicle is equipped with digital compass, odometer, and an OBU device, which are commonly available within the vehicles for V2I communications. The overall architecture relies on deployment of two RSUs on the same side of the road mounted with directional antenna. The directional antenna [10] is used to approximate the pattern in form of a conical section and apex angle θ ($\theta \in [0, \pi/2]$). The orientation of directional antenna mounted on RSU is kept fixed. The RSU is installed at distance of d from the other RSUs. It does not require any extra installations as all three units are available in standard vehicles. The beacon message from RSUs and the information from compass and odometer are used to compute the location of vehicle moving on the road.

2.2 GPS-Free Localization Algorithm

Now, we explain the proposed vehicle localization technique and the proposed algorithm (see Algorithm 1) in this section.

Proposed GPS-free localization is based on some assumptions as follows. The RSUs (say R_i and R_j) are installed along one (same) side of the road. The orientation

Fig. 1 Proposed system model

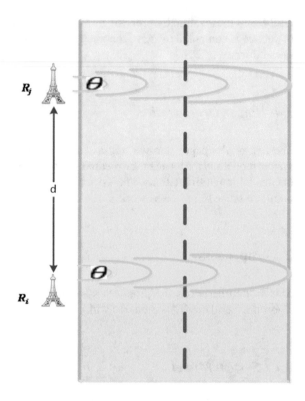

angles of directional antennas of R_i and R_j are φ_i and φ_j, respectively, with respect to east direction (see Fig. 2); $(0 < |\varphi_i|, |\varphi_j| < \pi/2)$. The candidate vehicle (denoted as V) is moving in northward direction.

The core of the algorithm is based on the exchange of beacon messages from RSUs (M_i and $M_{j,}$) to the vehicle. Using M_i and $M_{j,}$ the equations for two straight lines L_i and L_j are defined, and from these lines, the position of vehicle V (X_v, Y_v) is computed. These lines do not intersect on the road so, and it is essential to translate the line L_i accurately. Further, the vehicles do not move following straight line; hence, we propose to define motion vectors within the time-period between reception of two beacon messages (M_i and M_j). Then, the overall displacement p of the vehicle V from all 'N' moving vectors $(p_1, p_2, \ldots, p_N)$ is computed. The information from the digital compass and odometer is utilized to define the moving vectors. Our next challenge is that the radio patterns are not confined in straight line but, in conical form. Hence, the position (say S_0) at the time instant of reception first message M_i from RSU (say R_i) cannot be selected as starting point of moving vector p_1. Then, to select the start position for first motion vector from the set of location list SL $= \{S_l \mid l=1,2,\ldots,S_M\}$. S_l denotes current position with respect to S_0 and M denotes number of beacon messages). Also, a timer is set in the vehicle to wait for the reception of further beacon messages until the timer expires. In case, the timer expires, and

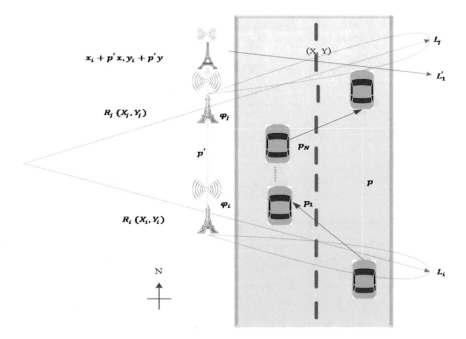

Fig. 2 Proposed GPS-free localization method

no message is further received, then the proposed method is used to compute the position of vehicle V as a point at which it intersects line L_i, i.e., V (X_v, Y_v).

Algorithm 1: GPS-Free Localization

Assumptions- RSUs(say R_i and R_j) are installed on one (same)side of the road; the orientation angles of directional antennas of R_i and R_j are ϕ_i and ϕ_j, respectively, with respect to east direction (see Fig. 2); $(0 < |\varphi_i|, |\varphi_j| < \pi/2)$; vehicle (denoted as V) moving in northward direction.

Step 1.   **a.** Vehicle moves in the coverage area of R_i and receives beacon message $M_i|$ {absolute position of $R_i(X_i, Y_i)$ and orientation angle φ_i}
        b. Vehicle moves in the coverage area of R_j and receives beacon message $M_j|${absolute position of $R_j(X_j, Y_j)$ and orientation angle φ_j}

Step 2.   Using M_i and M_j, two straight lines L_i and L_j are defined and from these lines

The position of vehicle V (X_v, Y_v) is computed (Fig. 3).

a.   Translate L_i along the motion of vehicle during the time interval of reception of M_i and M_j, because L_i and L_j do not intersect on the road.

b.   Utilize the readings from digital compass and odometer to compute all moving vectors obtained by the direction of motion and distance covered for accurate translation of L_i.

Fig. 3 Translating L_i to compute V (X_v, Y_v)

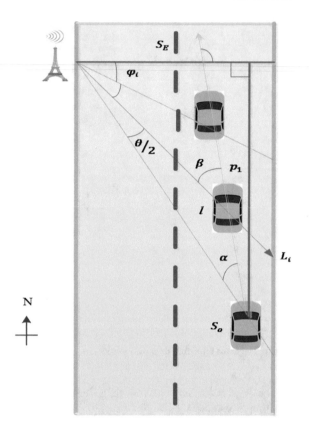

c.   Compute the overall displacement p of the vehicle V from all 'N' moving vectors $(p_1, p_2,...,p_N)$:

$$p = [p_x, p_y] \sum_{J=1}^{N} P_j = \sum_{j=1}^{N} p_{j,x} , \sum_{j=1}^{N} p_{j,y}$$

where p_x is x component and p_y is y component of p.

d.   Select the start position of first moving vector p_1: S=M * $\frac{1S_0}{S_N S_0}$ =M * $\frac{\sin(\alpha+\theta)}{2*\cos\left(\frac{\theta}{2}\right)\sin(\beta)}$ where S_1 denotes current position wrt S_0 and M denotes number of beacon messages. $\angle\alpha = \pi - \alpha 1 - \varphi_i - \theta 2$, $\angle\beta = \pi - \alpha 1 - \varphi_i$

e. $\alpha 1 = \arctan(m_{1,x}/m_{1,y})$

f.   Compute $p' = [p'_x, p'_y] = p'_1 + \sum_{j=2}^{k} p_j$, where p' is the final position for the first motion vector.

g.   Compute two straight line equations L'_1(translated L_1) and L_2 for vehicle V:

$$y - y_i - p'_y = \tan(\varphi_i)\left(x - x_i - p'_j\right) \text{ and } y - y_j = \tan(\varphi_j)\left(x - x_j\right)$$

h.   Compute the position of vehicle V (X_v, Y_v)

$$x = \frac{y_i - y_{j-} \tan(\varphi_i)_{x_i} \tan(\varphi_j)_{x_j} - \tan(\varphi_i) p'_x + p'_y}{\tan \varphi_j - \tan \varphi_i} \text{ and}$$

$$y = y_j - \tan(\varphi_j) x_j + \tan(\varphi_j) x$$

3   Experimental Results and Discussion

The proposed method is simulated using ns-2 simulator for experimental study. SUMO is used to simulate the road traffic [11]. The distance between two adjacent RSUs is kept d meters. The road is supposed to be two-lane, and width of a lane is taken as 3 meters. The traffic is generated using MOVE [12]. The moving speed is set as Spdkmph. For experimental study, RSUs are deployed on one side of the side. The distance d is set with the value of one. Vehicles are adjusted to move at a distance of 15 m from the RSU as in the near vicinity. The beacon messages are transmitted periodically at the time interval of 100 ms. The beam width of directional antennas is set as θ. The orientation is set as-φ_i or $+ \varphi_i$. The parameter-value pair used for the simulations is given in Table 1.

For each value of parameter, 500 runs are performed over changing traffic. The performance evaluation criteria are mean error, i.e., the mean of differences between localized position and actual position of the vehicle. The evaluation is made at the confidence level of 95%.

3.1   Result Analysis for Antenna Orientation

The mean positional error is computed for different antenna orientation and plotted as Fig. 4. We observe that the mean of the error values in localization decreases with increase in orientation angle of directional antenna. The steepest point is 1.496 at 37 degree of orientation angle. Hence, δ is assigned ade fault value of 37 degrees for further experiments.

Table 1 Parameters-value pair for simulations

Name of the parameter	Value(s)
Separation between adjacent RSU (d)	1 km, 2 km, 3 km, 4 km, 5 km
Beam spread (θ)	10, 12, 14, 16, 18°
Error in estimation of motion vector (E)	3%, 6%, 9%, 12%, 15%
Speed of vehicle (Spd)	60, 65, 70, 75, 80 kmph

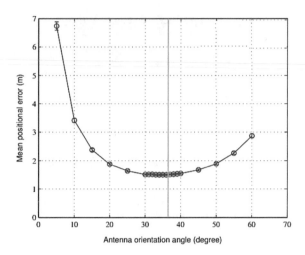

Fig. 4 Finding best antenna orientation

3.2   Result Analysis for Antenna Beam Width

The mean error in localization of vehicle at different beam width is plotted as Fig. 5. It is observed that the antenna beam width of 10 degrees shows the best results with error of 1.75 m.

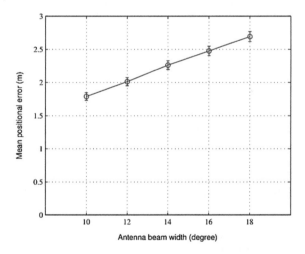

Fig. 5 Finding best antenna beam width

Fig. 6 Comparative analysis of proposed method

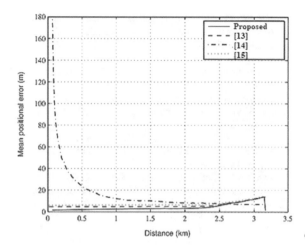

3.3 Comparison with State-of-Art Methods

The localization performance of the proposed scheme is compared with state-of-art localization methods, namely Single RSU [13] and RIALS [14], and GPS-assisted GeoLV [15] shown in Fig. 6.

It is clear from the plot that the proposed method gives best results over the state-of-the-art methods over the range of distances from zero to 2.5 kms.

4 Conclusion

In this paper, GPS-free localization method has been proposed, and experimental analysis has been made to confirm the usefulness of proposed technique. The proposed method is simulated using ns-2 simulator for experimental study. It is very inexpensive and better technique to localize the moving vehicle especially when GPS signals are not available. We infer that the propose GPS-free RSU-based technique using directional antenna can be implemented in real life. We further propose to extend this work to adapt the real-life bidirectional road traffic.

References

1. Boukerche A, Oliveira HA, Nakamura EF, Loureiro AA (2008) Vehicular ad hoc networks: a new challenge for localization-based systems. Comput Commun 31(12):2838–2849
2. Kaplan ED, Hegarty CJ (2006) Understanding GPS principles and applications, 2nd edn. Artech House, Norwood, MA

3. Eze EC, Zhang S, Liu E (2014) Vehicular ad hoc networks (VANETs): current state, challenges, potentials and way forward. In: 20th IEEE international conference on automation and computing (ICAC), pp 176–181, September 2014

4. Günay FB, Öztürk E, Çavdar T et al (2020) Vehicular Ad Hoc Network (VANET) localization techniques: a survey. Arch Computat Methods Eng. https://doi.org/10.1007/s11831-020-094 87-1

5. Lobo F, Grael D, Oliveira H, Villas L, Almehmadi A, El-Khatib K (2019) Cooperative localization improvement using distance information in vehicular ad hoc networks. Sensors 19(23):5231

6. Abd el aalAfifi W, Hefny HA, Darwish NR, Fahmy I (2020) Relative position estimation in vehicle Ad-Hoc Network. In: IoT and cloud computing advancements in vehicular Ad-Hoc Networks, pp 48–83. IGI Global

7. Fascista A, Ciccarese G, Coluccia A, Ricci G (2017) A localization algorithm based on V2I communications and AOA estimation. IEEE Signal Process Lett 24(1):126–130

8. An X, Zhao S, Cui X, Shi Q, Lu M (2020) Distributed multi-antenna positioning for automatic-guided vehicle. Sensors 20(4):1155

9. Ham Y, Yoon H (2018) Motion and visual data-driven distant object localization for field reporting. J Comput Civ Eng 32(4):04018020

10. Ou CH (2011) A localization scheme for wireless sensor networks using mobile anchors with directional antennas. IEEE Sens J 11(7):1607–1616

11. Behrisch M, Bieker L, Erdmann J, Krajzewicz D (2011) SUMO—simulation of urban mobility: an overview. In: Proceedings SIMUL, pp 63–68

12. Karnadi FK, Mo ZH, Lan K-C (2007) Rapid generation of realistic mobility models for VANET. In: Proceedings IEEE WCNC, pp 2506–2511

13. Khattab A, Fahmy YA, Wahab AA (2015) High accuracy GPS-free vehicle localization framework via an INS-assisted single RSU. Int J Distr Sensor Netw 11(5):1–16

14. Zarza H, Yousefi S, Benslimane A (2016) RIALS: RSU/INS-aided localization system for GPS-challenged road segments. Wirel Commun Mobile Comput 16(10):1290–1305

15. Kaiwartya O et al (2018) Geometry-based localization for GPS outage in vehicular cyber physical systems. IEEE Trans Veh Technol 67(5):3800–3812

An Improved Scheme in AODV Routing Protocol for Enhancement of QoS in MANET

Amit Kumar Bairwa and Sandeep Joshi

Abstract Mobile ad hoc networks (MANETs) are communication from the past few decades. Mobile ad hoc networks (MANETs) are alive as a wide research area in related investigators communities, intended to increase the connectivity of data network, principally in an area where formations of traditional networks are unfeasible. However, this network is capable to constitute and heal itself without having a predetermined infrastructure, but with high mobility and battery-operated functionality of network node, the founding of a proficient routing protocol always remains as a challenge for related field researchers. This article suggested a naive revamped variant of the AODV algorithm to improve the QoS by eliminating routing overheads in MANETs. Experimental results confirm the acceptability of the build method over the handy version of the AODV routing protocol.

1 Introduction

Mobile ad hoc networks (MANETs) are challenging area of wireless networks that were developed in the 1970s, to address the inadequacies of open communication networks and to offer express assistance to the defense system [1, 2]. It is mostly created for the locations where the construction of established network infrastructure became too costly or difficult to deploy [3]. Typically, in order to overcome the risk of exiting communication schemes [4], MANETs deliberately set up contact pathways between end nodes without any centralized control system and/or pre-existing network infrastructure [5]. Look at the typical scenario of MANET in Figs. 1 and 2.

Nevertheless, MANETs also supported the customers in a multitude of areas, but a range of problems also emerged due to a set of peculiar features, such as the usage of cell network knowledge nodes and limited power restrictions [6]. Developing a highly effective routing algorithm that can easily route with low overhead data

A. K. Bairwa · S. Joshi (✉)
Manipal University Jaipur, Jaipur-Ajmer Express Highway, Dehmi Kalan, Jaipur 303007, Rajasthan, India

© The Author(s), under exclusive license to Springer Nature Singapore Pte Ltd. 2022
P. Nanda et al. (eds.), *Data Engineering for Smart Systems*, Lecture Notes in Networks and Systems 238, https://doi.org/10.1007/978-981-16-2641-8_17

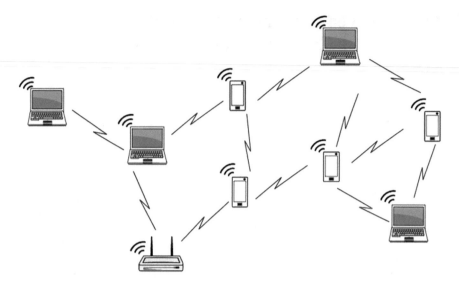

Fig. 1 Scenario of mobile ad hoc network

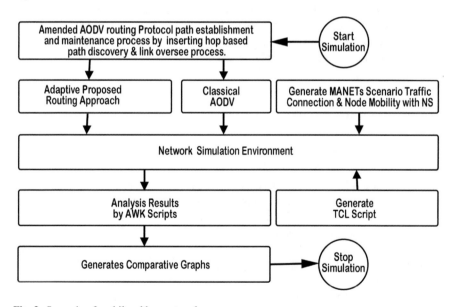

Fig. 2 Scenario of mobile ad hoc network

packets is a hot challenge in MANETs [7]. To overcome such problems, a naive routing procedure based on the AODV routing algorithm was discussed in this paper [8], which effectively improves the network throughput by reducing the amount of overhead routing [9].

1.1 AODV Routing Protocol

With high suitability for unicast /multicast routing and integrate the functionality of reactive and proactive procedures, this routing algorithm has smashed an act of another existing routing method [10, 11]. Among several, the key advantages and shortfalls of this routing practice are

1. More flexible structure of the network.
2. Joint functionality of reactive and proactive routing.
3. Power of diminishing the amount of flooding?
4. Expertly handle unicasting and multicasting of messages.
5. Utilize periodic indication, this leads to unnecessary consumption of bandwidth.
6. Path finding procedure is not extremely competent in this routine practice.
7. For a single broadcasted RREQ packet a network node possibly takes delivery of multiple RREP packets.

To pick up the routing act of AODV huge efforts [12, 13] has rapidly intended by researchers from past years but still, the implementation of an efficient routing method for MANETs environment is a tricky task in front of related field research community [14].

2 Related Works

Task assignment in the MANET is having a wide variety of research in which multi-objective optimization is one of its addon areas. It can be divided into major classes as class 1 and class 2 based on the individual objectives of the system. Class 1 is the situation in which there are several program goals for social health, but there are no specific objectives to achieve, whereas in Class 2 is the situation in which there are several program goals for global objectives, but there is also the specific objective for individual goals, which can contribute to or impede global welfare[15].

Our area of research is inspired by class 2 as they find the existence of malicious nodes executing destructive attacks and colluding to dominate resources to be the main priorities of social welfare. Class 1 involves the implementation of a mission assignment MOO problem to optimize global wellbeing, but specific nodes do not have particular goals.

3   Proposed Work

The proposed approach is an optimized edition of a customary AODV algorithm that puts its key consideration on diminishing several routing overheads and link breakages. To reduce routing overheads, an incorporated mechanism of the proposed approach has made and endeavors by condensing the broadcasting number of route request packets in the network. The approach utilizes a naive added verification method in the path searching procedure of a traditional AODV algorithm that efficiently explores the existence of the desired node in network previous to traveling of one host or can say broadcasting of route request packet. Also, the build mechanism is concerned with reshaping the effective path between communicative nodes as soon as the naive path takes place within the network and the recent path does not crack.

The evaluative results of the proposed approach confirm its suitability in comparison with the classical AODV routing algorithm.

3.1   Performance Evaluation Parameters [16, 17]

Several metrics are accessible to analyze an act of routing algorithms. Among several, only three matrices has taken in an account under this investigation to evaluate an act of offered method in comparison with the classical AODV protocol.

1. Throughput (THR): Number of data packets that have been successfully transmitted to the appropriate destination during a defined time frame. It is typically expressed in bits or in packets per second as in Eq. 1.

$$THR = PR/(ET - ST) * (\frac{8}{100})   \tag{1}$$

   where PR represents the total number of packets received, ET is the end time of simulation, and ST is the start time of the simulation.

2. Normalized routing load (NRL): Control the transfer ratio of packets to the network to deliver a packet as in Eq. 2. High NRL performance demonstrates the usefulness of the virtual system and helps to explain overhead routing.

$$NRL = RPF/RP(3)   \tag{2}$$

   where RL presents routing load, RPF is the number of routing packet, and RP is an amount of total received packet.

3. Packet delivery ratio (PDR): This matrix represents the number of packets successfully reached to the respective nodes compared to the total packets sent by the sender as in Eq. 3. It should have been computed as

$$PDR = \frac{\sum \alpha}{\sum \beta} \tag{3}$$

where α is the total number of received packets at the source node and β is an amount of packet that has forwarded by message inventor node.

4 Simulation and Result Discussion

Huge handy efforts in an area of MANETs have signified that variation in network parameters such as node numbers and alteration in their paces extremely affects an act of exploited routing procedure. Such a constraint was taken into consideration in the assessment of the proposed method.

The algorithm has implemented in a network area of 1000×1000 m with an alteration of nodes number in the network. Simulation has performed with node size of 10, 20, 40, and 50. Furthermore, results have measured with the three matrices parameters throughput, PDR, and normalized routing overhead.

To depict the effectiveness of the building approach, it has simulated with the classical handy routing algorithm of AODV over the same network parameters as the building approach has implemented. Table 1 denoted the details of network parameters under which build and classical AODV routing algorithm has implemented.

The same as Figs. 3, 4, and 5 network has implemented with the nodes number of 10, 20 30, 40, and 50. The following figure explains an evaluation act of build and classical AODV routing algorithms with each implemented network scenario. Every comparative graph has demonstrated the efficiency of build routing procedures in comparison with the classical AODV routing algorithm.

Table 1 Simulation parameters

Parameter	Values
Simulation time	200 s
Number of nodes	05, 10, 20, 40, 50
Simulation area	1000 m × 1000 m
Routing protocol	Modified AODV, Classical AODV
Maximum packet	50
Max speed of node	10m/s
Data packet size	512
Pause time	1.0
Initial nodes power (Jules)	100

Fig. 3 Attained throughput
proposed versus classical
AODV routing procedure

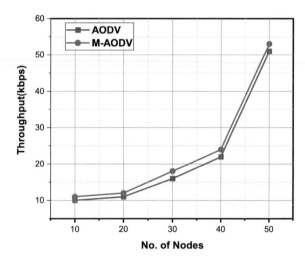

Fig. 4 Attained PDR
proposed versus classical
AODV routing procedure

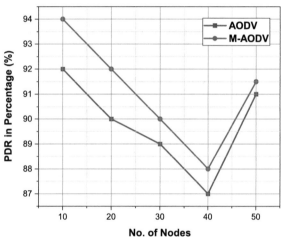

5 Conclusions and Future Work

With the adjustment in route discovery of AODV routing practice, a naive effective routing algorithm has converse in this paper. To improve QoS in MANET, build approach has put a key focus on trimming the amount of RREQ packet generation in the network and incorporate an extended searching mechanism that effectively and speedily discovers respective node existence with less transformation of RREQ packets. Furthermore implemented algorithms energetically set up a more effective route between communicative nodes by eliminating the existence of nodes that are not required to be a part of the active path. Such incorporated alteration in classical AODV routing protocol effectively improves routing QoS in MANET that attains

Fig. 5 Routing overheads proposed versus classical AODV routing procedure

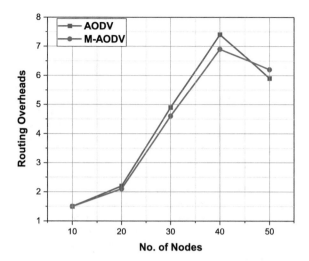

high throughput and PDR ratio with less amount of routing overheads in comparison of classical handy routing algorithm of AODV. Every relative graph denotes that the build algorithm is more suitable in place of classical AODV routing protocol.

References

1. Liu C, Li Y, Cheng W, Shi G (2019) An improved multi-channel AODV routing protocol based on Dijkstra algorithm. In: 2019 14th IEEE conference on industrial electronics and applications (ICIEA), Xi'an, China, pp 547–551. https://doi.org/10.1109/ICIEA.2019.8833838
2. Paranavithana P, Jayakody A (2017) Compromising AODV for better performance: improve energy efficiency in AODV. In: 2017 6th national conference on technology and management (NCTM), Malabe, pp 201–204. https://doi.org/10.1109/NCTM.2017.7872854
3. Aslam N, Xia K, Ali A, Ullah S (2017) Adaptive TCP-ICCW congestion control mechanism for QoS in renewable wireless sensor networks. IEEE Sens Lett 1(6):1–4. Art no. 7501004. https://doi.org/10.1109/LSENS.2017.2758822
4. Swami N, Bairwa AK, Choudhary M (2017) A literature survey of network simulation tools. Int J Creative Res Thoughts (IJCRT), ISSN: 2320-2882, Dec 2017
5. Mukherjee T, Bairwa AK, Joshi S (2017) Preventing MANET from black hole attacks within appropriate area and time. Int J Creat Res Thoughts (IJCRT), ISSN: 2320-2882, Dec 2017
6. Kumari U, Bairwa AK (2017) Paper titled "optimized routing algorithm for improvement of link connection and route optimization" published in Int J Engi Manag Sci (IJEMS) 4(9), ISSN-2348 –3733
7. Pradittasnee L, Camtepe S, Tian Y (2017) Efficient route update and maintenance for reliable routing in large-scale sensor networks. IEEE Trans Indus Inform 13(1):144–156. https://doi.org/10.1109/TII.2016.2569523
8. Kaler V, Bairwa AK (2017) Performance analysis of traffic oriented PCF in wireless local area network. Int J Creat Res Thoughts (IJCRT), ISSN: 2320-2882
9. Alamsyah E, Ketut SI, Eddy P, Hery Pumomo M (2018) Performance comparative of AODV, AOMDV and DSDV routing protocols in MANET using NS2. Int Seminar Appli Technol Inform Commun. Semarang,286–289. https://doi.org/10.1109/ISEMANTIC.2018.8549794

10. Bhagyalakshmi, Dogra AK (2018) Q-AODV: a flood control Ad-Hoc on demand distance vector routing protocol. In: 2018 first international conference on secure cyber computing and communication (ICSCCC), Jalandhar, India, pp 294–299. https://doi.org/10.1109/ICSCCC. 2018.8703220

11. Mai Y, Rodriguez FM, Wang N, "CC-ADOV: An effective multiple paths congestion control AODV," (2018) IEEE 8th Annual Computing and Communication Workshop and Conference (CCWC). Las Vegas, NV 2018:1000–1004. https://doi.org/10.1109/CCWC.2018.8301758

12. Kumar R, Tripathi S, Agrawal R (2018) A secure handshaking AODV routing protocol (SHS-AODV). In: 2018 4th international conference on recent advances in information technology (RAIT), Dhanbad, , pp 1–5. https://doi.org/10.1109/RAIT.2018.8389029

13. Varshney RK, Sagar AK (2018) An Improved AODV protocol to detect malicious node in Ad hoc network. In: 2018 international conference on advances in computing, communication control and networking (ICACCCN), Greater Noida (UP), India, pp 222–27. https://doi.org/ 10.1109/ICACCCN.2018.8748359

14. Srivastava SK, Raut MRD (2019) Enhancing the performance of average throughput, end-to-end delay, drop packets and packet delivery ratio by using improved AODV (AODV+) routing protocol in ad-hoc wireless networks. In: 2019 third world conference on smart trends in systems security and sustainability (WorldS4), London, United Kingdom, pp 266–269. https:// doi.org/10.1109/WorldS4.2019.8904032

15. Wang Y, Chen I, Cho J, Tsai JJP (2017) Trust-based task assignment With Multiobjective optimization in service-oriented Ad Hoc networks. IEEE Trans Netw Serv Manag 14(1):217–232. https://doi.org/10.1109/TNSM.2016.2636454

16. Gautam H, Bairwa AK, Joshi S (2017) Paper titled "Performance Evaluation of NANET in AODV Routing Protocol Under Wormhole Attack Using NS3" published in Int J Eng Manag Sci (IJEMS) 3(11), ISSN-2348 –3733, January 2017

17. Gautam H, Bairwa AK, Joshi S (2016) Paper titled "Routing Protocols under Mobile Ad-hoc Network" published in Int J Eng Manag Sci (IJEMS) 3(11), ISSN-2348 –3733

Knowledge Management Framework for Sustainability and Resilience in Next-Gen e-Governance

Iqbal Hasan and Sam Rizvi

Abstract Efficiency in knowledge management is crucial for the delivery of services in quality, time, and effectiveness. Since the introduction of ICT tools for e-government in the 1990s, technological interventions are increasingly playing important role in the functioning of governments. Worldwide, the governments are relying on digital technologies to manage the delivery of public services. ICT tools are imperative in managing the enormous amount of knowledge generated at different levels of government. The next generation e-Governance is citizen centric and based on the concept of the use of next-gen tools to involve multi-stakeholders in decision making and making government open and accountable. The governments are adopting next-generation technologies such as artificial intelligence, cognitive technologies, speech recognition, robotics, blockchain, IoT, cloud computing, grid and fog computing, automation, and smart technologies to catalyze the delivery of services to citizens and interact within and outside of the organization. In this paper, we present a conceptual knowledge management framework based on Indian Government initiatives for next-generation e-Governance aligned with the UN sustainable development goals. The proposed knowledge management framework will enhance sustainability and resilience in next-generation e-Governance. The paper also presents the challenges in adopting next-generation technologies and provides suitable recommendations to implement e-Governance in the Indian context.

Keywords Information and communication technology · Knowledge management · Next-gen technology · e-Governance · E-government · Digital transformation · Sustainability · Resilience

I. Hasan (✉) · S. Rizvi
Department of Computer Science, Jamia Millia Islamia University, New Delhi 110025, India
e-mail: ihasan@gov.in

S. Rizvi
e-mail: sarizvi@jmi.ac.in

© The Author(s), under exclusive license to Springer Nature Singapore Pte Ltd. 2022   191
P. Nanda et al. (eds.), *Data Engineering for Smart Systems*, Lecture Notes in Networks and Systems 238, https://doi.org/10.1007/978-981-16-2641-8_18

1 Introduction

Digital technologies are playing a pivotal role in the social and economic development of countries and the functioning of governments around the world through the deployment of information and communication technology (ICT) systems and tools. New technologies and novel initiatives day-to-day are changing the governing system by empowering the governments and citizens for effective and efficient delivery of public services. In this regard, ICT tools are increasingly being used for communicating and exchanging information between government and citizens and delivering services to citizens in a hassle-free and timely manner. However, sustainability and resilience are two important challenges in this technological arena. For sustainable development, governments or organizations need to build capabilities to uphold governance network, harness innovative technological solutions to the unresolved problems, and digitally transform the public delivery of services. This may contribute toward achieving the sustainable development goals vision 2030 proposed by the United Nation to transform the world in the areas of critical importance for people, planet, prosperity, humanity, and strengthening peace and global solidarity [4]. Digitization or the use of digital technologies to support governance and public delivery of services is an important driving force toward sustainable development. Indeed, digitization fosters all facets of our daily life.

The ICTs have empowered countries to establish electronic governance (e- governance) facilities that have transformed the way citizens interact with the government and the government organizations deliver various services to citizens and departments. This has helped to reduce the cost of delivering services, improved the function of government, and increased the speed and efficiency of the delivery of services. Governments possess citizen-centric data and resources that are mass respiratory of useful information and being used to provide services to citizens and manage various internal and external affairs. Knowledge inherent in government documents, reports, references, and other data and resources is needed to be managed effectively to mobilize resources, share information, provide effective services, and integrate the function of different organizations. ICT has an important role in the management and sharing of such inherent knowledge and information to the intended people at right time [17]. Moreover, ICT has the potential to enhance the transparency of government services and improve the efficiency of democratic governance through the adoption of next-generation technological solutions. Recent trends of adopting new technologies such as artificial intelligence, cognitive technologies, cloud computing, grid and fog computing, block-chain technologies, smart devices, Internet of Things (IoT), and robotic are strengthening the functioning of governments and transforming the government–citizen interaction and delivery of services.

The government of India launched a flagship program, namely *Digital India*[1] with an idea of institutionalizing its IT infrastructure and transforming India into a digitally empowered society by empowering its citizens with the power of technology. In this

---

[1] https://digitalindia.gov.in/.

regard, National Informatic Center[2] (NIC) a premier agency of Govt. under Ministry of Electronics and Information Technology (MEITY) has been instrumental in setting of a nation-wide ICT Infrastructure, design, and development of solutions/platforms for various sectors for the national, state, and local governments, facilitating wider transparency, data-driven management, planning and control, and enhanced quality of services.

In this paper, we propose a knowledge management framework for next-generation e-Governance. In particular, we consider the adoption of next-gen technologies for knowledge management and delivery of public services through e-Governance in India. We also present the different digital trends adopted by governments in shaping the citizens' experience and delivery of services.

2   Background

Developing countries, such as India once struggled for basic communication systems, have witnessed exceptional growth in information and communication technologies in recent decades. Access to ICT paved the way for citizens' participation in governance, educational and economic development, and effective delivery of public services, which are critical for effective, efficient, and transparent governance.

2.1   e-Governance in India

e-Governance in India is traced back to the 1970s with the establishment of the Department of Electronics, followed by NIC in 1977 [19]. However, e-Governance gained its momentum after the launch of the National Satellite-Based Computer Network (NICNET) in 1987 followed by the District Information System of the National Informatics Centre (DISNIC) to computerize all the districts [19]. In 2006, government of India approved a `National e-Governance Plan`[3] with multiple mission mode projects to promote e-Governance and develop supportive infrastructures [16]. State data centers (SDCs), common service centers (CSCs), state wide area network (SWAN), and middleware gateways are important initiatives toward the development of core infrastructure components. The e-Governance in India progressed through different phases of transitions. The first phase of e-Governance (e-Governance 1.0) includes *computerization* of government offices with the use of word processing and data processing. In the second phase (e-Governance 2.0), the expansion of Internet connectivity resulted in the interaction of government and citizens through Web that reduced the personal interface with the government bodies and allowed the provision of downloadable forms, acts, rules, and instructions. This facil-

---

[2] https://www.nic.in/.

[3] https://www.meity.gov.in/ (NeGP) divisions/national-e-governance-plan.

itated the interaction between government-to-government, government-to-citizen, government-to-business, and government-to-employees through various tools and techniques. The third phase of e-Governance's (e-Governance 3.0) core consideration is the use of ICTs and neighboring scientific and technological domains for solving societal problems and optimization of resource utilization through technological adoption for citizens' welfare and technological collaboration of civic and enterprise bodies at local, national, and international levels.

In e-Government development index (EGDI[4]) ranking for the year 2020, India is ranked 100 out of 193 UN member countries[5] [14]. EGDI ranking is based on the scope and quality of online services, telecom infrastructure, and human capacity building and how the countries promote ICT access and inclusion of its people. Through *Digital India* program, India has taken a significant step forward toward its e-Governance infrastructure with e-Governance services in almost every sector. E-court, e-district, e-office, e-hospital, aadhar-enabled payment system (AEPS), Dig-iLocker, direct benefit transfer (DBT), and MyGov are some of the major initiatives by the Government of India toward e-Governance. Moreover, to create a knowledge society and interconnect all research and knowledge institutions and promote collaborative research, Government of India has initiated National Knowledge Network[6] (NKN), which is a Pan-India network for the knowledge community and mankind at large.

2.2 Emerging Trends and Technologies

The technological changes have necessitated governments and organizations to harness the benefits of emerging trends and technologies for effective governance. To track new technologies, Gartner has tried to put a graphical depiction of new technology patterns, trends, and ideas through the hype cycle[7] for emerging technologies. Hype cycles describe the advancement of innovations from a period of disillusionment to their relevance understanding and role in a particular domain. Figure 1 presents the Gartner hype cycle for emerging technologies in delivering a high degree of competitive edge over the next five to ten years [15]. The hype cycle consists of five stages (shown in the *time* axis) through which a technology passes during its life cycle from innovation trigger toward its path to productivity and mainstream adoption of technology. The line graph depicts how expectations from a technology surge and contracts in its future expected value as it progresses. The hype cycle, 2020 (Fig. 1) has entries for AI-specific technologies for the first time. Moreover,

---

[4] https://publicadministration.un.org/egovkb/en-us/About/Overview/-E-Government-Development-Index.

[5] https://publicadministration.un.org/egovkb/en-us/Data/Country-Information/id/77-India.

[6] http://nkn.gov.in/en/about-us-lt-en.

[7] https://www.gartner.com/en/research/methodologies/gartner-hype-cycle.

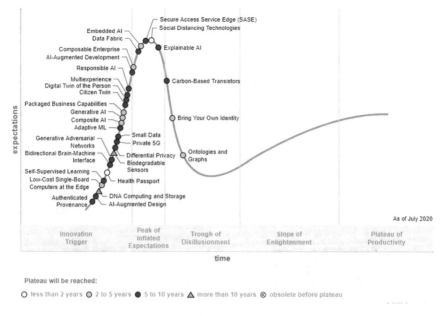

Fig. 1 Hype cycle for emerging technologies, 2020. *Source* [15]

the `health passports` made a transformational impact in less than two years whereas `social distancing` technologies reached made at the top of the *Peak of Inflated Expectations* because of the huge amount of media coverage.

3   Knowledge Management Framework

Knowledge is information in action and focused on results. Knowledge management involves the transformation of knowledge, information, and various intellectual assets into value for an organization. It is an important aspect for successful e-Governance as e-Governance involves domain knowledge, resources management, process reform management, change management, and project management [6]. It helps to transfer knowledge continuously and systematically from the knowledge generator to the decision makers at the top of the organization. Knowledge management (KM) helps an organization to perform its functions efficiently and use innovations to improve organizations' performance. Moreover, KM benefits citizens with better services and greater accountability and transparency of how their money being spent. As KM involves generating, acquiring, capturing, sharing, using, and reviewing knowledge, it requires the collection of technologies and ICT tools to perform such tasks that can help to improve the performance and increase the productivity of organizations [1, 7, 12]. Governments need to manage knowledge as it is their central resource and their

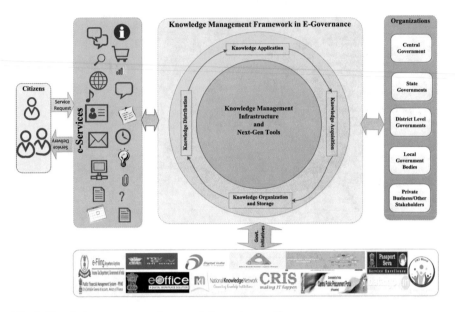

Fig. 2 Knowledge management framework for next-gen e-Governance

functioning rests on acquisition and dissemination of knowledge [13]. Moreover, due to the distributed enterprise nature of governments, the requisite knowledge across central, states, districts, and local governments are of similar kind [13]. Further, the *knowledge drain and demand* because of regular transfer of knowledge employees across government departments have necessitated the management of knowledge to address technological challenges [13].

The four basic components of knowledge management include *people*, *process*, *technology*, and *governance*. People are required to lead and support knowledge sharing. Similarly, defined processes are needed to manage knowledge flows. Technology helps connect the right people to the right content. A well-planned strategy and governance are required for using knowledge management to manage the needs of any organization.

Figure 2 presents the knowledge management framework for next-gen e-Governance. It presents that the knowledge can be acquired from the data collected from citizens in receiving service requests and delivery of services. Various government agencies at different levels of organizational hierarchy from the central to the local bodies provide services directly or indirectly to the citizens with the help of ICT tools. The ICT systems and tools generate lots of data from the receiving of requests to the delivery of services to citizens, organizations, and employees. Extracting knowledge collected from various data sources would help to enhance the effective management and delivery of services. With the use of next-generation tools and techniques, the domain and scope of data collection will tremendously increase that will further require proper data analysis through the use of recent trends and tech-

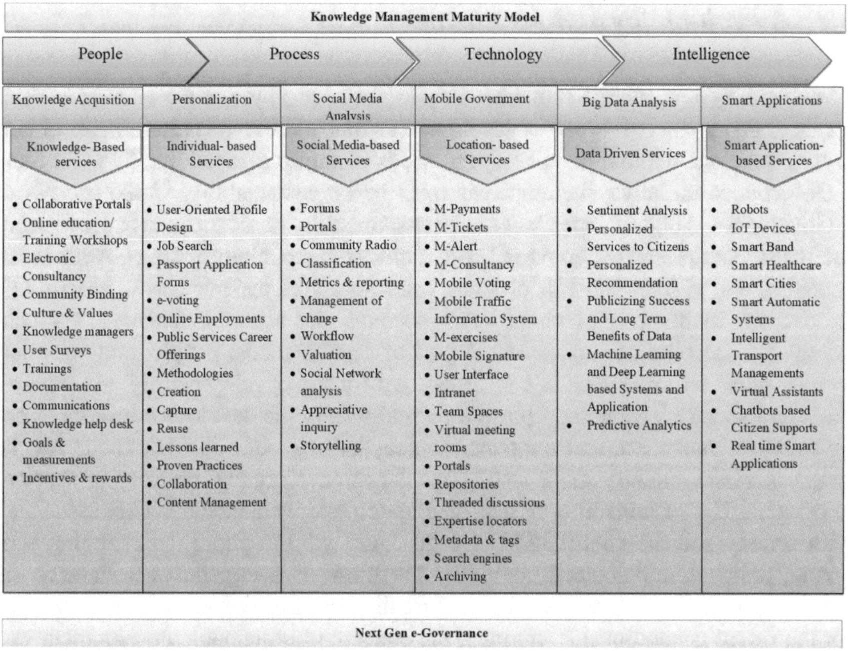

Knowledge Management Maturity Model					
People		Process		Technology	Intelligence
Knowledge Acquisition	Personalization	Social Media Analysis	Mobile Government	Big Data Analysis	Smart Applications
Knowledge-Based services	Individual-based Services	Social Media-based Services	Location-based Services	Data Driven Services	Smart Application-based Services
• Collaborative Portals • Online education/ Training Workshops • Electronic Consultancy • Community Binding • Culture & Values • Knowledge managers • User Surveys • Trainings • Documentation • Communications • Knowledge help desk • Goals & measurements • Incentives & rewards	• User-Oriented Profile Design • Job Search • Passport Application Forms • e-voting • Online Employments • Public Services Career Offerings • Methodologies • Creation • Capture • Reuse • Lessons learned • Proven Practices • Collaboration • Content Management	• Forums • Portals • Community Radio • Classification • Metrics & reporting • Management of change • Workflow • Valuation • Social Network analysis • Appreciative inquiry • Storytelling	• M-Payments • M-Tickets • M-Alert • M-Consultancy • Mobile Voting • Mobile Traffic Information System • M-exercises • Mobile Signature • User Interface • Intranet • Team Spaces • Virtual meeting rooms • Portals • Repositories • Threaded discussions • Expertise locators • Metadata & tags • Search engines • Archiving	• Sentiment Analysis • Personalized Services to Citizens • Personalized Recommendation • Publicizing Success and Long Term Benefits of Data • Machine Learning and Deep Learning based Systems and Application • Predictive Analytics	• Robots • IoT Devices • Smart Band • Smart Healthcare • Smart Cities • Smart Automatic Systems • Intelligent Transport Managements • Virtual Assistants • Chatbots based Citizen Supports • Real time Smart Applications

Next Gen e-Governance

Fig. 3 Knowledge management maturity model in e-Governance

niques such as machine learning, deep learning, and artificial intelligence. Figure 2 also presents four components of the knowledge management process that include *knowledge acquisition*, *knowledge organization and storage*, *knowledge distribution*, and *knowledge application*. With the use of proper knowledge management infrastructure and next-gen tools and techniques, we can acquire knowledge from the citizen-oriented data to obtain from performing government functions to public delivery of services. The acquired knowledge can be organized and stored to have most of the values out of it. The knowledge thus organized needs to be distributed to various stakeholders to mobilize the knowledge and expertise for the better citizens' services in terms of reach and effectiveness. The knowledge application through the use of ICT and next-gen tools and techniques would help the inclusiveness of citizens in governance. Knowledge management would further help to identify and address the knowledge gaps. Figure 3 presents the knowledge management maturity from the people, process, and technology to intelligence depicting knowledge acquisition, personalization, analysis, and smart applications. Figure 3 also depicts the progression of various services in e-Governance as the technologies evolved and being adopted.

3.1   Sustainability and Resilience

Sustainability is a mainstream idea in policy-making by governments at various levels of governance. It involves developments that address the requirements of current generations without hampering the needs of future generations [5, 9]. For any e-Governance initiative, the emphasis must be on sustainability as the success of e-Governance is highly dependent on the sustainability of solutions to succeed eventually [8]. For the next generations, the dynamics of technology among governments, societies, and citizens must incorporate future-oriented perspectives to sustain and survive any fundamental, technological, societal, and economic changes. As technology has become an indispensable part of organizations, people, and societies, technological sustainability is very important for the development of any government and society. Technology plays a pivotal role in the development of any public sector organization, and the technological infrastructure is a backbone for the progress of any country as its role is crucial in enabling and delivery of services to citizens. The sustainability of next-generation e-Governance is highly associated with society and the contribution of e-gov. projects to uphold the changing economic, political, and societal goals [9]. Therefore, it is imperative to balance the socioeconomic development while assuring that the natural resources are preserved for the future generations[3]. Initially, the concept of sustainable development was proposed by World Commission on Environment and Development in the United Nations general assembly to propose long-term environmental strategies and achieve common and mutually supportive goals considering the interrelationships between people, resources, environment, and development [5]. However, the technological advancement and the use of ICT in government and public interaction has necessitated the requirement of sustainability in the technological advancement to facilitate e-Governance to the next generations.

Resilience is the capability of any system to overpower modifications due to some troubling components and recoup and continue its underlying operation or states despite adversities. The system should be able to cope with all types of risks such as cyber-attacks, security breaches, data theft, loss of confidential information, and any kind of cybercrimes. The system should be able to adapt and adjust to the changing circumstances or environments based on requirements. Disruptive events such as natural disasters can damage infrastructure and critical functioning of the governments and can badly impact the decision making and stall major tasks and initiatives [2, 18]. In such circumstances, the ability of the government to withstand, recover, and adapt to the environment comes under resilience, which is a critical factor for sustainability [10].

Emerging technologies such as artificial intelligence with its highly uncertain benefits, risks, rapid pace of development and changes, broad range of applications that involve many stakeholders, regulatory agencies, and industries pose significant challenges in governance [11]. Resilience provides many potential advantages to deal with the unpredictable and highly uncertain risks associated with next-gen technologies. It is based on real risks and harms, and favors forwarding with the demanding technologies, and then identifies any malfunction or hazard that has occurred.

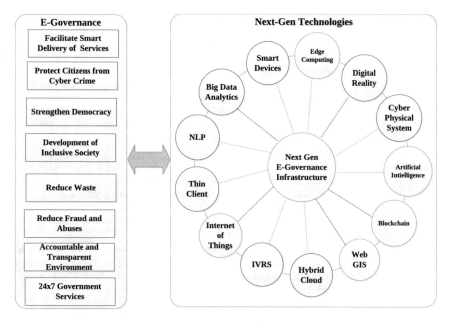

Fig. 4 e-Governance and the next-gen e-Governance tools and techniques

3.2 Next-Gen Technologies in e-Governance

The exploration and use of next-gen technologies such as big data, artificial intelligence, machine learning, deep learning, advanced analytics, cloud-based scalable applications, cybersecurity, and many other technologies would of great help to build sustainable and resilient societies through technological transformation. The ICT tools would be helpful toward reducing risks and adverse impact during disasters to help decision making for prevention, reduction, preparedness, response, and recovery of risks and maintaining sustainable development.

Figure 4 presents the e-Governance benefits and various next-gen tools and techniques. The next-gen e-Governance can facilitate seamless and smart delivery of government services at a reduced cost of time and resources. It will reduce waste, fraud, and abuse of power, and increase the participation of citizens, make accountable and transparent governance, and hence, strengthen democracy. Important technologies that would be the driving force for the next-generation e-Governance and shape the future include *digital reality, blockchain, hybrid cloud, edge computing, artificial intelligence, Internet of Things, chatbots, cyber-physical systems, natural language processing, thin client, web GIS*, etc. We have summarized and presented the probable applications of these next-gen technologies in Table 1 for e-Governance.

Table 1 Next-generation technologies and their probable applications

Technologeis	Applications
Digital reality	• Simulate environment to create real scenario for training military, law enforcement agencies, pilots, and astronouts • Create immersive experiences at museum, national parks, monuments • Visualize and interact with data • Enable maintenance of machines, infrastructures, and resources • Enhance education by providing access of resources anywhere, creating 36-degree videos for training • Improve public health and safety
Block chain	• Verification and sharing of certificates, degrees, and diplomas issued by governmental organizations and academic institutions • Publishing voting results and disseminating publicly available documents and copyrights. • Securely exchanging documents among organizations for records, registries, and inventories
Hybrid cloud	• Development/Testing in public cloud and Production in private cloud as production data is highly sensitive and confidential • New Applications to be developed in Public Cloud, Steady-State Apps in Private Cloud • Development of disaster recovery applications
Edge computing	• To prepare soldiers for modern warfare • For 3D site mapping and surveying, search and rescue operations, and big data collection • To develop `smart cities` by enabling autonomous cars for real-time decisions making based on traffic and/or road conditions
Artificial intelligence	• Maintaining public infrastructure • Rapid disaster response, search, tracking and rescue operations by grabbing real-time data • Managing traffic systems by synchronizing traffic data using AI • Reducing paperwork by automating the admin tasks • Improving query response mechanism through AI-enable applications • Assisting public interaction with governments • Precision farming in agriculture to maximize productivity
Internet of things	• Enhance infrastructures in healthcare, transportation, and power sectors • IoT based devices for national defense, border safety, and enhance security • Smart town and cities development • Analyzing factors for planning, control, and management of cities and towns • IoT based sensors for quick response to emergencies, reporting crimes, alerting police and government bodies
Chatbots	• Rapid access to public data and reporting citizens complaints • Form submission, tax, and bill payments • Delivering public services and providing multilingual support to citizens

(continued)

Table 1 (continued)

Technologeis	Applications
Cyber-physical systems	• For precision agriculture, intelligent water management, and efficient food distribution • Classroom eduction • Smart grids and energy management • Monitoring of surrounding environments, and disaster response • Intelligent transportation and cyber-transportation systems • Healthcare industries and medical-cyber physical systems • Industrial process control through sensors, actuators, and processors • Smart cities, smart home, and smart manufacturing
Natural language processing	• Human–computer interaction, machine translation, speech recognition, entity identification, information retrieval, information extraction, language understating and generation, sentiment and discourse analysis, and optical character recognition
Thin Client	• Cloud-based architecture and virtualization of servers • to provide higher level of security and ensure regular backup and recovery of all sessions • protect user data from any hazard or disaster
Web GIS	• to provides global reach, can be used by a large number of users simultaneously • better cross-platform compatibility, and low cost per user • to allow government departments and citizens to manage properly all their geographical data

3.3 Challenges in Adopting Next-Gen Technologies in e-Governance

e-Governance aims to enable seamless access and flow of information across different organizations, departments, and citizens. The next-gen technologies though a backbone for e-Governance also pose many challenges in their adoption. There are technological and organizational barriers that may hinder the adoption of emerging and next-generation technologies. Important challenging issues include the legal, economic, social, political willpower, security, privacy, infrastructure, usability and acceptability, and accessibility challenges. Lack of government policies, unclear revenue streams, inappropriate tendering, the unwillingness of funding by governments, no clear road maps for planning and implementation, and responsibility and accountability issues are important issues that may hinder in planning, development, and implementation of e-Governance services. Moreover, interoperability among ministries and departments to capture, process, and share data are also important issues. Besides these issues, the authenticity, security, and privacy issues of citizens' valuable information are important challenges that need to securely handled while adopting or implementing the next-gen technologies. To minimize the cost of operation and implementation, the governments need to focus on the development of applications and technologies that are reusable, portable, and easily maintainable to meet the citizens' requirements. In addition to these challenges, there are social issues in developing countries like India with diverse languages, cultures, religions, and illiteracy of a large section of the rural population. Social issues such as accessibility and usability of technologies due to the language barrier because of many regional languages and inadequate infrastructure in rural areas important challenges to be tackled while adopting next-gen technologies.

4 Conclusion

As the technological trends mature from innovation to productivity, they are being adopted by organizations and the governments in their mainstream functionality and business. Governments are taking initiatives to harness the benefits of next-gen technologies for e-Governance and public delivery of services and adapting themselves to the changing needs of the technology aware citizens. Organized initiatives and management of knowledge as a valuable resource are essential for effective delivery of services. This paper presents the next-gen technologies and their efficient use through knowledge management for sustainability and resilience of e-Governance and effective delivery of public services. The robust next-gen technological infrastructure along with effective knowledge management and sharing could play a key role in next-gen e-Governance.

References

1. Al Rawajbeh M, Haboush A (2011) Enhancing the egovernment functionality using knowledge management. In: World academy of science, engineering and technology. Citeseer
2. Cutter SL, Ahearn JA, Amadei B, Crawford P, Eide EA, Galloway GE, Goodchild MF, Kunreuther HC, Li-Vollmer M, Schoch-Spana M et al (2013) Disaster resilience: a national imperative. Environ Sci Policy Sustain Dev 55(2):25–29
3. Estevez E, Janowski T, Dzhusupova Z (2013) Electronic governance for sustainable development: how egov solutions contribute to sd goals? In: Proceedings of the 14th annual international conference on digital government research, pp 92–101
4. Ga U (2015) Transforming our world: the 2030 agenda for sustainable development. Division for sustainable development Goals: New York, NY, USA
5. Imperatives S (1987) Report of the world commission on environment and development: our common future. Accessed 10
6. India G (2008) Promoting e-governance–the smart way forward. Second Administrative Reform Commission, The Committee Report on Administrative reforms in India, Government of India, pp 1–186
7. Islam MS, Avdic A (2010) Knowledge management practices in e-government: a developing country perspective. In: Proceedings of the 4th international conference on theory and practice of electronic governance, pp 73–78
8. Klischewski R, Lessa L (2013) Sustainability of e-government success: an integrated research agenda. In: E-government success factors and measures: theories, concepts, and methodologies, pp 104–123. IGI Global

9. Larsson H, Grönlund Å (2014) Future-oriented egovernance: the sustainability concept in egov research, and ways forward. Govern Inform Quart 31(1):137–149
10. Lebel L, Anderies JM, Campbell B, Folke C, Hatfield-Dodds S, Hughes TP, Wilson J (2006) Governance and the capacity to manage resilience in regional social-ecological systems. Ecol Soc 11(1)
11. Marchant GE, Stevens YA (2017) Resilience: a new tool in the risk governance toolbox for emerging technologies. UCDL Rev 51:233
12. Misra D (2007) Ten guiding principles for knowledge management in e-government in developing countries. In: First international conference on knowledge management for productivity and competitiveness, pp 11–12. National Productivity Council New Delhi (IN)
13. Misra D, Hariharan R, Khaneja M (2003) E-knowledge management framework for government organizations. Inform Syst Manag 20(2):38–48
14. Nations U (2020) United nations e-government surveys, 2020: digital government in the decade of action for sustainable development. Department of Electronics and Social Welfare
15. Panetta K (2020) 5 trends drive the gartner hype cycle for emerging technologies, 2020. Technical report, https://www.gartner.com/smarterwithgartner/5-trends-drive-the-gartner-hype-cycle-for-emerging-technologies-2020/. August 18 2020
16. Singhai AJ, Faizan D (2016) Transition of Indian ICT processes to smart e-services-way ahead. In: 2016 international conference on advances in computing, communications and informatics (ICACCI), pp 1479–1486. IEEE
17. Subashini R, Rita S, Vivek M (2011) The role of ICTS in knowledge management (km) for organizational effectiveness. In: International conference on computing and communication systems, pp 542–549. Springer
18. Wakeman T, Contestabile J, Knatz G, Anderson WB (2017) Governance and resilience: challenges in disaster risk reduction. TR News (311)
19. Yadav N, Singh V (2013) E-governance: past, present and future in India. arXiv preprint arXiv:1308.3323

Enriching WordNet with Subject Specific Out of Vocabulary Terms Using Existing Ontology

Kanika, Shampa Chakraverty, Pinaki Chakraborty, Aditya Aggarwal, Manan Madan, and Gaurav Gupta

Abstract WordNet is a huge repository being used as a tool in various fields. With an increasing number of applications referring to WordNet as a dictionary, several attempts have been made to update it. The paper proposes to extend the huge repository by adding words and relationships derived from students' class notes through wikidata. These terms can be phrases, technical terms, or any subject specific terminology appearing in students' notes of a specific subject. Although various WordNet enriching techniques are available, it is for the first time that subject specific terminology is being added. The resulting version of WordNet has some very common phrases and technical terms along with the generic terms. Making subject specific and generic terms available in a hierarchy can improve the accuracy of various applications like text summarization and clustering for text belonging to a specific domain.

Keywords English WordNet · Hyponym enrichment · Wikidata

1 Introduction

WordNet is a recognized source of conceptual information for all kinds of linguistic processing in English [1]. It is a lexical resource extensively used as a research tool in natural language processing (NLP). In the database, nouns, verbs, adjectives, and adverbs are grouped into sets of cognitive synonyms called synsets. Each synset represents a unique concept [2]. The words are linked with respect to a particular sense in which they are used. Hence, we can say that the database also labels the semantic relations among words [3]. The huge repository is a useful tool for computational linguistics and NLP because of its structure. Improved general word sense disambiguation [4] and domain specific word sense disambiguation [5] are two most common applications where WordNet is directly used. It is a great resource to facilitate text categorization [1, 6], text summarization [7], and document clustering [8] too.

Kanika (✉) · S. Chakraverty · P. Chakraborty · A. Aggarwal · M. Madan · G. Gupta
Netaji Subhas Institute of Technology, Delhi, India

© The Author(s), under exclusive license to Springer Nature Singapore Pte Ltd. 2022   205
P. Nanda et al. (eds.), *Data Engineering for Smart Systems*, Lecture Notes in Networks and Systems 238, https://doi.org/10.1007/978-981-16-2641-8_19

The use of WordNet is not just confined to text and documents, but the repository is also used to improve the online searching and learning experiences. Search engines utilize the information stored in hierarchy to improve the precision of search query results [9]. It also helps in building dynamic learner profiles by extracting learners' interest [10]. The online dictionary is also used in extracting aspect terms from online reviews efficiently [11]. Computing semantic similarity between concepts [12] and learning a well-founded domain ontology [13] are a few other areas where WordNet plays a crucial role.

For almost two decades now, the artificial intelligence community working on computational linguistics and other fields has been using WordNet. While the resource has proved to be an excellent knowledge base, since it is a dictionary, it needs to evolve just like human language [14]. The enrichment and expansion of the online dictionary are also required to make it the best research tool for the newly emerged domain specific applications using it. In short, to prevent WordNet from failing to catch up with new applications and relations of various concepts and techniques [14], enriching it with "out of vocabulary" terms is an effective solution. Utilizing wiktionary [15] Wikipedia [16], topic signatures [17], and using the meta properties of concepts [13] are some of the ways one can improve the lexical database.

In this paper, we try to utilize the enormous knowledge base available through wikidata to incorporate subject-related terms into WordNet. The idea is to collect terms specific to a subject from notes, find more related terms present at wikidata, and place them at an appropriate place in the hierarchy. The proposed inclusion of subject specific terms will lead to an enriched version of WordNet that may efficiently assist in several e-learning applications. The paper proposes a novel hyponym enrichment in WordNet by enriching the database with subject related technical terms. This is important since the database has not been updated for a long time now [18]. To the best of our knowledge, we are the first to use wikidata for adding subject specific terms and relationships to WordNet.

The remaining paper is structured as follows. In Sect. 2, we discuss the various approaches used for WordNet enhancement and expansion in the past. Section 3 expounds the methodology adopted to add the new terms to the database. Section 4 presents the results and analysis of our attempt to enrich WordNet enrichment. We conclude and give future possibilities in the next section.

2 Related Work

With an enormous increase in the applications relying on the huge repository, a lot of attempts have been made to enrich, extend, and improve the contents of WordNet. Many automatic and semi-automatic approaches that improve WordNet in one way or the other exist. One way proposed almost two decades ago was the use of immense information already present on the World Wide Web (WWW). Since WordNet presents hierarchical information, researchers believe that spotting and fixing errors and ambiguities at different levels of taxonomy can result in varying betterments. [19] attempted to detect existing anomalies in the WordNet. The authors target to improve the quality of the database by using an automatic method to propagate domain information. The study revolved around assigning labels to unlabelled synsets in WordNet 3.0 and new domain labeling for synsets with variations. On comparison through a word sense disambiguation task, clear improvements were observed with the new labeling mechanism.

Verdezoto and Vieu [20] aimed to improve the top level of the taxonomy. For this, they proposed a semi-automatic error detection approach to spot errors in the lower levels of taxonomy. The system proposed detected errors automatically; however, there was a need for intervention from lexicographers and domain experts to decide on the part of solving the error. Slightly different approach was adopted by [13]. Rather than fixing the lower level errors and expecting its impact on top levels of taxonomy, the authors mapped the noun synsets to top level constructs of Unified Foundational Ontology (UFO). The semantically enriched WordNet thus generated had a wide scope of domain specific improvements with the philosophical meta properties of concepts available [13]. Apart from improving the quality, spotting errors and dealing with ambiguities, several proposals for extending the database also exist.

Community enRiched Open WordNet (CROWN) is one such example that improves quality of WordNet with technical words and idioms. The authors try to grow the size of WordNet by adding hypernym and antonym relations. For adding information at appropriate places, they used wiktionary. Wiktionary is a collaboratively constructed online dictionary. With CROWN, one can end up growing WordNet to double its size [15]. Not only by increasing its size by adding a variety of lemmas or establishing new relationships, but researchers have also attempted to improve the dense and complicated structure by enhancing the visualization. There are arguments that by improving the visualization, the understanding of WordNet connections can improve. The authors embed the concept of tag clouds into the synonyms rings present in WordNet. The results were improved human recognition of different senses in which words are used [21].

Considering the fact that WordNet for English is with us for more than two decades now, and the same version with slight variations was used by various applications relying on it, setting up new guidelines to update English WordNet is going to give the much required upgrade to the lexical database [22]. The introduction of new synsets and senses is underway. Developing standard guidelines for this addition and the integration of contributions from other projects are some recent developments in the enrichment of WordNet [18]. Some studies have tried to exploit the structure of wikipedia to reach correct synsets. However, the study maps those synsets to 14 languages other than English [16]. The aim of this paper is to improve the WordNet in English language. Since there are evidences that we can develop the existing WordNet using wiki resources such as wikipedia [23], and using Wikipedia to build and improve WordNet is not yet explored in depth [16], we try to analyze the potential of aligning the subject specific terms from wiki resources with WordNet. However, instead of using wikipedia, we used wikidata an ever evolving resource for structured data, relationships, and taxonomies [24].

3 Methodology

In order to add some more subject specific terms in a meaningful manner to WordNet, we try to exploit the data arranged in hierarchical manner in wikidata. Wikidata is an open source for structured data utilized by other resources like wikipedia [24]. Being one of the fastest growing wiki resources, one can rely on wikidata for updating crucial online dictionaries like WordNet. For demonstration purposes, we add terms related to artificial intelligence a computer science subject to WordNet. In order to collect relevant terms, we use classroom notes. We took notes on the introduction to artificial intelligence (AI) from undergraduate students of Netaji Subhas Institute of Technology studying computer engineering. These notes are processed, and keywords are extracted using the following steps.

3.1 Preprocessing the Data

In order to break down the entire text into meaningful sentences and syllables, we first remove/replace the pronouns first. During this process, it is made sure that the meaning the text conveys remains unchanged. For this, the data is given as input to a co-referencing tool. After co-referencing the entire text, it is now broken into sentences using an NLTK sentence tokenizer. To the words of these sentences, the system checks for any spelling mistakes. In case a particular word is not matching any word in the dictionary, it is converted to the closest match available.

3.2 Extract Informative Words

To the corrected set of words, part of speech tagging is applied, and this assigns each word to a lexical category. This is done to give each word a tag that helps in identification of the word in categories such as noun or adjective. Since it is believed that nouns carry most of the valuable information in any text, from the collection of tagged words, the system selects all nouns and nouns followed by adjectives as keywords. These informative words are then treated as nodes/vertices, and a graph is created using these nodes.

3.3 Generate Graph Using Wikidata

We have a set of vertices represented by important keywords. The next step is to establish linkages between these vertices. For this, for every informative word/ phrase, a closest equivalent called generic name existing on wikidata is used. Let us assume this generic name represents a wiki entity. Along with every generic name extracted, the corresponding wikidata id is also stored. For example, the informative word "maths" is converted into mathematics. The system ends up storing mathematics and its corresponding wikidata id which is Q395. Using SPARQL, a query language for accessing the data available on wikidata, the system stores the ancestor and descendant of every wiki entity. For illustration, the process of associating ancestors and descendants is repeated up to 3 levels. We now have a directed cyclic graph, G with each node representing a wiki entity related to a particular subject somehow. So, G = {T, E} where T is the set of subject specific terms present as wiki entities on wikidata and E is the set of edges between these terms. Each edge in G represents one of the two types of relations (i)"is a part of" and (ii) "is a subclass of." The graph thus generated represents the collection of wiki entities in a hierarchy. Due to overlap in topics, each entity can belong to, as you go up in hierarchy, G can contain more than just one subject/domain.

3.4 Identify Out of Vocabulary (OOV) Terms in WordNet

The system then traverses the graph G to reach every node present in T as concept. On reaching a particular vertex representing a term T_i in the graph, it searches for T_i in the WordNet corpus. The search can result in following outcomes: (i) T_i from graph G is available on WordNet and (ii) It is unavailable. The system searches for the child nodes of T_i in the WordNet database. This search can result in one of the three scenarios. The first case is when no child node of Ti is present in WordNet: This is the case when T_i but none of its children are present in the database. The term T_i alone is present in the existing database. Another scenario is when a node is present but not connected. The two nodes under consideration are present in the database, but are not directly related to each other through a hyponym/hypernym relation. Present and connected is the third situation. In such a case both T_i and the corresponding child node, T_j are present and are connected through any relation. This means both the terms as well as relationship exist. A slightly different case is when the current node, T_i, is not available on WordNet. This is possible only for the root node of G as all other nodes will be treated as child nodes, and the cases are listed above. This is the case when the root node of the Graph G is not present in the WordNet database.

3.5 Assign a Place in WordNet Hierarchy

All the terms present in G fall into one of the five categories listed above. According to the scenario, following actions are taken:

$T_i$present but not the child $T_j$from G: If T_j is not present in the database, the same is appended to the hierarchy in such a way that it is a hyponym T_i.

Present but not connected: In such cases, we propose a relation between the vertex- T_i and its child node- T_j in accordance to the wiki graph G.

Present and connected: If both T_i and T_j are present and are connected through any relation, we move on to the next vertex.

Root of the graph is not found: If the root of the graph is not found, the system searches for any of its child nodes in WordNet. If a child node is present, the root node is added as a hypernym to the child node. Else, add both the root node and its child to WordNet connected with a hyponym-hypernym relation.

4 Results

The proposed method ends up improving the WordNet by adding several subject specific terms. We take artificial intelligence as a subject for demonstration purpose. From the notes of one class of artificial intelligence, a total of 3.76% of the terms were either not present on WordNet or an appropriate relation according to wikidata was missing. In Fig. 1, the key term artificial intelligence was present on the online dictionary. However, "expert systems" is not available on the huge dictionary. According to the hierarchy provided by wikidata, expert system is a subclass of artificial intelligence.

Hence, it is added as a node following the relation determined by wikidata. Although there may be many other connected terms for the parent term, Fig. 1 shows only the newly added term. Similarly, chess theory is added as a new node under the term "chess" (see Fig. 2). It was observed that terms as common as tree data structure and binary tree were not available on WordNet. So, adding such terms can increase the vocabulary.

Fig. 1 Addition of a new node under artificial intelligence

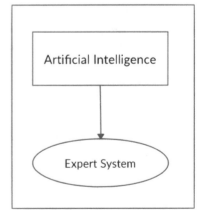

Fig. 2 Addition of new term related to the term chess

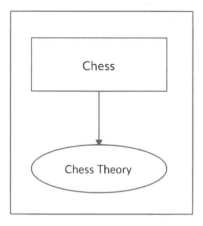

5 Conclusion

The paper proposes an extended version of WordNet that incorporates technical terms and subject specific words/ phrases. These phrases are extremely common in texts belonging to certain subjects. However, it was observed that not many of them were present on the existing WordNet database. WordNet is an online dictionary, and just like human language, there is a need to update it from time to time. Apart from adding new out of vocabulary terms, one way to increase the accuracy of applications revolving around a domain is to populate the database with terms belonging to that domain. We aimed at enriching WordNet with subject specific terms. For experimental purposes, we extracted terms from notes on a topic of artificial intelligence. Out of 186 subject specific extracted from the notes, we ended up adding 3.6% of the terms to WordNet.

The experiment is based on a small data of 186 words. It can be scaled up by taking into account notes of various other topics. In future, we will gather data from notes of an entire course. This will give more wiki entities covering almost the entire subject along with the relations. Even though we have considered a technical subject, the same method can be employed to incorporate terms related to any subject like social science or biology into WordNet.

References

1. Elberrichi Z, Rahmoun A, Bentaalah MA (2008) using wordnet for text categorization. Int Arab J Inform Technol 5(1):16–24
2. Miller GA (1995) WordNet: a lexical database for English. Commun ACM 38(11):39–41
3. Miller GA (1998) WordNet: an electronic lexical database. MIT press
4. Wang Y, Ming W, Fujita H (2020) Word sense disambiguation: a comprehensive knowledge exploitation framework. Knowl Based Syst 190

5. Lopez-Arevalo I, Sosa-Sosa VJ, Rojas-Lopez F, Tello-Leal E (2017) Improving selection of synsets from WordNet for domain-specific word sense disambiguation. Comput Speech Lang 41:128–145
6. Rodriguez M, Hidalgo J, Agudo B (2000) Using WordNet to complement training information in text categorization. In: Proceedings of 2nd international conference on recent advances in natural language processing II, pp 353–364 Bulgaria
7. Pal AR, Saha D (2014) An approach to automatic text summarization using WordNet. In: IEEE International advance computing conference, pp 1169–1173, India
8. Shehata SA (2009) WordNet-based semantic model for enhancing text clustering. In: IEEE international conference on data mining workshops, pp 477–482, USA
9. Moldovan DI, Mihalcea R (2000) Using WordNet and lexical operators to improve internet searches. IEEE Int Comput 4(1):34–43 (2000).
10. Sheeba T, Krishnan R (2020) A semantic approach of building dynamic learner profile model using wordnet. In: Advanced computing and intelligent engineering, pp 263–272. Springer, Singapore
11. Tao J, Zhou L (2020) A weakly supervised wordnet-guided deep learning approach to extracting aspect terms from online reviews. ACM Trans Manag Inform Syst 11(3):1–22
12. Zhang X, Sun S, Zhang K (2020) A new hybrid improved method for measuring concept semantic similarity in WordNet. Int Arab J Inform Technol 17(4), 433–439
13. Leão F, Revoredo K, Baião F (2019) Extending WordNet with UFO foundational ontology. J Web Semant 57:100499
14. Rusert J (2017) Language evolves, so should WordNet-automatically extending wordnet with the senses of out of vocabulary lemmas
15. Jurgens D, Pilehvar MT (2015) Reserating the awesometastic: an automatic extension of the WordNet taxonomy for novel terms. In: Proceedings of the conference of the north american chapter of the association for computational linguistics: human language technologies, pp 1459–1465, ACL, Colorado
16. Haziyev F (2019) Automatic wordnet construction using wikipedia data. Master's thesis, Fen Bilimleri Enstitüsü
17. Agirre E, Ansa O, Hovy E, Martinez D (2001) Enriching WordNet concepts with topic signatures. In: Proceedings of the SIGLEX workshop on "WordNet and other lexical resources": applications, extensions and customizations
18. McCrae JP, Rademaker A, Rudnicka E, Bond F (2020) English WordNet 2020: improving and extending a Wordnet for English using an open-source methodology. In: proceedings of the LREC 2020 workshop on multimodal WordNets (MMW2020), pp 14–19 ELRA, France
19. González A, Rigau G, Castillo M (2012) A graph-based method to improve WordNet domains. In: International conference on intelligent text processing and computational linguistics, pp 17–28, Springer, Berlin
20. Verdezoto N, Vieu L (2011) Towards semi-automatic methods for improving WordNet. In: Proceedings of the ninth international conference on computational semantics, pp 275–284, United Kingdom
21. Caldarola EG, Rinaldi AM (2016) Improving the visualization of WordNet large lexical database through semantic tag clouds. In: IEEE international congress on big data (BigData Congress), pp 34–41. IEEE, Washington
22. Khodak M, Risteski A, Fellbaum C, Arora S (2017) Extending and improving wordnet via unsupervised word embeddings. linguistic issues in language technology—LiLT 10(4):1–17
23. Oliver A (2020) Aligning Wikipedia with WordNet: a review and evaluation of different techniques. In: Proceedings of the 12th language resources and evaluation conference, pp 4851–4858 Marseille
24. Vrandečić D, Krötzsch M (2014) Wikidata: a free collaborative knowledgebase. Commun ACM 57(10):78–85

Automatic Detection of Grape, Potato and Strawberry Leaf Diseases Using CNN and Image Processing

Md. Tariqul Islam and Abdur Nur Tusher

Abstract Day-by-day the cultivation of plants and albumen are increased speedily in order to fulfill the demand of human being and all the animals in this universe. Recently, the production rate of crop is abated due to different crop diseases. Agricultural scientists tried hard to finding the medication for the plant disorder. But the manual identification takes huge amount of time and less efficient. For the quick detection of plant disease different types of new technologies involvement with the cultivation sector bring as blessing. In this research work, deep learning process is used to diagnose the affliction and finding its cure through the images of transited leaf of "grape" and "strawberry". In modern world, researchers can develop more accurate and efficient system for object detection and recognition using deep learning-based process. Here, we used convolutional-neural-network (CNN) algorithm to train the dataset where the accuracy rate is 93.63%. The farmer all over the world especially in Bangladesh can get the facilities form this work to the increment of production rate of grape and strawberry fruits through the reduction of disease and attack of insects.

Keywords Machine learning · Deep learning · CNN · Computer vision · Plant disease

1 Introduction

In Bangladesh, most of the people are engaged with the agricultural industry and about 80% people's life are directly or indirectly dependent with the agronomical work. The agricultural department plays an important role in the economic sector in Bangladesh. The cultivation of most of the crops and fruits remain whole year long which helps our economic condition moving forward. Rice and wheat are the most

Md. Tariqul Islam (✉)
Khulna University of Engineering and Technology (KUET), Khulna, Bangladesh

A. N. Tusher
Daffodil International University (DIU), Dhaka, Bangladesh
e-mail: abdur15-11632@diu.edu.bd

© The Author(s), under exclusive license to Springer Nature Singapore Pte Ltd. 2022   213
P. Nanda et al. (eds.), *Data Engineering for Smart Systems*, Lecture Notes in Networks and Systems 238, https://doi.org/10.1007/978-981-16-2641-8_20

popular crop in Bangladesh. The cultivation of 'Grape' and 'strawberry' are rose day-by-day and the popularity of this fruit is skyrocketed. As a result, the farmer shows great interest to cultivate these shorts of fruits than any previous decay. At the end of winter, the cultivation of 'Grape' is started. The grape plant started to grow from spring to summer season, and it started to ripe at the end of summer and early of fall. The season of strawberry production is varied with the variation of location. In our country, the season of strawberry production started from April and lasted in July.

The major problems of the production of these fruit are the infection of the leaf of these fruits plant. To cope with the demand of our countries growing people the increment of fruits is must. As a result, different types of pesticides are used which causes a huge damage in our environmental ecosystem. By using artificial intelligence and image processing technique, the disease identification from the disorder plant leafs is possible. In this research work, the grape and strawberry leap disease are detected, and the possible cure is provided with the attachment of deep learning technology. Mainly, CNN is used for the development of the system to train and testing the system. The desire of this paper is to determine the grape and strawberry leaps disorder and provide possible cure. The disease for 'Grape' is: Black measles, Black rot, Leaf blight and for 'Strawberry': Leaf scorch. This system is also capable to detect the healthy leaf.

Black measles is also called Spanish measles or Grapevine measles or Esca. When the spot is superficial, it is known as measles. There are seven types grapevine diseases names "pierce's disease, phylloxera, downy mildew, powdery mildew, gray mold, Black rot, vine trunk diseases". The main responsible for most of the grapevine diseases are bacteria and fungi. Insects and environmental condition are also responsible for spreading these types of disease. This disease is infected the plant's all airy part and the warm and humid weather helps to grow fungus. The temperature range 60–90° fahrenheit and rainy climate allow to increases the infection. At first, the leaves and shoots show round brown spots and then die. The fungus attacks the berries part of fruits. The moisture is loss after obtain gray spots from brown reddish than black blue and black dots obtained.

The fungus is responsible for leaf blight disease. The sorghum leaves are mainly produced by this disease. It is mainly occurred under the condition of humid. During this disease, the leaves show reddish purple or tan spots. This disease mainly attack on the seeding and older plants. Strawberry leaves have different types of disease for instance leaf spot, leaf scorch, leaf blight and so on. The fungus 'Diplocarpon earliana' is responsible for leaf scorch disease. When leaf scorch disease is appear the leaves upper surface shows many small, irregular, purplish spots or blotches and the blotches middle point become brownish. Fungus mainly attacked the leaves during whole winter season and in spring the spore forming structure appears in dead leaves. In midsummer, the structure produces spores. These spores infected the plant within 24 h with the presence of free water. The infection rate of older leaves is larger than the middle aged leaves.

The system which we design is capable to detect the above disease perfectly and efficiently provide the steps for farmer to get read those diseases. The education level of farmers in our country is not good enough for detecting the disease using scientific technologies, as a result our country's farmers are not used CNN anywhere. Most of the farmer use hand-made and measurement tools and for harvesting and detecting methods is non-scientific. They made their decision using eye view and blind guesses. Around the world, there have lots of developed countries for instance USA, Denmark, China, Germany are using image processing, AI (artificial Intelligence), CNN (Convolutional-Neural Network) for detecting diseases on harvesting field.

2 Literature Review

In our research work, some source paper reviewing is done and their summary is added in this section as literature review. For plant disease detection, classification and surveying properly lots of innovative techniques are established by researcher. In this section, these researcher works have been describe below.

In 2018, an international conference paper is published by Umar Ayub through which he tried to find out crop diseases in the Pakistan using DATA MINING model [1]. In this conference paper, they denoted the Pakistan farmer losses in their agricultural industry due to different types of diseases occurring because of the insect attract. They mainly introduced different types of procedure such as neural network, supporting vector machine, decision tree and so on and all these model are involve in data mining process. In 2006, an international conference paper is published by Esker in order to detect disease named 'Pantoea Stewartii subps' and this disease is found in corn plant [2]. In this paper, they introduced three models which are working with prediction, "Stevens", "Stevens-boewe" and "Iowa state" are these three model name. In the agricultural sector, a huge amount of pesticides are used in the department of corn and its estimation was given using Economical Threshold Concept of some USA personal named Craig Osteen, Johnson and Dowler [3] and this research work provides massive benefit to the farmer of the corn department.

The main target of our work is to fulfill the demand of detecting and providing appropriate cure of different types of diseases for Grape, Potato and Strawberry plant using their leafs picture in order to making the difficult task for the farmer to detect disease of plant into easy task and increase the accuracy of detection rate. We mainly used image processing techniques for manipulate the data and used CNN model for identification of the diseases and improves its accuracy. This technique is capable to provide real-time result and suitable cure.

3   Proposed Methodology

Image processing is a very powerful technique for analysis a picture pixel wise. The system is constructed through image processing for detecting disease from affected leafs where the disorder is not found with eye. With the proper knowledge of grape and strawberry diseases of a person, the use of technologies is still very useful and easy. The healthy and affected leaf image are applied as input of the system, than the system provides output after processing the input picture. The system is detected disease and healthy picture successfully.

3.1   Flowchart of the Methodology

The flowchart is shown below in Fig. 1.

Image Accretion: Though dataset is the most essential issue to complete any research work, the researcher needs to arrange as many data as possible at the beginning of the work. The accuracy of the research result becomes more expectable when the collected data is sufficient. At the beginning of our work, we successfully collected more than 5000 images (Grape, Strawberry). Most of the source of aggregated data is from grape and strawberry field which is harvested currently and rested data collected by our self. For making the system more convenient any type of image formats for instance .gif, .bmp, .jpg can be applicable as input for the system.

Primary Image Processing: The dataset which is collected before starting the work is subdivided into some small folders. We Select 5400 valid usable data for this

Fig. 1 Flowchart

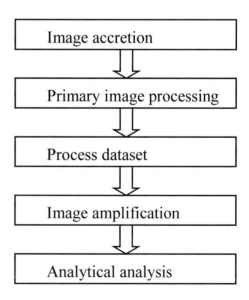

model. The dataset is designed with eighty (80) percent training data and twenty (20) percent testing data. We fixed Alternaria (67 images), Anthracnose (80 images), Downy Mildew (94 images), Healthy (80 images), powdery mildew (80 images) as well as Black Rot (233 images) during the testing performance. Apart from this, we fixed Alternaria (850 images), Anthracnose (670 images), Downy Mildew (1120 images), Healthy (720 images), powdery mildew (920 images) as well as Black Rot (877 images) at the time of training operation. We reshape every single image in a unique pixel value and the selected pixel value is 265×256. After reshaping the image, it is essential to increase the image quality and de-noising the image and for doing these activities different types of image processing method is applied. The elected dataset which is gather together for doing our work perfectly is illustrated in the down of this section and denoted as Fig. 2.

System Architecture: Using single level or multilevel CNN model, the system can be designed. Since the multilevel model gives better performance, we preferred multilevel model and the model step-by-step description is introduced in this section. At the initial layer known as first layer we used ReLu activation function, Input_Shape, Filter_size, Kernel_size "1" (256, 256, 3), "64", (8×8), respectively, and the padding and strides are "SAME" and (1×1), respectively. There have no any considerable difference between first and second layer with the exception of Max_Pool and strides (2×2) and (2×2), respectively, (Fig. 3).

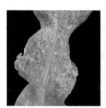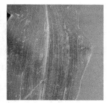

Fig. 2 Assembled dataset

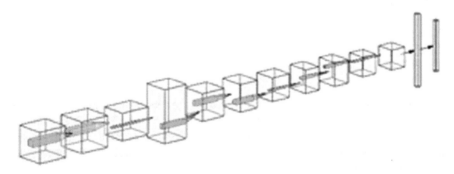

Fig. 3 Proposed convolutional neural network

$$ReLU(X) = MAX(0, X) \tag{1}$$

In third layer, we used ReLu activation function, Input_Shape, Filter_size, Kernel_size "1", (128,128, 3), "32", (5 × 5), respectively, and the padding and strides are "SAME" and (1 × 1), respectively. There have no any considerable difference between third and forth layer. In fifth layer, we used ReLu activation function, Input_Shape, Filter_size, Kernel_size "1", (64, 64, 3), "16", (5 × 5), respectively, and the padding and strides are "SAME" and (1 × 1), respectively. There have no any considerable difference between fifth and sixth layer. In seventh layer we used ReLu activation function, Input_Shape, Filter_size, Kernel_size "1", (32, 32, 3), "8", (3 × 3), respectively, and the padding and strides are "SAME" and (1 × 1), respectively. There have no any considerable difference between seventh and eighth layer. The ReLU activation function has been reduced into fifty percent of 512 units in the flatten layer [4]. The final result of the model is determined using SoftMAX activation function.

$$\sigma(Z) = \frac{e^{z_i}}{\sum\limits_{j=1}^{k} e^{z_j}} \, for \, i = 1, \ldots, k \tag{2}$$

The learning rate is 0.001 which is used in this proposed model as ADAM optimization.

Optimizer and Learning Rate: The result of the deep learning and computer vision shows sufficient change by selecting optimization method. Esoteric Adam paper, "Different subsamples data is evaluated by summing sub function and the subsample is produced by composing lots of objective functions; the optimization algorithm shows it's efficiency in increment manner on the assumption of taking gradient steps" [5]. At present, the Adam optimization model shows well adaption with most of the application of natural language processing. This algorithm is capable to compute adaptive learning rates individually using various parameters from gradients first and second moment estimation. In out proposed model used learning rate = 0.001 using ADAM optimizer amidst.

$$V_t = (1 - \beta_2) \sum_{i=1}^{t} \beta_2^{t-1} \cdot g_i^2 \tag{3}$$

Recently, the classification and prediction work are done by the neural network and cross-entropy provides better performance than classification and MSE. In general, the training does not stalls out because of the unable to get enough minor by cross-entropy error and the weight changes. The categorical cross entropy has been used as loss function in our proposed method "(4)".

$$L_i = \sum_{j} t_{i,j} \log(p_{i,j}) \tag{4}$$

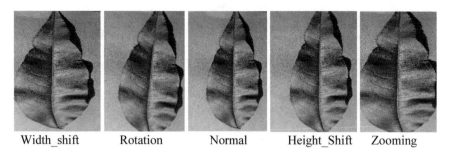

| Width_shift | Rotation | Normal | Height_Shift | Zooming |

Fig. 4 Example of image augmentation

Image Augmentation: Through augmentation process an image can partition into abundant augment. The main purpose of the augmentation of images is:

- To find the simplified and changing model of image representation.
- To changes the shape and angles of images for more data production.
- The image is rotated using maximum range is forty and the width and height shifting range and rescaling value is 1/5 and 1/155, respectively, 1/5 uses as the zooming range also. The horizontal flip consider as true during augmentation and for best accuracy finding the fill nearest mode is shown in Fig. 4
- The sheer range controls the angle in the direction of counterclockwise which allow our images to be sheared.
- At the beginning of processing the image data is multiplied with numeric value which known as 'Rescale'. The collected input images are RGB images and the range of each image coefficient is 0–255, but for our model these range of values is too high for processing. That is why zero and one are our target value range and this value is obtained by scaling the images with 1/255.

3.2 Training the Model

Around 30-batch size is used for training the model among different validation dataset. During processing time, the accuracy of validation and reduction rate is supervised by the method of learning rate reduction. After completing 35 epochs, the supervision between reduced learning rate and validation accuracy is worked manually. After that 10–25 epochs processed and the casual learning rate is set in certain time.

Layer Visualization: The layer visualization of softly changed image symbolizes and present visually. The multilayer image visualization is denoted into Figs. 5 and 6.

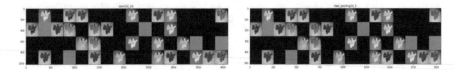

Fig. 5 3 × 3 matrix format layer visualization

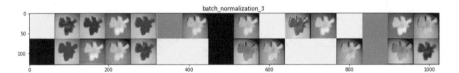

Fig. 6 2 × 2 matrix format layer visualization

4   Result and Discussion

The model which is introduced here has been worked with different data sets and obtained expected result after training, testing as well as validation and the detail description is introduced at the lower section of this part.

4.1   Analytical Analysis

This model provides 78.01% as training accuracy and 35.11% as validation accuracy. When one by one run is completed successfully, the result of the training model faced more accurate improved value. The training accuracy becomes 88.99% and the validation accuracy is 60.98% when the 10th run is completed. In this time the learning rate shows decreasing pattern and reached at 0.0005. The model observed 92.8% training accuracy and 94.89% validation accuracy and 3.124e-07 is the learning rate after thirty successful runs. While the final run is completed, the model provides us the final result as the training accuracy is **93.63%** and validation accuracy is 97.01%.

4.2   Accuracy Graph

The accuracy graph is designed by managing for factor such as training loss and accuracy and validation loss and accuracy. The huge amount of data used for this model is not noise free hence some distortion is taking place and for fitting and describing the loss and accuracy curve fitting is perfect choice. The accuracy graph is divided into two sections over and under part. The over fitting part gives description and reference for the loss of data and denoted as loss. On the other hand, the under

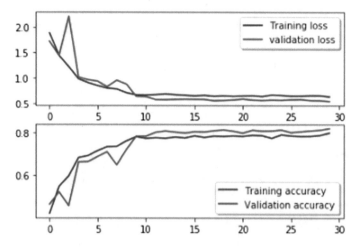

Fig. 7 Accuracy and loss graph for training and validation

fitting portion provides us the accuracy of the model. The training line is noted as blue color and the pink color is denoted the validation line and the graphical presentation is displayed below in Fig. 7.

4.3 *Confusion Matrix*

The confusion matrix is known as the error matrix table or error table through which the performance of the model is displayed. For this design error matrix, it is essential to find out the number of true and false images among all the images for individual diseases and these true and false images are represented into a table which is introduced below in Table 1.

From the confusion matrix, it is clear that the diagonal position value is grater than the any other position where the shape of the diagonal position is (4 × 4). The color of the diagonal position is dipper (Blue) than the any other position value which

Table 1 Accuracy and error data

Disease	True	False	Total
Altemaria blight	67	0	67
Anthracnose	41	45	86
Downy mildew	91	3	94
Powdery Milddew	44	36	80
Healthy	59	11	70
Black Rot	58	22	80

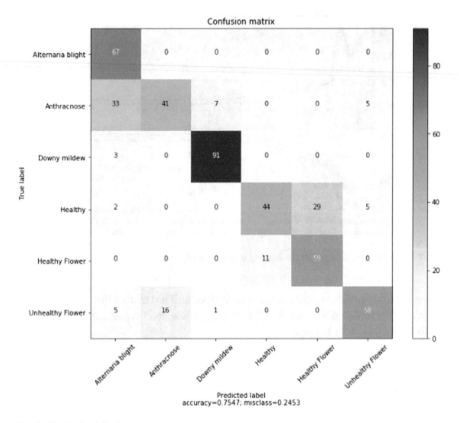

Fig. 8 Confusion Matrix

clearly indicate that this position maximize the data and provide best performance. The confusion matrix and the classification table are given in Fig. 8 and Table 2, respectively.

4.4 Result Analysis of Different Model

Our initial task to read and analysis some previous work which is directly or indirectly related to the work doing in this paper. In order to easily understand our model performance here some other researcher work model is compered in this section. From this comparison, it is clear that our model provide more accurate result which is denoted below in Table 3.

Table 2 Classification report

Disease	Precision	Recall	F1-score	Support
Alternaria blight	0.61	1.00	0.76	67
Anthracnose	0.72	0.48	0.57	86
Downy mildew	0.92	0.97	0.94	94
Healthy leaf	0.80	0.55	0.65	80
Healthy flower	0.67	0.84	0.75	70
Unhealthy flower	0.85	0.72	0.78	80
Accuracy			0.75	477
Macro avg	0.76	0.76	0.74	477
Weighted avg	0.77	0.75	0.75	477

Table 3 Accuracy comparison among different models

Work	Accuracy (%)	Work	Accuracy (%)
Kambale and Bilgi [6]	85.53	s.phadikar et al. [7]	79.50
Akila and Deepan [5]	89.93	Jyoti and tanuja [8]	93.00
Proposed model	**93.63**		

5 Conclusion

The farmer who is directly related to our agricultural industry specially in Grape, Potato and Strawberry department are always faced lots of unwanted damage of their fruits due to numerous diseases. The work which has been done here is capable to detect these diseases automatically and provides accurate cure so that the farmer can increment their production in a massive way. The model which we introduced here are capable to quick disease detection and for doing this work Image Processing techniques and CNN model is used. The affected portion of leafs or fruits are segmented and then analysis using very powerful CNN model which provides very accurate result within very short time. Therefore the ancient analog method which consumed a huge amount of time and less efficient for detecting plant disease is omitted and the farmer detect their plant disease using this automatic process. The model which we introduced is not associated with multimedia. That why, our main target is to associate this model with multimedia.

References

1. Ayub U, Atif Moqurrab S (2018) Predicting crop diseases using data mining approaches: classification. 2018 1st international conference on power, energy and smart grid (Icpesg). https://doi.org/10.1109/Icpesg.2018.8384523
2. Esker PD, Harri J, Dixon PM, Nutter FW Jr (2006) Comparison of models for forecasting of stewart's disease of corn in iowa. Plant Dis 90:1353–1357
3. Osteen C, Johnson AW, Dowler CC, Applying the economic threshold concept to control lesion nematodes on corn, natural resource economics division, economic research service, U.S. Department Of Agriculture. Technical Bulletin No. 1670
4. Janocha K, Czarnecki WM (2017) On loss functions for deep neural networks In: classification arXiv:1702.05659
5. Akila M, Deepan P (2018) Detection and classification of plant leaf diseases by using deep learning algorithm. Int J Eng Res Technol (Ijert) Issn: 2278-0181 Published By, Www.Ijert.Org Iconnect—2k18 Conference Proceedings
6. Kambale G, Bilgi N (2017) A survey paper on crop disease identification and classification using pattern recognition and digital image processing techniques
7. Prem Rishi Kranth G, Hema Lalitha M, Basava L, Mathur A (2018) Plant disease prediction using machine learning algorithms. Int J Comput Appl 182(25) (0975–8887)
8. Jyoti and Tanuja: Cotton plant leaf diseases identification using support vector machine. Int J Recent Scientific Res 8(12):22395–22398

Sentiment Analysis on Global Warming Tweets Using Naïve Bayes and RNN

Deborah T. Joy, Vikas Thada, and Utpal Srivastava

Abstract The concept is based on interpreting positive or negative sentiments expressed by human beings where the subject of climate change or global warming is concerned. The sentiment analysis was done using machine learning and as well been done using deep learning, with python. An existing dataset on climate change is used. After cleaning and processing the data, a dataset having the comments classified into positive and negative sentiment remains. This paper proposes to train the algorithm to interpret positive or negative sentiments expressed by human beings where the subject of climate change or global warming is concerned and thus using the multinomial Naïve Bayes algorithm and consecutively the long short-term memory algorithm to be able to classify the intentions given a new piece of data once the model is trained, tested and validated.

Keywords Sentiment analysis · Global warming · Multinomial naïve bayes (MNB) · Recurrent neural network (RNN) · Long short-term memory (LSTM)

1 Introduction

In the fact paced world of growing AI, human sentiments remain in question when a machine needs to understand it. For instance, interpreting sarcasm or irony can be challenging for an artificially intelligent machine. Sentiment analysis on global warming can help understand people's perspective and could even keep an AI from causing sentimental offence in dealing with the same.

Considering that it is well known that global warming has been a matter of deep concern for meteorologists and ecologists alike. What was once a speculation and a matter made light of, can no longer be ignored given its massive bearing over all sorts of existences. It is as well threatening to the very existence of mankind [1]. Defined much earlier, climate change or global warming connotated, the words

D. T. Joy (✉) · V. Thada
Amity University Gurgaon, Gurugram, Haryana, India

U. Srivastava
Banasthali University, Jaipur, Rajasthan, India

© The Author(s), under exclusive license to Springer Nature Singapore Pte Ltd. 2022
P. Nanda et al. (eds.), *Data Engineering for Smart Systems*, Lecture Notes in Networks and Systems 238, https://doi.org/10.1007/978-981-16-2641-8_21

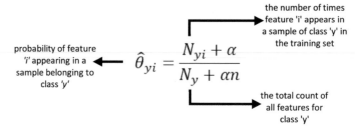

Fig. 1 Equation formulating the MNB [3]

themselves seem to bring down a cloud of dread and anxiety over anyone who interprets their meaning, besides the sense of the words being in any form. People express their thoughts and emotions on the frustration concerning the topic which in turn is actually harmful to hold worry about [2]. So, a look at people's views on the matter via Twitter statement's is taken.

1.1 Naive Bayes Classifier for Multinomial Models

The multinomial Naive Bayes classifier is apt for cataloguing with discrete features. The multinomial allotment normally requires integer feature counts [3] (Fig. 1).

1.2 Recurrent Neural Network

RNN are refined neural nets with learning from not only the past but also from prior inputs, and it is powerful when it comes to modelling sequential data [4] (Fig. 2).

The above-mentioned algorithms are the ones being used in this paper to classify global warming sentiments expressed by people via tweets.

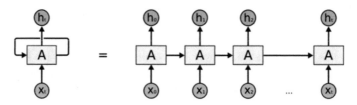

Fig. 2 Structure of unrolled RNN [5]

2   Literature Review

Sentiment analysis in artificial intelligence developing cognitive and emotional ability is a budding concept, but in no way novel. Though there is still much to discover, as technology has been on the up-rise people have started taking the view of reducing the trouble of having to understand and be able to classify human sentiments to predict dangers or even threats both personal and plural. Sentiment analysis, using machines, over Twitter data is no new concept either. There has been surplus study on the same with multidimensional aspects in view, given the fact that Twitter data has been the easiest access for sentiment-based data.

Kharde and Sonawane have put forward their theory of Twitter data as a key source of data where surveys are concerned, and they worked on the provision of a survey and comparative analysis of techniques involving the mining of Twitter data [5]. One of the methods that Kharde and Sonawane use is Naïve Bayes which has been extensively discussed in this paper. Considering just the source of data, Boyi Xie, in the paper Sentiment Analysis of Twitter Data, states that microblogging has particularly evolved to become a source of varied sorts of intel [6]. The authors have as well used Twitter data to build models for the categorization of tweets into their classes namely positive, neutral or negative [6]. One of the earliest works on Twitter data for sentiment analysis was done by Pak and Paroubek (2010). The focus laid on using Twitter which according to the authors happened to be the most popular microblogging arena, and they aimed at performing the linguistic analysis on the collected corpus [7]. One of the major reasons, why Twitter happens to be the most popular platform is, the availability of free expression, anyone can partake in a discussion of choice with no impede. Furthermore, using the Twitter corpus actually performs better with better efficiency, whilst sentiment classification [7]. The authors of Sentimental Analysis on Twitter Data using Naive Bayes, Wagh, Shinde and Wankhade, have compared different algorithms to detect the polarity in sentiment analysis. The unigram baseline as well other advantaged methods have been discussed on the performance of sentiment analysis on Twitter data particularly using Naïve Bayes [8]. Mucha in her paper highlights the analysis of people perception shifting over the years in the past decade where the topic of global warming is concerned, again using Twitter as the source of data [9]. The sentiment analysis has been done using machine learning algorithms Naïve Bayes and SVM classifiers [9]. The accuracy depends on perception.

3   Methods Adopted

The math working within sentiment analysis in this case is logistic regression, which is basically a supervised classifying algorithm. Although it is not a classifier, it acts as such based on probabilistic analysis. There is a binary form, a multi-form and ordinal form of logistic regression. The activation thus working is Sigmoid activation function, the binary form used here in RNN. Retrieving Sigmoid is of the form:

Fig. 3 Sigmoid activation function graph

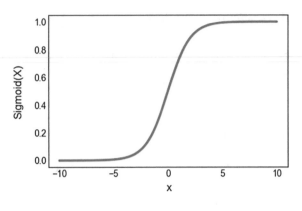

$$S(x) = \frac{1}{1 + e^{-x}}$$

Sigmoid is a binary activation function which represents very positive values as 1 and very negative value as 0, all values otherwise are found in between 0 and 1. The Sigmoid graph is plotted as given below (Fig. 3).

In this paper, a training algorithm was built to exercise a machine in understanding how sentiments on Global Warming can be categorized based on global warming tweets [10]. Subsequently, it was fed with labelled data for test in case of MNB algorithm and validation in case of LSTM algorithm. Once the system was readied to be able to classify the sentiments based on global warming, fresh data was supplied for it to categorize successfully, to pass.

The dataset used in this paper initially had three classes, 'Yes', 'No', and 'N/A'. For this specific work, only the absolute Yes and No were considered, N/A or Neutral was not taken into consideration, whilst training the algorithms. In other words, neutral tweets have been done away with altogether. The foremost step after importing the dataset is cleaning which includes, as mentioned before, the removal of neutral tweets followed by bringing all the tweets on the same target level and finally converting the target values into binary.

After the implementation of the two models, they are compared on the basis of their performance, measured in terms of accuracy and dependence; as well on their complexity, measured in terms of time and space (Figs. 4 and 5).

4 Experimental Results

Sentiment analysis can be done using multiple techniques. Consider multinomial Bayes aside, where machine learning is concerned there is support vector machine as well. As well instead of Keras and TensorFlow for LSTM in RNN, the torch module in python can be used to implement LSTM [11]. So, keeping the thought of effectiveness and accuracy in mind, multinomial Bayes is used, as it is efficient

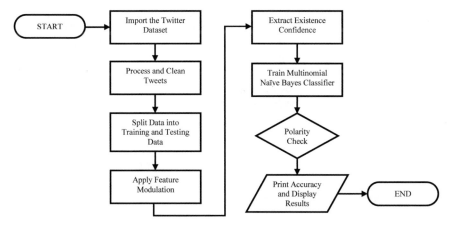

Fig. 4 Implementation of the MNB model

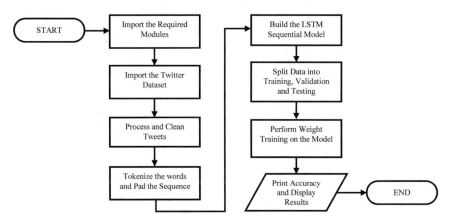

Fig. 5 Implementation of the LSTM model

when handling Twitter data, and the LSTM RNN is considered, as it holds on to the past learnings, Keras for simplicity.

As the final deciding factor on the accuracy a confusion matrix will be used for both the approaches. A confusion matrix has four segments that potentially represent the following: TP (True Positive): Actual Truth, Predicted True; FN (False Negative): Actual Truth, Predicted False; TN (True Negative): Actual False, Predicted False; FP (False Positive): Actual False, Predicted True. What is aimed to increase, so as to affect the accuracy, are the divisions of True Positives and True Negatives [12]. Accuracy is calculated as (Table 1):

$$\frac{TP + TN}{TP + FP + TN + FN}$$

Model type	Dataset dimensions	Dataset division
Naïve Bayes multinomial model	(4225, 2)	Training: 90% (3802) Testing: 10% (423)
RNN LSTM sequential model	(4225, 2)	Training: 67% (2830) Validation: 16.50% (697) Testing: 16.53% (698)

Table 1 Dataset dimensions and divisions

4.1 Multinomial Naïve Bayes Approach

Once the data is cleaned and split into training and testing, the feature modulating libraries are called to work on the data, whereby they convert the tweets into machine readable format. First, the count vectorizer does its job in producing a sparse matrix of token counts and their existence [13]. Once this is done, the TF-IDF transformer begins its walk and assigns the amount of impact that a token has on a certain tweet that causes a positive or negative classification [14] (Fig. 6).

In the above figure, true positives and true negatives are given in the locations (0, 0) as 316 and (1, 1) having 32. Basically, these two denote correct predictions of the apt labels. The other two scores, i.e. 6 and 69 denote false positives and false negatives, respectively, potentially incorrect predictions of the labels. The accuracy with MNB now is at 82.269%

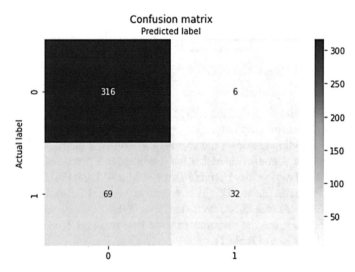

Fig. 6 Confusion matrix for MNB

4.2 Long Short-Term Memory Approach

Once again, the data is cleaned, and the tweets and targets are set. The next step involves a technique called feature padding. It includes encoding the data to form it into model compatible sequence. Even though padding does bring changes within the data, it does not affect the performance of the network, or rather, it should not [15]. After the aforementioned, a sequential model is built and the data split and fed into it (Fig. 7).

The training begins then and the training loss and validation loss that occur on the model are calculated via each epoch as depicted in Fig. 8. Further predictions are made and the accuracy of the model tested (Fig. 9).

```
Model: "sequential_1"

Layer (type)                   Output Shape            Param #
=================================================================
embedding_1 (Embedding)        (None, 28, 128)         256000

spatial_dropout1d_1 (Spatial   (None, 28, 128)         0

lstm_1 (LSTM)                  (None, 196)             254800

dense_1 (Dense)                (None, 2)               394
=================================================================
Total params: 511,194
Trainable params: 511,194
Non-trainable params: 0


None
```

Fig. 7 The RNN LSTM model structure

Fig. 8 Training and
validation loss by LSTM

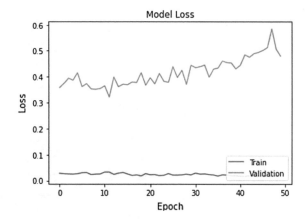

Fig. 9 Confusion matrix for
LSTM

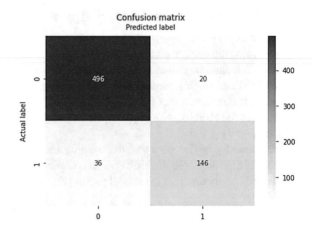

In the above figure, true positives and true negatives are given in the locations (0, 0) as 496 and (1, 1) having 146. Again, these two denote correct predictions of the original labels. The other two scores, i.e. 20 and 36 denote false positives and false negatives, respectively, as mentioned earlier, the incorrect predictions of the labels. So, the accuracy with LSTM is found at 91.977%.

4.3 A Brief Comparison

In case of Naïve Bayes, the accuracy can be improved by changing the feature length or shifting the sizes of the training and testing data, as is in case of RNN. The added advantage of RNN being running a greater number of epochs, whilst the model is being trained (Table 2).

Table 2 Comparison between MNB and LSTM

Points of difference	Naïve bayes multinomial model	RNN LSTM sequential model
Dependence	Existence confidence of words	Sequence and word interdependence
Accuracy	Lower: 82.27%	Higher: 91.97%
Precision	0.8421	0.8795
Recall	0.3168	0.8022
F1 score	0.4604	0.8391
ROC_AUC	0.6491	0.8817
Complexity	Low as compared to RNN	Higher, credited to time

Fig. 10 Chart–Comparison of MNB and LSTM

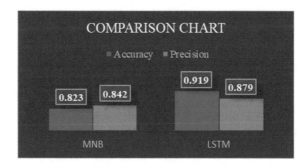

Although the conclusion from the table leads to believe that LSTM is performing better, though at the cost of time complexity, both models are quite independent with their own perks. The complexity both time and space are way lower in case of MNB making it a suitable choice for practical sentiment analysis on a smaller scale, with adequate accuracy. At the same time if the dealing itself is complex fitting it into the MNB may not be advisable. Both do work on sequential data though, and given the fact that today's world is filled with their desired data, none of the two models can be shunned (Fig. 10).

In the above chart, the comparison of the accuracy and precision is depicted. The blue bar is for accuracy, and the orange for precision, when the data split for training and testing is 90:10 in MNB and 67:33 in RNN.

5 Conclusion and Future Work

The sentiment analysis done using Naïve Bayes and recurrent neural network on the global warming tweet dataset has been a success. The performance gave an approximate accuracy of 82% with multinomial Naïve Bayes, but when the sentiment analysis is performed on the same dataset using recurrent neural network the accuracy has shifted to approximately 92%. The multinomial Naïve Bayes model and the LSTM RNN model are quite diverse in their working. One uses feature modelling the other uses feature padding, both work on tokenized data and have a set vocabulary. Now the future scope of this project ranges from being able to comprehend the sentiments with all irony and sarcasm all the way to being able to define the emotions in written words being interpretable to the further extent of detecting hostility via what humans tend to express in words with hidden thoughts or even the traces of emotions and anger displayed in what is written or how a person walks. The major future aspiration now relates to actually have a machine interpret the emotion in written words, whatever can be done beyond is all up to one's imagination.

References

1. Unitarian Universalist Association (2006) Threat of global warming/climate change. UUA action statements threat - global warming and climate change
2. Ro C (2019) BBC Future—the harm from worrying about climate change. BBC. Future article. How to beat anxiety about climate change and eco awareness
3. Scikit Learn. Sklearn.naive_bayes.MultinomialNB. https://scikit-learn.org/stable/modules/generated/sklearn.naive_bayes.MultinomialNB.html
4. Colah's Blog. Understanding LSTM Networks. Colah.Github.io—posta (2015–08) Understanding LSTMs
5. Kharde VA, Sonawane S (2016) Sentiment analysis of twitter data: a survey of techniques. Int J Comput Appl 139(11):0975–8887. http://arxiv.org/ftp/arxiv/papers/1601/1601.06971.pdf
6. Agarwal A, Xie B, Vovsha I, Rambow O, Passonneau R (2011) Sentiment analysis of twitter data. Department of computer science columbia university, New York, NY 10027 USA
7. Pak A, Paroubek P (2010) Twitter as a corpus for sentiment analysis and opinion mining. In: Proceedings of the seventh conference on international language resources and evaluation, 1320–1326
8. Wagh B, Shinde JV, Wankhade NR (2016) Sentimental analysis on twitter data using naive bayes. Int J Adv Res Comput Commun Eng 5(12). ISO 3297:2007 Certified
9. Mucha N (2018) Sentiment analysis of global warming using twitter data. https://library.ndsu.edu/ir/bitstream/handle/10365/28166/Sentiment%20Analysis%20of%20Global%20Warming%20Using%20Twitter%20Data.pdf?sequence=1
10. DataWorld. Sentiment of climate change. Dataset in IBM Watson AI Xprize—environment. https://data.world/xprizeai-env/sentiment-of-climate-change
11. Hana L, Sentiment_Analysis_rnn_LSTM. https://github.com/lamiaehana/Projects/blob/master/Sentiment_Analysis_rnn_LSTM/Sentiment_Analysis.ipynb
12. Ting KM (2017) Confusion matrix. In: Sammut C, Webb GI (eds) Encyclopedia of machine learning and data mining. Springer, Boston, MA. https://doi.org/10.1007/978-1-4899-7687-1_50
13. Edpresso Editor. A shot of dev knowledge. CountVectorizer in python. http://educative.io/edpresso/countvectorizer-in-python
14. Scikit-Learn. sklearn.feature_extraction.text.TfidfTransformer. http://scikit-learn.org/stable/modules/generated/sklearn.feature_extraction.text.TfidfTransformer
15. Reddy DM, Reddy NVS (2019) Effects of padding on LSTMs and CNNs. arXiv:1903.07288v1 [cs.LG]

Novel and Prevalent Techniques for Resolving Control Hazard

Amit Pandey, Abdella K. Mohammed, Rajesh Kumar, and Deepak Sinwar

Abstract In a standard pipelined RISC processor, the succeeding instruction in the program enters the instruction fetch (IF) stage before the branch condition is evaluated in the instruction decode (ID) stage. A wrong instruction may enter the processer in this hustle, causing control hazard. Earlier in single-core processors, techniques of static and dynamic branch predictions were used to predict the branch for fetching the correct instructions. Later, the processor technology enhanced and brought multi-threading and multicore techniques into play. Current study analyzes the novel and prevalent techniques, such as reconstituting processor's pipeline organization, speculative execution using multithreading, and fine-grained multithreading, for resolving the control hazard.

Keywords Control hazard · Branch prediction · Speculative multithreading · Fine-grain multithreading · Out of order execution · Thread-level parallelism

1 Introduction

The initial RISC style processor architectures were introduced by the microprocessor industry in the late 1980s. Induction of instruction level parallelism in the pipelined architecture and the use of multilayered caches were the key features used to enhance the performance in those architectures for next two decades [1].

Induction of pipelined architecture has undoubtedly enhanced the performance, but also has introduced a new type of hazard known as control hazard. Earlier various static and dynamic branch prediction techniques were used to handle this hazard, although none were 100% accurate. Later in 2019, Pandey A. has proposed a novel

A. Pandey (✉) · R. Kumar
College of Informatics, Bule Hora University, Bule Hora, Ethiopia
e-mail: amit.pandey@live.com

A. K. Mohammed
Faculty of Informatics, IOT, Hawassa University, Hawassa, Ethiopia

D. Sinwar
Department of Computer and Communication Engineering, Manipal University, Jaipur, India

© The Author(s), under exclusive license to Springer Nature Singapore Pte Ltd. 2022
P. Nanda et al. (eds.), *Data Engineering for Smart Systems*, Lecture Notes in Networks and Systems 238, https://doi.org/10.1007/978-981-16-2641-8_22

approach to resolve these type of hazards by refactoring the pipeline organization [2].

By the end of the second millennium, all the giants in the field of processor manufacturing had already realized that because of issues like heat dissipation it is not possible any further to enhance the performance of single core processors. Therefore, techniques of multithreading and multicore processors were induced. Speculative execution using multithreading [3–6] and fine-grained multithreading [7–12] are the techniques used to handle the control hazards in such architectures.

2 Established Techniques for Resolving Control Hazard

2.1 Handling Control Hazard by Re-forming Processor's Pipeline Organization

This approach uses a restructured pipeline organization to handle the control hazard [2]. Here, the program counter register (PC) in instruction fetch stage (IF) is initialized to minus 1, and by end of first clock cycle the first instruction from the program memory is loaded in to PC and the instruction fetch/instruction decode (IF/ID) inter-stage register. As the second clock cycle starts, the branch detector in the instruction decode stage (ID) takes persistent inputs from the general purpose register set (GPR) and the IF/ID inter-stage register. The branch condition is evaluated, and the persistent outcome is directly forwarded to the multiplexer in the IF stage. This multiplexer takes persistent inputs from the branch detector and the PC to load the appropriate instruction in the IF stage, before the end of second clock cycle (see Fig. 1). Further, the complete process can be conceptualized as expressed using Eqs. 1, 2, 3, 4, 5, and 6 (Fig. 1).

Processing in IF stage:

$$
\begin{aligned}
&(\text{if} \begin{pmatrix} (\text{IF} - \text{ID.InstructionRegister} \; = \; \text{Branch Instruction}) \\ \text{and (Branch.Taken} \; = \; \text{Taken}) \end{pmatrix} \\
&\{\text{InstructionMemory}[\text{PC} \; + \; \text{SignExtended (IF} - \text{ID.Value)}]\} \\
&\text{else } \{ \text{InstructionMemory}[\text{PC} \; + \; 1]\}) \rightarrow \text{IF} - \text{ID.InstructionRegister}
\end{aligned}
\tag{1}
$$

$$
\begin{aligned}
&(\text{if (PC.SetInitialize} \; = \; \text{True) } \{-1\} \\
&\text{else (if} \begin{pmatrix} (\text{IF} - \text{ID.InstructionRegister} \; = \; \text{Branch Instruction}) \\ \text{and (Branch.Taken} \; = \; \text{Taken}) \end{pmatrix} \\
&\{\text{PC} \; + \; \text{SignExtended (IF} - \text{ID.Value)}\} \text{ else } \{\text{PC} \; + \; 1\})) \rightarrow \text{PC}
\end{aligned}
\tag{2}
$$

Processing in ID stage:

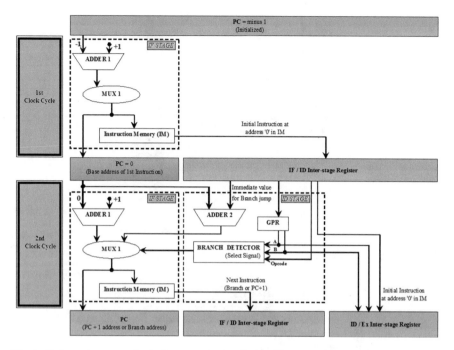

Fig. 1 Reformed processor's pipeline organization for handling control hazard [2]

$$\text{GPR}\left[\text{IF} - \text{ID.AddressOperandOne}\right] \rightarrow \text{ID} - \text{EX.OperandOne.Value} \quad (3)$$

$$\text{GPR}\left[\text{IF} - \text{ID.AddressOperandTwo}\right] \rightarrow \text{ID} - \text{EX.OperandTwo.Value} \quad (4)$$

$$\text{IF} - \text{ID. InstructionRegister} \rightarrow \text{ID} - \text{EX. InstructionRegister} \quad (5)$$

$$\text{SignExtended (IF} - \text{ID.Value)} \rightarrow \text{ID} - \text{EX.Value} \quad (6)$$

2.2 Fine-Grain Multithreading

In fine-grain multithreading, the threads are interchanged at every clock cycle, inducing a gap of one clock cycle between any two consecutive instructions from the same thread. This gap of one clock cycle provides the opportunity window to evaluate the branch condition and to select the correct instruction to be fetched in the next clock cycle [1]. Thus, the fine-grain multithreading is designed to handle the control hazards by default (see Fig. 2).

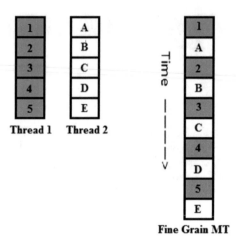

Fig. 2 Fine-grain multithreading

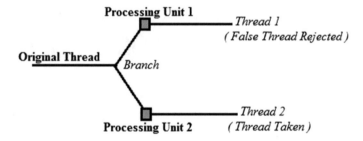

Fig. 3 Fine-grain multithreading with latencies induced in threads

Further, if any latency is induced in the first thread, then the instruction will be fetched from the second thread only. As in this case, the consecutive instructions from the same threads will be fetched. Hence, no gap will be induced between them, and in this particular scenario a control hazard may happen (see Fig. 3).

2.3 Speculative Execution Using Multithreading: A Case of Out of Order Processor

In this approach when a branch condition is encountered, two separate threads will take on both the possibilities simultaneously. Further, when the branch condition evaluation result is available, then the correct path is chosen and the other thread is discarded (see Fig. 4). Currently, this technique is supported by out of order processors and can handle even multilayered nested branches with the help of proper memory support [1].

Fig. 4 Speculative
execution using
multithreading

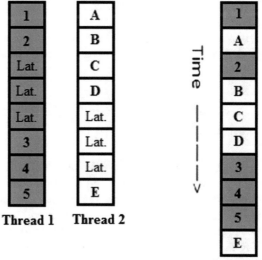

In an out of order processor (OoO), the instructions are executed on the basis of data-flow graph instead of the actual program order. Figure 5a shows the two alternative program orders in a case of a branch condition.

Fig. 5 Speculative
execution in OoO processor.
a Two alternative program
orders in case of a branch
condition. **b** The out of order
execution of both alternatives
in parallel. **c** Discarding the
wrong alternative

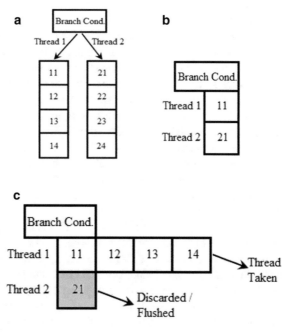

In this approach, while the branch condition is fetched and resolved, the other two possibilities will also start their out of order execution in parallel (see Fig. 5b).

Further, as the branch condition evaluation outcome is available, the correct thread is opted, and the other thread is discarded and flushed out (see Fig. 5c).

2.4 Accurate Branch Prediction for Short Threads on Multicore Processor

The average occupancy time of a single thread will be reduced when executed on a multicore processor, as compared to single core. In speculative multithreaded processor, the performance of any single thread is enhanced by partitioning its execution into various speculative threads, which will be executed on various available cores. A speculative thread will begin its execution as the processor fetches the first starting address from the couple of spawning instructions. A couple of spawning instruction basically has two instructions which are denoted as spawning point (SP) and control quasi-independent point (CQIP). When the spawning point is reached during the execution, the processor initiates the new speculative thread. This thread will keep on executing till the control quasi-independent point is reached. Further, the control quasi-independent point will initiate the next speculative thread. The control quasi-independent point is executed as the first instruction of the newly created speculative thread. A speculative multithreaded processor executes these speculated threads in parallel on various cores to exploit the Intra-thread parallelism. Putting this in a practical scenario when a function call is made becomes the SP and the return point for that function becomes the CQIP. In similar fashion, when dealing with iterations the branch target is treated as SP and CQIP both [13, 14].

As the intercore communication latencies in a multicore processor decreases, the architecture becomes more enticing for thread migrations. It was observed that during migration of shorter threads the substantial part of the warm-up cost on new core was imposed by the processes for setting up a branch predictor and for obtaining the appropriate Global History Register (GHR). For obtaining a valid GHR, various techniques are possible, which can further be categorized in to two main classes. In first category lies those techniques which will re-create or predict the estimated GHR using the existing historical data. While the second category consists of those techniques which will try to offer a steady initial point for the branch predictor, whenever the same thread is executed.

In speculative multithreading, the speculated thread starts its execution before all other previous branches are finished. Thus, it is not possible to obtain the outcome of these early branches in the GHR of the speculated thread. Even though it is not possible to get the actual values of real GHR, but it is possible to estimate a worthy prediction which can serve as positive proxy. Pre-spawning policy is the first approach that we will discuss here. Under this policy, the GHR of the parent thread will be inherited by the speculated thread. Although at the time of inheritance the GHR of the parent thread may not be complete, as it may not have finished its execution. Hence, it may not contain future predictions about all branches, but still may be indicative. One drawback of this approach is that it creates alias between the parent thread and the speculated thread.

The second approach is the Concat policy. In this policy, the GHR of the parent thread at SP is concatenated with the branch prediction history of this thread from last execution. This gives the policy an upper hand over the pre-spawn policy. As by concatenating the previous branch history, this approach becomes capable of making better predictions about the branches.

The last CQIP policy is the third approach in the first category. This approach directly uses the GHR from the history, when last time this thread has reached its CQIP. Thus, it uses the complete GHR history till CQIP from last execution. Both Concat policy and the last CQIP policy have shown promising results.

Further there are some approaches that fall into second category, which work by availing a consistent starting point for obtaining the correct GHR. The first zero policy proceeds by simply erasing the GHR whenever new thread arrives. This policy starts with little warm-up but outputs precise predictions once sufficient history is buildup for distinguishing this thread from other threads starting point.

The second PC policy in this category proceeds by mapping the PC of the speculated thread's CQIP value with the GHR. This gives each thread spawned from the same SP a uniquely identifiable consistent starting point.

Finally, the third XOR policy proceeds by XORing the PC with the GHR of the parent at SP for providing a consistent starting point [14].

2.5 TAGE-SC-L Branch Predictors

Nowadays, multicore technology is a pervasive processor technology that prominently exploits the thread level parallelism. To maintain the workload on each processor core, various load balancing techniques have been implemented. Although, while performing this load balancing when a thread is migrated from one core to another the information related to its branches is not transferred to the new branch predictor. Which induces a "cold startup phase" for the thread at new core. During this phase, the new branch predictor is not having enough information about the branches and its performance is diminished. To avoid this, a good approach is to migrate the branch history together with the branch. This will not only enhance the prediction accuracy but will also reduce the "warm-up cost" of the branch predictor.

The TAGE-SC-L predictor is combination of three basic predictors, namely the TAGE predictor, loop predictor, and statistical corrector predictor (see Fig. 6).

In TAGE-SC-L predictor, the TAGE predictor plays the role of initial predictor. The TAGE predictor incorporates a basic predictor for alternative predictions and a set of tagged predictor components which are indexed using the value obtained by hashing PC with various history length following a geometric progression obtained by using Eq. 7,

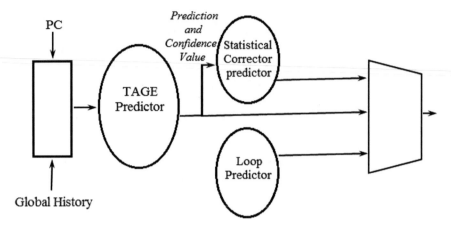

Fig.6 Overview of TAGE-SC-L predictor

$$Len(i) = (int)(\alpha^{i-1} * L(1) + 0.5) \tag{7}$$

Initially, the base predictor and the tagged components of the TAGE predictor are accessed concurrently. Then, the tagged component that uses the longest history is used to make the final prediction. If no match is found in tagged components, then the base prediction is used as the final prediction [15–17].

Although, TAGE predictor is an efficient predictor (Fig. 7). But in case of statistically biased branches, which do not show any strong correlation with the previous history, its performance diminishes. To handle such cases, the statistical corrector predictor comes in role. The statistical corrector predictor takes the TAGE prediction, branch address, local history, global path and global history as input to decide whether or not to invert the earlier prediction made by TAGE predictor. Usually, TAGE predictor is quite efficient, and rarely statistical corrector predictor comes in role [17].

Further, the loop predictor present in TAGE-SC-L predictor handles all the finite loops. The loop predictor starts making predictions about any loop after its seven successful runs with same number of iterations. At this time, the age counter for that loop is set to seven. This age value is incremented for every useful prediction and decremented otherwise. An entry can only be replaced when its age is null [17].

3 Conclusion

The current study discusses about the novel and prevalent techniques for handling control hazard by making efficient branch predictions. The fine-grain multithreading is well-designed technique to handle the control hazard until there is no latency induced in the threads. As in the case of latencies, any individual thread may be

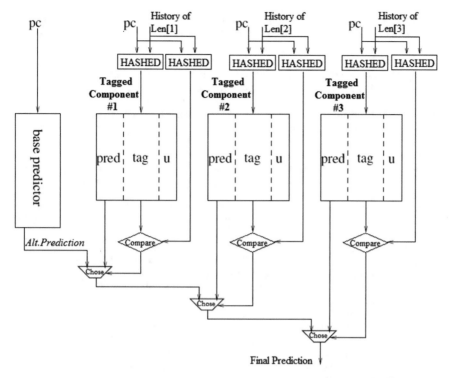

Fig.7 TAGE predictor: A base predictor with three tagged predictors arranged in increasing order of history length

forced to continue without the gap between its instructions, causing the control hazard.

The speculative execution of threads using multithreading is also a prominent technique for handling control hazard. This technique is capable of handling multilayered nested branches, but accounts for a complex hardware design.

The control hazard can also be handled by reconstituting the pipeline organization in the IF and ID stages of a standard RISC style processor.

Further, effective usage of GHR with speculative multithreading and implementation of predictors like TAGE-SC-L has also shown promising results.

References

1. Hennessy JL, Patterson DA (2011) Computer architecture: a quantitative approach. Elsevier
2. Pandey A (2019) Resolving control hazard by reconstituting processor's pipeline organization. In: Proceedings of 2019 Third International conference on inventive systems and control (ICISC). IEEE, pp 201–206

3. Tubella J, Gonzalez A (1998) Control speculation in multithreaded processors through dynamic loop detection. In: Proceedings of 1998 Fourth International symposium on high-performance computer architecture. IEEE, pp 14–23
4. Marcuello P, González A, Tubella J (1998) Speculative multithreaded processors. In: Proceedings of the 12th International conference on supercomputing, pp 77–84
5. Marcuello P, González A (1999) Clustered speculative multithreaded processors. In: Proceedings of the 13th International conference on supercomputing, pp 365–372
6. Marcuello P, Tubella J, González A (1999) Value prediction for speculative multithreaded architectures. In: MICRO-32 Proceedings of the 32nd Annual ACM/IEEE International symposium on microarchitecture. IEEE, pp 230–236
7. Pascoe J, Welch P, Loader R, Sunderam V (2002) Cache-affinity scheduling for fine grain multithreading. In: communicating process architectures 2002: WoTUG-25: Proceedings of the 25th WoTUG Technical meeting, Sep 2002, vol 60. University of Reading, IOS Press, UK, pp 15–18, p 135
8. Loikkanen M, Bagherzadeh N (1996) A fine-grain multithreading superscalar architecture. In: Proceedings of the 1996 Conference on parallel architectures and compilation technique. IEEE, pp 163–168
9. Sohn A, Kodama Y, Ku J, Sato M, Sakane H, Yamana H, Sakai S, Yamaguchi Y (1997) Fine-grain multithreading with the EM-X multiprocessor. In: Proceedings of the ninth annual ACM symposium on parallel algorithms and architectures, pp 189–198
10. Shrestha S, Manzano J, Marquez A, Feo J, Gao GR (2014) Jagged tiling for intra-tile parallelism and fine-grain multithreading. In: Proceedings of International workshop on languages and compilers for parallel computing. Springer, Cham, pp 161–175
11. Madriles C, López P, Codina JM, Gibert E, Latorre F, Martínez A, Martínez R, González A (2009) Boosting single-thread performance in multi-core systems through fine-grain multi-threading. ACM SIGARCH Comput Arch News 37(3):474–483
12. Daněk M, Kafka L, Kohout L, Sýkora J, Bartosiński R (2012) UTLEON3: Exploring fine-grain multi-threading in FPGAs. Springer Sci Bus Media
13. Zhang T, Zhou C, Huang L, Xiao N (2017) Branch prediction migration for multi-core architectures. In: Proceedings of International conference on networking, architecture, and storage (NAS). IEEE, pp 1–2
14. Choi B, Porter L, Tullsen DM (2008) Accurate branch prediction for short threads. In: Proceedings of the 13th International conference on Architectural support for programming languages and operating systems, pp 125–134
15. Seznec A (2011) A 64 kbytes ISL-TAGE branch predictor
16. Seznec A (2006) A case for (partially)-tagged geometric history length predictors. J Instr Lev Parallelism
17. Seznec A (2016) Tage-sc-l branch predictors again

Sound Classification Using Residual Convolutional Network

Mahesh Jangid and Kabir Nagpal

Abstract In this paper, we proposed a new architecture for environmental sound classification on the ESC-50 and urban sound dataset. The ESC-50 dataset is a collection of 2000 labeled environmental audio recordings and the urban sound dataset is a collection of 8732 labeled sound records. The Mel frequency cepstral has been used to obtain the power spectrum of the sound wave. The resulting matrix, made possible the use of the convolutional neural network architecture over the dataset. The new architecture extracts far more complex features repeatedly, while being able to carry it along a greater depth using a ResNet type architecture. After the fine-tuned network, we achieved 89.5% validation accuracy on environmental classification dataset and 96.76% on urban sound dataset.

Keywords Convolutional neural network · Residual neural network · Environmental sound classification · Mel frequency cepstral

1 Introduction

Recently, there have been many breakthroughs for image classifications using both supervised and unsupervised methods. Even natural language processing has been through some incredible changes. With the upcoming GPT-3 by Open-AI, trained on 40 terabytes of data, the bar is set to rise higher. But sound has remained an unexplored domain yet and not much research has been performed. An important application of understanding waves (sound as well others) is word independent voice recognition which can potentially revolutionize the way we use technology. These can be further used in understating characteristics of other waves as well.

Prior research based on traditional machine learning models were not able to achieve a decent score for classifying sounds. Most companies that deal with sound as their data are dependent on either similarity scores or in case of human speech generate transcripts manually or using a deep recurrent neural network [5] in the form of automatic speech recognition. These transcripts are then analyzed to find

M. Jangid (✉) · K. Nagpal
Department of CSE, Manipal University Jaipur, Jaipur, Rajasthan, India

© The Author(s), under exclusive license to Springer Nature Singapore Pte Ltd. 2022   245
P. Nanda et al. (eds.), *Data Engineering for Smart Systems*, Lecture Notes in Networks and Systems 238, https://doi.org/10.1007/978-981-16-2641-8_23

features. The reason for this is that the sound wave being captured is a set of 22,050 points per second (may vary depending on the sound rate for capturing device) in the form of a sinusoidal wave.

As these waves can stretch over a long duration, variable lengths of this data form most fails most machine learning models. This research was further carried out using several feature extraction techniques to reduce dimension of the audios. These features include zero-crossing, Mel frequency cepstral coefficients [1–3], and wavelet transformation.

Exceptional research performed by Ergün Yücesoy and Vasif Nabiyev in their paper, "Gender identification of a speaker using MFCC and GMM"[6] achieved 97.76% accuracy and similar models have since been used for a variety of different tasks. Although the accuracy achieved by them was mind-blowing, an issue with our (humans) voice is that pronunciation of a word depends on our accent and language. To extract deeper features, I decided to research with state of the art, convolutional neural networks. To avoid complexity due to human voice, environmental sound classification dataset [7] provided by Karol J. Piczak was used. In this dataset, the goal is to recognize the event type of a specific sounds.

It consists of 5-s-long recordings organized into 50 classes (with 40 examples per class) loosely arranged into 5 major categories: Animals, natural soundscapes and water sounds, human non-speech sounds, domestic sound, and urban noises. A smaller version of the same dataset is also available with 10 classes. The sound in this dataset is 5 s each, 44.1 kHz, mono (single channel) sounds. Another dataset of urban sound by Salamon et al. [8] was used which consists of 10 sound classes divided into 10 folds for evaluation purposes consisting of similar classes.

2   Related Work

CNNs have improved over time to a great extent, and with larger and wider models the capability of complex feature extraction has also improved. For ESC tasks similar feature extraction methods [9] were first proposed by Piczak which improved the classification accuracy over 13%.

Wei Dai and his team proposed to use deep convolutional neural networks (DCNNs) [10] to extract more discriminative features, which outperformed the shallow convolutional neural network model. Since then multiple models and filter bank techniques have been used to increase the performance of the model, the latest one being "ESResNet: Environmental Sound Classification Based on Visual Domain Models" by Andrey Guzhov and team [11], in which a pre-trained model was used to boost the accuracy of ESC-50 dataset to 91%. This however requires greater time to first train on a different dataset and fine tune on another. The proposed architecture is able to achieve similar accuracy in far less time due to extensive advantage of inspiration from residual networks.

3 Data Preparation

3.1 Dataset

ESC-50 Dataset defines the properties of the sound being Mono and 44.1 kHz and of a duration of 5 s. Few audios which were smaller were given a zero padding to maintain size input for the architecture. The urban sound dataset however does not define such properties and hence was assumed not to be mono and of a frequency of 22.05 kHz. Due to a huge variation of audio durations in this dataset, several audio samples were given zero-padding, while others were truncated. These tasks were performed using Librosa [12]. The dataset is divided into 10 folds with an equal proportion of each label to measure robustness at the time of testing.

3.2 Feature Extraction

Sound files are difficult to deal with due to their enormous size. The sound properties described in this dataset is 5 s, 44.1 kHz which means that the device collected 44.1 K data points in one second or approximately 200,000 data points for the complete file which when combined forms the sinusoidal sound wave. Due to this, traditional machine learning models failed when applied on raw audio. A raw form of audio is given in Fig. 1.

As observed longer audios will have a longer wave and hence a limitless dimension. To avoid this, special feature extraction methods are used to. The method used in this paper is Mel frequency cepstral coefficient (MFCC).

Mel Frequency Cepstral: [13] is the short-term power spectrum of a sound, based on a linear cosine transform of a log power spectrum on a nonlinear Mel scale

Fig. 1 Audio envelope of a sound wave for 1 s

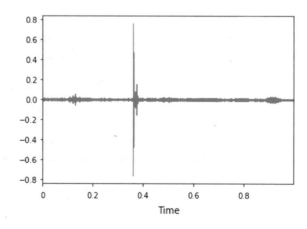

Fig. 2 Conversion of audio to mel frequency cepstral

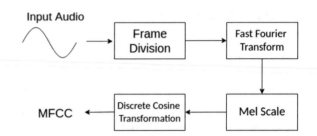

of frequency. Coefficients of these are known as MFCC. The block diagram for conversion of speech to Mel cepstral is given in Fig. 2.

Frame Division: The audio is cut into frames of fixed size. The next frame usually overlaps the previous frame to capture maximum information. The process is repeated until the end of the file and in case of presence of a shorter frame, zero paddings are added to it.

Fast Fourier Transform: These audio splits are converted to frequency domain from the time domain. The spectrum obtained is called the amplitude spectrum. Equation 1 defines the conversion with x representing the time.

$$\int_{-\infty}^{\infty} f(x)e^{-2\pi ikx}dx \tag{1}$$

Mel Scale: Stevens and Volkman in their research," A scale for the measurement of the psychological magnitude pitch" [14]. Experimentally proved that humans can perceive frequencies linearly up to 1 kHz and logarithmically above it. The powers of the spectrum obtained are mapped to Mel scale using Eq. 2.

$$M(f) = 1125 \ln\left(1 + \frac{f}{700}\right) \tag{2}$$

Discrete cosine transform: Discrete cosine transform is applied to the mapped powers. Discrete cosine transformation (DCT) [15], first proposed by Nasir Ahmed, is the sum of cosine functions oscillating at different frequencies. Coefficient of the spectrum thus obtained is called MFCC. For this network, 40 MFCCs are chosen.

Deltas: MFCC defines the power spectral of one frame only but over time these spectral coefficient change and hence it was observed that appending these trajectories of features to the power spectral significantly improved the performance of ASR. These are known as delta coefficients. To calculate the delta coefficients, Eq. 3 is used.

$$d_t = \frac{\sum_{n=1}^{N} n(c_{t+n} - c_{t-n})}{2\sum_{n=1}^{N} n^2} \tag{3}$$

Here, d_t is a delta coefficient, from frame t computed in terms of the static coefficients c_{t+n} to c_{t-n}. A typical value for N is 2 (derivative order).

4   Convolutional Neural Network

4.1   Architecture

As presented by the paper, "Deep Residual Learning for Image Recognition" [4] deeper convolutional networks are difficult to train due to vanishing gradients as well as the complexity of the network. As we add more convolutional layers, the accuracy tends to saturate and then degrades quickly. Residual networks provide a solution to this via shortcut connections. Features learned from one of the previous layers are stacked with a new layer. So as the layers increase, almost certainly new features can be learned due to residual features which are extending from layers before. A residual building block is visualized in Fig. 3. Here, X is the input, f(X) is the output generated by the layers.

As observed after every few stacked layers input of the primary layer gets added to the last layer. This enables features to be learned efficiently by deeper layers. A variety of operations can be performed on the two layers like addition, average, concatenation etc. Taking inspiration from this architecture, I used a block structure as shown in Fig. 4a. Additional details of the convolution layers are given in Table 1. This block is considered as the building block for this architecture and repeated several times. The pooling layer in this diagram changes after every two blocks and

Fig. 3 Architecture of residual block

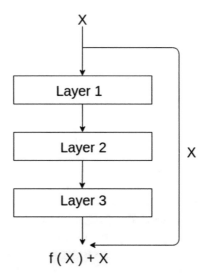

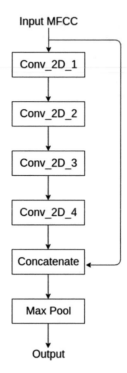

Fig. 4 Architecture of proposed residual block

Table 1 Properties of each convolution layer in the proposed architecture

Layer number	Properties kernel size	Filter	Activation
Conv_2D_1	(3,5)	32	Tanh
Conv_2D_2	(3,5)	64	Tanh
Conv_2D_3	(5,5)	128	Tanh
Conv_2D_4	(5,5)	256	Tanh
Concatenate	Along axis 1		
Max pool	Pool size = (2, 5)		

reduces pooling factor by 1 in dimensions two, i.e., for block 1 and 2 it is (2,5), for block 3 and 4 it is (2,4) and so on. The model appears as in Fig. 4b.

For the model, Stochastic Gradient Descent [18] has been used as an optimizer with the learning rate of 0.01. In SGD, gradients of the cost function are calculated for a single sample (chosen randomly). SGD was able to converge faster as compare to Adam [19] and hence was preferred for this architecture (Fig. 5).

Additionally, while training as the architecture fails to converge further, learning rate is dropped by a factor of 0.1 after observing for 7 consecutive increase in validation loss. The weights of the best-known model are used again to fine tune the network. The training was stopped if the loss did not decrease for 10 consecutive

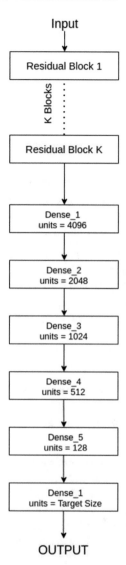

Fig. 5 Complete proposed architecture

epochs to prevent overfitting. The code was written using Keras [20]. The mean performance of the models is given in Table 2.

Table 2 Accuracy achieved by changing the number of blocks	Number of blocks	Accuracy for ESC 50	Accuracy for urban Sound 8 K
	3	55.5	87.19
	4	60.5	92.43
	5	47.50	92.8

4.2 Fine Tuning

As we keep training the model, it first converges and then the validation loss start increasing, while the training loss keep decreasing. This suggests that model is overfitting and learns the data instead of generalizing. A variety generalization/ regularizing technique is present to avoid this. Among the used ones are given below.

Regularizers: To reduce overfitting and generalize the model, regularizers are applied. Regularizers penalize the weight matrix by adding an extra term. Regularizers or weights are L1 and L2 [21] which are useful for updating the cost function by adding the term. For this paper, both regularizers were used.

L2 regularization is given in Eq. 4, and L1 regularization is given in Eq. 5.

$$Cost\ function = loss + \lambda * \|W\|^2 \tag{4}$$

$$Cost\ function = loss + \lambda * \|W\| \tag{5}$$

Here, Lambda is the regularizing factor, and W is the weight matrix. L1 tends to shrink coefficients to zero, whereas L2 tends to shrink coefficients evenly.

Batch normalization: Batch normalization is a method described by Sergey Ioffe and Christian Szegedy in their paper "Batch Normalization: Accelerating Deep Network Training by Reducing Internal Covariate Shift"[22], which normalizes the data in each batch. The advantage of this being each input is rescaled and recentered to have a value in a smaller range. It helps in faster computation and allow us to use higher learning which helps in finding global minima faster.

Dropout: Another regularization method is to drop some weight rows randomly [23]. This has been observed to reduce complex adaptation in the network, further helping the model to generalize.

Regularizers, batch normalizations and dropouts were added together in the architecture to fine tune the model. 33% dropout, batch normalization (done before and after the Residual Blocks Series) and regularization given as below were added to the network. Table 3 consists of the results after adding the three regularizing techniques.

- Kernel Regularizers: L1 regularization factor 10e-4, L2 regularization factor 10e-5
- Bias Regularizers: L2 regularization factor 10e-4
- Activity Regularizers: L2 regularization factor 10e-5

Table 3 Accuracies achieved after applying regularizers on the increased number of blocks

Number of n blocks	Accuracy for ESC 50	Accuracy for urban sound 8 K
5	81.5	93.35
6	85.5	94.87
7	**89.50**	**96.76**

Fig. 6 Testing accuracy of datasets

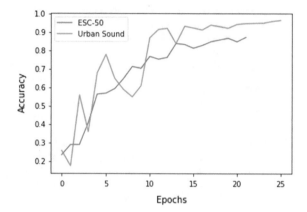

As observed in Table 3, accuracy increases giving us a result of 89.5% for ESC-50 Dataset and 96.76% for urban sound classification dataset. As seen in Fig. 6, we also observe that training time is much less (Best performance in 20–25 Epochs) compared to previous models on the same dataset.

5 Conclusion

In this paper, I proposed a new architecture adapted from ResNet for sound classification. Two public datasets, ESC-50, and urban sound 8 K are used to evaluate the model. The audios are converted to MFCCs along with its derivative to attain a two-dimensional input for the model. Using this architecture, the model learns quickly, achieving comparable accuracy on the public datasets. The model achieves 89.5% accuracy for ESC-50 Dataset and 96.76% for urban sound classification dataset accuracy which shows the potential of the architecture to perform better with further works. This also open the future capability of the network to be used for tasks like speaker identification and verification, and study of other similar wave forms.

References

1. Cotton CV, Ellis DPW (2011) Spectral vs. spectro-temporal features for acoustic event detection. In: Proceedings of the applications of signal processing to audio and acoustics. New Paltz, NY, USA, pp 16–19
2. Ntalampiras S, Potamitis I, Fakotakis N (2013) Large-scale audio feature extraction and SVM for acoustic scene classification. In: Proceedings of the applications of signal processing to audio and acoustics. New Paltz, NY, USA, pp 20–23
3. Automatic recognition of urban environmental sounds events (2008) In: Proceedings of the IAPR workshop on cognitive information processing cip. Santorini, Greece, pp 9–10
4. Kaiming H et al (2016) Deep residual learning for image recognition. In: Proceedings of the IEEE conference on computer vision and pattern recognition
5. Graves A, Mohamed AR, Hinton G (2013) Speech recognition with deep recurrent neural networks. In: Proceedings of 2013 IEEE International conference on acoustics, speech and signal processing. IEEE
6. Yücesoy E, Nabiyev VV (2013) Gender identification of a speaker using MFCC and GMM. In: Proceedings of 2013 8th International conference on electrical and electronics engineering (ELECO). IEEE
7. Piczak KJ (2015) ESC: dataset for environmental sound classification. In: Proceedings of the 23rd annual ACM conference on multimedia. Brisbane, Australia
8. Salamon J, Jacoby C, Bello JP (2014) A dataset and taxonomy for urban sound research. In: Proceedings of 22nd ACM International conference on multimedia, Orlando, USA
9. Piczak KJ (2015) Environmental sound classification with convolutional neural networks. In: Proceedings of 2015 IEEE 25th International workshop on machine learning for signal processing (MLSP). IEEE
10. Dai W, Dai C, Qu S, Li J, Das S (2017) Very deep convolutional neural networks for raw waveforms. In: Proceedings of 2017 IEEE International conference on acoustics, speech and signal processing (ICASSP). IEEE, pp 421–425
11. Guzhov A et al (2020) ESResNet: Environmental sound classification based on visual domain models. ArXiv abs/2004.07301:nPag
12. McFee B, Raffel C, Liang D, Ellis DPW, McVicar M, Battenberg E, Nieto O (2015) Librosa: audio and music signal analysis in python. In: Proceedings of the 14th python in science conference, pp 18–25
13. Wikipedia contributors (2020) Mel-frequency cepstrum. Wikipedia, The Free Encyclopedia. Wikipedia, The Free Encyclopedia, 21 Dec 2019, Web 16 Sep
14. Stevens SS, Volkmann J, Newman EB (1937) A scale for the measurement of the psychological magnitude pitch. J Acoust Soc Am 8(3):185–190
15. Ahmed N, Natarajan T, Rao KR (1974) Discrete cosine transform. IEEE Trans Comput 100(1):90–93
16. Davis S, Mermelstein P (1980) Comparison of parametric representations for monosyllabic word recognition in continuously spoken sentences. IEEE Trans Acoust Speech Signal Process 28(4):357–366
17. Huang X, Acero A, Hon H (2001) Spoken language processing: a guide to theory, algorithm, and system development. Prentice Hall
18. Bottou L (1991) Stochastic gradient learning in neural networks. Proc Neuro-Nımes 91(8):12
19. Kingma DP, Ba J (2014) Adam: a method for stochastic optimization. 1412.6980
20. Chollet F et al (2015) Keras. GitHub. https://github.com/fchollet/keras
21. Wikipedia contributors (2020) Tikhonov regularization. Wikipedia, The Free Encyclopedia. Wikipedia, The Free Encyclopedia, 15 Sep 2020. Web
22. Ioffe S, Szegedy C (2015) Batch normalization: accelerating deep network training by reducing internal covariate shift. arXiv:1502.03167
23. Srivastava N et al (2014) Dropout: a simple way to prevent neural networks from overfitting. J Mach Learn Res 15(1):1929–1958

Dimensionality Reduction Using Variational Autoencoders

Parv Dahiya and Sahil Garg

Abstract In the past few years, the problem of processing of big data in less time and with great accuracy has been a major concern in the world of machine learning. Problems like overfitting and underfitting have also been on the troubling side of projects in the years leading to the current stages of artificial intelligence. This model works on the principle of dimensionality reduction, a potential solution to the above problems, with the growth of artificial intelligence and machine learning, data has become one of the most important assets and yet the biggest challenge. Over the years fast and accurate processing of data has been a constant pursuit, and this is where dimensionality reduction comes into play. In dimensionality reduction, authors reduce the dimension of the data keeping the essence and all the important points of the data intact, this leads to faster and equally accurate result. In my project, I have shown the concept of dimensionality reduction on images using the fonts dataset. Variational autoencoders are used to reduce dimension of the images from larger dimensions to smaller dimensions. The model was then checked for the training loss and the validation loss. Also, GPU based approach is used for the upscaling of the model and obtaining competent and satisfying results.

Keywords Artificial neural networks · Deep learning · Dimensionality reduction · Variational autoencoder

1 Introduction

Dimension reduction is not only an important but also a very essential concept in the field of machine learning. Big data has been an asset for the artificial intelligence world, but only if we know how to use it optimally. This is where dimension reduction comes into play. Dimension reduction helps us process huge data in smaller time and good accuracy. Dimension reduction has achieved state-of-art accuracy results with impressive performance on the validation error scale and the training error stage. Dimensionality reduction always keeps the essence of the data intact. The modern

P. Dahiya (✉) · S. Garg
Amity University Haryana, Gurugram, Haryana, India

© The Author(s), under exclusive license to Springer Nature Singapore Pte Ltd. 2022
P. Nanda et al. (eds.), *Data Engineering for Smart Systems*, Lecture Notes in Networks and Systems 238, https://doi.org/10.1007/978-981-16-2641-8_24

world problems such as autonomous driving and many other models which have to process the real time data and give an accurate output uses the concept of dimensionality reduction since it reduces the computation time significantly and saves the model from problems like overfitting or underfitting. Using variational autoencoders provide us with precise control over our latent representation and what you would like them to represent. Precise modelling helps us to capture better representation. This concept of dimension reduction is going to be a live saver in the very near future or the authors say it already is. In this project, I have converted a 2500 dimensions image to a 32-dimension image without losing the essence of it or the important data points or labels, now this concept when applied to a huge amount data will reduce the computation time significantly so that in various fields, we can be earning valuable time. For example, in medical field, it can save someone's life with the saved time or can train our models and create technologies years earlier in the field of Engineering.

2 Related Work

Variational autoencoder for deep learning of images, labels and captions a research by Pu et al. in 2016 mentioned development of a novel variational autoencoder which models images and the related features and captions. He used deep generative deconvolution neural network (DGDN) as the decoder of the latent image features and a deep convolution network (CNN) was used as the encoder for the image latent features [1]. Roweis et al. in his report on Nonlinear Dimensionality Reduction by Locally Linear Embedding mentioned that a huge number of areas of science are now dependent on exploratory data and the data's analysis and visualisation. The need of the hour is to analyse large amounts of data and that has raised the need of the concept of dimensionality reduction [2]. Tenenbaum et al. in his research on A Global Geometric Framework for Nonlinear Dimensionality Reduction in the year 2000 said that dimensionality reduction is nothing new to the human race in fact the human brain does dimensionality reduction every day, the information that human brain extracts from all it's 30,000 auditory nerve fibres or it's 10^6 optic nerve fibres, it reduces all that information to a manageable small number of relevant features [3]. Maaten et al. in his experiment, Dimensionality Reduction: A Comparative Review laid the emphasis on the concept and gave out the limitations of the conventional techniques like PCA and classical scaling [4]. In 2002, Rabbani et al. in his publication JPEG2000: Image Compression Fundamentals, Standards and Practice mentioned about the JPEG200 as the best compression scheme at the time and talks about its acceptance by the ISO Committee [5], it is still considered as the standard compression scheme for the lossless and lossy compression. In recent developments, Theis et al. submitted his research on lossy image compression with compressive autoencoders, in 2017, the author claimed that his way of doing lossy compression on images was competitive with the standard JPEG2000. It said that the autoencoders have the potential to address this solution and outperform the recently proposed approaches that used RNNs as well [6].

Fig. 1 A sample of image of numeric digit '7' in the dataset

3 Methodology

3.1 Dataset

A huge variety of datasets is readily available for dimensionality reduction. The dataset I have used is the font dataset. The font dataset contains 62 samples sets with 1016 images in each of the sample set, making it a data set of 62,992 images. Any images dataset can be used to demonstrate the concept of dimensionality reduction in this model, but I used the fonts dataset due to the comparative smaller size and hence the less time required to load the data in the model and also the reduction in processing time (Fig. 1).

3.2 Model Generation

Model Generation using a Variational Autoencoder

A variational encoder is an autoencoder which is regularised during the training period as well to avoid the problem of overfitting, and also it ensures that the good properties are chosen for the latent attributes which enable the regeneration process. It is a Compound of both an encoder and a decoder and it is trained to minimise the error of reconstruction between the initial data and reconstructed data (Fig. 2).

This model was trained on the fonts dataset to perform dimensionality reduction on the images of the dataset and it is done in four basic steps of encoding, latent attribute generation, decoding, image generation.

Encoding

The authors build the encoder model to get a set of possible values, a statistical distribution. Out of this set of possible values, the authors will randomly choose and feed sample into the decoder. It is a simple step which takes the data converts it into much smaller encoded form which contains enough processed information for the next steps of the model to process the information. It gives the desired output

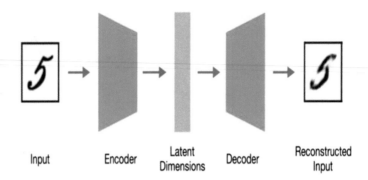

Fig. 2 A basic autoencoder

format according to the next steps of the model. Traditionally, the encoder and all the other parts of the network are trained together and is then optimised via the process of back-propagation, to produce the encodings for the specific task at hand. In variational autoencoders, the encoder is sometimes also called as the recognition model.

Latent Attribute Generation

In this step, the authors process the information given by the encoder, the encoder encodes the information into some latent values. Supposedly, the authors train the autoencoder model with the encoding dimension as 6 on a large faces dataset. The autoencoder will try and learn the descriptive features of the faces such as hair colour, if the person is wearing a beard or not etc., and then attempt to represent the observation in a compressed form. In the example below, we describe the input image in terms of the latent attributes, whilst only using just one value for each attribute. The authors may prefer to use a range of possible values for the representation of each latent attribute. Authors can use the variational autoencoder for the description of the latent attributes in probabilistic terms (Fig. 3).

Decoding

The decoder than trains on the input data. It learns from the latent attributes and tries to reconstruct the image or the information in an exact manner. Whilst doing dimensionality reduction, authors can ask the decoder to only decode or to only generate the image up to a specific number of layers and not to do it the whole way back. The decoder reads the latent attributes and then starts reconstructing the image according to the values of each latent attributes provided and does it only to the exact number of layers the authors want the original image to be reduced to. The decoder reads the probabilistic distribution amongst the latent attributes and then reconstructs the information. In the example below, the decoder decodes the latent attributes generated after encoding the information of the faces data set and produces the exact same data (Fig. 4).

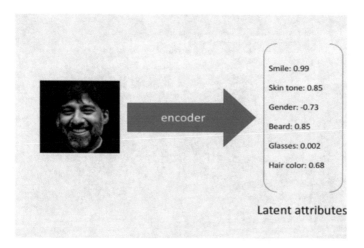

Fig. 3 Latent attribute generation after encoding [7]

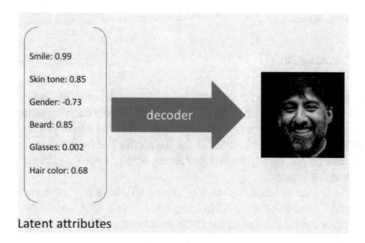

Fig. 4 Decoding [7]

Image Generation

After building the model, authors fit the model using the training dataset. The model is made to run for 1000 epochs, the encoder encodes the data into latent attributes and then the decoder reads the latent attributes and then generates an image with the required number of dimensions and then the dimensionally reduced image is produced [8].

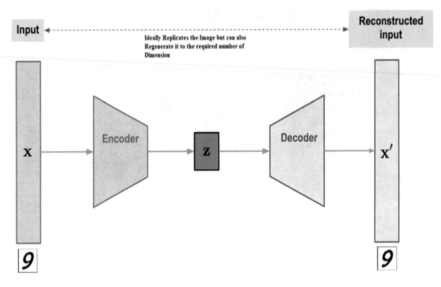

Fig. 5 Model architecture

4 Implementation

The authors have done the implementation of the model with the help of Jupyter Notebook environment which is provided to us by Anaconda. The model is based on the framework of variational autoencoders which have been implemented by the use of PyTorch. It also uses Matplotlib for the plotting of the training loss and the validation loss graph. The Python image Library of PIL is used for the processing of the images in the dataset.

To define the structure of the model, authors will be using the variational auto encoder model from Functional API. It consists of two major parts: Encoder—The feature extracted from the image has 2500 dimensions, with a dense layer, we will reduce the dimensions to 32. Decoder—By learning from the encoded input or the latent attributes, authors will process it to make the final prediction. The final layer will contain the images with the required number of dimensions that is 32. Visual representation of the complete model is as follows (Fig. 5).

5 Results

After the successful implementation of the model, the results are generated as shown in Fig. 6. All the font images in the upper row are the original images with 2500 dimensions and the images in the bottom row are the reduced images with just 32 dimensions. After 950 epochs, there is not much of a difference to be seen between

After 950 epochs:

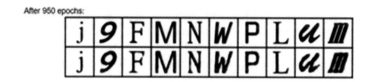

Fig. 6 Dimensionally reduced images

Fig. 7 Graph—Epochs versus losses

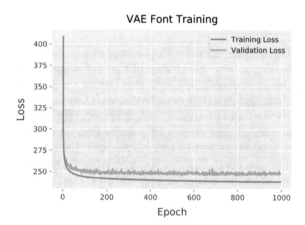

the original image and the dimensionally reduced images. The bottom row of the compressed images has been reproduced using the variational auto encoder, they have been decoded and reconstructed with only 32 dimensions according to the requirements of the authors.

In Fig. 7, the authors produced a graph showing the validation loss and the training loss against the Epochs the graph clearly shows the declining training loss and the validation with each epoch and within a range of 1000 epochs the validation loss falls from an approximate 320 to a near value of 250 and the training loss falls from 400 to something around 150.

Also, the above graph shows that the model is not overfit since the training loss is not much lower than the validation loss, also it shows that the model is not underfit since the training and validation losses are not equal.

6 Conclusion

In this paper, the authors have successfully developed a dimension reduction model with the use of variational autoencoders, the reduction model has then been trained and tested, further the authors have calculated the training loss and the validation loss. The model has successfully reduced the dimensions of the images of the fonts dataset without losing the important bits or the essence of the data. The model used a variational autoencoder setup, learning and improving after each epoch and reducing both the validation loss and training loss functions to a great extent. The usage of the python language has simplified the model. The algorithm has been scaled up with a GPU-led approach, achieving state-of-the-art results which are competitive and satisfying.

7 Future Scope

This project discusses the concept of dimensionality reduction which is already proving to be a vital concept of the field of machine learning, projects ranging from Tesla's Autonomous Car to Hand writing recognition are all using the concept of dimension reduction to make their data smaller and more understandable and hence getting faster processing with approximately equal accuracy and at the same time preventing their models from under training or over training. In the future, it can be used in the field of space exploration since the data collected from the rovers is huge and often difficult to process. Also, the authors plan to improve the project by ingesting live data in the code and giving live outputs for real time statistics.

References

1. Pu Y, Gan z, Henao R, Yuan X, Li C, Stevens A, Carin L (2016) Variational autoencoder for deep learning of images, labels and captions. http://papers.nips.cc/paper/6528-variational-aut oencoder-for-deep-learning-of-images-labels-and-captions
2. Sam T, Roweis, Saul LK (2000) Nonlinear dimensionality reduction by locally linear embedding. Science 290(5500):2323–2326
3. Tenenbaum JB, Silva VD, Langford JC (2000) A global geometric framework for nonlinear dimensionality reduction. Science 290(5500):2319–2323
4. Maaten LVD, Postma E, Herik JVD (2009) Experiment dimensionality reduction: a comparative review. https://members.loria.fr/moberger/Enseignement/AVR/Exposes/TR_Dimensiereductie. pdf
5. Rabbani M (2002) JPEG2000: Image compression fundamentals. Standards and practice. J Electron Imaging 11(2). https://doi.org/10.1117/1.1469618
6. Theis L, Shi W, Cunningham A, Huszár F (2017) Lossy image compression with compressive autoencoders. https://arxiv.org/abs/1703.00395
7. Verma, Kumar V, Tiwari PK (2015) Removal of obstacles in Devanagari script for efficient optical character recognition. In: 2015 International conference on computational intelligence and communication networks (CICN). IEEE
8. Variational Autoencoders—Jordan J. https://www.jeremyjordan.me/variational-autoencoders

Computational Intelligence for Image Caption Generation

Sahil Garg and Parv Dahiya

Abstract The problem of generating a description just by seeing an image has been of great interest to the developers. In recent years due to the advancements in the fields of deep learning and natural language processing (NLP), we are able to solve this problem. This problem can be divided into two sub problems. First, to retrieve the image feature vectors to create a vocabulary of all the objects in the image and second, to generate a meaningful caption by using those objects. We have used convolutional neural networks (CNN) and recurrent neural network (RNN) to achieve the solution to these problems. CNN architecture helps to extract features from the images and then these features are further passed to a RNN which uses long short-term memory (LSTM) units which generates an accurate and meaningful description for that image. The dataset used for training and testing of this model is Flickr8k, and the evaluation of the generated caption is done using the BLEU metric. The BLEU metric is an algorithm used for the evaluation of a machine translated text by comparing it with natural language texts for the same image.

Keywords Deep learning · Convolutional neural network (CNN) · Recurrent neural network (RNN) · Long short-term memory (LSTM) · BLEU

1 Introduction

Caption generation is the problem where we generate descriptive sentence for an image. This problem can be solved by using a deep learning model which consists of a CNN which helps in understanding the objects in the image and an RNN which is used to arrange those objects into words in such a way that it makes a meaningful sentence. Using these techniques, we have shown that these models are able to produce such fine captions, which can also be compared to the ones generated by a person. We also do not need to work on any sort of complex data to accomplish this task, we just need a dataset containing images which will be converted to numeric form for further processing. In order to evaluate the performance of the model, BLEU

S. Garg (✉) · P. Dahiya
Amity University Haryana, Gurugram, Haryana, India

metric is used. Image captioning has extensive set of applications and can be used in numerous ways like it can be an aid for blind peoples–to help them get a better understanding of what is in front of them and in self-driving cars. It can be used to improve the Google Search Images and can also be enhanced to work even for videos.

In the subsequent sections, we will be comparing the approach that we used to the already existing ones and how they are different from each other. We will also be explaining the in-depth approach that we used to create our model, followed by the implementation of our model and the results we achieved from the model. In the final section, we will be talking about the limitations and the scope of this technology.

2 Related Work

Image captioning has existed for a, while now and many researchers have been working on it and came up with various solutions. Multimodal neural language models, a research paper by Kiros et al. in which he implemented a feed forward neural network to predict the next word using the image and the previously predicted words [1]. Sutskever et al., in his publication, ImageNet classification with deep convolutional neural networks have shown the implementation of a neural network which consists of five convolution layers, followed by max-pooling layers, three fully connected layers and a 1000-class softmax layer. They used non-saturating neurons to make the training process faster [2]. Their network was successful enough to overcome the limitation of overfitting by introducing a regularization method known as dropout. Deshpande et al. proposed a convolutional network model in addition with RNN's LSTM units which performs machine translation and conditional image generation [3]. Toshev et al., in his paper-Show and Tell: A neural image caption generator, proposed a caption generation model which makes use of techniques such as machine translation and computer vision and is based on deep recurrent architecture [4] which generates natural sentences for a given image by selecting the words with the highest probability so that the generated caption matches the target objects for the given training image. Socher et al. introduced the database known as ImageNet which consists of millions of images and are much more accurate than the current image datasets [5]. It is organized according to the WorldNet hierarchy. Wang et al. proposed a multimodal neural network which can generate novel descriptions for an image [6]. It consists of two sub-networks: the first one is a deep recurrent neural network for caption generation, and the second one is a deep convolutional neural network for images. In a recent study, Fei-Fei proposed a multimodal recurrent neural network architecture which uses a dataset containing images and their respective sentence descriptions to learn about the inter-modal correspondences between visual data and languages [7]. Cho et al. introduced an attention model which learns to describe the content of an image automatically. The model is trained using standard back propagation techniques by maximizing a variational lower bound [8]. They made use of visualization technique to demonstrate how the model is able to detect the boundaries of objects, while generating the caption for the given image.

3 Methodology

3.1 Dataset

A large variety of datasets are available for the task of image captioning out of which most commonly used datasets are Flickr8k, Flickr30k and MSCOCO.

For this model, we have used Flick8k dataset which consists a total of 8092 images in which there are 6000 training, 1000 dev and 1000 test images. This dataset is made by the pictures which we see or use on a daily basis. There are five different captions present for a single image. The model will be trained on these captions for 6000 different images resulting in 30,000 different captions for training. If we face the problem of overfitting, then we can use a larger dataset like Flickr30k or MSCOCO, but for a larger dataset we require extensive memory and processing power.

3.2 Evaluation Metrics

For the evaluation of the model, we have used BLEU (Bilingual Evaluation Understudy) metric [9] which is used to evaluate the performance of machine translation. The model generates a caption for an image which it has never seen before and that generated caption is compared with another caption from the actual results for the same image. This comparison is done by using the BLEU metric.

3.3 Model Generation

Image Feature Extractor

We need to extract the feature vectors from the images as we cannot pass the image as it is to the model as it cannot interpret what it is. So, we need to convert the images to a fixed size vector by using the Inception-v3 model (Fig. 1).

This model can carry out classification on 1000 different classes of objects. But our goal is not to classify the objects but to convert each image to a fixed length vector. In order to do so, we need to remove the last layer, i.e., the softmax layer from the inception-v3 model.

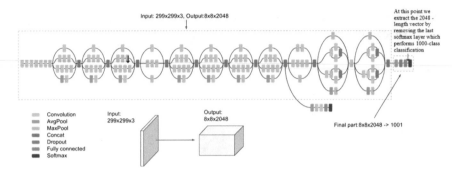

Fig. 1 Inception-v3 model [10]

Text Preprocessor

As the name suggests, text pre-processing is done to clean the captions before feeding it to the machine. After cleaning the captions, all the unique words are extracted from the training captions to generate a vocabulary. But as machine do not understand English words, each word in the vocabulary is mapped with a unique index value, and then those words are encoded into a fixed sized vector.

Output Predictor

The final model that we have created is the result of a series of Sequential models containing only one layer per model. Therefore, each sequential model can also be referred as a single layer for the final model.

Fitting the Model

When the model is prepared, it is trained using the training dataset. The model is made to run for a total of 90 epochs. Here, our goal is to minimize the below-mentioned loss function.

The loss function for our model can be defined as [4]:

$$L(I, S) = -\sum_{t=1}^{N} \log p_t(S_t) \tag{1}$$

where
  I: Input image
  S: Sentence describing the input image
  N: Length of the generated caption
  p_t: Probability of the words at time t
  S_t: Predicted word at time t

Fig. 2 LSTM model
combined with a CNN image
embedder and word
embeddings [4]

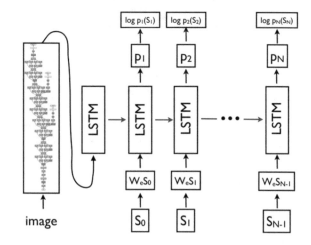

Caption Generation

After training the model, it will be tested by passing an image from the test dataset.
The image undergoes through the process explained earlier (Fig. 2).

At each time-step, the LSTM considers the previous cell output and returns a
prediction for the most likely next value in the sequence by selecting the word
which has the maximum weight for that particular iteration. The indexed word is
then converted to word and then added together to generate the final caption. The
generated image caption should not only contain the image object names, but also
their properties, relationships and functions.

4   Implementation

The implementation of the model is done using the Jupyter Notebook environment
which is provided by Anaconda. The framework on which the model is based upon
is Keras as we need to work with the images. It uses TensorFlow as its backend
for creating and training the neural network. In order to retrieve the image feature
vectors, we need to clean the descriptions so that the model do not create different
embedding vectors for insignificant words or symbols and also for the alphabetic
form of the same word. The feature vectors are extracted by passing the training
and testing images through a pretrained model called inception-v3. After that a new
model is defined as shown in Fig. 3 below and is trained on those feature vectors.
When the training is completed for the defined number of epochs, the model is used
to generate captions for the unseen data, i.e., the testing data. The generated caption
will be evaluated by comparing it against the already defined captions present in the
testing data.

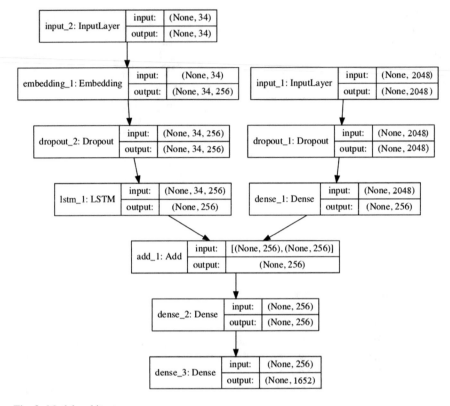

Fig. 3 Model architecture

5 Results

After the successful implementation of the model, the results are generated as shown in Fig. 4. We can clearly see that the captions generated for the images are pretty much relevant and are close to how a human being will define these images.

The BLEU score lies in the range of 50–60 which means that our model has generated high-quality translations.

But it will not be the case for every image. The model will sometimes predict a caption which is not the correct depiction of the image as shown in Fig. 5.

6 Conclusion

In this paper, the authors have created a model which is based on a CNN that encodes an image into dense feature vectors which are further passed on to an RNN that generates corresponding captions based on the learned image features. It performed

Caption: man on motorcycle rides dirt road   Caption: white bird flies over water

Fig. 4 Captions generated by the model

Fig. 5 Imprecise caption
generated by the model

Caption: boy in red shirt is walking on the beach

quite well on the test images. But also, there are some limitations in this model. It might not work precisely with the images having distant objects. So, there is still some room for improvement.

The model is using a greedy approach for the prediction of the next word as it selects the word with the maximum probability. This can also be improved by using Beam Search which selects a group of words with the maximum likelihood and parallel searches through all the other sequences. This might help increase the accuracy of the predictions.

7 Future Scope

Image captioning can be used for a variety of use cases. The first and the most important one in our perspective is that it can be an aid for the blind peoples. An application can be developed which convert the frames from a video to images and then convert them from text to speech so that the person will know which path to follow or toward what they are heading. Also, in robotics, image captioning can help the robot in gaining the insight of its environment by converting the visual feed through the camera into captions and further to machine language so that the robot can get a better understanding of its surroundings. If every image is captioned on the internet then it can also lead to faster searching and indexing.

References

1. Kiros R, Salakhutdinov R, Zemel R (2014) Multimodal neural language models. In: Proceedings of the 31st International conference on machine learning PMLR, vol 32(2), pp 595–603
2. Sutskever I, Krizhevsky A, Hinton GE (2017) ImageNet classification with deep convolutional neural networks. In: Proceedings of communication of the ACM
3. Deshpande A, Aneja J, Schwing A (2018) Convolutional image captioning. In: Proceedings of 2018 IEEE/CVF Conference on computer vision and pattern recognition
4. Toshev A, Vinyals O, Erhan D, Bengio S (2015) Show and tell: a neural image caption generator. In: Proceedings of 2015 IEEE Conference on computer vision and pattern recognition (CVPR)
5. Socher R, Deng J, Dong W, Li LJ, Li K, Fei-Fei L (2009) ImageNet: a large-scale hierarchical image database. In: Proceedings of 2009 IEEE Conference on computer vision and pattern recognition
6. Wang J, Yang Y, Huang Z, Xu W, Mao J, Yuille A (2015) Deep captioning with multimodal recurrent neural networks (M-RNN). In: Proceedings of A conference paper at ICLR
7. Fei-Fei L, Karpathy A (2017) Deep visual semantic alignments for generating image descriptions. In: Proceedings of IEEE Transactions on pattern analysis and machine intelligence, vol 39, issue 4
8. Ba JL, Cho K, Courville A, Kiros R, Xu K, Salakhutdinov R, Bengio Y, Zemel RS (2015) Show, attend and tell: neural image caption generation with visual attention. In Proceedings of the 32nd International conference on machine learning, PMLR, vol 37, pp 2048–2057
9. Papineni K, Roukos S, Ward T, Zhu WJ (2002) BLEU: A method for automatic evaluation of machine translation. In Proceedings of IBM TJ Watson research center yorktown heights, NY 10598, USA
10. Advanced guide to inception v3 on cloud TPU–Google cloud. https://cloud.google.com/tpu/docs/inception-v3-advanced

An Empirical Statistical Analysis of COVID-19 Curve Through Newspaper Mining

Shriya Verma, Sonam Garg, Tanishq Chamoli, and Ankit Gupta

Abstract In India, thousands of suspected cases have been tested positive of corona virus resulting in more than 10 M confirmed corona virus cases. The newspaper played a vital role in supplementing abundant and accurate information through corona-related articles. This research paper presents a detailed analysis toward the rising coverage of COVID-19 and its related terms aiming at newspapers like "THE HINDU" using the base technology text mining. The PDFs of the newspaper are extracted through automation and converted to text files free from any ambiguous characters. The graphs are plotted for data visualization depicting the percentage coverage of COVID-19-related words and lines from March 2020 to July 2020 in the newspaper. The objective is that the given idea can be employed on various such topics and provide us refined information that can be easily used or represented graphically.

1 Introduction

Newspapers have been a great resource for data analysts as they provide a lot of information and go back to the very old era as well. With the boom of the Internet, huge volumes of unorganized data (also called "big data") are available online. Electronic newspapers also called e-newspapers are increasingly being read by users from anywhere and anytime. According to Reuters, Institute for the Study of Journalism [1] surveyed that the 56% of the population consider online as their primary source and only 16% consider print newspapers but it was easily overtaken by the 28% people who identify social media such as Facebook, Twitter or Youtube as their main source of information and daily news. The readership of e-papers soared post the lockdown and the major reason was the problem caused due to the lockdown in the newspaper distribution system which put a spotlight on the e-editions of dailies. According to a joint study by Broadcast Audience Research Council of India (BARC) and Nielsen India [2], in March, the platform of news aggregation had concluded that around

S. Verma · S. Garg · T. Chamoli (✉) · A. Gupta
Chandigarh College Of Engineering And Technology (Degree Wing), Chandigarh, India

© The Author(s), under exclusive license to Springer Nature Singapore Pte Ltd. 2022
P. Nanda et al. (eds.), *Data Engineering for Smart Systems*, Lecture Notes in Networks and Systems 238, https://doi.org/10.1007/978-981-16-2641-8_26

40,000–50,000 downloads were done per day, which went up from 20,000–25,000 downloads in February. The data could be extracted and gathered manually from the newspaper but this process remains error prone; the solution to the problem can be text mining. Thus, this research paper aims to make an Open Source Intelligence Tool (OSINT) for information gathering on the basis of the newspaper.

Text mining is the process of extracting useful knowledge from the text documents [3]. It helps users find useful information from a large amount of digital text data [4]. It is a challenging issue to seek out the most appropriate and accurate knowledge (or features) in text documents [3]. Many text-mining techniques have been developed in order to achieve the goal of retrieving useful information for users [4–8]. These techniques are developed to unravel the matter of text mining that is nothing but the relevant information retrieval according to user's requirement.

In this research paper, the data is collected through web scraping and the PDFs are downloaded. The data from the PDF file is extracted in the form of a text file where multiprocessing is used to reduce the time for the extraction. The text files are cleaned, processed, and visualized in the form of graphs to understand trends.

Rest of the paper is as follows: Sect. 2 discusses about the related work on newspaper mining. Section 3 provides an insight to the experimental process and analysis. Section 4 concludes the paper.

2 Motivation and Some Earlier Work

Newspaper mining is the application of data mining techniques for data analysis/interpretation. In past few years, interest and activity have been seen in applying computational methods to extract knowledge from collection of documents. For readers, there is a unique, personal value in news articles that are written and published by people from within the country. However, the value cannot be easily extracted from these articles. These archives are typically unstructured [9]. While traditional attempts to analyze these archives are limited in processing large amounts of data, text mining presents a set of approaches that allow researchers to explore large-scale collections of texts in an efficient manner [10]. Various researches have been carried out in this domain.

Kim et al. [11] used newspapers as dataset and proposed an approach to investigate keywords related to medical tourism in daily and medical newspapers and to analyze networks between keywords.The purpose of the study was to understand the perspectives of the public and the medical community on medical tourism using text mining.

Matto et al. [12] used newspapers, social media, and other similar platforms as a source of crime data that are not necessarily reported in police stations. The objective was to mine frequently reported crimes, to investigate on the distribution of crimes per regions, and to generate association rules between the mined crimes. Shahzad et al. [13] display studies clandestine proliferation of diseases among favorable inhabitants, propagation speed, and surreptitious properties may potentiate a disease to emerge as pandemic such as COVID-19, which due to high contig-stimulus and furtiveness, emulated as contagion and risked immortality of mankind. Liu et al. [14] adopted the WiseSearch database to extract related news articles about the corona virus from major press media between January 1, 2020, and February 20, 2020. The data was sorted and analyzed using Python software and Python package Jieba. The number of articles published per day has been listed.

Hanumanthappa [15] used the technique of information retrieval to extract useful and relevant information from e-newspapers because they are becoming more easy to access nowadays and are a source of timely information and news about what happens all around the world. They preferred to use e-newspapers because many different newspapers present news on the same topic but with different perspectives.

Hanumanthappa [16] used e-newspapers by analyzing that from past few years, the use of e-newspapers has significantly increased. The system focuses mainly on extracting PDF documents into text format for easy reading.

Jung et al. [17] proposed a study that aims to find out the pattern of COVID-19-related news to minimize the pandemic. The concept of big data is utilized to focus on the relevant news in the given period. The results of the analysis are as follows. The articles of COVID-19 rose more than 100 indicated that our society is watching with great interest in the government's response to the disease.

3 Experimental Setup

3.1 Methodology

The research is based on text mining to automatically process data and generate valuable insights, enabling to make data-driven decisions. Text mining serves basis for the report as it provides a complete pathway to enrich and analyze content. According to Hotho et al. (2005), three different perspectives can be differed on the basis of text mining: information extraction, data mining, and a knowledge discovery in databases (KDD) process. Uma [18] told all about knowledge discovery in databases (KDD) which helps various organizations to convert their collection of data into various datasets according to the requirements of valuable and relevant information [19]. Text analysis involves: Information retrieval (IR), lexical analysis, pattern recognition, tagging/annotation, and information extraction.

3.2 Choice of Data

Among newspapers of all sorts of publications and languages, "The Hindu" [20] was the first choice as it had a big number of readers. As per the data released by the MRUC, "The Hindu" had an all-India total readership of 62,26,000 in the IRS 2019.

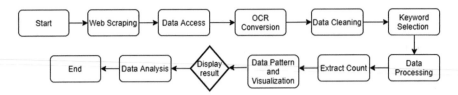

Fig. 1 Flowchart

So, this was chosen and searched for the site which could provide free PDFs of the newspaper and found the site "dailyepaper.in".

The data is collected through web scraping. In order to save time from downloading each newspaper manually by clicking the link on web, automation technique is used which parses through each URL and searches for keywords provided in the code by generating an API request and stores links in a text file. The library beautiful soup is used to make the web scraping process handy.

3.3 Brief Overview

This project aims at first collecting data through web scraping. The files are downloaded using the wget module which refers to the API links stored in text files. Data is cleaned to bring efficiency in the data for better results. The text files thus obtained are free from UTF-8 characters and contain simple text. These cleaned files are sent for data processing. The percentage of total words and sentences are calculated which gives a proper understanding of rising and declining COVID-19-related articles. The data is visualized in the form of graphs to understand trends, outliers, and patterns. The flowchart Fig. 1 given below describes the whole project.

3.4 Data Extraction

Web Scraping

Ashiwal et al. [21] developed a method of retrieving all the relevant information using the library of Python called as beautiful soup. They concluded that most of the information present on the web is in the form of unstructured data.

This method just makes the use of two parameters which are URL path of the file to be downloaded and second is local path of the file where it is stored in PC which has made extracting the links more facile.

Our motive for this program was simple and straight forward, i.e., to extract the data from the PDFs into a text file and then being able to use that data as many times for the analysis. So the program was made to do this task.

The mundane task includes downloading each file manually where each file takes approximately thirty seconds to download plus the extra time taken to visit through each link. The intelligent and faster way is to use the functionalities of Python.The extracted URLs requested by the web servers are stored in the text file. These links are accessed and referred for downloading files. A special Python library called *w*get is used. *G*NU Wget is a computer program that retrieves content from web servers.

This process is very efficient when data to be downloaded is very large.

The code also checks if the file is already present in the designated folder; it is done using the os.path module. If it is there, it prints exists otherwise downloads it using *w*get library. It matches with the link name. In the end, the name of the downloaded file is printed on the console.

3.5 Data Cleaning

In our research paper, the text files thus obtained are free from UTF-8 characters and contains simple text. The text files obtained are UTF-8 encoded capable of encoding all 1,112,064 valid character code pointed in unicode using one to four one-byte (8-bit) code units. The newspapers consist of all form of characters which are not required in text file. Moreover, the text files contain irrelevant data that has no connection with analysis.

In the given project, the text file obtained after web scraping is opened in read mode which contains links that have been parsed through the mentioned link. To get the name of the newspaper from the parsed link, split function in Python is used and the link is being split from/and name of the newspaper is printed. Folder in which files have been extracted is opened in read binary mode and it is further decoded in "UTF-8." The output is converted into strings and then new lines and the specified characters in the code are replaced by empty space so that text is cleaned properly. Data after cleaning is written into new text files and saved in a folder, and at last, the file is closed.

3.6 Data Pre-Processing

The information is not couched in a manner that is amenable to automatic processing. That is why, the cleaned files are sent for data processing. Data processing is the collection and manipulation of items of data to produce meaningful information. Data processing involves various processes, including:

1. Validation—The newspaper chosen "The Hindu" was the most trusted source of information.
2. Sorting—The PDFs of every newspaper were collected in different folder, and after the conversion of them into text file,they were stored in different folder.

3. Summarization—Removing redundant words reduced detailed data to its main points.
4. Analysis—the "collection, organization, analysis, interpretation, and presentation of data" was performed on later stages.

The cleaned data is analyzed to count the keywords that have occurred with respect to total words in every newspaper dating from March 01, 2020, to July 20, 2020. Keywords chosen are COVID-19, corona, pandemic, corona warriors, incubation period, community spread, n95, quarantine, isolation, epidemic, flattening the curve, comorbidity, social distancing, hydroxychloroquine, aarogya setu app, lockdown, lockdown extension, virus, infection, airborne, cough, fever, masks. Keywords represented the frequently and most queried words throughout the crisis. Choice of keywords increased the credibility of the research. Rest, whole process of opening, reading, processing the files remains same.

Total number of keywords are counted and percentage is calculated as: [(total number of keywords)/(total number of words)]*100.

Any number of keywords occurrence pattern can be found on any current debating issue. Result obtained can help in fostering data in another research.

3.7 OCR Conversion

OCR conversion is a useful process which is used for transforming the data that is available in the form of image into encoded data and well-defined machine readable form. The data on an image is identified using detection algorithms.

Hamad et al. [22] gave detailed analysis of OCR technology. OCR is a technology that extracts the text from the images. In our project, this is used for those PDFs that are found to be in the form of images rather than text form. Pytesseract library is used for extracting the text from the images and PIL-Python imaging library is used for manipulating and enhancing the images.

The library *p*df2image was used as well in this so that tesseract can be used on the images extracted from the PD.

The code snippet which converts the PDF into a number of images (Fig. 2)-

In this code, the PDF is converted using the convert from path to an image of every page with only 100 dots per cm cube. The number of dots was tried and had to choose better quality to obtain good results but will take more than 5 min to process a single page. Various values were tested, and it was found that 100 dots per inch took only 15–18 s to process the image of size 300–400 kb. So images were created of the PDF and data was extracted from it, and on moving to the next PDF, the previous images were deleted to keep the space constant.

```
pages = convert_from_path(Path+file_name, 100)
counter=0
for page in pages:
    counter+=1
    print("image : ",counter)
    if os.path.isfile(Path+"out"+str(counter)+".jpg"):
        os.remove(Path+ "out"+str(counter)+".jpg")
        page.save(Path+"out"+str(counter)+'.jpg', 'JPEG')
```

Fig. 2 Code Snippet

3.8 Data Export

One of the libraries of Python-xlwt is being used to export the listed data to an excel file.

The tempfile module in Python is used for working with temporary files, folders, and directories. After importing xlwt, a temporary file is made in excel and data is stored in that file. To prevent overwriting into the excel sheet, "cell overwrite ok" is used. One of the major disadvantage is that exported data can only be added but cannot be modified. Further, Pandas library is used to convert the excel file to csv file.

R language has influenced text mining with various statistical approaches and has improved graphics which further helps in analysis and integration of data at a large scale.

The csv file is accessed in R, and tables and graphs are concluded on the basis of imported data. Process of graphical representation of data is called data visualization.

Therefore, by using visual elements like charts/graphs, data visualization tools provide an accessible way to see and understand trends, outliers, and patterns in data. The tables are shown in Figs. 3 and 4.

Date of month	Selected words	Total words
23/03/2020	326	48700
24/03/2020	378	50632
..	..	..
..	..	..
14/07/2020	121	41074
15/07/2020	102	39680

Fig. 3 Dataset of count of words

Date of month	Selected lines	Total lines
23/03/2020	302	4113
24/03/2020	348	5861
..	..	..
..	..	..
14/07/2020	116	5344
15/07/2020	104	5277

Fig. 4 Dataset of count of lines

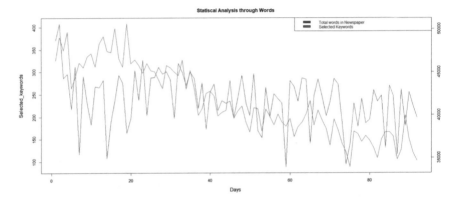

Fig. 5 Graph of count of words

The table in Fig. 3 shows the dates of every month with words which are related to "COVID-19" and total no. of words excluding the bad words and after cleaning of text in each newspaper since March 20, 2020, till July 15, 2020.

Graph made in Fig. 5 shows the trend of COVID-19 words found in newspapers and total no. of words that were present. Red line shows words that were related to COVID-19 and black line shows total number of words in the newspaper.

The table in Fig. 4 includes dates of each month with all the sentences which include words related to "COVID-19" and total number of sentences in each newspaper since March 20, 2020, till July 15, 2020.

Similarly, graph in Fig. 6 shows the trend of sentences that were found containing words related to COVID-19 and total sentences.

3.9 Analysis

The dataset visualized after the formation of graphs depicts an overall decrease in the newspaper articles related to COVID-19 from the month of February 2020 till July 2020. When the lockdown was declared in India, there was a lot of panic in

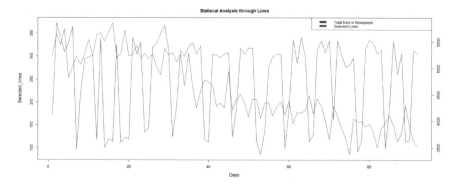

Fig. 6 Graph of count of lines

the public and this can clearly be seen, when there was a sudden increase in the number of mentions in the newspapers in the month of March. This curve did go up a lot of times like when PM Narendra Modi interacted with the people in Mann ki Baat or during the Tabilighi Jammat where 4,500 people assembled and many more incidents . This curve did rise and fall but as time passed, the citizens and the news about COVID-19 kept on falling down as the people also got more accustomed and started adapting the new locked lifestyle.

4 Conclusion

The designed project analyzed the increasing coverage of COVID-related words initially, and later, the decrease was observed. The given project will not only help in analyzing the COVID-19 trend in news but on any other issue as well. The given project strives to bring result out of the text in a form that is suitable for consumption by computers directly, with no need for a human intermediary. It will help to get real insight about various domains. The programmer of the research has to just change the keywords related to any issue and can easily analyze the trend of that issue. It gives an overlay of when/how/to what extent the issue was discussed. The project has the potential to be applicable in various fields. Various other research topics could be administered in different areas for instance in the case of lawyers or cyber security analysts, etc. keeping the project code as the base.

References

1. https://scroll.in/article/937657/the-future-of-indias-newspapers-has-to-be-digital-and-it-has-to-be-now

2. https://www.financialexpress.com/brandwagon/coronavirus-impact-digital-news-outlets-see-growth-spurt/1928767/
3. Zhong N, Li Y, Wu S (2012) Effective pattern discovery for text mining. IEEE Trans Knowl Data Eng 24(1):30–44, 30. https://doi.org/10.1109/TKDE.2010.211
4. Inje B, Patil U (2014) Operational pattern revealing technique in text mining. In: 2014 IEEE students' conference on electrical, electronics and computer science, Bhopal, pp 1–5. https://doi.org/10.1109/SCEECS.2014.6804509
5. Sujatha GS, Poonguzhali E (2014) Text mining using pattern taxonomy model for effective pattern discovery. Int J Eng Res Technol (IJERT) NCRTS—2014 2(13)
6. Aas K, Eikvil L (1999) Text categorization: a survey. Technical report. Norwegian Computing Center, Raport NR 941
7. Edda L, Jorg K (2002) Text categorization with support vector machines. How to represent exits in input space? Mach Learn 46:423–444
8. Lam W, Ruiz ME, Srinivasan P (1999) Automatic text categorization and its application to text retrieval. IEEE Trans Knowl Data Eng 1(6):865–879
9. Rahman HMT, Sherren K, van Proosdij D (2019) Institutional innovation for nature-based coastal adaptation: lessons from salt marsh restoration in Nova Scotia Canada. Sustainability 11:6735
10. Ritala P, Schneider S, Michailova S, Innovation management research methods: embracing rigor and diversity. R&D Management. 50. 10.1111/radm.12414
11. Kim S, Lee W, Network text analysis of medical tourism in newspapers using text mining: the South Korea case. Tour Manag Perspect 31:332–339. https://doi.org/10.1016/j.tmp.2019.05.010
12. George M, Joseph M (2017) Int J Knowl Eng Data Min 4(2)
13. Shahzad MA, NVIVO based text mining on COVID-19 studies: precision technologies and smart surveillance may help to discover, and coup COVID-19 transmogrify. SSRN Electron J. https://doi.org/10.2139/ssrn.3623499
14. Liu Q, Zheng Z, Zheng J, Chen Q, Liu G, Chen S, Chu,B, Zhu H, Akinwunmi B, Huang J, Zhang C, Ming W-K, Health communication through news media during the early stage of the COVID-19 outbreak in China: a digital topic modeling approach (Preprint). J Med Int Res 22. 10.2196/19118
15. Hanumanthappa M (2014) A study of information extraction tools for online English newspapers (PDF): comparative analysis. In: An ISO 3297: 2007 certified organization, vol 2, Issue 1, Oct 2014
16. Hanumanthappa M (2015) Identification and extraction of different objects and its location from a Pdf file using efficient information retrieval tools. In: International conference on soft computing and network security (ICSNS-2015), 25–27 Feb 2015
17. Jung Ji-Hee, Shin Jae-Ik (2020) Data cleaning: current approaches and issues. Int J Environ Res Publ Health 17(16):5688
18. Uma K, Hanumanthappa M (2017) Data Collection methods and data preprocessing techniques for healthcare data using data mining. Int J Sci Eng Res 8(6)
19. Agnihotri D, Verma K, Tripathi P (2014) Pattern and cluster mining on text data. In: 2014 fourth international conference on communication systems and network technologies, Bhopal, pp 428–432. https://doi.org/10.1109/CSNT.2014.92.
20. https://dailyepaper.in/the-hindu-pdf-newspaper-free-download/
21. Pratiksha A, Tandan, Tripathi P, Miri R (2016) Web information retrieval using python and beautifulsoup. Int J Res Appl Sci Eng Technol (IJRASET) 4(VI). ISSN: 2321-9653
22. Karez AH, Mehmet K (2016) A detailed analysis of optical character recognition technology. Int J Appl Math Electr Comput

Wearable Monopole Antenna with 8-Shaped EBG for Biomedical Imaging

Regidi Suneetha⃝ and P. V. Sridevi

Abstract In this paper, Electromagnetic Band Gap (EBG) backed dual band monopole antenna that is radiating efficiently is being presented. A planar monopole antenna with a single EBG unit operating at microwave frequencies between 1–10 GHz is proposed, to be useful for wearable biomedical applications. The EBG structure resembles number 8. The gap between the EBG unit and the ground plane is adjusted to get the required results. The dimensions of the antenna are 50 × 56 × 1.6 mm^3 fabricated on FR4 material having 4.4 relative permittivity and 0.02 loss tangent. The obtained results of various performance parameters from simulations and fabrications prove that the antenna can be used for wearable biomedical applications.

Keywords Electromagnetic band gap (EBG) · Microwave frequencies · Microwave imaging monopole antenna · Wearable biomedical applications

1 Introduction

Cancer is one among the major deadly diseases and the complete process starting from diagnosis to treatment is a very expensive and painful process filled with trauma. In few cancers like breast cancer, the survival rate is high, up to 97% with early detection and treatment. Computed tomography (CT) scan, magnetic resonance imaging (MRI), ultra-sonic imaging (US) are few highly expensive, highly efficient and reliable techniques used in cancer diagnosis. Apart from being costly, these techniques also have few limitations like missed detection, sometimes false positive and it is uncomfortable, time-consuming for the deep-lying or solid tumor due to compression and ionization. From the literature, it is obtained that the highest rate of accuracy is 75.6% only, so it is pretty much required to have a complementary technique that must be portable, efficient and non-ionizing at a comparatively low cost. Over last

R. Suneetha (✉) · P. V. Sridevi
Andhra University College of Engineering (A), Andhra University, Visakhapatnam, AP, India
e-mail: rsuneetha@rocketmail.com

© The Author(s), under exclusive license to Springer Nature Singapore Pte Ltd. 2022
P. Nanda et al. (eds.), *Data Engineering for Smart Systems*, Lecture Notes in Networks and Systems 238, https://doi.org/10.1007/978-981-16-2641-8_27

few years, the microwaves operating up to 10 GHz in the fields of biomedical applications like biomedical imaging is gaining a considerable amount of interest, by 2021, 6 GHz is going to rule and the European Commission regulations [1–3] may permit 5925–6425 MHz (US U-NII-5 band) band operations using very low-power devices, and the Federal Communications Commission [1–3], U.S. defined ultra-wide band [4] frequency range of 3.1–10 GHz. Particularly during diagnosis of tumor and ruling out the type of the tumor whether it is malignant or not in breast and lung cancer detection, medical diagnostic imaging [7] is used to detect the damage, location or movement within the body such as bone fracturing, skin lesions, brain stroke as well. Microwave imaging also finds a wide range of applications in treatment like hyperthermia and various other treatments. In this paper, a planar monopole antenna and a monopole antenna with EBG [5] structure are designed and fabricated. A single unit of EBG structure in the shape of 8 is used, that can be placed on human body commonly used in wireless body area network applications, ultra high frequency (UHF), ultra wide band (UWB) applications etc.

2 Antenna Designs

The monopole antennas are designed and simulated using HFSS Software and fabricated on Fr4 substrate with dielectric constant of 4.4 and loss tangent of 0.02 with 50Ω microstrip line feeding. Fabricated monopole antennas are tested and results are obtained from Anritsu Vector Network Analyzer. EBG structure is used to enhance the performance of the antenna, the proposed antenna EBG structure is backed in the shape of 8. The antenna size [6] is reduced by imprinting the substrate either sides, with monopole antenna [8–10] on one side and partial ground [11, 12] along with EBG structure on the other side.

2.1 Design of Multi Band Monopole Antenna

The dimensions of the substrate are $50 \times 56 \times 1.6 \text{ mm}^3$, the structure and dimensions [6] of a monopole antenna that is imprinted on upper side and the ground on the lower side of the substrate are as shown in Fig. 1. The structure designed for multi band frequencies between 1 to 10 GHz and the fabricated monopole antenna is as shown in Fig. 2.

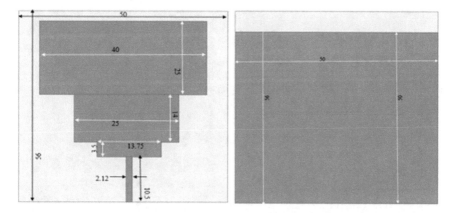

Fig. 1 Top and bottom view of a monopole antenna with dimensions

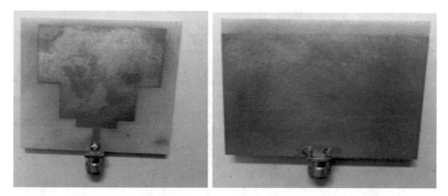

Fig. 2 Fabricated monopole antenna

2.2 Design of Double Band Monopole Antenna with EBG Structure

EBG structure with dimensions of 45 × 44 mm^2 in the shape of 8 is used and the partial ground with dimensions of 50 × 4.5 mm^2 using a layer of copper on same side of the substrate with monopole patch on the other side are as shown in Fig. 3. The antenna is radiating with dual band, impedance bandwidth obtained is from 1.5 to 2.2 and 5.8 to 6.9 GHz which is 700 and 1100 MHz, respectively, find applications in ISM, UWB and 6 GHz biomedical applications. The fabricated monopole antenna with EBG structure is as shown in Fig. 4.

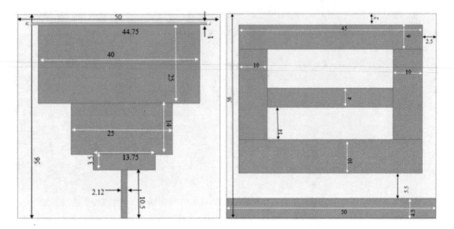

Fig. 3 Top and bottom view of a monopole antenna with EBG structure

Fig. 4 Fabricated monopole antenna with EBG structure

3 Discussion of Results

The simulated and measured results of monopole antenna and monopole antenna with EBG are presented in this section. They both agree to a maximum extent and the difference in the results may be due to the increase in permittivity value of the substrate material as FR4 is a lossy material, but it is mostly preferred due to its several advantages like ease of availability and low cost etc.

3.1 Return Loss (S₁₁)

Figures 5 and 6 show S_{11} versus frequency plot of monopole antenna and monopole antenna with EBG structure, from the observed results of S_{11} it can be concluded that with the inception of the EBG structure the antenna impedance bandwidth is increased and the S_{11} value has decreased, compared to other antennas it can be used in ultra-wideband wireless applications and microwave imaging in biomedical applications.

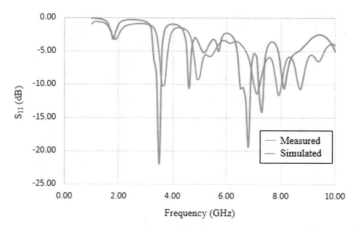

Fig. 5 S_{11} versus frequency plot of simulated and measured results of monopole antenna

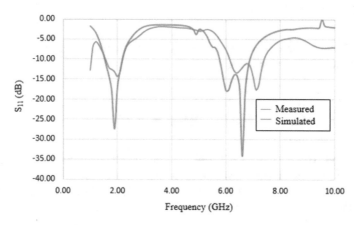

Fig. 6 S_{11} versus frequency plot of simulated and measured results of monopole antenna with EBG structure

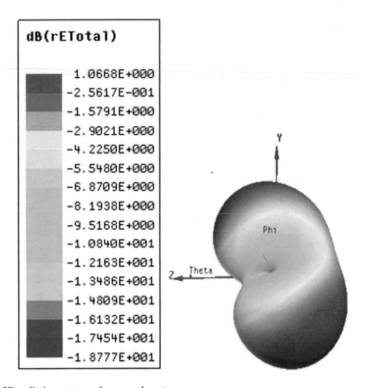

Fig. 7 3D radiation pattern of monopole antenna

3.2 Radiation Performance

Figures 7 and 8 show the radiation pattern of monopole antenna and monopole antenna with EBG, respectively.

3.3 VSWR(Voltage Standing Wave Ratio)

Figures 9 and 10 show the VSWR versus frequency plot of simulated and measured values of monopole antenna and monopole antenna with EBG structure. It can be seen in the regions of operation the VSWR value is low and <2 which is a prominent requirement of an antenna to perform well.

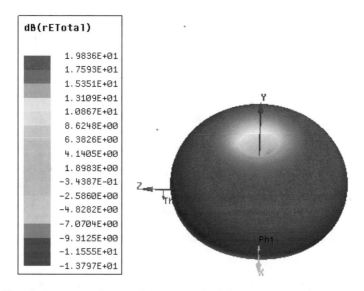

Fig. 8 3D radiation pattern of monopole antenna with EBG structure

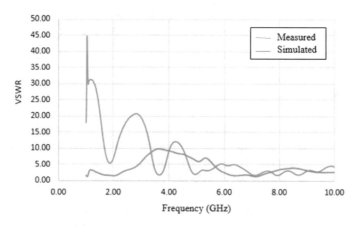

Fig. 9 VSWR versus frequency plot of simulated and measured results of monopole antenna

3.4   Comparison of Monopole Antenna and Monopole Antenna with EBG Structure

When the results of the monopole antenna and monopole antenna with EBG structure are compared as shown in Table 1, the monopole antenna results obtained are with multi band, but when compared with the dual band monopole antenna with EBG structure, there is an increase in bandwidth, gain and decrease in S_{11}. Radiation pattern obtained is also omni directional which is much preferred in biomedical antennas.

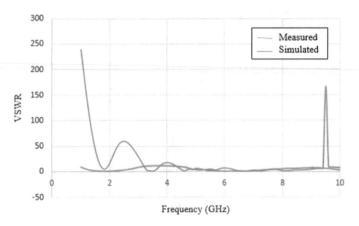

Fig. 10 VSWR versus frequency plot of simulated and measured results of monopole antenna with EBG structure

Table 1 Comparison table of monopole antenna without and with EBG structure

Parameter	Monopole antenna	Monopole antenna with EBG structure
Bandwidth	100, 500 MHz	700, 1100 MHz
Return loss(S_{11})	−21.94, −19.36 dB	−27.48, −34.21 dB
Gain	2.35 dB	6.35 dB

4 Conclusion and Future Work

As the human body is a combination of different layers and materials like blood, bones, muscles, skin etc., due to the non-homogeneous nature of the body it is obvious of having different dielectric constants. The significant variation of permittivity affects the interface between skin and free space. To study and maximize the coupling effect between the antenna and human body parts like head, chest, hand etc., for good reception of data, the antenna is placed in mediums with different values of permittivity and simulation results are obtained as shown in Figs. 11 and 12. The antenna performance can be analyzed for various permittivity values to improve the antenna design for better results. The minimum value of S_{11} is increased as the permittivity of the medium is increased also the impedance bandwidth is varied. It can be concluded that the impedance bandwidth of the antenna when placed in a medium, is varied randomly as the value of the dielectric constant of the medium is increased. This monopole antenna with EBG is more advantageous and can be used for microwave imaging as it covers the required frequency range of operation. The antenna performance can be varied with respect to the placement of the antenna and good results can be obtained if placed at values of low-dielectric constant. By making slight variations in ground structure [11, 12] and EBG structure, also considering the

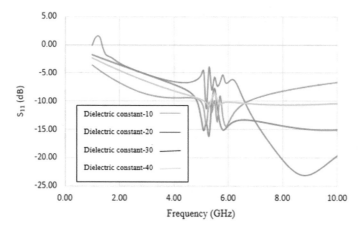

Fig. 11 S_{11} versus frequency plot of simulated results of monopole antenna in different medium

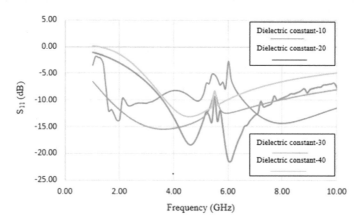

Fig. 12 S_{11} versus frequency plot of simulated results of monopole antenna with EBG in different medium

medium and the value of the dielectric constant of the medium multi band can be attained for UWB and microwave imaging applications.

Acknowledgements The author with Unique Awardee Number MEITY-PHD-2245 would acknowledge Visvesvaraya Ph.D Scheme, DeitY, New Delhi for providing the fellowship and Andhra University College of Engineering (A), Andhra University, Visakhapatnam for providing facilities to conduct research activity.

References

1. FCC opens 6 GHz band to Wi-Fi and other unlicensed uses. FCC.gov
2. Europe prepares to harmonize the 6 GHz spectrum band for Radio Local Area Networks. ECC newsletter. Aug 2019
3. Ofcom UK to make 6GHz band available for faster home Wi-Fi." ISP review. 17 Jan 2020
4. Sheikh S, et al (2013) Directive stacked patch antenna for UWB applications. Int J Antennas Propag vol. 2013, Article ID 389571, p 6
5. Rahmat-Samii Y (2008) Electromagnetic band gap (EBG) structures in antenna engineering: from fundamentals to recent advances. Asia-Pacific microwave conference, Macau, pp 1–2
6. Balanis CA (2016) Antenna theory: analysis and design. Wiley India edition, third edition, pp 811–876
7. Scapaticci R, Di Donato L, Catapano I, Crocco L (2012) A feasibility study on microwave imaging for brain stroke monitoring. Prog Electromagn Res B 40:305–324
8. Ray KP (2008) Design aspects of printed monopole antennas for ultra-wideband applications. Int J Antennas Propag
9. Thakkar Y, Lin X, Chen Y, Yang F, Wu R, Zhang X (2019) Wearable monopole antennas for microwave stroke imaging. In: 13th European conference on antennas and propagation, no EuCAP, pp 1–5
10. Yan S et al (2015) Wearable dual-band magneto-electric dipole antenna for WBAN/WLAN applications. IEEE Trans Antennas Propag 63(9):4165–4169
11. Peng S, Zheng H (2015) Design of coplanar-waveguide-feed antenna. Int J Eng Res Technol 4(7):1171–1177
12. Samal P et al (2014) UWB all-textile antenna with full ground plane for off-body WBAN communications. IEEE Trans Antennas Propag 62(1):102–108

Speech Signal Processing for Identification of Under-Resourced Languages

Shweta Sinha

Abstract The present research paper addresses the problem of language identification (LID) in short-time utterances. The success of support vector machine in acoustic modeling for speech recognition and accent recognition motivated the author to extend its applicability to LID. Acoustic model and the classifier play an important role in system performance. The paper presents the LID task by the use of state-of-the-art i-vector system using MFCC-SDC as acoustic feature and support vector machine (SVM) as backend classifier. Both the systems have been implemented for the identification of two Indian languages (Hindi and Punjabi). The speech utterances are divided into short segments of 5, 10, 20 and 35 s. The performance of the system is measured in EER (%), and the results obtained highlights the efficiency of the model in short duration utterances.

Keywords LID · i-vectors · Support vector machine; MFCC-SDC

1 Introduction

Automatic language identification deals with identification of language in the uttered speech utterances by the means of computer algorithms and logics. The purpose is to identify which of **L** languages does the spoken utterance **S** belong to. With the rise of globalization trends need for multilingual systems have come into existence. Even the human communication with machines have shifted from mouse based to voice based [1]. The demand for multilingual speech recognition poured in for the demand for language identification system to make the system performance better.

The need for LID is rising at a very fast pace. In this era of globalization, the service sectors be it travel, healthcare or hospitality deals with multilingual translation systems. The translation systems that deal with multiple languages would use LID system at the first stage to identify the language of the spoken utterance and direct to language-specific translation/recognition system. Apart from this, the LID systems

S. Sinha (✉)
Amity University, Gurugram, Haryana, India
e-mail: meshweta_7@rediffmail.com

© The Author(s), under exclusive license to Springer Nature Singapore Pte Ltd. 2022   291
P. Nanda et al. (eds.), *Data Engineering for Smart Systems*, Lecture Notes in Networks and Systems 238, https://doi.org/10.1007/978-981-16-2641-8_28

can be used by police for emergency assistance, in forensic studies, in the tourism sector and also in call routing services for request routing. In a nation like India with multiple languages spoken across the country, the demand for multilingual speech/language ID system is essential.

This paper presents a robust techniques for LID. The prominence of i-vector technology in automatic speaker identification works as a motivating factor for its application in LID. Several pieces of research highlight the application of i-vector front end features with advance classification mechanism that takes into consideration the speaker and session variability [2–4]. The i-vector is obtained by the use of factor analysis technique that maps the sequence of frames into a low-dimensional fixed-length vector space and is the compact representation of a whole utterance. This low-dimension space is known as total variability space. This is a date-driven technique and has been configured for LID from Indian multilingual (Hind and Punjabi) corpus to achieve reasonable performance for spoken utterances. Equal number of utterances have been used for both the languages.

The rest of the paper is organized as follows: Sect. 2 presents the background research done for LID and analyses the speech features and techniques used to date. Section 3 presents a description of the corpus used in the work. Section 4 presents the description of i-vector based system. Results and evaluation are discussed in Sect. 5, and the paper completes with a conclusion and future work in Sect. 6.

2   Background Research

Automatic LID is carried out in two remarkable phases. The front-end being the first phase, and the backend classifier constitutes the second phase. The front-end processing deals with the training of the system. The training can be done in the simplest way using the supervised learning approach or it can follow a complex methodology that demands for a phonetic transcription or furthermore orthographic transcription along with pronunciation dictionary. The digital processing presents the speech signal as vectors that contain all the acoustic, prosodic and phonetic features [5] engrossed in the signal.

The review paper [6] highlights the use of spectral features for sample representation in initial research and the identification of the language was done using template matching [7]. After series of researches were performed it was perceived that MFCC, LPC, PLPs are a few prominent acoustic features used for this purpose [8]. Further to capture additional temporal information about the speech vectors that is embedded in the signal, researchers have included shifted delta cepstral (SDC) coefficients. The inclusion of SDC over large frames are motivated by the phonetic concept of using multiple frame span [9]. In the recent past, i-vector has given promising results in speech and speaker recognition tasks, and this has motivated its inclusion for the LID task. The backend utilizes the acoustic models generated using the features representing language specific fundamental characteristics of input data [10, 11]. Research in [12, 13] highlighted the interdependency of speech features and feature vector sequences through HMM. Sooner, the importance of phonetic contents was realized and parallel phone recognition and language modeling was used in LID [14, 15].

In recent years, deep learning techniques have been used for a few LID system developments. Lopez-Moreno et al. [16] used feed-forward DNN(FF-DNN) for language classification using NIST LRE 2009 and also 5 M Google corpus. These techniques are highly data centric and hence not ready to be used for under-resourced languages at this stage.

3 Identification Process and the Corpus

The architecture of the generalized LID system can be represented as a source-channel model. Almost all LID systems have similar components shown in Fig. 1.

The input speech signal is obtained as a source and are pre-processed to extract the feature vectors. The language scoring block has the speech decoder as one of the important unit. The decoder uses the feature vectors and generates acoustic model based on all sound units. The second important component, the language decoder uses the sequence generated by speech decoder for evaluating over the language set **L.** The role of the backend classifier is to combine multiple scores to reach to a final decision. The classifiers can work as a linear or non-linear models.

3.1 Corpus Used for LID

In India, there are 22 official languages and most of the Indian languages are scarce in digital resources. As language technology is a very data-intensive, it is a challenge to handle Indian languages [17]. In this work, we have used Hindi and Punjabi languages for the identification purpose. India has a rich diversity in terms of languages and these languages borrow words from one another. The samples were manually annotated and the foreign words used in the samples were also marked concerning the language they belong to. For further processing, we segment the samples from our corpus into 35, 20, 10 and 5 s duration. We aim for reasonable performance for the short time

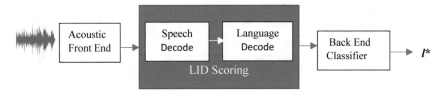

Fig. 1 Schematic diagram of the LID system

utterances. The training data is obtained from the 35 s segment and the rest of the segments together with some of the 35 s segment is used for system performance evaluation.

The performance of the system is evaluated in terms of equal error rate (**EER**) measured in percentage. It is the error rate at a point where the false acceptance rate and the false rejection rate are equal. The identification task is analogous to the binary classification task for each of the languages, hence the performance of the system is the average of its performance for each of the two languages.

4   The i-Vector System

The first approach used for LID task is through the use of i-vectors. This section discusses the baseline system briefly. The approach used for i-vector extraction, the feature set and the classifier are discussed.

4.1   Total Variability Modelling for i-Vector

The i-vectors have shown their prominence in the field of speaker verification for long [18] and now they have been used in LID task too [19]. The i-vectors are extracted in total variability space represented as low-dimension fixed-length feature vector. The idea is to provide compact representation to GMM super vectors that are composed of the channel and language-dependent super vectors. The model represents the super vector of utterances M as:

$$M = m + Tw \quad (1)$$

where
     m: Language and channel-independent UBM-GMM super vector.
     T: A rectangular low-rank matrix that that captures the super vector space variability.
     W: Normally distributed low-dimension hidden variable.

The total variability matrix **T** is obtained assuming every utterance for a given language class to belong to a different class. In this work, the dimension of i-vector is varied from 200 to 600.

4.2 Feature Extraction

The speech utterances are segmented into successive overlapping frames of 20 ms with an overlap rate of 10 ms. From each frame 13 MFCC coefficients including C0 are computed. Variance and cepstral mean normalization are applied on these feature sets and then shifted delta coefficients (SDC) are computed over the combination of multiple frames using 13–1-2–3 configuration [20]. Finally, these features are stacked to obtain 52-dimensional vector every 10 s. These feature sequences are converted into single i-vector.

4.3 Backend Classifier

The backend classifier used in the present experiment is the support vector machine (SVM), a discriminative linear classifier with RBF (Radial Basis Function) kernel. The RBF kernel works with overlapping data and finds the support vector classifiers in infinite dimension. For the tuning of RBF Kernel, the parameters of interest are the variance of the kernel, $\mathbf{Y}$ and the parameter $\mathbf{C}$ used to penalize the errors associated with training. The best value of $\mathbf{Y}$ is determined by cross-validation. The goal is to obtain $(\mathbf{C}, \mathbf{Y})$ pair, such that any unknown data can be accurately predicted. The classifier is trained based on one-vs-all strategy (Fig. 2).

Fig. 2 Flow-graph of DNN-LID system

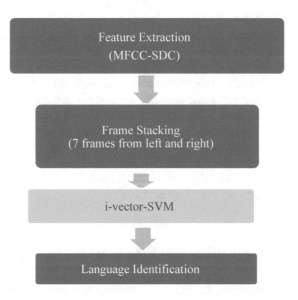

Table 1 Performance measured in EER (%) for various i-vector dimensions and test speech durations

Speech utterance	5 s	10 s	20 s	35 s
Dimensions				
200	13.40	10.55	7.21	5.80
300	12.64	8.87	4.43	3.42
400	**12.10**	**7.64**	**3.65**	**2.94**
500	14.03	9.21	5.02	4.01
600	14.76	10.02	7.11	4.96

5 Evaluation and Results

This section presents the performance of the i-vector system discussed above in the given experimental setup. The results of different size utterances are presented here. The evaluation is done for different penalty values. The accuracy of the system with varying penalty values and utterance sizes is also evaluated.

5.1 Results of the i-Vector System

The i-vector system explained here uses SVM with RBF kernel as a backend classifier. The value of C and Y has to be fixed for experimenting. Cross-validation is the mechanism to find out the optimal (C, Y) pair, where the value of Y is fixed and C is varied and this is repeated for several Y values to obtain the optimal pair. Also, here the i-vector dimension is varied from 200 to 400 to measure the performance. After several cross-validation cycles, the best value of Y is obtained as **0.1** for **$C = 1$.** Table 1 presents the performance of the i-vector system for varying vector dimension ranging from 200 to 600. It has been observed that the best performance of the system at this (C, Y) pair is obtained for dimension = 400.

The best performance for any duration of speech utterance was observed with i-vector dimension 400. We also present here the performance of this system with a varying value of penalty C, while keeping the Y fixed at 0.1. Figure 3 presents the performance in this context.

6 Conclusion and Future Work

In this paper, we presented an extensive study of acoustic–phonetic features and recent techniques for automatic language identification. The paper presents a probabilistic approach to LID and discusses the acoustic–phonetic aspects that influence language characteristics. The MFCC-SDC features have been used to explore the

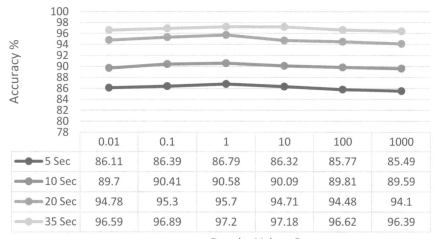

	0.01	0.1	1	10	100	1000
5 Sec	86.11	86.39	86.79	86.32	85.77	85.49
10 Sec	89.7	90.41	90.58	90.09	89.81	89.59
20 Sec	94.78	95.3	95.7	94.71	94.48	94.1
35 Sec	96.59	96.89	97.2	97.18	96.62	96.39

Penalty Value: C

Fig. 3 i-vector system accuracy for various values of penalty C of RBF kernel and test speech durations. The number of dimension 400 and $\mathbf{Y} = 0.1$

capability of i-vector with SVM as a backend classifier. It has been observed that for short utterances also the i-vector systems can capture many discriminative features. Looking into the potential of DNN system we believe that DNN-LID can perform much better and aim to use other feature sets and network configurations in future.

References

1. Sinha S, Agrawal A, Singh A, Raj P (2020) Transforming interactions: mouse-based to voice-based interfaces. Telecommun Radio Eng 79(14):1259–1271
2. Brummer N, Cumani S, Glembek O, Karafiát M, Matejka P, Pesan J et al (2012) Description and analysis of the Brno276 system for LRE2011. In: Proceedings of Odyssey 2012: the speaker and language recognition workshop. International speech communication association, Singapore, June 25–28, pp 216–223
3. Li H, Ma B, Lee KA (2013) Spoken language recognition: from fundamentals to practice. Proc of the IEEE 101(5):1136–1159
4. Lopez-Moreno I, Gonzalez-Dominguez J, Plchot O, Martinez D, Gonzalez-Rodriguez J, Moreno P (2014) Automatic language identification using deep neural networks. In: Proceeding IEEE international conference on acoustics, speech and signal processing (ICASSP), Florence, Italy, May 4–9, 5337–5341
5. Ambikairajah E, Li H, Wang L, Yin B, Sethu V (2011) Language identification: a tutorial. IEEE Circuit Syst Mag 11(2):82–108
6. Muthusamy YK, Barnard E, Cole RA (1994) Reviewing automatic language identification. IEEE Signal Process Mag 11(4):33–41
7. Leonard G (1980) Language recognition test and evaluation. https://apps.dtic.mil/dtic/tr/fulltext/u2/a084752.pdf. Accessed 12 Feb 2020
8. Hanna J, Webb N (2019) Deep learning for low-resource automatic language identification. http://jonathan-hanna.com/publication/lid/lid.pdf.2019. Accessed 12 May 2019

9. Torres-Carrasquillo PA, Singer E, Kohler MA, Greene RJ, Reynolds DA, Deller Jr (2002) Approaches to language identification using Gaussian mixture models and shifted delta cepstral features. In: Proceedings of seventh international conference on spoken language processing, Denver, USA, Sept 16–20, pp 89–92

10. Foil J (1986) Language identification using noisy speech. In: Proceeding ICASSP'86 IEEE international conference on acoustics, speech, and signal processing, Tokyo, Japan, Apr 7–11, vol 11, pp 861–864

11. Ives RB (1986) A minimal rule AI expert system for real-time classification of natural spoken languages. In: Proceeding of 2nd artificial intelligence advanced computer technology, May 337–40

12. Nakagawa S, Ueda Y, Seino T (1992) Speaker-independent, text-independent language identification by HMM. In: Proceedings of second international conference on spoken language processing, Oct 13–16, Banff, Canada, pp 1011–1014

13. Zissman MA (1993) Automatic language identification using Gaussian mixture and hidden Markov models. In: Proceedings of IEEE international conference on acoustics, speech, and signal processing. Minneapolis, USA, Apr 27, vol 2, pp 399–402

14. Zissman MA (1996) Comparison of four approaches to automatic language identification of telephone speech. IEEE Trans Speech Audio Process 4(1):31

15. Singer E, Torres-Carrasquillo PA, Gleason TP, Campbell WM, Reynolds DA (2003) Acoustic, phonetic, and discriminative approaches to automatic language identification. In: Proceedings of eighth European conference on speech communication and technology, Geneva, Switzerland, Sept 1–4, pp 1345–1348

16. Lopez-Moreno I, Gonzalez-Dominguez J, Martinez D, Plchot O, Gonzalez-Rodriguez J, Moreno PJ (2016) On the use of deep feedforward neural networks for automatic language identification. Comput Speech Lang 1(40):46–59

17. Feng K, Chaspari T (2019) Low-resource language identification from speech using transfer learning. In: Proceedings of IEEE 29th international workshop on machine learning for signal processing (MLSP), Pittsburgh USA, Oct 13, pp 1–6

18. Dehak N, Dehak R, Kenny P, Brümmer N, Ouellet P, Dumouchel P (2009) Support vector machines versus fast scoring in the low-dimensional total variability space for speaker verification. In: Proceedings of tenth annual conference of the international speech communication association. Brighton, UK, Sept 6–10, pp 1559–62

19. Dehak N, Torres-Carrasquillo PA, Reynolds D, Dehak R (2011) Language recognition via i-vectors and dimensionality reduction. In: Proceedings of twelfth annual conference of the international speech communication association, Florence, Italy, Aug 27–31, pp 857–860

20. Sinha S, Jain A, Agrawal SS (2019) Empirical analysis of linguistic and paralinguistic information for automatic dialect classification. Artif Intell Rev 51(4):647–672

A Unique ECG Authentication System for Health Monitoring

Kusum Lata Jain, Meenakshi Nawal, and Shivani Gupta

Abstract Remote well-being checking has turned into a significant part of the present life because of expanding populace and huge number of sick individuals. While observing the patients remotely, confirmation of the patient and the relating information assumes a fundamental job in exchanging the patient's information for breaking down and ensuing treatment guidance. Much of the time, it is required to remotely screen patient's electrocardiogram (ECG) signal that can likewise fill the need of human recognizable proof. Presently in biometrics look into numerous scientists are working after demonstrating uniqueness of ECG signal with reference to a person. This paper shows a novel individual distinguishing proof framework that utilizes 12 lead electrocardiogram and support vector machine (SVM). The 12 lead ECG signal has been handled through principal component analysis to get three sign (Vx,Vy,Vz). Wavelet highlights were have out and connected to SVM for preparing and testing. The detection accuracy of 89.69% is accomplished for test information, while with preparing information, it could achieve the discovery precision of 100%.

Keywords ECG · Wavelet features · SVM · PCA · Remote health monitoring · Patient authentication

1 Introduction

The reason for this examination is to demonstrate an inconvenience free patient validation arrangement that may be utilized to monitor patients remotely. The checking component needs to make positive that the records are drawing nearer from the

K. L. Jain (✉)
Manipal University Jaipur, Jaipur, RJ, India
e-mail: kusumlata.jain@jaipur.manipal.edu

M. Nawal
Poornima College of Engineering, Rajasthan Technical University, Jaipur, RJ, India

S. Gupta
Vellore Institute of Technology, Chennai, TN, India
e-mail: shivani.gupta@vit.ac.in

right person before capacity and look at of gained physiological or well-being information. License-based confirmation frameworks, for example username and pass-
words, authentications and so on, are not proper for remote observing as patients
may handover to another person. Likewise, one-time verification utilizing license
or highlight-based biometrics; face, fingerprints and so on do not coat the total
observing stage and may prompt unlawful post-validation use. The patients who
is on high hazard with perpetual clutters checked remotely assume a significant
job as the information of patient can be gotten to or dissected remotely to deal with
patient opportune. All other expensive treatment like high resolution ECG of medical
clinics, deficiency of master specialists are likewise can be kept away from by remote
patient checking. Confirmation from human body parameter is extremely helpful in
remote patient observing of patients experiencing basic illnesses and needs ceaseless
checking. It is additionally helpful for heart patients on the off chance that the individual not physically ready to give the license-based data. Interest of patient should be
dynamic if there should arise an occurrence of biometric, which may not be feasible
for a basic unhealthy patient and may prompt dissatisfaction. The proposed system
uses ECG-based biometric scheme to authenticate the patient.

- What is Electrocardiogram?

The electrocardiogram is utilized to analyze and monitor practically a wide range
of basically unhealthy patients. It is an electrical sign created through beat of blood
spilling out of heart to different pieces of the human body. There are some particular
areas of the body where these electrical heartbeats are recorded and the ECG is
recorded. The quantity of areas the sign to be recorded relies upon the prerequisite of
exact heart movement. It might be 3 lead, 12 lead or different assortments of the types
of gear accessible to record it and watch the movement of heart. It displays a regular
example as appeared in Fig. 1. This example is normal for every person as far as
varieties; yet, it contains some run of the mill contrasts from every person. Specialists
perceive this with reference to watch and dissect the working of heart and stream

Fig. 1 Shape of an
electrocardiogram

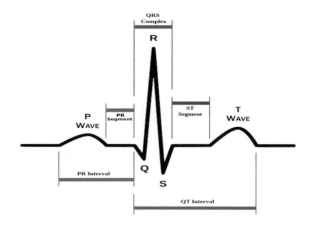

of blood to recognize heart related illnesses. The commonplace estimations of those are seen from ECG are estimation of P, Q, R, S and T. These are named as waves having diverse scope of amplitudes for ordinary working of hearts. The typical scope of variety is chosen by thinking about the varieties among various people of shifting age, weight and different parameters. At the point when this example is considered as mix of various frequencies, there happens to be assorted variety from individual to individual as far as different time and recurrence area segments. For specialists, QRS mind boggling and different interims may be of run of the mill significance to identify the illness and treat the patients. For the proposed target, the ECG is considered as an electrical sign having creation of various frequencies having partition in time and recurrence space. The key goal would decide one of a kind element that speak to every person and can be speak to for validation in verified manner. The security here issues as the proposed framework is focused for remote patient observing.

2 Existing Work

Many of the researchers work upon biometric authentication through retina, finger print, gait, etc., but very few researchers work upon ECG- and EEG-based authentication which is continuous and difficult to lie. Biel et al. [1] worked upon 12 lead ECG signals with some diagnostic characteristics and achieve 100% of accuracy by using SIMCA classifier, but it is difficult to rely on diagnostic data rather than using feature extraction technique. Shen et al. [4] used the combination of two classifiers and achieved 100% of accuracy. Although there have been very good results but limited to typical set of data, and long-term ECG recordings have been rarely considered.

3 Data Collection

Information is gathered for 25 subjects with 12 lead accounts of electrocardiogram from MIT-BIH database. The recurrence of ECG recording is 1000 examples for every second, and every one of subjects has 60,000 example esteems with every one of 12 lead ECG.

The chronicle of 12 lead ECG made up of I, II, III three appendage drives three avL, avR and avF expanded leads and six V1, V2, V3, V4,V5, V6 precordial drives. The enlarged leads are gotten from same I, II and III leads, in any case, they see the heart from various edges on the grounds that the negative anode for these leads is a change of Wilson's focal terminal.

Text recording of one subject with 12 lead ECG is as following Fig. 2.

Elapsed time hh:mm:ss.mmm	i (mV)	ii (mV)	iii (mV)	avr (mV)	avl (mV)	avf (mV)	v1 (mV)	v2 (mV)	v3 (mV)	v4 (mV)	v5 (mV)	v6 (mV)
0:00.000	0.274	-0.035	-0.308	-0.119	0.291	-0.172	0.68	0.333	0.391	0.054	-0.384	-0.198
0:00.001	0.287	-0.037	-0.324	-0.125	0.305	-0.18	0.669	0.33	0.413	0.077	-0.36	-0.18
0:00.002	0.292	-0.032	-0.324	-0.131	0.308	-0.178	0.664	0.33	0.431	0.097	-0.341	-0.169
0:00.003	0.312	-0.018	-0.329	-0.147	0.321	-0.173	0.652	0.319	0.441	0.109	-0.33	-0.162
0:00.004	0.328	-0.018	-0.345	-0.154	0.336	-0.181	0.641	0.316	0.45	0.121	-0.319	-0.153
0:00.005	0.331	-0.022	-0.352	-0.154	0.342	-0.187	0.63	0.314	0.468	0.138	-0.305	-0.141
0:00.006	0.332	-0.015	-0.346	-0.159	0.34	-0.181	0.623	0.311	0.481	0.151	-0.293	-0.136
0:00.007	0.353	0	-0.353	-0.177	0.353	-0.176	0.615	0.305	0.491	0.161	-0.286	-0.132
0:00.008	0.361	0.016	-0.344	-0.189	0.352	-0.164	0.612	0.307	0.509	0.177	-0.272	-0.124
0:00.009	0.354	0.025	-0.329	-0.19	0.342	-0.152	0.61	0.311	0.53	0.199	-0.253	-0.111
0:00.010	0.367	0.036	-0.331	-0.202	0.349	-0.147	0.598	0.303	0.543	0.213	-0.242	-0.101
0:00.011	0.382	0.05	-0.332	-0.215	0.356	-0.141	0.588	0.298	0.552	0.222	-0.233	-0.102
0:00.012	0.392	0.07	-0.322	-0.231	0.356	-0.126	0.594	0.308	0.577	0.246	-0.212	-0.085
0:00.013	0.411	0.09	-0.321	-0.251	0.366	-0.115	0.589	0.314	0.604	0.27	-0.191	-0.07
0:00.014	0.419	0.105	-0.313	-0.263	0.366	-0.104	0.57	0.302	0.617	0.284	-0.175	-0.062
0:00.015	0.423	0.127	-0.296	-0.276	0.36	-0.085	0.575	0.311	0.638	0.306	-0.157	-0.043
0:00.016	0.438	0.143	-0.294	-0.291	0.366	-0.075	0.58	0.324	0.661	0.325	-0.141	-0.033
0:00.017	0.448	0.161	-0.286	-0.304	0.367	-0.063	0.577	0.326	0.679	0.341	-0.126	-0.015
0:00.018	0.453	0.176	-0.276	-0.315	0.364	-0.05	0.575	0.328	0.696	0.357	-0.113	-0.017
0:00.019	0.448	0.172	-0.276	-0.309	0.361	-0.052	0.564	0.315	0.702	0.363	-0.11	-0.024
0:00.020	0.443	0.177	-0.266	-0.31	0.354	-0.044	0.556	0.307	0.711	0.376	-0.098	-0.015
0:00.021	0.436	0.177	-0.259	-0.306	0.347	-0.041	0.542	0.295	0.718	0.384	-0.087	-0.009
0:00.022	0.448	0.185	-0.262	-0.317	0.354	-0.038	0.517	0.28	0.722	0.387	-0.083	-0.005
0:00.023	0.461	0.206	-0.254	-0.334	0.357	-0.024	0.491	0.266	0.732	0.399	-0.072	0.002
0:00.024	0.454	0.213	-0.239	-0.334	0.346	-0.013	0.471	0.245	0.741	0.412	-0.058	0.011
0:00.025	0.448	0.227	-0.222	-0.337	0.335	0.003	0.453	0.229	0.748	0.424	-0.045	0.015
0:00.026	0.446	0.233	-0.212	-0.339	0.329	0.011	0.417	0.199	0.75	0.43	-0.041	0.027
0:00.027	0.46	0.252	-0.208	-0.356	0.334	0.022	0.4	0.18	0.754	0.438	-0.035	0.034
0:00.028	0.48	0.284	-0.196	-0.383	0.339	0.044	0.393	0.176	0.766	0.453	-0.017	0.056
0:00.029	0.485	0.291	-0.193	-0.389	0.339	0.05	0.383	0.169	0.773	0.461	-0.011	0.058

Fig. 2 ECG text file sample

4 Proposed Patient Identification System

The proposed patient identification system appeared in Fig. 3 forms on ECG sign of 12 leads. The 12-dimensional ECG is changed over into three-dimensional and a short time later one-dimensional utilizing preprocessing. The extracted statistical and wavelet features are applied as an input to the SVM for training and recognition.

4.1 Preprocessing

Step1: Transformation of 12-dimensional ECG to three-dimensional data.
Fusion of three frank leads Cx,Cy,Cz by 12 ECG leads is as follows:

Fig. 3 Block diagram of proposed patient identification system

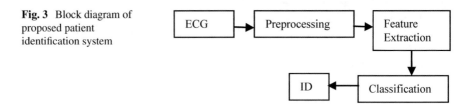

$$Cx = 0.4 * II - 0.8 * (II + III)/3 + 0.2 * V5 + 0.5 * V6 + 0.1 * V4 \qquad (1)$$

$$Cy = 0.3 * III + 0.8 * II + 0.5(II + III)/3 - 0.2 * V5 - 0.3 * V6 \qquad (2)$$

$$\begin{aligned} Cz = &-0.1 * III - 0.2 * II + 0.4 * (II + III)/3 - 0.3 * V1 - 0.1 \\ &* V2 - 0.1 * V \quad 3 - 0.2 * V4 - 0.1 * V5 + 0.4* \end{aligned} \qquad (3)$$

Step2: Principal component analysis (PCA).

On applying PCA to (Cx,Cy,Cz), the correlated variables are converted to uncorrelated variables(Ux,Uy,Uz).

Step3: Transformation of three lead (Ux,Uy,Uz) data to one lead.

The customized three leads Ux,Uy,Uz are changed into one lead by evaluating spatial magnitude (SpMag) as follows:

$$SpMag = \sqrt{Ux^2 + Uy^2 + Uz^2} \qquad (4)$$

4.2 Feature Extraction

In order to examine the distinctness of ECG signals, recorded wavelet features were taken out using Daubechie's mother wavelet and decomposing the single lead electrocardiogram into different levels.

Wavelets are obtained from a single prototype wavelet $\Psi(t)$ called mother wavelet by dilations and shifting:

$$\varphi_{a,b}(t) = \frac{1}{\sqrt{a}} \varphi\left(\frac{t-b}{a}\right) \qquad (5)$$

where a is the scaling parameter and b is the moving parameter. The 1 lead ECG signal was separated into little casings, and the Daubechies mother wavelet of different requests with dimension of decay shifting from 3 to 15 has been utilized. The point-by-point coefficients and inexact coefficients got through disintegration, and for remade waves were utilized to decide factual parameters such as

- Maximum (MAX)
- Minimum (MIN)
- Standard deviation (STD)
- Mean (Mean)
- Variance (VAR)
- Mode (Mode)

On applying Debauchees order 3 on signals, approximate coefficients CA3 and and CD1-CD3 are obtained for decomposed signal, A3 and D1-D3 are obtained for reconstructed signal. This eight feature matrix are used to calculate six statistical features standard deviation, variance, mean, mode, maximum, minimum and lead to 48 values for every frame. To analyze the association between different feature matrix, the scatter plots are plotted. For five subjects, Fig. 4 exhibits correlation between CD7MODE and CD2MIN. The overlapping behavior in both the figures proves the requisite of a nonlinear classifier, and therefore, SVM is opted for classification.

In addition to the above wavelet features, 18 statistical features were also derived for each frame. Frame energy, standard deviation of signal, standard deviation of frequency modulation, maximum of signal, maximum of frequency modulation, mean of frequency spectrum with gradient and slopes are some of these parameters.

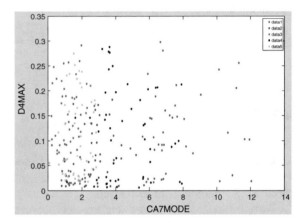

Fig. 4 Scatter plot for D4Max versus CA7Mode for five subjects

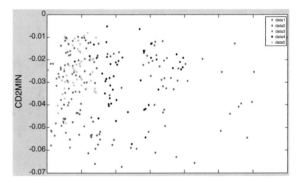

Fig. 5 Scatter plot for CD2Min versus CA7Mode for five subjects

4.3 Support Vector Machine as a Classifier

SVM is a restrictive classifier that is prominent because of its colossal job in zone of example recognization. The SVM is actualized utilizing the bit Adatron calculation. This calculation maps contributions to a high-dimensional component space and after that ideally isolates information into their separate classes by disengaging those sources of info which falls near the information limits. Along these lines, the Kernal Adatron is particularly viable in isolating arrangements of information which share complex boundaries.

4.4 Performance Parameters

The performance parameters such as mean square error (MSE), mean detection accuracy (MDA), normalized mean square error (NMSE) are measured to evaluate the simplification ability of SVM classifier. However, accuracy is most essential parameter for classifiers.

- The parameters MSE and NMSE are obtained using the following formulae:

$$MSE = \frac{\sum_{j=0}^{P} \sum_{i=0}^{N} \left(d_{ij} - y_{ij}\right)^2}{N.P} \tag{6}$$

where P = number of output neurons, N = number of exemplars in the dataset, yij = network output for exemplar i at neuron j, dij = desired output for exemplar i at neuron j.

$$NMSE = \frac{P.N.MSE}{\sum_{j=0}^{P} \frac{N \sum_{i=0}^{N} d_{ij}^2 - \left(\sum_{i=0}^{N} d_{ij}\right)^2}{N}} \tag{7}$$

The classification error is measured which defines incorrectly classified samples from the overall samples.

- Mean detection accuracy (MDA) is the mean detection accuracies of different classes in terms of percentage. The MDA has been reflecting as a particular classification performance parameter.
- Correlation coefficient (r).
  The calculation r is obtained by substituting estimates of the covariance's and variances based on a sample into the formula.
- Mean absolute error (MAE) is used to compute how close the prophecies are with the actual results.

Table 1 Size detail of input feature matrix

Feature file	Total (M X N)	Training feature (P X Q)	CV matrix (L X N)	Test data (S X V)
DB3_L3	66	66	66	66
DB3_L4	78	78	78	78
DB3_L5	90	90	90	90
DB3_L6	102	102	102	102
DB3_L7	114	114	114	114
DB3_L8	126	126	126	126
DB3_L9	138	138	138	138
DB3_L10	150	150	150	150
DB3_L11	162	162	162	162
DB3_L15	210	210	210	210

- Minimum absolute error: During training phase of SVM, the minimum value of absolute error is obtained.
- Maximum absolute error.

  During training phase of SVM, the maximum value of absolute error is obtained.

5 Results and Discussions

5.1 Input Feature Matrices

Daubechies wavelet order 3(DB3) is applied with the levels 3–15 to extract features of the ECG signal. An input feature file of the ECG signal consists 119 frames of 25 subjects with 500 samples. The input feature matrix is used for training, testing and cross-validation in which 25% of input feature matrix has been used for training, 50% for testing and 25% is used for cross-validation. Size detail of input matrix is given in Table 1. The total size is of M X N where M = 2975, training feature if P X Q where P = 1487, CV matric, and test data is of size L X N and S X V where L = S = 744.

5.2 Analysis and Study of SVM for DB3_Level_3

When the data is trained using SVM for 1000 epoch values, it is observed that the mean detection accuracy could be 100% for patients with training data set whereas

Table 2 Analysis for DB3_L3 feature file

Performance measure	Test data results	Cross-validation	Training data results
MDA%	70.87654209	70.7865009	**100**
MSE	**0.47830704**	0.62270893	0.500493363
NMSE	13.89833	**12.159778**	13.06954231
MAE	0.682996	0.55114315	**0.538782539**
Max Ab Error	3.78956074	6.027612775	**2.215549028**
Min. Ab. Error	0.0065784	0.001220747	**0.000664875**
Correlation Coefficient	0.30456787	0.330364964	0.522014339

approximately 70% for testing and cross-validation of data. The average values of errors are also obtained minimum for training data set as shown in Table 2.

5.3 Testing Data Comparison Among Different Levels of DB-3

As per previous discussion of DB-3 level 3, we found the performance parameters through the deviation of decomposition level's 3–15. As Table 3 shows the maximum mean detection accuracy is attained with DB-3 level 7, whereas the other best results of performance parameters could be obtained by DB-3 level 15.

The feature matrix of DB-3 level 15 is more complex as compared to other feature files. If we compare both feature files of DB-3 level 7 and DB3 level 15, the variations of remaining performance parameters values can be negotiate as the mean detection accuracy of DB-3 level 15 is 74% only."

Table 4 demonstrates the mean detection accuracy of patients is 100% for all training feature set files, whereas the good results of MSE, MAE, etc., could be obtained by DB3 level 15. Table 5 demonstrates the cross-validation data set results. As per the table, the mean detection accuracy of patients could be attained by DB-3 level 9 which is approximately 79%, whereas the good results of other performance parameters are attained from DB-3 level 15. On comparing the size of both feature files of levels 9 and 15, the performance parameter values can be negotiate as the mean detection accuracy of DB3-level 15 is 73% only.

5.4 Graphical Analysis of Patient Detection Accuracy

Figure 6 exhibits the mean detection accuracy(MDA) for training, testing and cross-approval data sets on using the different wavelet feature files of Daubechies 3 with the levels of 3–15. The figure shows the detection accuracy is approximately 100% for

Table 3 Comparative performance analysis of SVM for various input features (test data)

Feature	MSE	NMSE	MAE	Min Ab Error	Max Ab Error	R	ADA %
Db3-L3	0.47830704	13.89833	0.682996	0.0065784	3.78956074	0.30456787	70.87654209
Db3-L4	0.7374305	19.84232	0.651223	0.00183625	5.024501039	0.28351698	70.94319531
Db3-L5	0.81687256	21.4698	0.694175	0.00189095	4.750186748	0.30529231	69.62777987
Db3-L6	0.66437334	17.49812	0.626403	0.001228722	3.09581315	0.31470639	69.49638528
Db3-L7	0.68880645	18.38301	0.624462	0.001182298	5.365114792	0.2856135	**89.69186143**
Db3-L8	0.58916714	15.69657	0.572857	0.001151973	5.893494054	0.30132607	70.04756569
Db3-L9	0.6393676	16.52176	0.584275	0.001176183	5.594951884	0.27936676	87.03107416
Db3-L10	0.57850877	15.28277	0.573863	0.001220659	**2.697316167**	0.32082832	71.89675952
Db3-L11	0.62656442	16.91975	0.582088	0.001462788	6.11399942	**0.28124604**	72.8869188
Db3-L15	**0.43412158**	**11.52931**	**0.496486**	**0.000926669**	3.909474476	0.30990807	74.229267

Table 4 Comparative performance analysis of SVM for various input features (Training Data)

Feature	MSE	NMSE	MAE	Min Ab Error	Max Ab Error	Correlation coefficient	MDA %
Db3-L3	0.500493363	13.06954231	0.538782539	0.000664875	2.215549028	0.522014339	100
Db3-L4	0.593926551	15.48903746	0.58038559	0.001021635	2.51304102	0.505377051	100
Db3-L5	0.732588753	19.24077873	0.635959191	0.000760843	2.804835026	**0.482882537**	100
Db3-L6	0.641262957	16.89172637	0.587568665	0.00073358	2.673404281	0.493867144	100
Db3-L7	0.558311313	14.6509324	0.552225085	0.000748539	2.357482031	0.512784791	100
Db3-L8	0.492508705	12.93473784	0.510959832	0.000850657	2.340463229	0.525902087	100
Db3-L9	0.47968282	12.59534671	0.500592242	0.000652719	2.269014984	0.513941336	100
Db3-L1	0.516521282	13.80512987	0.520337411	0.000754931	2.351122471	0.520855404	100
Db3-L11	0.498958965	13.0667774	0.51643403	0.000459713	2.458487314	0.520348309	100
Db3-L15	**0.338149579**	**8.973049616**	**0.425594616**	**0.000337272**	**2.03996751**	0.547787047	100

Table 5 Comparative performance analysis of SVM for various input features (CV data)

Feature	MSE	NMSE	MAE	Min As Error	Max Ab Error	Correlation coefficient	MDA %
Db3-L3	0.62270893	12.159778	0.55114315	0.001220747	6.027612775	0.330364964	70.7865009
Db3-L4	0.66564515	17.657223	0.62899585	0.001411391	5.191881997	0.313777338	67.72139764
Db3-L5	0.86322833	23.111401	0.69208464	0.001046685	5.813453731	**0.271632938**	69.23882681
Db3-L6	0.69562332	18.573574	0.64556745	0.001509878	3.380297835	0.318948314	70.44327859
Db3-L7	0.65055197	17.001584	0.60999926	0.001104473	5.407333115	0.290948633	68.40820019
Db3-L8	0.60054043	15.826742	0.58387794	**0.000915978**	5.800332778	0.294327124	67.44838641
Db3-L9	0.57567028	15.58044	0.5871764	0.001771897	2.959514771	0.302388363	**78.98153845**
Db3-L10	0.54210463	14.373163	0.56196158	0.001115322	**2.358436127**	0.33506983	73.14879755
Db3-L11	0.5651567	14.961294	0.57984464	0.00115583	3.163221025	0.31820984	69.94620242
Db3-L15	**0.41748038**	**10.934788**	**0.48822415**	0.001099631	4.681593045	0.31838745	73.14461941

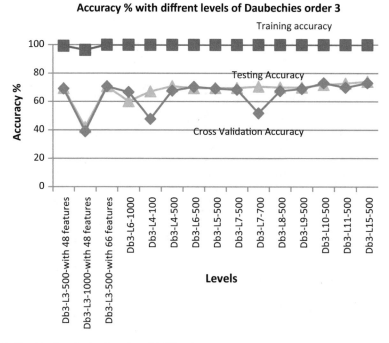

Fig. 6 Graphical analysis of results with 25 patients

training with all the feature files. Same as for testing, maximum detection accuracy is obtained with the feature file of decomposition level 7 which is roughly 90%.

Decomposition level 9 is also fetching roughly same detection accuracy, while remaining feature files is also fetching 70% accuracy.

6 Conclusion

In the proposed system of patient authentication, we have used 12 lead ECGs of 25 patients recorded under different activity states. The 12 lead signals were converted to 3 leads and finally to 1 lead data. One lead ECG data has been processed through Daubechies mother wavelet of order 3 with decomposition level varying from 3 to 15. The SVM is trained and tested for all the varying feature sets. It is found that if the support vector machine is trained and then investigated for identical data, it provides 100% patient detection accuracy. The robustness of the system could be tested for CV and test datasets and observed that the maximum accuracy of 89.69% could be obtained for test data of DB3_L7 feature set. In future work, record the ECGs of the same patients at different times and situations performing different activities will be considered.

References

1. Biel L, Pettersson O, Philipson L, Wide P (1999) ECG analysis: a new approach in human identification. In: Proceedings of the 16th Instrumentation and measurement technology conference. IEEE, vol 1
2. Conway JCD, Coelho CJN, Andrade LCG (2000) Wearable computer as a multi-parametric monitor for physiological signals. In: Proceedings of IEEE International conference on bioinformatics and biomedical engineering, pp 236–242
3. Dasrao MS, Hock YJ, Sim EKW (2001) Diagnostic blood pressure wave analysis and ambulatory monitoring using a novel non-invasive portable device. In: Proceedings of International conference on biomedical engineering, pp 267–272
4. Shen TW, Tompkins WJ, Hu YH (2002) One-lead ECG for identity verification. In: Proceedings of the 24th annual conference on engineering in medicine and biology and the annual fall meeting of the biomedical engineering society, vol 57, issue 2
5. Johannessen EA, Wang L, Tang TB, Cui L (2003) Implementation of multichannel sensors for remotebiomedical measurements in a microsystems. IEEE Trans Biomed Eng 51(3):525–535
6. Chen W, Wei D, Cohen M, Ding S, Toxinoya S (2004) Development of a scalable healthcare monitoring platform. In: Proceedings of Fourth International conference on computer and information technology (cit'04)
7. Israel SA, Irvine JM, Cheng A, Wiederhold MD (2005) ECG to identify individuals. Pattern recognit virtual reality medical center, USA, vol 38(1)
8. Alajel KM, Yousuf KB, Ramji AR, Ahmed ES (2005) Remote electrocardiogram monitoring based on the internet. Kmitl Sci J 5:493–501
9. Kim HK, Biggs SJ (2006) Continuous shared control for stabilizing reaching and grasping with brain-machine interfaces. IEEE Trans Biomed Eng 53(6):1164–1173
10. Lin CH, Young ST, Kuo TS (2007) A remote data access architecture for home-monitoring. J Med Eng Phys 29:199–204
11. Wang Y, Agrafioti F, Hatzinakos D, Plataniotis KN (2008) Analysis of human electrocardiogram for biometric recognition. EURASIP J Adv Signal Process
12. Chan DC, Hamdy MM, Badre A, Badee V (2008) Wavelet distance measure for person identification using electrocardiograms. Trans Instrum Meas IEEE 57(2)
13. Chiu C, Chuang C, Hsu C (2008) A novel personal identity verification approach using a discrete wavelet transform of the ECG signal. In: Proceedings of the International conference on multimedia and ubiquitous engineering, vol 6, issue 4
14. Janani S, Minho S, Tanzeem C, David K (2009) Activity-aware ECG-based patient authentication for remote health monitoring. In: Proceedings of International conference on mobile systems. ACM
15. Can Y, Miguel C, Vijaya K BVK (2010) Investigation of human identification using two lead electrogram signals. In: Proceedings of the 4th International symposium on applied sciences in biomedical and communication technologies, vol 50. IEEE
16. Fahim S, Ibrahim K, Jiankun H (2010) ECG-based authentication, vol 1. Springer
17. Arthur RM, Wang S, Jason WT (2011) Changes in body-surface electrocardiograms from geometric remodeling with obesity. IEEE Trans Biomed Eng 58(6)
18. Abdelraheem MMT, Selim H, Abdelhamid TK (2012) Human identification using the main loop of the vectorcardiogram. Am J Signal Process 2012 2(2):23–29
19. Nawal M, Purohit GN (2014) ECG based human authentication: a review. Int J Emerg Eng Res Technol 2(3):178–185
20. Meenaksh N, Bundele M, Purohit GN (2015) An Exhaustive CHAID based authentication approach for remote health monitoring. Int J Appl Eng Res 10(17)
21. Nawal M, Sharma MK, Bundele MM (2016) Design and implementation of human identification through physical activity aware 12 lead ECG. In: 2016 international conference on recent advances and innovations in engineering (ICRAIE). IEEE

Assamese Dialect Identification From Vowel Acoustics

Priyankoo Sarmah and Leena Dihingia

Abstract Assamese language is spoken in the state of Assam in a geographical area spread over 400 km. The Assamese language, spoken in different areas of the state, is known to differ by geographically defined dialects. In this work, we collect speech samples of Assamese speakers from five different geographical regions in Assam, producing the Assamese vowels in various contexts and examine the acoustic characteristics of the vowels. The regional variations of Assamese language were captured in terms of formant frequencies, their discrete cosine transforms, F0 and duration of the vowels. A random forest was trained to classify an vowel as belonging to one of the five regions. The accuracy of classifying unseen test data was 94.0%. The confusion matrix shows that geographically close regions have greater tendency to be confused. The results show that a reasonable accuracy in dialect identification can be attained only by using vowel specific acoustic properties.

1 Introduction

As it happens in several languages around the world, Assamese also has acoustic differences in vowels spoken in different regions where the language is spoken. This work reports an attempt at dialect identification in Assamese using acoustic-phonetic features of vowels in the language, with a random forest (RF) classifier. In this work, we consider five regional variations of Assamese, namely Tinsukia, Jorhat, Nagaon, Barpeta and Nalbari. According to Goswami & Tamuli [1], there are three distinct dialects of Assamese, Western Assamese (WA), Central Assamese (CA) and Eastern Assamese (EA). The Tinsukia and Jorhat regions are considered to be under the EA variety, Barpeta and Nalbari regions under the WA, and the Nagaon region is considered to be under CA.

P. Sarmah (✉) · L. Dihingia
Indian Institute of Technology Guwahati, Guwahati, India
e-mail: priyankoo@iitg.ac.in

L. Dihingia
e-mail: l.dihingia@iitg.ac.in

© The Author(s), under exclusive license to Springer Nature Singapore Pte Ltd. 2022   313
P. Nanda et al. (eds.), *Data Engineering for Smart Systems*, Lecture Notes in Networks and Systems 238, https://doi.org/10.1007/978-981-16-2641-8_30

Vowels are attested to be indicators of dialectal variation in several languages. In the seminal work by Labov [2], it is shown how vowel variation is one of the distinguishing features in American English varieties [3]. Williams & Escudero [4] undertook an acoustic study of British English and compared the formant frequencies of vowels in Northern and Southern British English and found the vowels to be considerably different. Similarly, vowel variation is attested in dialects of Dutch [5], Swedish [6] and Portuguese [7].

The importance of incorporating vowel specific features in automatic dialect identification was discussed in several early works [8]. In Ferragne & Pellegrino [9], automatic dialect identification was attempted on 15 British English varieties using only vowels with an overall accuracy of 90%. Biadsy et al. [10] also underlined the importance of vowels, specifically the emphatic vowels in Arabic in achieving higher accuracy in Arabic dialect classification. Chittaragi et al. [11] used acoustic-phonetic features, namely formant frequencies (F1–F3) and prosodic features such as, energy, pitch (F0) and duration. In classifying five Kannada dialects, an accuracy of 76% is reported by the authors. Recently, Devi & Thaoroijam [12] also used F1, F2, F3, duration, energy and F0 features from vowels to classify three dialects of Meeteilon (Manipuri) language. They report the best accuracy of 61.57% in Meeteilon dialect identification. In case of Assamese, a previous attempt at automatically classifying Assamese dialects using vowels as input segments achieved a dialect recognition accuracy of 89.32% [13].

In this work, we attempt dialect identification in five Assamese regions using acoustic-phonetic features derived from vowel phonemes of the language. We provide a brief acoustic analysis of vowels produced in these five areas and then proceed to the dialect identification experiment using RF classifier. In the immediately following Sect. 2, we present a brief description of the Assamese language and its vowels. In Sect. 3, we detail the methodology used for the classification experiment using a random forest (RF) classifier and provide an analysis of the acoustic features of Assamese vowels. In Sect. 4, we report the results of our experiments, and in Sect. 5, we conclude the paper.

2   Assamese Language and Vowels

Assamese is one of the official languages of the state of Assam in India and it is spoken by almost 15 million speakers. It is also one of the languages included in the eighth schedule of the Indian constitution as a scheduled language. Earlier studies on the Assamese language have shown that there are noticeable variations in the language based on the geographical distribution of the speakers. Goswami & Tamuli [1] divide Assamese into three distinct varieties, EA, CA and WA. Among these varieties, the Eastern variety, spoken around Jorhat and Sibsagar area is considered the standard variety of Assamese.

There is no general agreement on the types of vowels found in standard Assamese. Mahanta [14] provides an inventory of eight vowels in Assamese, namely /i, e, ɛ, u, ʊ, o, ɔ, ɑ/. Nevertheless, the realization of these vowels may be different in the varieties of Assamese spoken in various geographic locations. For example, Taid [15] and Deka [16] claim that the /o/ vowel is neutralized to /u/ in the speech of Eastern Assamese speakers and in case of the Kamrupi variety, /o/ is neutralized to /ɔ/. The vowel inventory proposed for Assamese in Mahanta [14] is representative of the standard Assamese and hence followed in the rest of this paper.

Considering the reports that vowel inventory of Assamese dialects may be different from each other, in this work, we explore if vowels can be used as inputs to model dialectal variation for Assamese. In the section immediately following this, we provide a brief summary of the database used in this work and a discussion on the random forest classification experiments we conducted.

3 Methodology

In this section, we provide a brief description of the database used in this work. We also give a brief explanation of the acoustic-phonetic features extracted from the speech database and the parameters used for the RF classification.

3.1 Database

The database used in the random forest classification is of 14289 vowel tokens produced by 72 speakers of 5 regions of Assam. The database contained the eight Assamese vowels produced in isolation and in sentence frames. The vowels are embedded in grammatical words in Assamese in monosyllables with consonants on both onset and coda positions. The distribution of the speakers and the vowel tokens in the database is summarized in Table 1. A more detailed account of the database used may be found in Dihingia [17].

Table 1 Summary of the database

Region	No. of speakers		No. of vowel tokens	
	Female	Male	Female	Male
Barpeta	06	04	1261	996
Jorhat	09	12	1592	1599
Nagaon	05	05	1288	1303
Nalbari	10	11	2664	1210
Tinsukia	05	05	1088	1288

3.2 Acoustic Phonetic Features

In order to capture vowel specific spectral features, we extracted the values of the first three formants of each vowel token. Considering the variation assumed in vowel production across the Assamese varieties, we hope that the first three formants will be able to capture dialectal differences, apart from the vowel specific vocal tract features. Hence, the first three formant values of the vowels were averaged from the mid-20% of total duration of the vowels as this area is shown to be more effective in robust characterization of vowel features [4]. Apart from the formant features, duration of vowels and fundamental frequency at the vowel midpoint were also extracted. In order to remove speaker specific values, the format and F0 values were normalized using the Lobanov normalization method [18].

Goswami [19] has mentioned that certain monophthongs may be pronounced as diphthongs in the Western varieties of Assamese. Considering that, in order to capture vowel inherent dynamic changes, the formant frequency trajectory within a vowel was transformed using discrete cosine transform (DCT), and the first three coefficients of the DCT, namely C0, C1 and C2, were extracted for each formant. Hence, the final set of features consisted of 14 features, namely F1, F2, F3, and the first three DCT coefficients of the 3 formant trajectories ($3 \times 3 = 9$), F0 and duration. Apart from the feature vectors, the data grid included information about the speakers, vowels, gender, contexts (sentence or isolation), onset and coda consonants.

3.3 Acoustics of Assamese Vowels

While the primary focus of this work is dialect identification, in order to justify the acoustic-phonetic features, we present a brief acoustic analysis of the vowels in the five regions of interest in this work. The systematic relationship between vowel quality and the first three formants has been established quite early, as reported by McKendrick [20] in 1901. Since then, lower formant frequencies, namely F1, F2 and F3 are considered to be distinctive acoustic measures for vowel characterization. Hence, in the current study, we have extracted the formants of the vowels spoken in the five regions of Assamese and plotted the vowels in Fig. 1. This figure shows the distibution of the eight Assamese vowels across five regions of Assam. Figure 1 also shows the dialect-wise variation of the vowels, particularly in case of /ɛ/, /ʊ/, /u/ and /o/.

In order to see the distinctiveness of the acoustic-phonetic features in terms of the geographical regions, we conducted a series of two-way analysis of variance (ANOVA) tests on each of the 14 features with interaction of region and vowel types as the factor. The results confirms significant interaction of all the features ($p < 0.001$) except for the C3 of F3. Hence, 13 of the 14 features have the probability of varying depending on the geographical regions.

Fig. 1 Average speaker normalized formants of Assamese vowels, color coded by the five regions

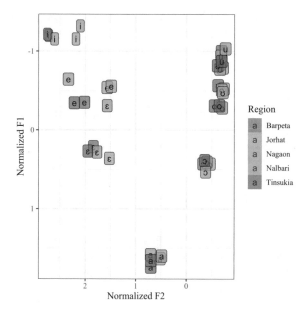

Region

a Barpeta
a Jorhat
a Nagaon
a Nalbari
a Tinsukia

Normalized F1

Normalized F2

4 Identification of Assamese Dialects Using Random Forest

In this section, we present the results of various experiments carried out using RF to identify the geographical region of the speakers based on vowel spoken by them.

4.1 Random Forest Classification

Random forest (RF) is an ensemble method used for classification or regression. When it is used for classification, a trained RF gives an output which is the class label of an input data. The classification or regression task is carried out independently by several decision trees employed by the RF. The label set by a majority of the decision trees is decided as the class label of a new instance of input. Randomness in an RF is introduced through several schemes. In general, increasing the number of decision trees in an RF improves accuracy of classification without resulting in over-training [21]. At the same time, uncorrelated trees predict classes more accurately. Hence, each tree is trained with a dataset that is obtained by randomly sampling the entire training data with replacement, known as the bootstrap procedure. Additional randomness is introduced by random feature selection during the construction of a decision tree. The size of this random subset of features is set by the user and is known as mtry parameter in various RF classification program on the R software [22].

The other parameters to be set by the user are nodesize, maxnode and ntree. The splitting of a node of tree is repeated until each cell of the tree contains less than `nodesize` data points. The parameter `maxnode` sets maximum number of terminal nodes/cells in a tree. Finally, the effectiveness of a random forest classification depends on the number of uncorrelated trees, `ntree`, that are used in computing the aggregate results. To summarize, the architecture of a RF is characterized by four parameters, namely `mtry`, `nodesize`, `maxnode` and `ntree` [23]. The values assigned to these parameters determine the accuracy obtained and computational cost incurred. In the following section, we discuss how the optimal values for the four parameters mentioned in this section were determined to be used in dialect classification of Assamese using vowels.

4.2   Experimental Setup and Parameter Tuning

The random forest classification experiment was set up using the R software [24] with the `randomForest`, `caret` and `e1071` packages [22, 25, 26]. In case of the experiment reported in Sect. 4.3, the classifier was trained on 80% of the data, i.e., 11431 randomly selected vowel tokens. In Sect. 4.4, data from 80 to 20% of the speakers from each dialect were used for training and testing the RF, respectively. In both experiments, `mtry`, `ntree` and `maxnode` parameters were changed stepwise to derive the optimum value for training the models.

4.3   Region-Wise Classification

The RF classifier was trained using randomly chosen 80% of the data, i.e., 11432 vowel tokens. The classifier was tested with 20% of the remaining data, i.e., 2857 vowel tokens. Initially, the training set was subjected to training using 5 fold validation. In the initial training, the `mtry` parameter was varied from 1 to 70, keeping `nodesize` at 14 and `maxnode` at 24. The optimal results were obtained with `mtry = 56`. Following that we varied the `maxnode` parameters from 20 to 2000 and found the highest accuracy when `maxnode = 452`. Finally, the `ntree` parameter was varied from 250 to 1100 at steps of 50, and at `ntree = 1000`, the best accuracy was obtained. Hence, for training the RF classifier for regions, the following parameters were set: `nodesize` = 14, `mtry` = 56, `maxnode` = 452 and `ntree` = 1000.

The system derived accuracy is calculated as the percentage of total number of tokens correctly identified out of the 2857 vowel tokens. However, for each of the five regions in the test set, we also calculated overall accuracy as an average of the individual accuracy (we refer to as 'manually calculated' hereafter). The accuracy of dialect classification for both system generated and manually calculated increased to 99.7% when the model is trained with all the 14 features. As the train and test datasets were not mutually exclusive in terms of the speakers, we assumed that this

Table 2 Confusion matrix for region identification (values in %)

	Barpeta	Jorhat	Nagaon	Nalbari	Tinsukia
Barpeta	**22.2**	0	0	77.8	0
Jorhat	0	**100**	0	0	0
Nagaon	0	0	**100**	0	0
Nalbari	2.6	0	0	**97.4**	0
Tinsukia	0	100	0	0	**0**

high classification score is possibly due to speaker resemblance rather than dialect resemblance.

Considering the possibility of speaker effects, another classification was conducted with 80% speakers in each dialect included in the training set and 20% speakers in each dialect included in the test set. This resulted in 12644 vowel tokens in the training set and 1645 vowel tokens is the test set. The average accuracy of classification in the five regions reduced to 63.9% (system generated, 60.4%). As seen in Table 2, the classification results indicate that the Barpeta and Tinsukia regions have poor accuracy and they are confused with regions within their own dialectal regions, i.e., Western Assamese (WA) and Eastern Assamese (EA), respectively. The classification of Tinsukia is biased toward the Jorhat region and Barpeta is biased toward the Nalbari region which is possibly due to class imbalance arising due to less amount of data available for the Tinsukia and Barpeta regions (see Table 1).

4.4 Dialect-Wise Classification

We also conducted a classification attempt by dialect areas as identified for Assamese dialects in Goswami & Tamuli [1]. In this classification experiment, the Jorhat and Tinsukia regions were subsumed under the EA group, Nagaon under the CA group and Barpeta and Nalbari regions under the WA group. A fivefold experiment was run by dividing the speaker set in each of the 3 dialect regions into 5 mutually exclusive subsets of speakers while maintaining gender balance. Data from the first subset of speakers of all regions formed the test set in fold1 experiment. The train set was formed with the remaining data. In order to explore the variation of classification accuracy due to randomness inherent in the RF classifier, the fold1 experiment was run 10 times. The results of the 10 trials, (T1 – T10), of the first fold are shown in Table 3. The highest accuracy is 81.8% while the average of 10 accuracy is 70.6%.

In order to determine the optimal values of the parameters, a series of experiments was conducted. The parameters nodesize, mtry, maxnode and ntree were set at 14, 4, 34 and 248, respectively. Fivefold classification was conducted with alternating speaker exclusive train and test sets. The classification accuracies (in %) of the fivefold experiments were 81.8, 68.0, 76.7, 72.4, 71.8; the average accuracy

Table 3 Classification accuracy (in %) in 10 trials of the first fold of training-testing

T1	T2	T3	T4	T5	T6	T7	T8	T9	T10
68.5	65.0	81.8	70.3	68.6	68.7	64.8	67.1	71.7	79.8

Table 4 Confusion matrix for Assamese dialect classification (values in %)

	Eastern	Central	Western
Eastern	**94.9**	00.0	05.1
Central	07.9	**61.7**	30.4
Western	11.4	00.0	**88.6**

Table 5 Confusion matrix for region identification using balanced dataset (values in %)

	Barpeta	Jorhat	Nagaon	Nalbari	Tinsukia
Barpeta	**96.4**	0	0	3.6	0
Jorhat	0	**83.7**	0	0	16.3
Nagaon	0	0	**100**	0	0
Nalbari	9.4	0	0	**90.6**	0
Tinsukia	0	0.5	0	0	**99.5**

was 74.1%. Table 4 shows the confusion matrix of the classification in the firstfold. While EA and WA dialects are classified with an average accuracy of 91.8%, the CA dialect has lower accuracy and tends to be misclassified as WA. The low accuracy in case of CA can be attributed to data imbalance arising due to lower sampling points.

4.5 Region-Wise Classification with Balanced Dataset

We assumed in Sects. 4.4 and 4.3 that the errors in classification were a result of imbalanced datasets. In order to confirm that, we conducted a final experiment, where, for each region, the number of speakers was kept constant in the test and the train sets, i.e., for each region, 2 speakers were in the test set and 8 speakers were in the training set. This resulted in 7684 and 1454 vowel tokens for training and testing, respectively. The training parameters were same as described in Sect. 4.3. When the test set was used to evaluate the model, it yielded an accuracy of 94.0% in correctly recognizing the region of the speakers. The confusion matrix in Table 5 shows that with the balanced dataset, the accuracies for region identification are higher. Compared to the accuracy in Table 2, the accuracy in classifying the Barpeta and Tinsukia regions is significantly high.

5 Conclusion

This study showed that vowels do have enough information that can be exploited for dialect or region classification. The results also showed that geographically closer areas have greater tendency to be confused. Moreover, it was also noticed that random forest classification is prone to errors arising due to imbalanced classes that need to be addressed for more reliable classification. This problem has been discussed extensively in the literature as in More & Rana [27]. A future direction of this work will be address the issues arising due to data imbalance to achieve more robust classification.

Acknowledgements This work is part of the Ph.D. research work of the second author submitted at the Indian Institute of Technology Guwahati, Assam, India. The authors would like to thank Dr. K Samudravijaya for his insightful comments and clarifications regarding the statistical modeling used in this work.

References

1. Goswami, GC, Tamuli J (2003) Asamiya. In: D Jain, G Cardona (eds) The Indo-Aryan languages. Routledge, pp 391–443
2. Labov W, Ash S, Boberg C (2008) The Atlas of North American English. Berlin. De Gruyter Mouton, New York
3. Labov William (1991) The three dialects of english. New ways of analyzing sound change 5:1–44
4. Williams Daniel, Escudero Paola (2014) A cross-dialectal acoustic comparison of vowels in Northern and Southern British English. The Journal of the Acoustical Society of America 136(5):2751–2761
5. Adank Patti, Van Hout Roeland, Smits Roel (2004) An acoustic description of the vowels of Northern and Southern Standard Dutch. The Journal of the Acoustical Society of America 116(3):1729–1738
6. Leinonen Therese (2008) Factor analysis of vowel pronunciation in Swedish dialects. International Journal of Humanities and Arts Computing 2(1–2):189–204
7. Escudero P, Boersma P, Rauber AS, Bion RA (2009) A cross-dialect acoustic description of vowels: Brazilian and European Portuguese. J Acoust Soc Am 126(3):1379–1393
8. Higgins A, Benson P, Li KP, Jack P (1999) Dialect identification. Technical report, ITT Aerospace/Comminications Division
9. Ferragne E, Pellegrino F (2007) Automatic dialect identification: A study of British English. In: Speaker classification II. Springer, pp 243–257
10. Biadsy F, Hirschberg J, Habash N (2009) Spoken Arabic dialect identification using phonotactic modeling. In: Proceedings of the EACL 2009 Workshop on computational approaches to Semitic languages, pp 53–61
11. Chittaragi NB, Koolagudi SG (2019) Acoustic-phonetic feature based Kannada dialect identification from vowel sounds. Int J Speech Technol 22(4):1099–1113
12. Devi TC, Thaoroijam K (2021) Vowel-based Meeteilon dialect identification using a Random Forest classifier. arXiv preprint arXiv:2107.13419
13. Sarma M, Sarma KK (2016) Dialect identification from Assamese speech using prosodic features and a neuro fuzzy classifier. In: 2016 3rd International Conference on Signal Processing and Integrated Networks (SPIN), pp 127–132

14. Mahanta Shakuntala (2012) Assamese. Journal of the International Phonetic Association 42(2):217–224
15. Taid T (1988) A note on the synchronic description of Assamese sounds in Kakati's AFD. In: Taid T, Goswami RD (eds), Banikanta Kakati, the man and his works. Publication Board Assam, pp 70–83
16. Deka KS (2007) Kamrupi upabhaxa. In: Bhasar Asamiya (ed) Dipti Phukan Patgiri. Gauhati University, Guwahati, Upabhasa. University Publication Department
17. Dihingia L (2020) Vowel acoustics of Assamese spoken in five geographical locations. PhD thesis, Indian Institute of Technology Guwahati
18. Lobanov BM (1971) Classification of Russian vowels spoken by different speakers. J Acoust Soc Am 49(2B):606–608
19. Goswami GC (1982) Structure of Assamese. Department of publication, Gauhati University
20. McKendrick JG (1901) Experimental phonetics. Nature 65(1678):182–189
21. Ho TK (1995) Random decision forests. In: Proceedings of 3rd International Conference on Document Analysis and Recognition, vol 1. IEEE, pp 278–282
22. Liaw A, Wiener M (2002) Classification and regression by randomforest. R news 2(3):18–22
23. Scornet E (2017) Tuning parameters in Random Forests. ESAIM: Proc Surv 60:144–162
24. R Core Team (2019) R: A language and environment for statistical computing. R Foundation for Statistical Computing, Vienna, Austria
25. Kuhn M (2015) Caret: classification and regression training. Astrophysics Source Code Library, ascl–1505
26. Dimitriadou Evgenia, Hornik Kurt, Leisch Friedrich, Meyer David, Weingessel Andreas (2008) Misc functions of the department of Statistics (e1071), tu wien. R package 1:5–24
27. More AS, Rana DP (2017) Review of random forest classification techniques to resolve data imbalance. In: 2017 1st International Conference on Intelligent Systems and Information Management (ICISIM), pp 72–78

A Review on Metaheuristic Techniques in Automated Cryptanalysis of Classical Substitution Cipher

Ashish Jain, Prakash C. Sharma, Nirmal K. Gupta, and Santosh K. Vishwakarma

Abstract Between the year 1993 and 2019, a considerable new and different meta-heuristic optimization techniques have been presented in the literature for automated cryptanalysis of classical substitution cipher. This paper compares the performance of these new and different metaheuristic techniques. Three main comparison measures are considered to assess the performance of presented metaheuristics: efficiency, effectiveness, and success rate. To the best of author knowledge, first time this kind of review has been carried out. It is noteworthy that among the presented meta-heuristics, the performance of genetic algorithm technique is best with respect to effectiveness and success rate.

Keywords Genetic algorithm · Scatter search · Simulated annealing · Tabu search · Cryptanalysis

1 Introduction

Combinatorial optimization is an approach to deal with a given problem and locate the best answer out of a very large set of possible solutions. The problems for which one need to find the best solutions are mostly come under the umbrella of NP-hard and NP-complete combinatorial problems. The problem associated related to solving these problems is that the time and/or memory increases drastically with the increase

A. Jain · P. C. Sharma (✉) · N. K. Gupta · S. K. Vishwakarma
School of Computing and Information Technology, Manipal University Jaipur, Jaipur, India
e-mail: prakashchandra.sharma@jaipur.manipal.edu

A. Jain
e-mail: ashish.jain@jaipur.manipal.edu

N. K. Gupta
e-mail: nirmalkumar.gupta@jaipur.manipal.edu

S. K. Vishwakarma
e-mail: santosh.kumar@jaipur.manipal.edu

in size of problems [1]. Branch and bound and simplex methods are examples of exact optimization techniques that can be used to speed up the search. However, often these techniques have prohibitive complexity requirements (time and/or memory) which makes the use of these techniques impractical [2]. In such cases, approximate techniques, i.e., metaheuristics are utilized to determine an adequate solution to the problem [2]. This paper presents four different metaheuristic techniques in solving the problem related to the classical substitution cipher.

Cipher provides information security. Basically, ciphers are used to transform one form of text called "plaintext" into another form of text called "ciphertext" which is tough to break if the secret key is not known. Cryptanalysis is the procedure to find weakness in the construction of ciphers. Cryptanalyst performs cryptanalysis. One of the most difficult tasks of the cryptanalyst is to discover (detect or search) the secret key of the cipher by knowing only some of the ciphertext characters. In terms of information security if cryptanalyst or attacker able to discover the secret key of the cipher, then we say that the cipher has been successfully attacked. Attacking cipher comes in the class of NP-complete problem [3–9]. If the exhaustive search is carried out to detect secret key in the keyspace, then the whole keyspace required to be examined in the worst case that will take significant number of years [3–10]. However, automated attacks can be formed using metaheuristic techniques that can search the secret key of classical ciphers in acceptable amount of time [5, 6, 9, 10].

2 Performance Measurement Criteria

Substitution cipher is considered for attacking using metaheuristics. For the substitution, cipher refer [9]. One can ask what the weakness in the substitution cipher is so that it can be attacked. The answer is—the encryption process used in the substitution cipher does not altered the character frequency distribution significantly. Therefore, the metaheuristics are capable to match the known language statistics with the character frequency statistics (n-grams) of the encrypted message (a standard strategy to automatically attack the classical ciphers).

There are three criteria based on which the performance of metaheuristic techniques can be assessed with regard to automated attacks: (1) number of ciphertext characters available for the attack (effectiveness measurement criterion); (2) number of key elements detected correctly (success rate measurement criterion); (3) time required to recover the key (efficiency measurement criterion). Based on these three main criteria, we will assess the performance of different metaheuristic techniques in the result section.

3 Literature Review

The application of metaheuristic techniques in automated attacks of classical ciphers was first reported in 1993 (e.g., [11–13]), and the outcomes have demonstrated that metaheuristic strategies are exceptionally efficient and effective. With this inspiration, numerous metaheuristic techniques have been reported for mounting automated attacks on the substitution cipher, for example, genetic algorithm (GA), scatter search (SS), simulated annealing (SA), and tabu search (TS). Among all these algorithms the GA strategy recently proposed by Jain and Chaudhary [9] has shown the best performance with respect to effectiveness and success rate.

For automated cryptanalysis of the "classical substitution cipher, hereinafter, substitution cipher" multiple metaheuristic techniques have been used in the past that have been mentioned above. Below we describe the standard form of these techniques in brief. In 1960s, Holland and his students [14, 15] proposed a popular population-based metaheuristic, namely GA. This method starts by haphazardly creating a population of individuals. Three operators, namely selection, crossover, and mutation, control the population and to generate the new population from the old population. In each generation, a cost function, namely fitness function, assesses the suitability of individuals. After some number of iterations, the individual with best cost provides the adequate answer to the associated issue. For point-by-point depiction on the GA, the reader can refer [14–17].

SS method is a population-based metaheuristic technique which start by a population of some scattered and good solutions. After every iteration, probably the best solutions are extricated and remembered for a reference set. From the extricated solution, a new solution is obtained by applying a linear combination. Afterward, the quality of the new solution is improved using the local search technique. Finally, the final solution is comprising in the reference set. For point-by-point depiction on the SS method, the reader can refer [18].

SA method is a metaheuristic technique that handles and updates a single solution during optimization. Kirkpatrick et al. [19] mimicked the annealing process with respect to combinatorial optimization. The SA strategy starts with a haphazard solution for the issue to be solved and a beginning temperature. At every temperature, various endeavors are made to bother the current solution. For point-by-point depiction on the SA, the reader can refer [19].

TS method is a direction-based metaheuristic technique which gives a way to deal with search to find an ideal arrangement of the given issue. A separate list, namely a tabu list, is preserved by the tabu strategy during the hunt of the solution [20]. The additional arrangement stays in the tabu for a characterized number of iterations. For point-by-point depiction on the TS, the reader can refer [20, 21].

In the literature, GA, SS, SA, and TS methods have been utilized to tackle many optimization issues. These methods have also utilized for the optimization problems related to cryptology. The cryptology problems solved using these methods and their applications are given in Table 1.

Table 1 Applications of cryptology problems solved using GA, SS, SA, and TS

Authors [References]	Metaheuristics	Problem solved	Application
Jain and Chaudhari [9] Matthews [12] Spillman et al. [13] Clark [22] Dimovski and Gligoroski [23] Garg and Sherry [24] Verma et al. [25] Omran et al. [26] Mudgal et al. [27]	GA	Automated cryptanalysis of classical substitution cipher	Modern substitution ciphers uses functions of classical substitution cipher in a complicated way [8]
Garici and Drias [28]	SS		
Forsyth and Naini [11] Clark [22]	SA		
Clark [22] Garg and Sherry [24] Verma et al. [25]	TS		
Clark [22] Giddy and Safavi-Naini [29] Toemeh and Arumugam [30] Song et al. [31] Muhajjar [32] Al-Khalid et al. [33] Garg [34]	GA	Automated cryptanalysis of classical transposition cipher	Modern transposition ciphers uses functions of classical transposition cipher in a complicated way [8]
Clark [22] Giddy and Safavi-Naini [29] Song et al. [31] Garg [34] Mishra and Kaur [35]	SA		
Clark [22] Garg [34]	TS		
Clark [22] Spillman [36] Yaseen and Sahasrabuddhe [37] Garg et al. [38] Ramani and Balasubramanian [39]	GA	Automated cryptanalysis of knapsack cipher	Knapsack cipher is a reasonable alternative, particularly for security of little implanted gadgets, e.g., cellular devices [8]

(continued)

Table 1 (continued)

Authors [References]	Metaheuristics	Problem solved	Application
Song et al. [40] Vimalathithan and Valarm-athi [41] Sathya et al. [42] Sharma et al. [43] Al Adwan et al. [44] Dworak and Boryczka [45]	GA	Automated cryptanalysis of data encryption standard (DES)	DES is a modern block cipher used for encryption of confidential information [8]
Nalini and Rao [46] Nalini [47]	SA		
Nalini and Rao [46] Soyjaudah [48]	TS		
Cowan [49]	SA	Automated cryptanalysis of short playfair ciphers	Modern substitution ciphers use functions of playfair substitution cipher in a complicated way [8]
Clark et al. [50]	SA	Design of substitution-boxes	Substitution-boxes are nonlinear elements that are used in block cipher for encryption of confidential information [8]

4 Comparative Analysis

Recall from Sect. 2, the standard strategy for escalating attacks on the substitution cipher is the matching of the known language statistics with the observed n-gram statistics of the decrypted message. Through matching, we determined the cost of the candidate key. A candidate key is a key which is evolved using metaheuristic technique during the hunt of original secret key.

Fitness Function. The input of this function is the candidate key. This function determines the "quality" of the candidate key. For example, from the population of the evolved candidate keys, a key K is selected. Using K, a known ciphertext is decrypted. Afterward, an examination is carried out between n-gram statistics of the decoded ciphertext and the known language statistics (for instance, for English language statistics refer [9]). Thusly, the fitness of K is determined. Formally, Eq. (1) is utilized for statistics comparison.

$$\text{Cost}_k = \alpha \left(\sum_{i \in \varsigma} \left| k_i^u - d_i^u \right| \right) + \beta \left(\sum_{i,j \in \varsigma} \left| k_{i,j}^b - d_{i,j}^b \right| \right)$$
$$+ \gamma \left(\sum_{i,j,k \in \varsigma} \left| k_{i,j,k}^t - d_{i,j,k}^t \right| \right) \tag{1}$$

For clarification on Eq. (1), the reader can refer [9]. As clear, the estimation of n in the term n-gram ought to be higher in number to play out a precise appraisal of candidate keys. In any case, in the writing, it has demonstrated that typically the best operational reason for a fitness function utilized in automated cryptanalysis of substitution ciphers is the bigrams only, i.e., $n = 2$ [9, 11, 22, 28]. If the fitness function utilizes trigrams just than the bit of leeway over the bigrams is minimal (see Fig. 1), while the computational multifaceted nature of trigrams is high, roughly, (key-size)^3. These realities persuade us to utilize the fitness function which is mentioned in Eq. (2) that depends on just the bigrams.

$$\text{Cost}_k = \left(\sum_{i,j \in \varsigma} \left| k_{i,j}^b - d_{i,j}^b \right| \right) \tag{2}$$

Experiment. For performing experiments, the considered metaheuristic techniques have been implemented in Java. We followed the guidelines reported in the respective papers during implementation of each of the metaheuristics. Given ciphertext, length of the ciphertext, and the English language, bigram statistics are input to every algorithm.

Analysis of Results. Regarding all the performance criteria, we mention the obtained results in Table 2. Note that the metaheuristic technique that takes a greater number of ciphertext characters are said to be less effective than the metaheuristic which takes lesser number of ciphertext characters. From the obtained results, we can observed that the following algorithms are most effective because taking only 800 number of ciphertext characters for successful recovery of key: GA proposed in [9, 22], SA proposed in [22], TS proposed in [22], and SS proposed in [28]. From the obtained

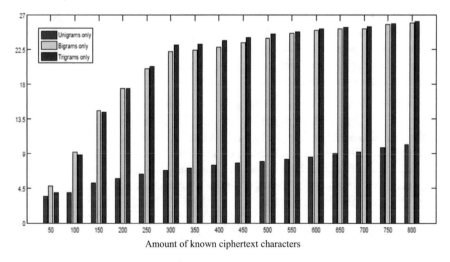

Amount of known ciphertext characters

Fig. 1 Plotting of average outcomes as for number of key components accurately recovered utilizing cost function dependent on unigrams just, bigrams just and trigrams as it were

Table 2 Cryptanalytic results obtained through various metaheuristic techniques on the substitution cipher

Year	Authors [Reference]	Metaheuristics used	Maximum number of ciphertext characters used	Average number of key elements correctly recovered out of 27	Mean performance time (in seconds) to recover the key
1993	Matthews [12]	GA	1000	24.43	0.513
1993	Forsyth and Naini [11]	SA	1000	24.96	0.421
1993	Spillman et al. [13]	GA	1000	24.52	0.496
1998	Clark [22]	GA	800	24.64	0.416
1998	Clark [22]	SA	800	25.08	0.350
1998	Clark [22]	TS	800	24.18	0.241
2003	Dimovski and Gligoroski [23]	GA	1000	24.74	0.527
2005	Garg and Sherry [24]	GA	1000	24.82	0.546
2005	Garici and Drias [28]	SS	800	25.13	0.712
2005	Garg and Sherry [24]	TS	1000	24.91	0.549
2007	Verma et al. [25]	GA	1000	24.87	0.603
2007	Verma et al. [25]	TS	1000	24.96	0.592
2010	Omran et al. [26]	GA	1000	24.82	0.546
2017	Mudgal et al. [27]	GA	1000	24.87	0.562
2019	Jain and Chaudhari [9]	GA	**800**	**25.72**	0.397

results, we can clearly observe that the GA proposed by Jain and Chaudhari [9] takes only 800 ciphertext characters and as an outcome able to recover 25.72 number of key elements out of 27. The time taken by the algorithm is also comparable which is 0.397 s. This study indicates that the GA is most effective and most successful in mounting automated attacks on the substitution cipher.

5 Conclusions

The efficient, effective, and successful utilization of various metaheuristic techniques in solving the substitution cipher is presented. Based on the experimentation, we noted the following outcomes: (1) in terms of success rate the GA proposed in [9] performs significantly better than the rest of the metaheuristic techniques. (2) in terms of efficiency, the GA proposed in [9] performs better than the GA proposed in [12, 13, 22–27], SA proposed in [12], TS proposed in [24, 25], and SS proposed in [28]. However, performs poorer than the SA and TS proposed in [22]. This study indicates that the GA technique proposed in [9] is a viable option for solving such kind of NP-complete problems.

References

1. Goldreich O (2010) P, NP, and NP-completeness: the basics of computational complexity. Cambridge University Press, pp 1–183
2. Du KL, Swamy MNS (2016) Search and optimization by metaheuristics: techniques and algorithms inspired by nature, Birkhäuser, pp 1–434
3. Menezes AJ, Van Oorschot PC, Vanstone SA (1996) Handbook of applied cryptography. CRC press, pp 1–780
4. Stinson DR (2005) Cryptography: theory and practice. CRC press, pp 1–593
5. Castro JCH, Viñuela PI (2005) Evolutionary computation in computer security and cryptography. N Gener Comput 23(3):193–199
6. Danziger M, Henriques MAA (2012) Computational intelligence applied on cryptology: a brief review. IEEE Lat Am Trans 10(3):1798–1810
7. Awad WS, El-Alfy ESM (2015) Computational intelligence in cryptology. Improving information security practices through computational intelligence, vol 28, pp 1–17
8. Holden J (2017) The mathematics of secrets: Cryptography from caesar ciphers to digital encryption. Princeton University Press, pp 1–373
9. Jain A, Chaudhari NS (2019) An improved genetic algorithm and a new discrete cuckoo algorithm for solving the classical substitution cipher. Int J Appl Metaheuristic Comput (IJAMC) 10(2):109–130
10. Bhateja AK, Bhateja A, Chaudhury S, Saxena PK (2015) Cryptanalysis of vigenere cipher using cuckoo search. Appl Soft Comput 26:315–324
11. Forsyth WS, Safavi-Naini R (1993) Automated cryptanalysis of substitution ciphers. Cryptologia 17(4):407–418
12. Matthews RA (1993) The use of genetic algorithms in cryptanalysis. Cryptologia 17(2):187–201
13. Spillman R, Janssen M, Nelson B, Kepner M (1993) Use of a genetic algorithm in the cryptanalysis of simple substitution ciphers. Cryptologia 17(1):31–44
14. Goldberg DE (2006) Genetic algorithms. Pearson Education India
15. Michalewicz Z (2013) Genetic algorithms+ data structures= evolution programs. Springer Science & Business Media
16. Gonzalez TF (Ed) (2007) Handbook of approximation algorithms and metaheuristics. CRC Press. https://doi.org/10.1201/9781420010749
17. Kramer O (2017) Genetic algorithm essentials. Springer. https://doi.org/10.1007/978-3-319-52156-5
18. Laguna M (2014) Scatter search. In: Search methodologies. Springer, Boston, MA, pp 119–141

19. Kirkpatrick S, Gelatt CD, Vecchi MP (1983) Optimization by simulated annealing. Science 220(4598):671–680
20. Glover F, Laguna M (2013) Tabu Search. Handbook of combinatorial optimization. Springer, New York, pp 3261–3362
21. Rego C, Alidaee B (eds) (2006) Metaheuristic optimization via memory and evolution: tabu search and scatter search. Springer Science & Business Media
22. Clark A (1998) Optimisation heuristics for cryptology, doctoral dissertation. Queensland University of Technology, Australia
23. Dimovski A, Gligoroski D (2003) Attack on the polyalphabetic substitution cipher using a parallel genetic algorithm. Swiss-Macedonian scientific cooperation trought SCOPES project
24. Garg P, Sherry AM (2005) Genetic algorithm & Tabu search attack on the mono-aiphanetic substitution cipher. Paradigm 9(1):106–109
25. Verma AK, Dave M, Joshi RC (2007) Genetic algorithm and tabu search attack on the mono-alphabetic substitution cipher i adhoc networks. J Comput Sci
26. Omran SS, Al-Khalid AS, Al-Saady DM (2010) Using genetic algorithm to break a mono-alphabetic substitution cipher. In: 2010 IEEE conference on open systems (ICOS 2010). IEEE, pp 63–67
27. Mudgal PK, Purohit R, Sharma R, Jangir MK (2017) Application of genetic algorithm in cryptanalysis of mono-alphabetic substitution cipher. In: 2017 International conference on computing, communication and automation (ICCCA). IEEE, pp 400–405
28. Garici MA, Drias H (2005) Cryptanalysis of substitution ciphers using scatter Search. In LNCS proceedings of international work-conference on the interplay between natural and artificial computation 2005. LNCS Springer Heidelberg, pp 31–40
29. Giddy JP, Safavi-Naini R (1994) Automated cryptanalysis of transposition ciphers. Comput J 37(5):429–436
30. Toemeh R, Arumugam S (2007) Breaking transposition cipher with genetic algorithm. Elektronika ir Elektrotechnika 79(7):75–78
31. Song J, Yang F, Wang M, Zhang H (2008). Cryptanalysis of transposition cipher using simulated annealing genetic algorithm. In: International symposium on intelligence computation and applications. Springer, Berlin, Heidelberg, pp 795–802
32. Muhajjar RA (2010) Use of genetic algorithm in the cryptanalysis of transposition ciphers. Basrah J Sci 28(1A english):49–57
33. Al-Khalid AS, Omran SS, Hammood DA (2013) Using genetic algorithms to break a simple transposition cipher. In: 6th International conference on information technology ICIT
34. Garg P (2009) Genetic algorithms, tabu search, and simulated annealing: a comparison between three approaches for the cryptanalysis of transposition cipher. J Theor Appl Inf Technol 5(4)
35. Mishra G, Kaur S (2015) Cryptanalysis of transposition cipher using hill climbing and simulated annealing. In: Proceedings of fourth international conference on soft computing for problem solving. Springer, New Delhi, pp 293–302
36. Spillman R (1993) Cryptanalysis of knapsack ciphers using genetic algorithms. Cryptologia 17(4):367–377
37. Yaseen IF, Sahasrabuddhe HV (1999) A genetic algorithm for the cryptanalysis of Chor-Rivest knapsack public key cryptosystem (PKC). In: Proceedings third international conference on computational intelligence and multimedia applications. ICCIMA'99 (Cat. No. PR00300). IEEE, pp 81–85
38. Garg P, Shastri A, Agarwal DC (2007) An enhanced cryptanalytic attack on Knapsack Cipher using Genetic Algorithm. Int J Comput Inf Eng 1(12):4071–4074
39. Ramani G, Balasubramanian L (2011) Genetic algorithm solution for cryptanalysis of knapsack cipher with knapsack sequence of size 16. Int J Comput Appl 35(11):17–23
40. Song J, Zhang H, Meng Q, Wang Z (2007) Cryptanalysis of four-round DES based on genetic algorithm. In: 2007 International conference on wireless communications, networking and mobile computing. IEEE, pp 2326–2329
41. Vimalathithan R, Valarmathi ML (2009) Cryptanalysis of S-DES using genetic algorithm. Int J Recent Trends Eng 2(4):76

42. Sathya SS, Chithralekha T, Anandakumar P (2010) Nomadic genetic algorithm for cryptanalysis of DES 16. Int J Comput Theory Eng 2(3):1793–8201
43. Sharma L, Pathak BK, Sharma RG (2012) Breaking of simplified data encryption standard using genetic algorithm. Glob J Comput Sci Technol
44. Al Adwan F, Al Shraideh M, Al Saidat MS (2015) A genetic algorithm approach for breaking of simplified data encryption standard. Int J Secur Appl 9(9):295–304
45. Dworak K, Boryczka U (2017) Genetic algorithm as optimization tool for differential cryptanalysis of DES6. In: International conference on computational collective intelligence. Springer, Cham, pp 107–116
46. Nalini N, Rao GR (2005) Cryptanalysis of simplified data encryption standard via optimization heuristics. In: 2005 3rd International conference on intelligent sensing and information processing. IEEE, pp 74–79
47. Nalini N (2006) Cryptanalysis of block ciphers via improved simulated annealing technique. In: 9th International conference on information technology (ICIT'06). IEEE, pp 182–185
48. Soyjaudah KMS (2012) Cryptanalysis of simplified-data encryption standard using tabu search method. In: International conference on information processing. Springer, Berlin, Heidelberg, pp 561–568
49. Cowan MJ (2008) Breaking short playfair ciphers with the simulated annealing algorithm. Cryptologia 32(1):71–83
50. Clark JA, Jacob JL, Stepney S (2005) The design of S-boxes by simulated annealing. N Gener Comput 23(3):219–231

An Empirical Study of Different Techniques for the Improvement of Quality of Service in Cloud Computing

Chitra Sharma, Pradeep Kumar Tiwari⊙, and Garima Agarwal

Abstract In a large, heterogeneous, and distributed environment, the computing infrastructure expands, and resource management becomes a challenging task. In a cloud world, one experiences problems of resource distribution, triggered by items like heterogeneity, dynamism, and errors, with uncertainty and distribution of resource. Unfortunately, to manage these environments, applications, and resource behaviors, current resource management techniques, frameworks, and mechanisms are insufficient. In recent years, the computer system is mostly based on cloud computing. Service level agreement (SLA) and quality of service (QoS) decrease by the minimum utilization of resources. Proper utilization of resources reduces the SLA violation and maximize QoS. Proper management of resources managed by service algorithms. This research paper analyzes the different types of resource management strategies which play the vital role mange the resources computing resources.

Keywords Quality of service · Service level agreement · Energy efficient · Load balancing

1 Introduction

The cloud service provider with SLA is able of handling its resources. QoS is one of SLA's key attributes. QoS attributes, e.g., reliability, security, availability, performance, data privacy, etc. In cloud SLA help with the QoS, the facilities like software as a services (SaaS), platform as a service (PaaS), infrastructure as a services (IaaS) are consumed by the end-users [9]. Cloud resources should be allotted not exclusively to fulfill quality of services (QoS) needs and service level agreements (SLAs), but also to minimize the energy usages and time to execute the client work.

C. Sharma · P. K. Tiwari (✉) · G. Agarwal
Manipal University Jaipur, Jaipur, India

G. Agarwal
e-mail: garima.agarwal@jaipur.manipal.edu

© The Author(s), under exclusive license to Springer Nature Singapore Pte Ltd. 2022
P. Nanda et al. (eds.), *Data Engineering for Smart Systems*, Lecture Notes in Networks and Systems 238, https://doi.org/10.1007/978-981-16-2641-8_32

According to scheduling and load balancing, strategies are very critical to maximizing the efficiency of the cloud with the setup [1–10].

1.1 Load Balancing

The main role of load balancing is to maximize the resource use, decrease the time of response, increases the throughput and the important point is to avoid the load on any single node [11]. Load balancing is an act when the server is overloaded. Two reasons for the server are overloaded. One, the variation in load distribution, and the other require additional resources is allocated to VM [12]. Load balancing part of cloud computing to performs a task. Proper distribution of the load managed by the help of a load balancer. A load balancer is improving the quality of cloud computing [13].

The main objectives of load balancing are:

- Expandability of the data centers and improve the energy efficiency.
- Increases the utilization of resource.
- Maximize user satisfaction.
- Stability maintains of the system.
- Make the fault-tolerant system.
- Increases the performance.
- Accommodate the modifications of the future.
- Increase service accessibility and consistency.
- Reduce response time [1–13].

2 Literature Review

Malekloo et al. [1] solved the biggest problems facing cloud services is how to do so. This balance between energy savings mechanism and ensuring the implementation of the mechanism is required to satisfy the prerequisites for QoS and SLA. In this work, authors suggested an energy and QoS-conscious approach to the positioning and consolidation of multi-objective VMs that point to a balance between energy productivity, implementation of the system, and administrative design. To classify the over-used and under-used hosts in the data center, Tarafdar et al. [2] suggested a Markov chain-based prediction method. Haghshenas [3] proposed a linear regression-based algorithm to minimize energy usage and maintain service efficiency and decrease SLA violations. This technique is used to compare another algorithm using the CloudSim and PlanetLab simulator. The suggested algorithm illustrates scalability to conserve data center efficiency and establishes expectations of size. Mandal et al. [4] proposed live migration-enabled dynamic consolidation of virtual machines result in significant energy savings. Excessive migration cause of SLA violation. Researchers recommend a power-aware VM selection strategy in this research paper to solve the issue of energy-efficient VM selection with minimal SLA violation. The proposed strategy was modeled and simulated with several flexibilities and limitations in a simulation setting. With increased SLA in terms of CPU and Gul et al. [5] suggested space-aware virtual machine consolidation systems reduce energy usage. The findings indicate that proposed energy-aware models, namely MaxCap and RemCap, improve the SLA violations treated in terms of energy use. To solve the issue of QoS service composition with SLA, Yaghoubi and Maroosi [6] suggested an enhanced multi-verse optimization algorithm for the Network service composition (IMVO) algorithm. Increase the QoS to 57%, while SLA is fulfilled. The consuming Markov chain was suggested by Rajabzadeh et al. [7] to reduce energy usage with a service level agreement. The key problem was broken into smaller sub-problems to control the management of the resource process and minimize energy usage, and an algorithm was implemented for each sub-problem.

Gupta et al. [8] suggested an advanced optimization algorithm to maximize the efficiency of QoS and reduce the cost of implementation. The genetic algorithm is considered in this paper to look for the best answer to the problem. The proposed balance of algorithms between QoS and the cost of deployment. Li et al. [9] suggested the improved task schedule algorithm (ITSA) to manage the problem of workload imbalance and excessive SLA violations. The greedy technique for executing the fixed activity with the lowest gain value and maximum gain value to name a task pair and implement schedule was adopted by this researcher. The suggested algorithm was used by the service provider and the customer to characterize the cloud-based framework for optimal QoS. To tackle the challenge of unpredictability, the authors used the queuing model. Several reports have attempted to describe the QoS exhibited by benchmarking in cloud implementation environments. In QoS modeling, mathematical characterizations of observational evidence are useful to estimate threats without the requirement for an ad-hoc assessment analysis to be carried out. They are important for estimating practical values for parameters of the QoS model, such as the variance of network bandwidth, startup times of VM.

3   Quality of Service (QoS)

Several reports have attempted to describe the QoS exhibited by benchmarking in cloud implementation environments. In QoS modeling, mathematical characterizations of observational evidence are useful to estimate threats without the requirement for an ad-hoc assessment analysis to be carried out. They are important for estimating practical values for parameters of the QoS model, such as the variance of network bandwidth, startup times of virtual machine (VM), and probabilities of starting failure. Output variability findings have been recorded for various types of VM instances [1–7]. The primary reason for such inconsistency is hardware heterogeneity. Other works describe the heterogeneity in VM startup times [7, 8], which is especially associated with the image size of the operating system [8]. Some tests have noticed high-performance contention in CPU-bound workers on Amazon EC2.

3.1 QoS Attributes Defined as Follows

(a) Accountability: confirms the clients that their knowledge is treated by the earner as per their prospects. It is predictable in terms of audibility, administration, consensus, sustainability, and ownership, etc.

(b) Agility: the cloud community illustrates how easily services are scaled with a minimal amount of investment. Flexibility, extensibility, elasticity, probability, and scalability are composed of it, etc.

(c) Assurance: guarantees that the efficiency of the distributed service is the same as that promised in the contract related to the service stage. Assurance is estimated in terms of usability, recoverability, conservation, the durability of resilience, etc.

(d) Expense: the most critical feature, the cost of which depends on all the attributes. It relates to the cost of growth, ongoing costs, and the expense of a service or collection of services for use.

(e) Performance: this defines the provider's roles and functionality. In terms of accuracy, reaction time, suitability, and functionality, efficiency is approximate.

(f) Security: identified as the controller of the employer's data service provider. Protection depends on control of entry, the geography of data, honesty of data and confidentiality of data, etc. [11].

3.2 Service Level Agreement (SLA)

SLA is the agreement with the consumer and service provider for a certain period. The service provider is using SLA to control its resources. SLA is a joint contract between cloud workers and their clients in cloud computing [14].

SLA violations are measured two metrics in IaaS model. The first metric is SLA violations time par active host (SLATAH).

$$SLATAH = \frac{1}{M} \sum_{i=1}^{M} \frac{Ts_i}{Ta_i} \tag{1}$$

Where M is the number of servers, Tsi is the total time of experiencing 100% CPU utilization by server i, resulting in an SLA violation and Tai is the total time of server i being in an active state [15].

The second metric is the overall performance degradation by VMs due to migrations (SLAVM)

$$SLAVM = \frac{1}{N} \sum_{j=1}^{N} \frac{Cd_j}{Cr_j} \tag{2}$$

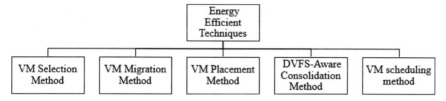

Fig. 1 Energy-efficient technique in cloud computing

where N denotes the number of VMs, Cdj denotes the performance degradation caused by migrating VM, and Crj denotes the total CPU utilization requested by VM j during its lifetime [18].

Third and the last metric are SLA violation (SLAV) which is the product of (1) and (2), defined as follows.

$$SLAV = SLATAH * SLAVM \tag{3}$$

4 Energy Efficient

Energy efficient is a new area of computer science that achieves successful computation, while using minimum energy on cloud networks for that computing. Energy efficient is characterized as a computation that is environmentally meaningful and friendly. It seeks to optimize the usage of electricity use and energy conservation and to reduce the expense of servers and therefore the emission of carbon dioxide. Each cloud network can have many data centers and each data center can have a vast range of services and various forms of resources that through their service can expend a massive amount of energy.

4.1 Energy-Efficient Techniques

Energy-efficient techniques are depicted in Fig. 1. Energy efficient techniques are such as VM selection, VM migration, VM placement, DVFS-Aware consolidation, VM scheduling, VM allocation method [12].

4.2 VM Selection Method

There might be several virtual machines on a physical computer (PM). The virtual machine has migrated the load to the other server, while a server is overwhelmed to reduce the host load and reduce the SLA breach. All virtual machines are migrated immediately, while the server is low load [20]. Many VM selection policies are available: Minimum migration period (MMT) policy, random selection (RS) policy, maximum correlation (MC) policy, minimum usage (MU) policy, host fault detection (HFD) policy [13].

4.3   VM Migration Method

One host to another is the migration of a virtual machine. The time required for and process of cloud computing migrates between virtual machines [15, 16]. This approach establishes a workload level for each system, and when a server is overwhelmed, the VM migrates the load from one server to another server to reduce the overloaded server's energy consumption. Without disregarding the QoS, this method will accomplish an effective decrease in power utilization [12]. There are a few migration strategies that are accessible for the virtual machine migration from one host to the next, such as the pre-copy, the post-copies, the flexible strain, the least recently used (LRU) and the spreading tree, accompanied by the checkpoint recovery and finally, the replays. In resources such as fault tolerance, load balancing, server consolidation, online servicing, this approach helps minimize the cost of use, improve electricity usage and very rapid resource utilization [17].

VM Placement Method: To save energy usage, VMs packed the hosts used to save energy consumption [13]. The aim is to look for hosts where those VMs will operate the least energy and minimal SLA violations [18]. All data center hosts are categorized into three types: overloaded hosts, underloaded hosts, middle loaded hosts. The middle-loaded hosts are the main VM placement containers chosen from underloaded and overloaded [16].

4.4   DVFS-Aware Consolidation Methods

The power control approach is commonly used for dynamic voltage and frequency scaling (DVFS) techniques. DVFS enables the processor's running speed and dynamic modification of the supply voltage, with working load conditions, set [19]. The researchers indicated that the two approaches are VM consolidation first, and power management focused on DVFS techniques second. The first strategy is the consolidation of VMs used to evaluate the complex frequency to assign QoS to the task. DVFS is the second approach used to use power usage with output loss and saves resources for complex load states by 39.14% [12].

4.5 VM Scheduling Method

The process of VM scheduling lowers the use of resources. This strategy is to run in multiple job frameworks to establish an enhanced service level for the heterogeneous machine. In cloud datacentres, one solution is to shut down an idle physical node [12]. Efficiency, time, and expense are the most common scheduling parameters. These criteria improve the effectiveness and decrease energy consumption in the cloud data center. Needed to execute a role in data centers without decreasing efficiency and utilizing less energy [20]. For the service level arrangement and the requested equipment, the data center classifies regular work tasks. Each assignment is then assigned to one of the servers provided at either cost. An answer or consequence is conveyed back to the customer. The purpose of this approach is to maximize its usage and decrease the implementation time of the job [21–24].

5 Conclusion

This study describes a methodical literature review of autonomous resource management in the field of cloud in general and QoS in particular, autonomous resource management. In cloud computing, the present condition of autonomous resource management is distributed into separate divisions. As developed by numerous industry and academic organizations, the methodical study of autonomous resource management in cloud computing and its approaches are defined. In cloud systems, resource pricing models play an important role. There is an increasing interest in understanding better markets for cloud spots, where bidding techniques for the acquisition of computational services are created. Approaches are currently being suggested to simplify competitive pricing and the collection of cloud services. We believe that these models may play a larger role in resource allocation systems in the coming years than they do today.

References

1. Malekloo M-H, Kara N, El Barachi M (2018) An energy-efficient and SLA compliant approach for resource allocation and consolidation in cloud computing environments. Sustain Comput Inf Syst 17:9–24
2. Tarafdar A, Debnath M, Khatua S, Das RK (2020) Energy and quality of service-aware virtual machine consolidation in a cloud data center. J Supercomput 1–32
3. Haghshenas K, Mohammadi S (2020) Prediction-based underutilized and destination host selection approaches for energy-efficient dynamic VM consolidation in data centers. J Supercomput 1–18
4. Mandal R, Mondal MK, Banerjee S, Biswas U (2020) An approach toward design and development of an energy-aware VM selection policy with improved SLA violation in the domain of green cloud computing. J Supercomput 1–20

5. Gul B, Khan IA, Mustafa S, Khalid O, Hussain SS, Dancey D, Nawaz R (2020) CPU and RAM energy-based SLA-aware workload consolidation techniques for clouds." IEEE Access 8: 62990–63003

6. Yaghoubi M, Maroosi A (2020) Simulation and modeling of an improved multi-verse optimization algorithm for QoS-aware web service composition with service level agreements in the cloud environments. Simul Modell Pract Theory 102090

7. Rajabzadeh M, Haghighat AT, Rahmani AM (2020) New comprehensive model based on virtual clusters and absorbing Markov chains for energy-efficient virtual machine management in cloud computing. J Supercomput 1–20

8. Gupta A, Bhadauria HS, Singh A (2020) SLA-aware load balancing using risk management framework in cloud. J Ambient Intell Humaniz Comput 1–10

9. Li Z, Xinrong Yu, Lei Yu, Guo S, Chang V (2020) Energy-efficient and quality-aware VM consolidation method. Futur Gener Comput Syst 102:789–809

10. Bharathi PD, Prakash P, Kiran MVK (2017) Energy efficient strategy for task allocation and VM placement in cloud environment. In: 2017 Innovations in power and advanced computing technologies (i-PACT). IEEE, pp 1–6

11. Goraya MS, Singh D (2020) Satisfaction aware QoS-based bidirectional service mapping in cloud environment. Clust Comput 1–21

12. Ali SA, Affan M, Alam M (2018) A study of efficient energy management techniques for cloud computing environment. arXiv:1810.07458

13. Haghighi MA, Maeen M, Haghparast M (2019) An energy-efficient dynamic resource management approach based on clustering and meta-heuristic algorithms in cloud computing IaaS platforms. Wirel Pers Commun 104(4):1367–1391

14. Raza MR, Varol A (2020) QoS parameters for viable SLA in Cloud. In: 2020 8th international symposium on digital forensics and security (ISDFS). IEEE, pp 1–5

15. Hsieh S-Y, Liu C-S, Buyya R, Zomaya AY (2020) Utilization-prediction-aware virtual machine consolidation approach for energy-efficient cloud data centers. J Parallel Distrib Comput 139:99–109

16. Saadi Y, Kafhali SE (2020) Energy-efficient strategy for virtual machine consolidation in cloud environment. Soft Comput 1–15

17. Kumar GG, Vivekanandan P (2019) Energy efficient scheduling for cloud data centers using heuristic based migration. Clust Comput 22(6):14073–14080

18. Khoshkholghi MA, Derahman MN, Abdullah A, Subramaniam S, Othman M (2017) Energy-efficient algorithms for dynamic virtual machine consolidation in cloud data centers. IEEE Access 5:10709–10722

19. Stavrinides GL, Karatza HD (2019) An energy-efficient, QoS-aware and cost-effective scheduling approach for real-time workflow applications in cloud computing systems utilizing DVFS and approximate computations. Futur Gener Comput Syst 96:216–226

20. Bhattacherjee S, Das R, Khatua S, Roy S (2019) Energy-efficient migration techniques for cloud environment: a step toward green computing. J Supercomput 1–29

21. Jain M, Priya A (2019) Energy efficient algorithms in cloud computing: a green computing approach. Int J Adv Eng Technol 47–52

22. Tiwari PK, Joshi S (2016) A review on load balancing of virtual machine resources in cloud computing. In: Proceedings of first international conference on information and communication technology for intelligent systems, vol 2. Springer, Cham, pp 369–378

23. Tiwari PK, Joshi S (2018) Effective management of data centers resources for load balancing in cloud computing. Int J Inf Retr Res (IJIRR) 8(2):40–56

24. Sisodia PS, Tiwari V, Dahiya AK (2015) Measuring and monitoring urban sprawl of Jaipur city using remote sensing and GIS. Int J Inf Syst Soc Change (IJISSC) 6.2:46–65

Contribution Analysis of Scope of SRGAN in the Medical Field

Moksh Kant, Sandeep Chaurasia, and Harish Sharma

Abstract This paper focuses on the concept of generative adversarial networks (GANs) and their scope in the medical field. "Generative adversarial networks" has been one of the most prominent research areas in the domain of machine learning in the past few years. It consists of two neural networks competing against each other. One of them is a generator model which generates fake samples of data and the other is a discriminator model which receives both real data (from the training data) and fake data (from the generator model) and tries to identify them as real or fake. Use of generative adversarial networks has been done here for the purpose of "super-resolution" because GANs work on the concept of generative modeling. Since applying super-resolution to an image means adding more data to the image which was not previously there (2017), it would require generation of data which might not be actually real data but is so close to the real one that one cannot know the difference. Hence, when we apply super-resolution using generative adversarial networks it gives us way better results in comparison to many other approaches such as "SRCNN".

Keywords GANs · Generator · Discriminator · Data · Resolution · SRGAN

1 Introduction

1.1 About GAN

Generative adversarial network is a type of neural network architecture for generative modeling. Two neural network models are used to train a generative model. The first is known as the "generator" or "generative network" model that learns to produce new samples. The second one is known as the "discriminator" or "discriminative network" and learns to distinguish between real and generated examples (2019). Training is done in a way that the generator tries to generate some examples and discriminator

M. Kant · S. Chaurasia (✉) · H. Sharma
Department of Computer Science and Engineering, Manipal University Jaipur, Jaipur, India
e-mail: sandeep.chaurasia@jaipur.manipal.edu

© The Author(s), under exclusive license to Springer Nature Singapore Pte Ltd. 2022   341
P. Nanda et al. (eds.), *Data Engineering for Smart Systems*, Lecture Notes in Networks and Systems 238, https://doi.org/10.1007/978-981-16-2641-8_33

is provided with both generated ("Fake") and real images. The discriminator tries to differentiate between real and fake whereas it is the work of the generator to fool the discriminator so as to increase its probability of making a mistake by producing various examples like the "real images".

The myriad GANs are deep convolutional GANs (DCGANs), conditional GANs (CGANs), StackGAN, Wasserstein GANs(WGAN), etc. Revent contributions of GANs have shown plausible results in many areas like semantic-image-to-photo translation, text-to-image translation, generate new human poses, and super-resolution. Their applications start from and are not limited to entertainment industry, fashion industry, medical field, science and research.

Although a lot has been done with GANs, there is still a lot of unexplored potential. Since there are a lot of applications of GANs, we restrict our focus to "super-resolution" and its possible applications and scope in the medical field. Super-resolution using GANs is an application, which apart from being a separate application, may also bring significant improvement in other applications as well as different types of researches that do not provide sufficient results due to "shortage of high-definition data" or "low-resolution results".

2 Related Work

From majority of the researches, it can be seen that these generative adversarial networks being comparatively new have a lot of unexplored potential. Their training can however be a little complicated in some cases such as a generator learning way faster than the discriminator and producing images that the discriminator cannot identify as fake. The reverse of this is also true, i.e., the discriminator may learn so fast that it starts identifying everything as fake (generator fails due to vanishing gradient problem). However, if the learning rate and other parameters are kept stable (learning rate can neither be too slow nor too fast), they have a lot of potential. Generative adversarial networks have shown commendable results and have outperformed many other approaches.

Now if we talk about the medical field, a lot of research is going on about how GANs can be useful, however, it is still relatively new. Two particular research papers have shown that there is a lot of potential for GANs in the medical field. If we talk about the challenges that are being faced in the medical field, two major challenges come up again and again and GANs have shown the potential to overcome them.

Firstly, having image data along with their corresponding explanatory verities is very expensive and takes a lot of time since data annotations are done manually by medical experts (2018). This causes many problems since for GANs and deep learning models in general, there is a requirement of large amounts of data. However, GANs being based on generative modeling produce the data themselves. One good example of this is the dataset SynthMed (2018). SynthMed is a dataset which consists of retinal images (91 pairs till date) which have been generated based on the original data from DRIVE database. This also results in overcoming of the second problem which is data privacy. Since another major reason of data shortage is data privacy, GANs solve this along with huge amounts of data, since the images do not belong to any actual person even though they are really close to them.

For GANs to be used by doctors in the medical field there is a requirement of instant result. Now the authors of the paper "Towards Real-Time Image Enhancement GANs" [1] propose and implement a method for real time image enhancement using GANs based on which an approach of Fast-SRGAN (Link) has been made. This Fast-SRGAN approach if used for medical images has tremendous potential.

3 Methodologies

A lot of techniques are there to implement SRGAN. Since to implement GANs the first step is to understand the structure of GANs, we have studied and implemented the UCGAN. Fashion MNIST dataset is used so as to produce various kinds of fashion images (2019) with UCGAN [2]. It can be considered as a five-step process (Fig. 1).

Now to understand super resolution, we have first implemented SRCNN [4] in which image degrading and other preprocessing functions have been applied on the high-resolution images of various varieties. Although there is considerable improvement and a good match in terms of structural similarity in a lot of cases, in some cases it is not clearly visible to the naked eye. When we compare GANs to CNNs, a lot of researches show that GANs outperform CNNs [5] and even some other approaches in a lot of tasks.

To perform super-resolution with GANs, different approaches can be implemented, some of them being Tensorlayer (2017), Pytorch (2018) and third of the Fast-SRGAN [6] which the author has based on the paper "Towards Real-Time Image Enhancement GANs" [1]. In Fast-SRGAN [7] inverted residual blocks rather than

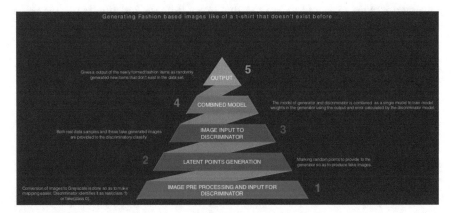

Fig. 1 It shows a pyramid structure of the various steps taken during implementation of unconditional GAN for Fashion MNIST dataset [3]

normal residual blocks are used to get parameter efficiency and speed the operation. The medical dataset used for testing is "NIH chest X- Ray Dataset" [8].

4 Experimental Work

Implementation of Unconditional GAN using Fashion MNIST dataset (2019).

First, we import Fashion MNIST dataset. We use NumPy and Keras. The dataset in a compressed form is of 25mb. We start by taking 100 images, hence, it does not require a sophisticated generator or discriminator. The image is converted into grayscale to make the color mapping easier. Random images are taken and latent points are generated which act as generator input and help in generating novel images.

We define two functions: first for the generator (generator function generates the fake images) and the second for the discriminator. Both of these are sequential model functions. Discriminator [9] takes as input a grayscale 28×28 image and gives the prediction identifying the image as real (class $= 1$) or fake (class $= 0$).

It is carried out as a simple CNN using finest practices for GAN such as the LeakyReLU [10] function having a slope of 0.2 along with the Adam(optimizer) version of stochastic gradient descent having a learning rate of 0.0002 and a momentum of 0.5.

A GAN model is designed that combines both the discriminator and generator model into one larger model which will be used to train the generator model weights with the help of output and error calculated by the discriminator model. Separate training of the discriminator model is done and the model weights are marked as not trainable in this larger GAN model to ensure that only the generator model weights are updated. The resultant change to the trainability of the discriminator weights is effective only when training the combined GAN model rather than when training the discriminator alone. Channels are added to make it a 3D set of images. In the generated images (the output), we see that every time the function is run we see a different set of images that are given as the result (Figs. 2a–c, 3 and Tables 1, 2).

Observations:

1. Increasing no. of epochs result in better images. Lesser epochs result in more random looking images and (Fig. 2a is better than Fig. 2c).
2. Some images are not complete (e.g., t-shirt missing sleeve).
3. Different types of images are created with no specific order

4.1 SRCNN Implementation

For this we import keras, openCV2, NumPy, Skimage, and matplotlib [11]. Some other necessary packages are also imported. We then define the various quality metrics like PSNR, MSE, and SSIM. First, we degrade the data images to generate

Fig. 2 **a** Output of newly generated images after 100 epochs. **b** Output of newly generated images after 25 epochs. **c** Output of newly generated images after 5 epochs

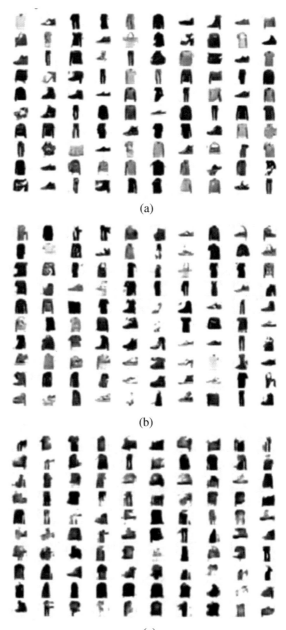

(a)

(b)

(c)

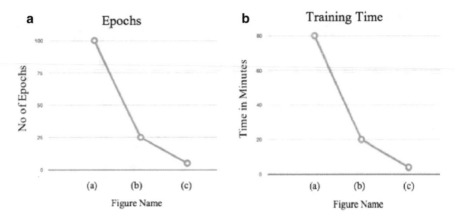

Fig. 3 **a** Epoch comparisons of all three outputs. **b** Training time comparisons of all three outputs

Table 1 Epochs along with respective training time for UCGAN using the Fashion MNIST dataset: **a** output with generator model trained on 100 epochs showing results better than 25 and 5 epochs; **b** output with generator model trained on 25 epochs showing results better than 5 epochs; **c** output with generator model trained on 5 epochs showing comparatively distorted images. All the training has been done on Google Colab along with accelerated hardware (GPU)

	(a)	(b)	(c)
Training time	80 min	20 min	4 min
Epochs	100	25	5

Table 2 Shows the comparison of the original images 4(a) and 4(b) with respect to degraded and SRCNN applied images based on the metrics PSNR, MSE, and SSIM

	PSNR	MSE	SSIM
4(a) Degraded	19.8378	2024.9522	0.8446
4(a) SRCNN	23.3545	901.0465	0.8722
4(b) Degraded	29.6993	209.0595	0.9407
4(b) SRCNN	31.1153	150.8933	0.9485

low-resolution images and then we apply super resolution and use the above-mentioned metrics to compare the original images with the newly generated SR images. It is a three-step process:

A. Degrading Images

Images are degraded by resizing them (changing of height and width). Then we expand them which result in a lower resolution and save them along with their comparison with the original image (bicubic downscaling is also tried to degrade images in some cases).

B. SRCNN Model

Adam optimizer is used with a learning rate of 0.0003, and the model is a sequential model. There are three Conv2D layers which are added. Prediction function is used in which first the degraded images are loaded. Image preprocessing is done and then super-resolution is applied. Basic shaving of image is done, and calculations are performed and comparisons of images are done (Fig. 4).

C. Outputs

Output is given as a comparison of original image, degraded image, and the super-resolution applied image (Fig. 5).

In case of its applications in the medical field, we can see an example mentioned below (Fig. 6). This type of example which may help on a computer will not help doctors/scientists in real life because it is not visible to the naked eye.

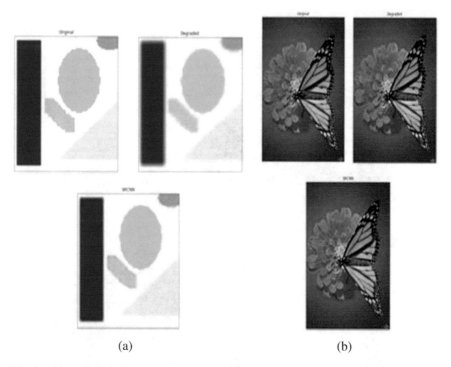

(a)   (b)

Fig. 4 **a** Shows the original, degraded, and SRCNN applied example of a.bmp shape image and **b** shows the original, degraded, and SRCNN applied example of a butterfly image. Now as we can see SRCNN shows results but they are not that clearly visible to the naked eye in some cases

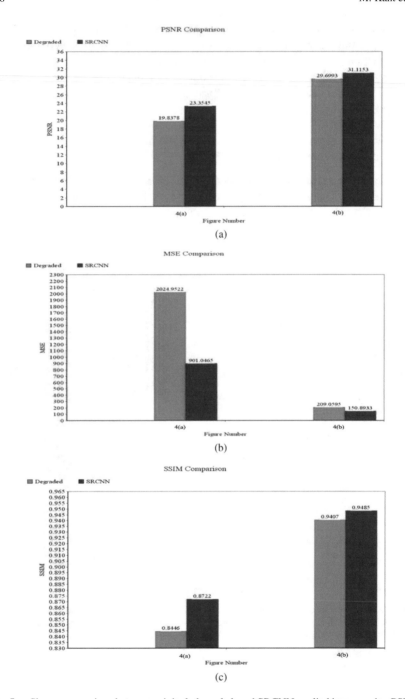

Fig. 5 a Shows comparison between original, degraded, and SRCNN applied images using PSNR metric; **b** shows comparison between original, degraded. and SRCNN applied images using MSE metric; **c** shows comparison between original, degraded, and SRCNN applied images using SSIM metric [12–14]

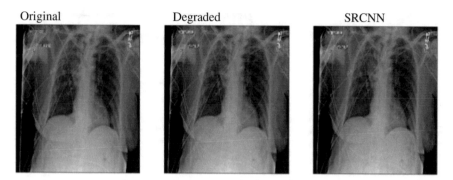

Fig. 6 Shows the comparison of original, degraded, and SRNNN on NIH chest X-Ray dataset

4.2 SRGAN Implementation

There can be many approaches for implementing SRGAN such as Tensorlayer (2017), Pytorch (2018), and Fast-SRGAN (2019).

In the discriminator, we take the LeakyReLU activation function (unlike in the generator which has ReLU activation function) to prevent dying state by allowing some negative values to pass through. The generator receives gradient values from the discriminator, and if the network is stuck in a dying state situation, the learning process will not happen (2018). We also have to ensure proper residual blocks, SubpixelConv2D layer for upsampling and batch normalization [15, 16].

The third approach is that of the Fast-SRGAN which the author has implemented based on the research paper "Towards Real-Time Image Enhancement GANs" [1]. The Author describes the structure as follows (Fig. 7).

In implementation of Fast-SRGAN, we first import various libraries like Tensor-Flow, OpenCV, Python, NumPy, etc. This approach is based on the normal SRGAN approach just that inverted residual blocks, rather than normal residual blocks, are used to get parameter efficiency and speed the operation. We start by optimizing and assigning HR and LR shapes. We have six inverted residual blocks. We take a learning decay rate of 0.1 and 100,000 steps for both the generator and the discriminator. We use the Adam optimizer. We use a pretrained VCC19 model to get image characteristics from the high resolution and the generated high-resolution images and minimize the use between them [17–19].

Content Losses (2019): Pixelwise MSE loss

$$l_{MSE} = \frac{1}{} \sum_{x=1}^{W} \sum_{y=1}^{H} \left(I_{x,y} - I_{x,y}^{R}\right)_{WH\cdot2} \tag{1}$$

Perceptual loss

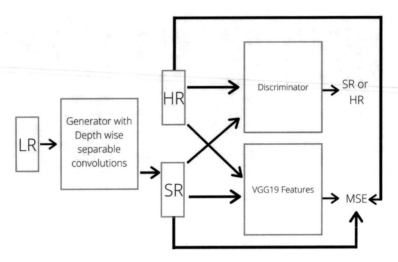

Fig. 7 Depicts the GAN structure for the Fast-SRGAN approach

$$l_p^2 = \frac{1}{WH} \sum_{x=1}^{W_f} \sum_{y=1}^{H_f} \left(\phi_j(I)_{x,y} - \phi_j(I^R)_{x,y} \right) \tag{2}$$

φj(I) indicates the activations of some jth layer of the pretrained network for an input image I, Wf represents the width and Hf represents the height of the feature maps. Adversarial loss.

Total loss of the generator is

$$l_{AR} = l_{MSE} + \lambda 1 l p + \lambda 2 l a d v \tag{3}$$

where adversarial loss is:

$$l a d v = -\log \left(D_\psi \left(I^R | I^C \right) \right) \tag{4}$$

We now calculate the output shape of discriminator and give 32 filters in the first layer of both the discriminator and generator. We calculate the contest loss and return the mean squared error. We then build and compile the discriminator followed by building and compiling of the generator for pretraining. In the residual blocks, we have Conv2D, batch normalization, and activation layers. We also have a deconvolutional layer. We create and compile the model and build the generator that will do the super-resolution task [20–22]. The output can be seen (Fig. 8).

Original   Degraded   SRGAN

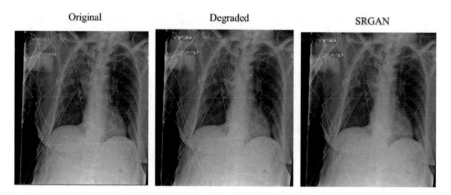

Fig. 8 Shows the comparison of original, degraded, and SRGAN on NIH chest X-ray dataset

5   Conclusion

Performance of a Fast-GAN is better than SRCNN in case of enhancement of medical images which can be seen with the naked eye.

GANs can be used for high-quality data generation which can be highly beneficial in medical field since it would result in overcoming of data privacy issue and data shortage.

We do not use the same metrics that we use for SRCNN in the case of SRGAN because it is not optimized for it. Even for a pair of identical images, a shift by only one pixel on one of the images causes a large increase in the distance of the metrics. Pretrained on natural images, GAN-based models may synthesize unrealistic patterns in medical images affecting clinical analysis and diagnosis. MOS is one metric which can be used for better visual comparison.

All training has been done on Google Colab. However, in the case of training, accelerated hardware (GPU) is a must. A lot of RAM is required so that the training and testing can be completed otherwise, and it may crash while training/testing. Google Colab offers a RAM of 12 GB which is not sufficient in case of the second approach especially since Pytorch itself reserves some of the memory which is not usable and hence CUDA/CPU does not have sufficient memory allocation. However, with pretrained generator model in the case of Fast-GAN (trained on DIV2k dataset), we can see that the results are better than SRCNN if we compare them visually.

6   Future Work

Training on specific types of medical images for, e.g., training on chest scans for enhancement of chest scans itself and similarly for other specific needs may give better results. Also, implementation of a metric that makes various medical mappings easier is also very important (2019), since it would help in complete automation of reports and data which would make the tasks of doctors and scientists very easy.

References

1. Galteri L, Seidenari L, Bertini M, Bimbo AD (2019) Towards real-time image enhancement GANs
2. Radford A, Metz L, Chintala S (2016) Unsupervised representation learning with deep convolutional generative adversarial networks
3. Ledig C, Theis L, Husz´ar F, Caballero J, Cunningham A, Acosta A, Aitken A, Tejani A, Totz J, Wang Z, ShiTwitter W (2017) Photo-realistic single image super-resolution using a Generative Adversarial Network
4. Wang T-C, Liu M-Y, Zhu J-Y, Tao A, Kautz J, Catanzaro B (2018) High-resolution image synthesis and semantic manipulation with conditional GANs
5. Iqbal T, Ali H (2018) Generative adversarial network for medical images (MIGAN)
6. Kazeminia S, Baur C, Kuijper A, vanGinneken B, Navab N, Albarqouni S, Mukhopadhyay A (2019) GANs for medical image analysis
7. Liu J, Spero M, Raventos A (2017) Super-resolution on image and video
8. Ramavat K, Joshi M, Swadas PB (2016) A survey of super resolution techniques
9. Naik S, Patel N (2013) Single image super resolution in spatial and wavelet domain
10. Takano N, Alaghband G (2019) SRGAN: training dataset matters
11. Kovalenko B (2017) Super resolution with generative adversarial networks
12. Goodfellow I (2017) NIPS 2016 tutorial: Generative adversarial networks
13. Guibas JT, Virdi TS, Li PS (2018) Synthetic medical images from dual generative adversarial networks
14. Che Z, Cheng Y, Zhai S, Sun Z, Liu Y (2017) Boosting deep learning risk prediction with generative adversarial networks for electronic health records
15. Brownlee J (2019) 18 impressive applications of generative adversarial networks (GANs)
16. Mwiti D (2018) Introduction to generative adversarial networks (GANs): types, and applications, and implementation
17. Hui J (2018) GAN—super resolution GAN (SRGAN)
18. Thomas C (2019) Deep learning based super resolution, without using a GAN
19. Birla D (2018) Single image super resolution using GANs—Keras
20. Brownlee J (2019) How to develop a conditional GAN (cGAN) From scratch
21. Nayak M (2018) Deep convolutional generative adversarial networks (DCGANs)
22. Wang X, Yu K, Wu S, Gu J, Liu Y, Dong C, Loy CC, Qiao Y, Tang X (2018) ESRGAN: enhanced super-resolution generative adversarial networks

Machination of Human Carpus

Sumit Bhardwaj, Bhuvidha Singh Tomar, Adarsh Ankur, and Punit Gupta

Abstract Along with its multiple benefits, advancements in technology brings the serious problem of sedentary lifestyle. A sedentary lifestyle is defined as a type of lifestyle where an individual does not receive regular amounts of physical activity. This physical inactivity is a leading cause of joint immobility in younger adults as well as adults. Joint immobility has serious impacts on a person's social, mental and physical well-being. The model presents an ideal initiation to the determination of dysfunction in the human wrist. The ideal is to bring about a design to better understand the range of motion of the human wrist at its two extremes, by plotting it on the co-ordinate axis. The idea is to bring about a reform in the domain of anatomical displacements.

Keywords Co-ordinates · Plot · Wrist · Web camera · Capture · Machination

1 Introduction

This paper overturns the old-school way of visiting an orthopaedician which was viewed as an ideal approach for diagnosing wrist dysfunction [1]. The past few months of lockdown in view of the alarming rise in the COVID-19 cases have thrown light upon the possible problems of revisiting hospitals at the drop of a hat [2]. The idea behind the model is to develop a viable monitoring system by combining miniaturized, durable and low-cost sensors (web cameras), with computer vision [3–6]. The system will find the co-ordinates for the two extreme positions of the wrist. The idea of using a web camera for plotting co-ordinates has long been floating in the domain of technology [7]. Applying this pre-existing ideology to the biomedical

S. Bhardwaj (✉) · B. S. Tomar · A. Ankur
Department of Electronics and Communication Engineering, Amity School of Engineering and Technology, Amity University, Noida, Uttar Pradesh, India
e-mail: sbhardwaj58@amity.edu

P. Gupta
Department of Computer and Communication Engineering, Manipal University Jaipur, Jaipur, India

field is the true inspiration behind the research [8]. The sole purpose is finding the co-ordinates of the wrist at its two extremes. The model deals with remote self-analysis of the range of motion of the wrist [9].

In the past, machination of human joints have been carried out to better understand human–computer interaction. Real-time hand gesture recognition using finger segmentation have been carried out for feature extraction, where the region of interest is extracted. The following method has been used for machination using data gloves and Kinect sensors. However, our method stands novel due to the real-time wide application it possess. Furthermore, no expensive technologies have been incorporated [10]. Local binary patterns and histograms algorithm [11] have been applied to face detection using a web camera. Our model takes inspiration from this very invention.

In geometry, a coordinate system is a system that uses one or more numbers, or co-ordinates, to uniquely determine the position of the points or other geometric elements on a manifold such as Euclidean space [12]. We have used this very co-ordinate system to plot the positions of the two extremes of the wrist.

The wrist is simply an ellipsoidal synovial joint that permits movements along two axis. All the movements at the wrist are carried out by the muscles in the forearm. A bi-axis movement means that all four movements, i.e. abduction, adduction, flexion and extension, can occur at the site of this joint. Our model deals with the co-ordinates of flexion [13, 14] and extension [15, 16]. The idea is simply to mathematically analyse the bi-axis movement at the wrist. For this very purpose we have determined the various co-ordinates. Further studies will help in using the co-ordinate system for biomedical fields.

2 Related Work

In this section, a survey of existing work is showcased. McClintock [20] represents in his paper HGPS mutant Lamin A primarily targets human vascular cells as detected by an anti-Lamin A G608G antibody by performing tests on rabbits. The methods used by him were characterization of anti-Lamin A, primary dermal fibroblast cells, indirect immunofluorescence and western blot analysis. Kashyap et al. [21] talk about HGPS in his case study in which a 3-year-old boy was observed. The boy showed all the common symptoms from the age of 6 months like hair loss, short stature, wrinkled skin, large head, etc. He noticed that the boy had some pigmentations on his body along with the similar IQ of a normal person. He studied the differences between a normal cell and a cell with Progeria. As his conclusion, he wrote that HGPS is a very different syndrome and should not be confused with other syndromes. Ullrich [22] in his paper about craniofacial abnormalities in HGPS says that the patients' data were taken from the Progeria Research Foundation medical and research database. He used neuroimaging studies which included MR, CT and radiographs of the skull, full face and neck. As his results, it was known that the disease-related abnormalities were thinning of calvarium, short mandibular rams, kinking of optic nerves, hypotelorism and prominent eyes. Guide [23] wrote in his study about a 22-year-old male who was admitted in a hospital. He named the study as 'the curious case of ageing'. When tests were conducted on the patient, it was found that he was way too old for his age. According to him, early diagnosis and suitable treatment for the same can increase life expectancy. Gungor [24] in his study 'Comprehensive dental management' writes about a 3-year-old girl with lower height and weight as compared to normal people. Thin sharp nose was also seen as another symptom among others. Cranial radiograph, complete blood count, biochemistry tests and intramural examination was done on her. A complete dental treatment plan was proposed for her. After giving her local anaesthesia, the required dental treatment was done. Kuru [25] in his paper about analysis of biomedical visual data in order to build an intelligent diagnostic decision support system in medical genetics which talks about image acquisition and face detection in dysmorphology. Image enhancement and feature extraction module were made. As his conclusion, automatic image processing and computational moiling can help to predict syndromes by reading facial images.

3 Proposed Methodology

The model initiates with the command to turn on the camera. Basic machine learning techniques have been implemented to enable the program to auto-detect the wrist via the web camera. The sole system pre-requisites are a personal computer or laptop enabled with a web camera. Once the camera has been turned on, the program auto-detects the wrist. The user will then be given 10 s to adjust his/her wrist to one of the two extreme positions. Post completion of the allotted time for making adjustments, the program captures the image of the wrist. The images thus captured are subsequently saved in a folder. This folder will be later used for creation of a database, for an in-depth study of the range of motion of the wrist. An algorithm was created using Python to convert these images into set of co-ordinate points. Python being an open-source user-friendly language simplified the task at hand. The library used for the model is OpenCV. This library of programming functions was selected for its real-time computer vision applications. Also, the library is free to use, thus, reducing the monetary cost of the project. The icing on the cake is its cross-platform.

The model is aimed at bringing a reform in the pre-existing studies of wrist mobility and morbidity (Fig. 1).

The subject was introduced to the task at hand. Next, he was asked to follow the steps requested by the program. The algorithm gives a novel real-time method for hand gesture recognition. In the given framework, the hand region is extracted from the background using background subtraction. Then the detected hand gesture is converted to yield its co-ordinates. Background subtraction is a methodology used for segmenting foreground elements from background ones. The idea is to extract the foreground elements. This is simply done by generating a foreground mask. This method is incorporated for exposing dynamically operating subjects from static image capturing equipment. The actual motive being the tracking of the subject.

Fig. 1 Flowchart

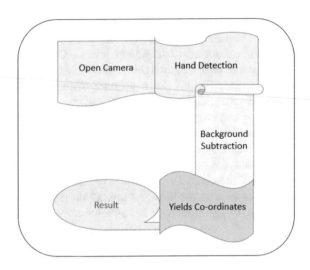

Figure 2 is the position of the wrist that was captured post the completion of the 10 s allotted for wrist adjustment. The latter position is at the other extreme of the range of motion of wrist in comparison to the initial one. The images obtained are post background subtraction, i.e. the background of the real-time hand has been eliminated from the captured image.

4 Results

4.1 Bone Density

The input that was taken in the form of Fig. 1 was then processed to yield the following results. In order to obtain the co-ordinates, the hand was captured in real-time using a web camera. This captured image underwent background subtraction, to yield the processed input images [11]. Later, these images were converted to co-ordinates.

Figure 3 is the image co-ordinate of Fig. 2.

Furthermore, the above co-ordinates were yielded on the screen (Fig. 4).

The above image represents the co-ordinates of the two extremities of the wrist of a given subject.

5 Conclusion

Across the globe, joint dysfunction has become a serious threat to healthy living. Musculoskeletal disorders result in significant morbidity and higher mortality rates

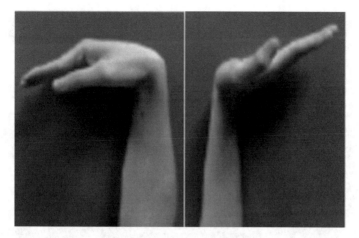

Fig. 2 Position of wrist

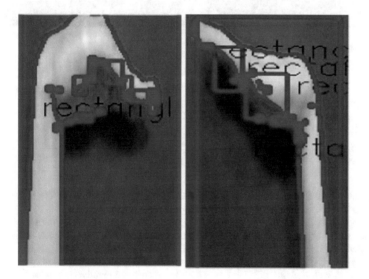

Fig. 3 Result

(in comparison to age and gender-matched peers). The impact on global disability burden due to joint malfunctioning is enormous [17]. Lack of time due to work overload combined with long waiting hours makes visit to an orthopaedician difficult. The optimal target is to bring the model into orthopaedic use as the same model can be universally applied to all subjects. Our study dealt with finding the co-ordinates for one given subject. However, the success of the model proves its utility and generalization. With a few more developments, the model can be bought into daily use

Fig. 4 Output

[18–26]. A similar model can also be created for all other human joints. A similar model can be created for other organisms as well.

References

1. Charnley J (1959) The lubrication of animal joints. In: Proceedings of the symposium on biomechanics, Institution of Mechanical Engineering, London, pp 12–22
2. El-Gohary M, Pearson S, McNames J (2008) Joint angle tracking with inertial sensors. In: Proceedings of the 2008 30th annual international conference of the ieee engineering in medicine and biology society. Vancouver, BC, Canada, pp 1068–1071
3. McCutchen CW (1962) The frictional properties of animal joints Wear 5:1–17
4. Baradel F, Wolf C, Mille J (2017) Pose-conditioned spatiotemporal attention for human action recognition. 1703.10106
5. Gkioxari G, Toshev A, Jaitly N (2016) Chained predictions using convolutional neural networks. In: European conference on computer vision (ECCV)
6. Herath S, Harandi M, Porikli F (2017) Going deeper into action recognition: a survey. Image Vision Comput 60(Supplement C):4–21
7. Insafutdinov E, Pishchulin L, Andres B, Andriluka M, Schiele B (2016) DeeperCut: a deeper, stronger, and faster multi-person pose estimation model. In European Conference on Computer Vision (ECCV)
8. Ionescu C, Papava D, Olaru V, Sminchisescu C (2014) Human3.6m: large scale datasets and predictive methods for 3d human sensing in natural environments. TPAMI 36(7):1325–1339
9. Iqbal U, Garbade M, Gall J (2017) Pose for action-action for pose. FG-2017
10. Jhuang H, Gall J, Zuffi S, Schmid C, Black MJ (2013) Towards understanding action recognition. In: The IEEE international conference on computer vision (ICCV). https://doi.org/10.1109/iccv.2013.396

11. Kokkinos I (2017) Ubernet: training a 'universal' convolutional neural network for low-, mid-, and high-level vision using diverse datasets and limited memory. Computer Vision and Pattern Recognition (CVPR)

12. Lifshitz I, Fetaya E, Ullman S (2016) Human pose estimation using deep consensus voting. Springer International Publishing, Cham, pp 246–260

13. Liu J, Shahroudy A, Xu D, Wang G Spatio-temporal lstm with trust gates for 3d human action recognition. In B Leibe, J Matas, N Sebe, M Welling (Eds.), ECCV, pp 816–833

14. Liu J, Wang G, Hu P, Duan L-Y, Kot AC (2017) Global context-aware attention lstm networks for 3d action recognition. In: The IEEE conference on computer vision and pattern recognition (CVPR)

15. Luvizon DC, Tabia H, Picard D (2017) Human pose regression by combining indirect part detection and contextual information. CoRR, abs/1710.02322

16. Luvizon DC, Tabia H, Picard D (2017) Learning features combination for human action recognition from skeleton sequences. Pattern Recognit Lett

17. Martinez J, Hossain R, Romero J, Little JJ (2017) A simple yet effective baseline for 3d human pose estimation. In: ICCV

18. Bhardwaj S, Kumar A, Yadava RL Approximation and analysis of single band FIR pass integrator centered around mid-band frequencies with degree $k = 1, 2, 3\ldots$ Period Polytech Electr Eng Comput Sci 64(4):66–373

19. Bhardwaj S, Kumar A, Yadava RL (2021) Approximation and analysis for fir based multiband pass integrator for frequency ?m ; 0<?m<?. Suranaree J Sci Technol

20. McClintock D, Gordon LB, Djabali K Hutchinson–Gilford progeria mutant lamin a primarily targets human vascular cells as detected by an anti-Lamin A G608G antibody. In: Proceedings of the national academy of sciences of the United States of America 103(7):2154–2159

21. Kashyap S, Shanker V, Sharma N (2014) Hutchinson-Gilford progeria syndrome: a rare case report. Indian Dermatol Online J 5:478–481

22. Ullrich NJ, Silvera VM, Campbell SE, Gordon LB (2012) Craniofacial Abnormalities in Hutchinson-Gilford Progeria Syndrome. Am J Neuroradiol 33(8):1512–1518

23. Gude D, Abbas A, Zubair M (2013) The curious case of ageing. Int J Health Allied Sci 2:43–45

24. Gungor OE, Nur BG, Yalcin H, Karayilmaz H, Mihci E (2015) Comprehensive dental management in a Haller-mann-Streiff syndrome patient with unusual radiographic appearance of teeth. Niger J Clin Pract 18:559–562

25. Kuru K et al (2014) Biomedical visual data analysis to build an intelligent diagnostic decision support system in medical genetics. Artif Intell Med

26. Verma VK, Jain T (2018) Machine-learning-based image feature selection. Feature dimension reduction for content-based image identification. IGI Global, pp 65–73

Building Machine Learning Application Using Oracle Analytics Cloud

Tarun Jain, Mahek Agarwal, Ashish Kumar, Vivek Kumar Verma, and Anju Yadav

Abstract The main objective behind doing this work is that the future is autonomous and these systems improve analytics in cloud by providing "machine to human" environments. Oracle Analytics Cloud introduces the ability to "explain" an attribute in context of the other attributes and metrics in the dataset! Uncover what drives your results, be transparent with your findings, and analyze key segments of customer behavior. Hence with the growing amount of data in today's world, an autonomous database would be perfect to handle it. The complete methodology adopted focuses on data preparation using the explain and other features of Oracle and then analyzing using different visuals after cleaning the data to get insights followed by creating different machine learning models and a small sample of natural language processing model. This entire paper focuses on bringing out the insights of sales that was further used to forecast the same for future, hence helping the company to strategize its sales all over. Through this paper, I have created several models which are further used on the autonomous database to answer the queries. Through this paper, I gained vast knowledge regarding various utilities of Oracle and how the technology is emerging which can help me in my future works. Oracle Analytics Cloud is seeking to go toe-to-toe with Microsoft (Power BI), Tableau and Qlik in particular. Why? All three vendors' offerings are popular for BI and analytics. Moreover, they address many of the same areas as Oracle's offering, including visual analysis and discovery, machine learning-driven data prep, and augmented analytics in the form of natural language queries. The tools used where Oracle Analytics Cloud, Oracle Analytics Desktop, Sql loader.

1 Introduction

Autonomous data warehouse provides an easy-to-use, fully autonomous data warehouse that scales elastically, delivers fast query performance, and requires no database administration. It is designed to support all standard SQL and business intelligence

---

T. Jain · M. Agarwal · A. Kumar (✉) · V. K. Verma · A. Yadav
SCIT, Manipal University Jaipur, Jaipur, India

(BI) tools and provides all of the performance of the market-leading Oracle Database in an environment that is tuned and optimized for data warehouse workloads. Oracle Machine Learning provides a notebook style application designed for advanced SQL users and provides interactive data analysis that lets you develop, document, share, and automate reports based on sophisticated analytics and data models [14]. In order to model this data, dimensional data modeling is used. A dimensional model is a data structure technique optimized for data warehousing tools. The concept of dimensional modeling was developed by Ralph Kimball and is comprised of "fact" and "dimension" tables. The key features of Oracle Analytics Cloud are best practice key performance indicators (KPIs) driven by insights from machine learning. KPIs have the capability for rapid deployment which facilitates a subsequent rapid return on investment and the ability for users to expand and contract their usage based on the needs of their business [1, 13].

The goal is to build up a few ML models with a superior exhibition than the visual determination of specialists and consequently at the same time limit the subjectivity. The dataset is a cigarette sales dataset over the years in different parts of the world by two major companies used comprised of three parts. The three major algorithms used are: Neural network, random forest (RF), and support vector machine (SVM).

2   Methodology

We acquired the cigarette sales dataset and utilized Oracle Analytics Desktop as the stage to code. Our approach includes utilization of arrangement systems like support vector machine (SVM), random forest, and neural network.

2.1   Dataset

The dataset used is comprises of three parts:

- VST-GPI Primary Data: Primary sales of two companies (GPI & VST) of cigarettes, various brands it sells on, its nature, its monthly sales over three years which is used as training set for the algorithms. The total sales are divided monthly, quarterly, total average sales from the year 2013–2015.
- VST-GPI Secondary Data: Secondary sales of two companies (GPI & VST) of cigarettes, various brands it sells on, its nature, its monthly sales over three years which is used as test set for the algorithms. The total sales are divided monthly, quarterly, total average sales from the year 2013–2015.
- Zone-wise Sales Distribution: Sales of two companies and its various branches over three years in different zones, in order to maximize, minimize sales in various zones along with its state-wise distribution in all regions.

2.2 Data Preprocessing

Information preprocessing is an information mining system that changes raw, improper information into a reasonable one providing better details. Proper information providing genuine results is usually less, conflicting, and hence contains numerous mistakes. Information preprocessing plans crude information for additional preparing [9, 12]. Here in this dataset, there where a lot of numerical factors having NA values that needed to be preprocessed, as well as when feature values are not in the same range abrupt results are obtained so using feature scaling they are bought on a similar scale.

2.3 Store in Autonomous Data Warehouse

In order to store the data in autonomous data warehouse, following steps needs to be followed:

1. Open SQL Developer and connect to your autonomous data warehouse database as user adwc_user.
2. Create a credential name. You reference this credential name in the copy_data procedure in the next step.
3. Specify the credentials for your Oracle Cloud Infrastructure Object Storage service.

After running this script, your object store's credentials are stored in your autonomous data warehouse adwc_user schema. All data load operations done using the PL/SQL package DBMS_CLOUD are logged in the tables dba_load_operations and user_load_operations. These tables contain the following:

- dba_load_operations: shows all load operations.
- user_load_operations: shows the load operations in your schema.

1. Query these tables to see information about ongoing and completed data loads.
2. Examine the results. The log and bad files are accessible as tables.

Now the entire data is loaded on cloud so we have to create a connection in order to work ahead which is done as follows. Menu ⇒ Data ⇒ Data flow ⇒ Create Data flow ⇒ Data flow ⇒ Add data.

2.4 Data Preparation

We can create a data flow and select the columns that we would want to feed for our algorithm. Data flow enables to organize and integrate data to produce a curated dataset that users can analyze. To build a data flow, we add steps. Each step performs a specific function, for example, add data, join tables, merge columns, transform data, save the data. Use the data flow editor to add and configure our steps. Each step is validated when we add or change it. When we have configured our data flow, we execute it to produce a dataset.

3 Model Selection

The most energizing stage in building any ML model is determination of calculation. We can utilize more than one sort of information mining strategies to huge datasets. In any case, at elevated level, each one of those various calculations can be characterized in two gatherings: regulated learning and unaided learning. Regulated learning is the technique where the machine is prepared on the information which the info and yield are all around marked. In unaided learning calculation, the machine is prepared from the information which is not marked or ordered making the calculation to work without legitimate guidelines [7]. In our dataset, we have the result variable or dependent variable for example Y having just two arrangement of qualities, either M (Malign) or B(Benign). So, classification calculation of directed learning is applied on it. We have picked five distinct sorts of arrangement calculations in machine learning [4, 10].

1. Support vector machine
2. Decision tree
3. Random forest

Once a particular model is approved, we can use the same model and apply to other data files as and when they are updated on the autonomous database; in this way, it would help us in future where we have a model ready now as more and more data comes we just need to apply that model on it, instead of doing everything from scratch for it [2].

3.1 Support Vector Machine

A discriminative classifier officially characterized by an isolating hyperplane is the support vector machine. The primary goal of this calculation is to discover a hyperplane in a N-dimensional space that particularly orders the information focuses. So as to characterize information focuses, we use hyperplanes as choice limits. These information focuses which spread in either sides can be credited to various classes. The straight separation which is determined between the nearest information focuses and the plane is alluded to as edges. Maximal marginal hyperplane is that ideal line that can isolate two classes in the line that has the biggest margin [11]. The margin is determined as the opposite good ways from the line to any nearest focuses. Just

Fig. 1 SVM classifier

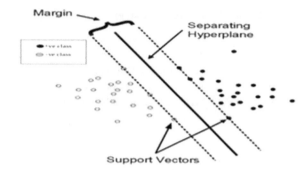

these focuses are significant in characterizing the line and in the development of classifier. These focuses are called support vectors which bolster or characterize the hyperplane (Fig. 1).

$$h_{w,b}(x) = g(w^T x + b)$$

Class labels are denoted as −1 for negative class and +1 for positive class in SVM.

$$y \in \{-1, 1\}$$

The final optimization problem that we get fitting the best parameters:

$$min \frac{1}{2} ||w||^2$$

$$s.t. y_i(w.x_i) >= 1, \forall x_i$$

Optimization problem that the SVM algorithm solves

3.2 Random Forest

Random forest is an adaptable, simple to utilize calculation that gives, even without hyperparameter tuning, an extraordinary after effect of the time. It is one of the most utilized calculations and its easy to utilize. It tends to be utilized as both order and relapse assignments. Random forest is a supervised (labeled data) learning algorithm. Random forest builds multiple decision trees and then merges them together to give a more accurate and stable and efficient prediction. There are two stages right now, first is random forest creation and the second is to make an expectation from the irregular random forest classifier made in the principal stage [15].

Random Forest creation strategy:

1. Arbitrarily pick K features from aggregate of N features where $K << N$.
2. Among the K features we compute the hub N utilizing the best part point strategy.
3. Split the hub into girl hubs utilizing the best part technique.
4. Rehash the means from a to c until l number of hubs has been come.
5. Build forest by repeating steps a to d for N number times to create N number of trees.

3.3 Neural Network

Neural networks are relatively crude electronic networks of neurons based on the neural structure of the brain. They process records one at a time and learn by comparing their classification of the record (i.e., largely arbitrary) with the known actual classification of the record. The errors from the initial classification of the first record are fed back into the network and used to modify the networks algorithm for further iterations. Neurons are organized into layers: input, hidden, and output. The input layer is composed not of full neurons but rather consists simply of the record's values that are inputs to the next layer of neurons. The next layer is the hidden layer. Several hidden layers can exist in one neural network. The final layer is the output layer, where there is one node for each class. A single sweep forward through the network results in the assignment of a value to each output node, and the record is assigned to the class node with the highest value [5]. The training process normally uses some variant of the Delta Rule, which starts with the calculated difference between the actual outputs and the desired outputs. Using this error, connection weights are increased in proportion to the error times, which are a scaling factor for global accuracy. This means that the inputs, the output, and the desired output all must be present at the same processing element. The most complex part of this algorithm is determining which input contributed the most to an incorrect output and how must the input be modified to correct the error. (An inactive node would not contribute to the error and would have no need to change its weights.) To solve this problem, training inputs are applied to the input layer of the network, and desired outputs are compared at the output layer. During the learning process, a forward sweep is made through the network, and the output of each element is computed by layer. The difference between the output of the final layer and the desired output is back-propagated to the previous layer(s), usually modified by the derivative of the transfer function. The connection weights are normally adjusted using the Delta Rule. This process proceeds for the previous layer(s) until the input layer is reached [3] (Fig. 2).

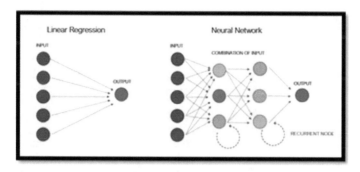

Fig. 2 Nerual network classifier

4 Data Analysis

Data analysis is a method of aggregation, remodeling, cleaning, and modeling information with the goal of discovering the specified data. The results thus obtained area unit communicated, suggesting conclusions, and supporting decision-making.

4.1 Model Evaluation

After you create the training model and run the data flow, you can review information about the model to determine its accuracy. Use this information to iteratively adjust the model settings to improve accuracy and predict better results.

1. Click the navigator icon and select mmachine learning.
2. Click the menu icon for a model and select inspect [8].

The inspect dialog is displayed. Browse the dialog's tabs for information about the model and to view the model's accuracy to determine if you need to adjust the model's parameters or select a more suitable training algorithm.

- Quality tab—This tab contains model quality details that include accuracy metrics like model accuracy, precision, recall, F1 value, false positive rate, and so on. Oracle Analytics provides similar metrics irrespective of the algorithm used to create the model thereby making comparison between different models easy. During the model creation process, the input dataset is split into two parts to train and test the model based on the Train Partition Percent parameter. The model uses the test portion of the dataset to test the accuracy of the model that is built.
- Related tab—Use to navigate to the datasets generated when you train a model. Depending on the algorithm, these datasets contain details about the model like: prediction rules, accuracy metrics, confusion matrix, key drivers for prediction, and so on. These parameters help you understand the rules the model used to determine the predictions and classifications. You can double click a related dataset to view

Table 1 Confusion matrix

	Class 1 (Predicted)	Class 2 (Predicted)
Class 1 (Actual)	TP	FN
Class 2 (Actual)	FN	TP

it or to use it in a project. If based on your findings in the quality and related tabs, you need to adjust the model parameters and retrain it, then close the information dialog, click the navigator icon, select data, click the data flows tab, locate the data flow, and click open. Here, we get to know the accuracy of the model, and we can fine tune model parameters further and achieve better results [6] (Table 1).

5 Results and Discussion

A confusion matrix is a tabulated version of prediction results on a problem which is of type classification, the number of correct and incorrect predictions with count values broken down by each class (Figs. 3, 4 and 5).

While it might seem the trend (as shown in Fig. 6) is driven by the chaotic market events taking a step back and looking at overall sales from regular and DSFT segments has shown interesting insights. We take the combined sales volume of RSFT and DSFT sales, and we find that this volume in spite of some fluctuation does not show significant increase or decrease. In fact, the loss in the sales from the fall in 54 RSFT

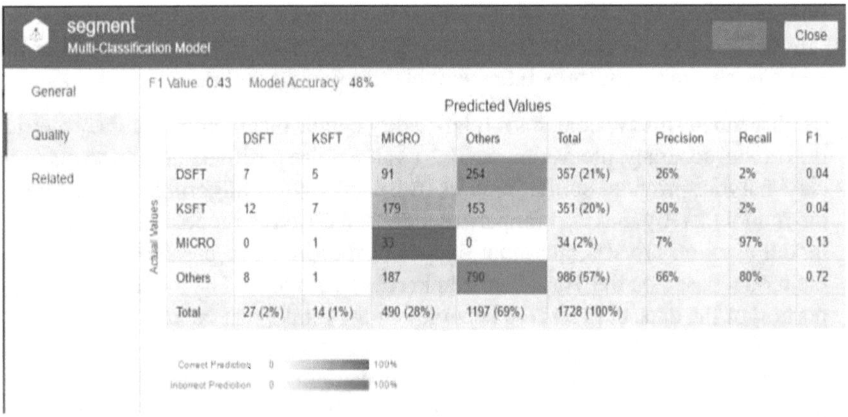

Fig. 3 Different types of cigarette brand sales over the years using decision tree

state model
Multi-Classification Model
Close
General
Quality
Related
F1 Value 0.39 Model Accuracy 47%
Predicted Values
Actual Values
Correct Prediction 0 100%
Incorrect Prediction 0 100%

	East	North	South	Total	Precision	Recall	F1
East	936	31	64	1031 (34%)	44%	91%	0.59
North	382	401	17	800 (27%)	56%	50%	0.53
South	572	26	88	686 (23%)	52%	13%	0.21
West	239	258	0	497 (16%)	0%	0%	0.00
Total	2129 (71%)	716 (24%)	169 (6%)	3014 (100%)			

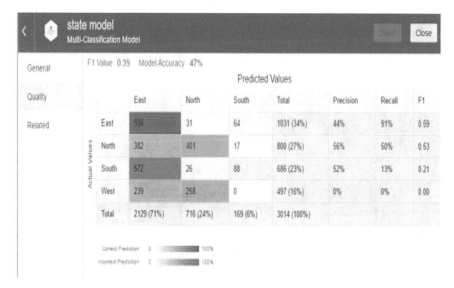

Fig. 4 Sales over the years in different states

ation Model
Close
F1 Value 0.28 Model Accuracy 40%
Predicted Values
Actual Values

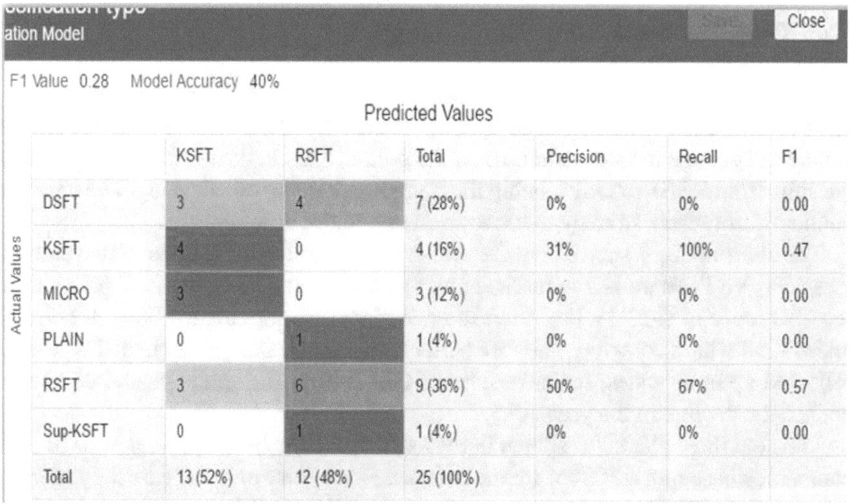

	KSFT	RSFT	Total	Precision	Recall	F1
DSFT	3	4	7 (28%)	0%	0%	0.00
KSFT	4	0	4 (16%)	31%	100%	0.47
MICRO	3	0	3 (12%)	0%	0%	0.00
PLAIN	0	1	1 (4%)	0%	0%	0.00
RSFT	3	6	9 (36%)	50%	67%	0.57
Sup-KSFT	0	1	1 (4%)	0%	0%	0.00
Total	13 (52%)	12 (48%)	25 (100%)			

Fig. 5 Different types of cigarette brand sales over the years using random forest

Overall Effect

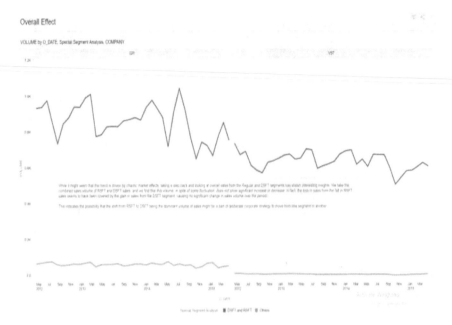

Fig. 6 Graph-based trends

sales seems to have been covered by gain in sales from DSFT segment, causing no significant change in sales volume over the period. This indicates the possibility that the shift from RSFT to DSFT being the dominant volume of sales might be a part of deliberate corporate strategy to move from one segment to other.

The chart in Fig. 7 shows how the year-wise difference is the sales of two companies GPI, VST. There is a numerical chart to see numerical difference, pie chart to see difference in % and a box chart to understand the numerical values. It helps to understand which company sales are better over a particular year, whether a trend is followed up in the sales. It shows sales of GPI is high during the years 2012–2014 but faces a decline in the year 2015.

The next three charts (as shown in Figs. 8, 9 and 10) tells about total sales of both companies based on different segments(types) year-wise and forecast the same for future years. This forecast will hence help the seller on focusing on the right type and make its profit better. The left part plots the data available and the right side makes predictions for three future years using the models made. DSFT has a major growth in the year 2013–2014 then suddenly faces a decline in 2015 and seeing the forecast then will become almost constant over the next three years. Sales of KSFT are however almost negligible.

The algorithms are further applied to make a comparison of net sales among the three years and hence produce outcomes for future as shown in Fig. 11.

Hence, the above ones are all the insights gathered from my models, its accuracy over the data. Oracle Analytics is producing amazing performance, AI, NLG,

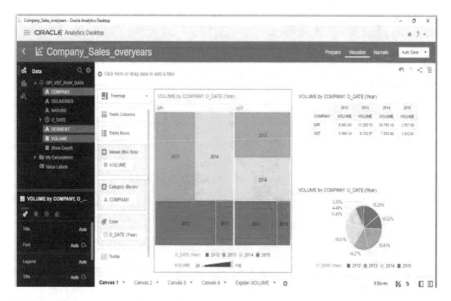

Fig. 7 Pie Chart based Trends

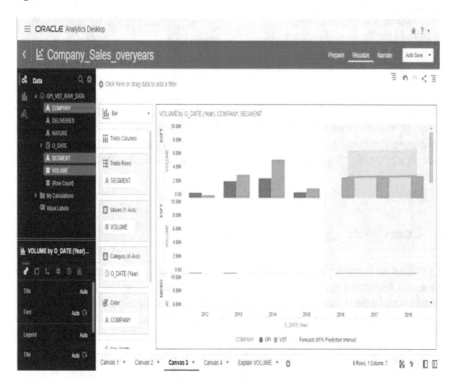

Fig. 8 Forecast sales for Micro, KSFT & DSFT in 2016, 2017, 2018

Fig. 9 Forecast sales of Micro, KSFT & DSFT in 2016, 2017, 2018

NLP and machine learning features. The data flow module allows data preparation, cleaning and transformation. The data visualizations are pixel perfects and stunning. Oracle Analytics is not just a data visualization tool but an entire enterprise business analytics platform. So, it is difficult to compare it with traditional data visualization tools.

6 Conclusion

The three pillars for Oracle Analytics are: Augmented, integrated, collaborative. Hence to conclude, there are a lot of benefits provided when the work is done on cloud as compared to on premises. This paper shows an utilization of cloud and hence it shows how cloud is far more beneficial than that on premises. Some benefits are as follows:

- Cloud is a robust product and its offerings like networking require no cost.
- Oracle provides a private cloud as well as database service to work on.
- Lot of limitations need not have to be handled anymore by the user, cloud takes care of those on its own. The security, information management all these factors

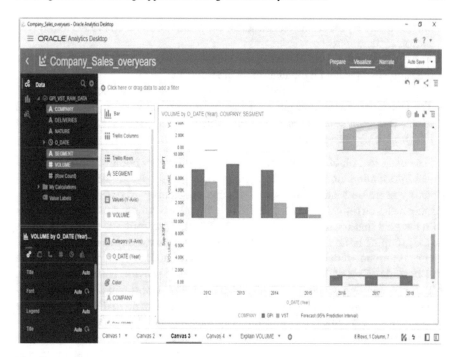

Fig. 10 Forecast sales of Sub-KSFT & Micro in 2016, 2017, 2018

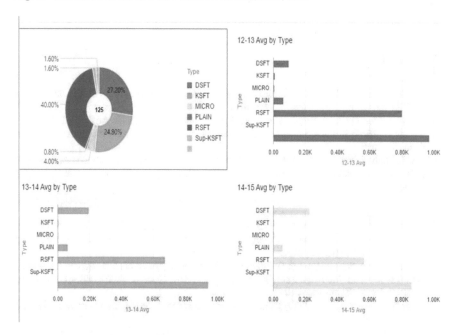

Fig. 11 Comparative study of different types among 3 years

are taken care by the cloud itself whereas on premise these factors have to be given a strong attention to.

- Various kinds of cost is reduced by putting the project on cloud. Maintenance cost, building cost, product depreciation all are reduced.
- Cloud performance is no doubt better than that of on premise project
- The user does not need to put on the server everything is managed by the cloud itself. This brings me to the end of the project and hence concluding why doing this project was beneficial.

The examination of the outcomes connote that the incorporation of multidimensional information alongside various grouping, highlight determination and dimensionality decrease systems can give favorable apparatuses to derivation right now. Further research right now be completed for the better execution of the order strategies with the goal that it can anticipate on more factors. Therefore, the factual measures on the grouping issue were likewise agreeable. This application has high range in the future, as amount of data is increasing tremendously handling it on premise platform will become more and more difficult and cloud will provide better performance. After implementing ML and NLP, next step to move forward will be building artificial intelligence models with it because of its growing demand in the industry. In the future course, a complete model using deep learning and AI can be built using the autonomous database where the user would just have to type in simple English and all his queries would be answered. This would provide high security and low cost to the application because of being on the cloud.

References

1. Abellera R, Bulusu L (2018) Business intelligence, big data, and the cloud. In: Oracle business intelligence with machine learning. Springer, pp 11–21
2. Bi Z, Cochran D (2014) Big data analytics with applications. J Manage Anal 1(4):249–265
3. Bisong E (2019) Building machine learning and deep learning models on Google cloud platform. Springer
4. Fei X, Shah N, Verba N, Chao KM, Sanchez-Anguix V, Lewandowski J, James A, Usman Z (2019) Cps data streams analytics based on machine learning for cloud and fog computing: a survey. Future Gener Comput Syst 90:435–450
5. Kim YM, Ahn KU, Park CS (2016) Issues of application of machine learning models for virtual and real-life buildings. Sustainability 8(6):543
6. Londhe A, Rao PP (2017) Platforms for big data analytics: trend towards hybrid era. In: 2017 International conference on energy, communication, data analytics and soft computing (ICECDS). IEEE, pp 3235–3238
7. Nazarov D, Nazarov A, Kovtun D (2020) Building technology and predictive analytics models in the sap analytic cloud digital service. In: 2020 IEEE 22nd conference on business informatics (CBI). vol 2. IEEE, pp 106–110
8. Priya ES, Suseendran G (2020) Cloud computing and big data: a comprehensive analysis. J Crit Rev 7(14):2020
9. Rosenbrock CW, Homer ER, Csányi G, Hart GL (2017) Discovering the building blocks of atomic systems using machine learning: application to grain boundaries. NPJ Comput Mater 3(1):1–7

10. Sharma S, Srivastava S (2019) Qcpw: a quality centric process workflow improvement approach for a legacy healthcare information system. IET Softw 14(2):129–137

11. Sharma S, Srivastava S, Kumar A, Dangi A (2018) Multi-class sentiment analysis comparison using support vector machine (svm) and bagging technique-an ensemble method. In: 2018 International conference on smart computing and electronic enterprise (ICSCEE). IEEE, pp 1–6

12. Silva JA, Faria ER, Barros RC, Hruschka ER, Carvalho ACD, Gama J (2013) Data stream clustering: a survey. ACM Comput Surv (CSUR) 46(1):1–31

13. Verma VK, Tiwari PK (2015) Removal of obstacles in devanagari script for efficient optical character recognition. In: 2015 International conference on computational intelligence and communication networks (CICN). IEEE, pp 433–436

14. Yadav N, Kumar A, Bhatnagar R, Verma VK (2019) City crime mapping using machine learning techniques. In: International conference on advanced machine learning technologies and applications. Springer, pp 656–668

15. Yeşilkanat CM (2020) Spatio-temporal estimation of the daily cases of covid-19 in worldwide using random forest machine learning algorithm. Chaos, Solitons & Fractals 140

An Empirical Analysis of Heart Disease Prediction Using Data Mining Techniques

Ashish Kumar, Sivapuram Sai Sanjith, Rajkishan Cherukuru, Vivek Kumar Verma, Tarun Jain, and Anju Yadav

Abstract Data mining refers to analyzing an existing dataset related to heart patients from which patterns are discovered within the dataset, and hence we are able to find out meaningful information from the dataset. Large amounts of data cannot be processed by traditional methods to predict heart diseases; hence, we use data mining concepts to resolve the issue. Accuracy achieved by predicting heart diseases using data mining techniques is better than prediction by doctors. In our study, we summarize the observation from multiple papers in which more than one technique (algorithm) of data mining is used for predicting heart diseases. Results from the study were found that using neural networks accuracy obtained is 100%, from classification accuracy obtained is 99.62%.

1 Introduction

Heart is an important part of the human body. Heart diseases are the reason for an increase in the death rates. The dataset used here includes patient details such as symptoms during the occurrence of heart disease and other signs, etc. It is estimated by WHO that by the year 2020 approximately 23.6 million might die from diseases related to heart [7]. In traditional methods, heart diseases are predicted through pulse, ECG, Troponin T levels, etc. But these methods are accurate up to a certain extent.

The accuracy can be increased by implementing data mining techniques and hybrid intelligent techniques. This can be done by collecting huge amounts of data from healthcare industries or universities and thus performing the necessary operations to predict the output which also helps us to discover additional information for effective decision making [28]. This helps us to easily predict complex heart patient cases, make doctors and practitioners work easy to take proper decisions for avoiding heart stroke cases which is totally impossible to do through traditional methods. But, from [8] it is seen that sometimes additional information and patterns divert from the main track.

A. Kumar (✉) · S. S. Sanjith · R. Cherukuru · V. K. Verma · T. Jain · A. Yadav
SCIT, Manipal University Jaipur, Jaipur, India

© The Author(s), under exclusive license to Springer Nature Singapore Pte Ltd. 2022
P. Nanda et al. (eds.), *Data Engineering for Smart Systems*, Lecture Notes in Networks and Systems 238, https://doi.org/10.1007/978-981-16-2641-8_36

Some of the risk factors that cause heart diseases are:

Smoking: Smoking is one of the major causes for heart attacks because when people smoke cigarettes, it contains a high proportion of tobacco in it which on taking adequately leads to causing damage to the functioning of the heart and also the blood vessels.

Cholesterol: Generally, cholesterol generates cells and hormones. When this is seen in excess amount, the path for the blood flow in the heart gets narrowed which slows down the blood flow and at one point gets blocked. Due to unequal proportions of blood and oxygen flowing through the heart, it leads to chest pain which is a symptom for heart disease. The cholesterol content in the body must in a proper proportion to allow efficient flow of blood and oxygen to the heart.

Blood-Pressure: Having high blood pressure in the body decreases the elasticity of the arteries which are responsible for the blood and oxygen flow. If the elasticity decreases automatically, it becomes difficult for the flow leading to heart strokes. Cholesterol and blood pressure show effect through the same arteries of the heart but in different ways.

Diabetes: It generally causes when insulin production in body reduces. Insulin generally increases when more than a certain amount of glucose enters into the blood stream, and then the insulin is released and converts the glucose into fats. When the insulin in body reduces in old age or due to hereditary reasons, then more glucose enters into blood, and when this process is prolonged, then the glucose in the blood will end up blocking the arteries in the heart causing heart attack—one of the major heart diseases.

C-Reactive protein: It is a protein released in your body when something is inflamed in your body. The release of c-reactive protein can be simply found by getting tested from a doctor. If c protein is present in your body, it may be because you might have an inflammatory artery in your heart. When these arteries are inflamed, then might cause a stroke.

Lack of physical exercise: One of the measures of this is composite risk factor. Researchers studied children averaging age of 9 years old, and children who are active had a lesser composite score than the children who are less active. Lack of adequate exercise is also reason for coronary artery disease (CAD)

2 Data Mining

Data mining techniques are used for analyzing chunks of data. It helps in finding patterns in the huge dataset that we use in order to predict future outputs efficiently. In this process, we collect heart disease-related data from various sources (discussed above), and we filter the data for removal of unnecessary attributes and send it for processing and hence to perform predictions. By using data mining techniques for heart disease prediction, we can reduce the chances of heart strokes as we can assess the situation well before doctors can assess it. Due to these advantages of data mining, it becomes a useful tool for analysis of various heart diseases.

There are various techniques in data mining; they are as follows:

2.1　Classification

It is one of the fundamental data mining process. In this, we classify the input features of the dataset in a predefined set of categories, and the future input features output prediction is done based on these categories. This method involves decision trees, neural networks, linear programming and some concepts of statistics. For example, say we have dataset of heart patients with outputs as positive (+ve) or negative (-ve) based on the input features. Then we can predict the future inputs based on the input features of the patient whether it is positive or negative [13, 30].

2.2　Clustering

Clustering is a technique in which we group data having similar characteristics into one category and assign it an output value. Clustering is different from classification; in classification, we have predefined value for the input data, but in clustering we decide the value for a data groups that we make. For example, say for a heart patients dataset we make groups based on the input features, i.e., by finding patterns in the dataset and hence give a value to the group, i.e., the cause of heart disease (or) risk factor [11, 12].

2.3　Association

In this technique, we discover patterns with respect to the relationship between the features in the dataset. This helps us to find the probability of the occurrence of a feature within a dataset. For example, by finding out all the attributes of heart patients dataset required analysis, we can filter all those patients having values for those attributes and them for training and to obtain prediction for future inputs [34].

2.4　Regression

Regression is the process of analyzing the relationship between different variables and building a mathematical model that can be used to predict the value of a variable(label) based on the values of remaining variables(features). Dependent variable: It is the variable to be predicted, also known as label, and usually represented by Y. Independent variable is the predictor or explanatory variable, also known as feature and usually represented by X [36] (Fig. 1).

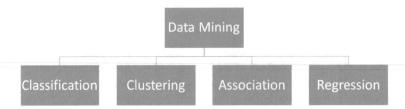

Fig. 1 Data mining techniques

3 Data Mining Techniques Used for Predicting Heart Diseases

Many data mining techniques are present in data science, out of which few were used on different dataset types by various people. They are mentioned below along with the accuracies obtained.

3.1 Naive Bayes

This technique works based on a common concept named Bayes Theorem. Here we consider a classification-type dataset. Now we divide the dataset into two parts named as featured matrix and response vector. All input features meaning the features of heart diseases are included in the feature matrix, and the outputs of the corresponding inputs are included in the response vector [23].

Fundamental concept of Naive Bayes is that each feature of the dataset makes independent and equal contribution for the output. This means that all our heart disease features have equal importance while training the data. Hence using this concept, we train the dataset and then our future inputs contain either one of the outputs from the dataset based on input features (Table 1).

3.2 Decision Trees

It is a model which is used to predict both continuous and discrete type of outputs. Here we train the dataset based on the input features through which we get a generalized tree which is used by future input features. In tree, all the inputs are organized in the root and child nodes of the tree and the leaf nodes are the outputs which are obtained based on the training of dataset. The split concepts of the features are done based on the concept of entropy and information gain.

When we consider this for the case of predicting heart diseases, the first major feature of all the features is placed on the root node, now which feature is placed

Table 1 Accuracies obtained using Naive Bayes

Author	Technique used	Year	Accuracy (%)
Mary et al. [18]	Naive Bayes	2019	78
Cherian et al. [5]	Naive Bayes	2017	86
Malav et al. [17]	Naive Bayes	2017	88
Sultana et al. [31]	Naive Bayes	2016	87
Dessai [10]	Naive Bayes	2013	84
Jesmin et al. [19]	Naive Bayes	2013	NA
Chaitrali et al. [7]	Naive Bayes	2012	90.74
Sundar et al. [32]	Naive Bayes	2012	78
Peter et al. [24]	Naive Bayes	2012	85.18
Srinivas et al. [29]	Naive Bayes	2011	52.33
Anbarasi et al. [2]	Naive Bayes	2010	96.5
	Clustering Naive Bayes		88.3

Table 2 Accuracies obtained using decision trees

Author	Technique used	Year	Accuracy (%)
Dessai [10]	Decision trees	2013	84.2
Chaitrali et al. [7]	Decision trees	2012	99.62
Srinivas et al. [29]	Decision trees	2011	52
Anbarasi et al. [2]	Decision trees	1999	99.2

in the root of the tree is dependent on the entropy of each independent variable in the data. The outputs of heart disease can be discrete or continuous type of outputs, and all these are placed in the leaf nodes based on the root and child node values (Table 2).

3.3 Neural Networks

Neural network is a better version of classification algorithms which can perform training for complex classification datasets. It is good for nonlinear training of models. The layers can be CNN or just a simple multilayer. This method focuses on how the data that is input as features is represented in a pack of layer before reaching the output layer. In this, as more as the hidden layer, as better as the training and understanding the data. There are different types of neural networks; they are CNN, PNN, BNN, RNN, etc. [3, 9] (Table 3).

Table 3 Accuracies obtained using neural networks

Author	Technique used	Year	Accuracy (%)
Latha et al. [16]	Neural networks	2019	76
Abdar et al. [1]	Neural networks	2015	80.23
Dessai [10]	Probabilistic neural networks	2013	94.6
	Bitwise neural networks	2013	80.4
Chaitrali et al. [7]	Neural networks	2012	100
Tanawut et al. [33]	Bitwise neural networks	2008	74.5
Matjaz et al. [14]	Neural networks	1999	74
	Neural networks	1999	85

Table 4 Accuracies obtained using SMO

Author	Technique used	Year	Accuracy (%)
Sultana et al. [31]	Sequential minimal optimization	2016	89
Jesmin et al. [19]	Sequential minimal optimization	2013	96.04

3.4 Sequential Minimal Optimization (SMO)

It is an algorithm used for solving problems based on quadratic programming. These kinds of problems arise because of support vector machines which are models that use two-group classification method. SMO is a commonly used training vector support system, and the popular LIBSVM tool is used to implement it. The SMO algorithm created a huge amount of excitement in the SVM community, as previous SVM trainings were much more complex and costly third-party QP solvers needed (Table 4).

3.5 IBK

It is known as instance-based learning. It is an algorithm based on K-NN technique and runs on WEKA tool [15]. Different from model-based approaches that first learn models from training samples and then predict test samples using the trained model, the model-free k nearest neighbors (kNNs) approach has no training process and performs classification tasks by first measuring the gap between the test sample and all the training samples in order to determine its closest neighbors and then performing kNN classification tasks. Due to its easy implementation and substantial

Table 5 Accuracies obtained using IBK

Author	Technique used	Year	Accuracy (%)
Jesmin et al. [19]	IBK	2013	95.05
John et al. [24]	K-NN	2012	85.55
Srinivas et al. [29]	K-NN	2011	45.67

Table 6 Accuracies obtained using AdaBoost

Author	Technique used	Year	Accuracy (%)
Jesmin et al. [19]	AdaBoost	2013	96.04

classification efficiency, kNN method is a very common tool in data mining and analytics and has therefore been voted as one of the top ten data algorithms to mine [37] (Table 5).

3.6 AdaBoost

It is a meta-algorithm which was used to get better performance when compared to other learning algorithms of this type. It is an upgraded version of the decision trees algorithm. This algorithm is used for performing binary classification. The main difference between the AdaBoost and the decision trees is that AdaBoost improves the classifiers which are weak in case of decision tree algorithms (Table 6).

3.7 J48

It is a classification-type algorithm and an extension of ID3. This algorithm is used for finding the missing values, decision trees, derivation purposes, etc. It is an open-source code based on Java programming language. It is built to implement C4.5 algorithm. This algorithm generates rules from which data identification is generated. This gives better generalization of the decision tree (Table 7).

Table 7 Accuracies obtained using J48

Author	Technique used	Year	Accuracy (%)
Sultana et al. [31]	J48	2016	86
Jesmin et al. [19]	J48	2013	96.04
John et al. [24]	J48	2012	85.18

4   Hybrid Intelligent Techniques Used for Prediction of Heart Diseases

Hybrid intelligent is a software system that uses mixture of many methods of artificial intelligence such as evolutionary neural network and genetic fuzzy hybridization.

The development and design of hybrid techniques are quite difficult as they have a large number of components that have many interactions. There is a great use to these hybrid intelligent techniques as the interactions between the current software cannot be managed [6]. Generally, interactions among these may occur at unforeseen times between unpredictable components for unpredictable reasons.

It is presented in the paper [22] that the genetic algorithm is based on Charles Darwin theory of natural selection, best known for this works in evolution of science who wrote many books like On the Origin of Species, etc. This genetic algorithm is inspired from natural selection, and the algorithm is heuristic in nature. Coactive neuro-fuzzy inference system (CANFIS) is a type of adaptive neuro-fuzzy inference system (ANFIS) which is an adaptive neuro-fuzzy inference system developed in 1990s. It uses Takagi sugeno fuzzy interference system to function as an artificial neural network. Generally, we use linear rule formation, but in case of CANFIS, it uses a nonlinear rule formation that gives more than one output, and this enables it to solve even a partially defined problem.

As discussed in the paper [8] that prediction is the process of finding missing data for new observation that could make the future little predictable. Predictive analytics is the processing of large amounts of useless data into an useful form which can help us understand or predict future events with a little certainty. A scoring system is used in this method, and these methods are used in many domains of diagnosis. A dataset was taken from federal medicines that used these input values and used in input layers with 16 nodes. Initially, neural network weights are randomized. Later on, this method helps us to measure the training and generalization accuracy.

Neural network was discussed in this paper [21] that it is an algorithm that goes through data in way that is similar to how human brain operates. Here each data or dataset is like a neuron, and the algorithm goes through the data by interlinking all the data known to it. Back propagation implies back propagation of errors which optimizes the neurons by giving weights to neurons by the process of gradient descent; this is also called as feedback step [27] (Fig. 2 and Table 8).

5   Heart Disease-Related Datasets

Dataset is an amalgamation of huge amounts of data. When a tabular form is used here, the columns represent a variable and every row represents a record of dataset. All the values in a dataset cell are known as datum, and datasets can also happen to constitute a collection of files or documents. These datasets can be obtained from various resources such as data.gov, European open data portal which allows you to

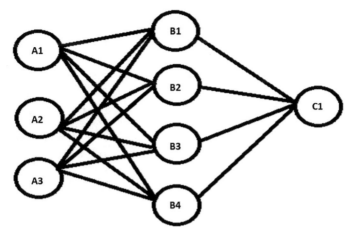

Fig. 2 Neural network example structure

Table 8 Accuracies obtained using hybrid intelligent techniques

Author	Technique used	Year	Accuracy (%)
Dangare et al. [8]	Prediction system	2012	NA
Olabode et al. [21]	Artificial neural network with back propagation error	2012	NA
Parthiban et al. [22]	CANFIS and genetic algorithm	2008	NA

use the data collected by various government agencies. These datasets can be of wide variety from health care to weather and climate.

Datasets in generally have data that can be classified broadly into two categories, first in categorical data and the other as numerical data. In the former, the data in datasets is categorized by observations like dog breed or its color, and many such categories and data are then enlisted in their respective tables, whereas in the latter data in the datasets are based on numerical values where data is ranked by its numerical values like the speed of the car, its mileage, etc. These classifications can help us identify one data from the other and can help you derive useful information from them with greater ease.

The first dataset is called "heart.csv", and it contains an astonishing 76 attributes and subset of 14 only used in experiments that are published. There is a goal field in this dataset that indicates the existence of heart disease. The goal field is basically an integer from values ranging from 0 to 4 where 4 is absent. This can only detect the existence of heart disease but cannot establish the severity of disease, and almost all these experiments are conducted using Cleveland database [25]. Before, the dataset also had social security number and names of the patients who recently bought both and are replaced by dummy values.

Table 9 Input datasets based on heart disease

Dataset name	No. of attributes	Web link
heart.csv	75	https://www.kaggle.com/ronitf/heart-disease-uci
framingham.csv	15	https://www.kaggle.com/dileep070/heart-disease-prediction-using-logistic-regression?select=framingham.csv
Data.csv	76	https://www.kaggle.com/sanskrutipanda/heart-disease-prediction

The second dataset is directly accessible on Kaggle website, and it is a study from Framingham, Massachusetts, of their local residents. This dataset has 15 attributes and used all of them. The main purpose of this study is to detect a 10-year future coronary heart disease prediction. Here attributes are classified into categorical data, and each attribute is a risk factor. The categories used are demographic, behavioral and medical risk factors. This study is more of sensitive than specific so the negative values are highlighted more than the positive values. The results of this study helped us to derive some conclusion, and they are men more inclined to get a heart disease than woman as they age and also as a result of their smoking habits and systolic blood pressure. Contrary to the popular notion of cholesterol and glucose causes increased chances of heart diseases, but it predicted a very low probability of these two factors contributing to eventual heart attack. The model is more specific than sensitive; this can be improved by more data (Table 9).

The third dataset that we chose is data.csv, and it contains 76 attributes, but only 14 of them are being and there is also an option to change or manage the number of columns to appear or be used. This directory consists of four databases relating to heart diseases treatment. All attributes that are used in this dataset are numerical value.

6 Feature Selection from the Dataset

The feature selection process is used for selecting the most important features or attributes from the independent variables in the dataset. This process is done by first splitting the independent and dependent variable from the datasets and then converting all the categorical inputs in the heart disease dataset to numerical from the concepts of dummy variables and Label Encoder to make their use in feature selection [20, 35]. Once converted, we use the sklearn functions to perform feature selection where we select based on the scores of each feature (independent variables)

Table 10 Feature selection used by authors

Author	Technique used	Actual attributes	Reduced attributes
Anbarasi et al. [2]	Feature selection using genetic algorithm	13	10
Shanthi [26]	Neuro-genetic approach	NA	NA

in the dataset. This helps in reducing complexity in the data and also the training and testing error thus increasing the efficiency of prediction.

Many of the authors have used the concept of feature selection before performing any algorithm on the dataset (Table 10).

7 Conclusion

One of the biggest causes of natural deaths today is heart diseases. Predicting it has become a big problem today as we are not able to predict it properly from the information, and we are able to predict only based on few symptoms though we have other information about the patients too. To predict heart disease through remaining data which doctors could not use at this point, data mining techniques come into play. Our aim in this review paper was to provide a detailed study of various data mining techniques that have been used for predicting the heart diseases.

From the analysis of different datasets and techniques from different papers along which large number of attributes, it is found that different techniques have different accuracies which are based on the type of dataset we use, the type of tools used and the number of attributes. For example, for a classification type dataset uses classification technique or decision tree technique. In the paper [4], it is mentioned that the accuracy obtained by using decision tree technique is 99.62% with the use of 15 features of the dataset. In some other papers, 100% accuracy is obtained by predicting heart diseases using neural networks. The ultimate advantage that we get by using data mining techniques is that we can avoid unnecessary creation of heart diseases and can try to cure the disease well before.

In some cases, using feature selection process before performing the algorithm has given better results. Also, many authors have used Naive Bayes the most out of all other algorithms.

References

1. Abdar M, Kalhori SRN, Sutikno T, Subroto IMI, Arji G (2015) Comparing performance of data mining algorithms in prediction heart diseases. Int J Electr Comput Eng (2088-8708) 5(6)
2. Anbarasi M, Anupriya E, Iyengar N (2010) Enhanced prediction of heart disease with feature subset selection using genetic algorithm. Int J Eng Sci Technol 2(10):5370–5376
3. Atkov OY, Gorokhova SG, Sboev AG, Generozov EV, Muraseyeva EV, Moroshkina SY, Cherniy NN (2012) Coronary heart disease diagnosis by artificial neural networks including genetic polymorphisms and clinical parameters. J Cardiol 59(2):190–194
4. Bhatla N, Jyoti K (2012) An analysis of heart disease prediction using different data mining techniques. Int J Eng 1(8):1–4
5. Cherian V, Bindu M (2017) Heart disease prediction using Naive Bayes algorithm and Laplace smoothing technique. Int J Comput Sci Trends Technol (IJCST) 5(2):68–73
6. Chitra R, Seenivasagam V (2013) Review of heart disease prediction system using data mining and hybrid intelligent techniques. ICTACT J Soft Comput 3(04):605–09
7. Dangare C, Apte S (2012) A data mining approach for prediction of heart disease using neural networks. Int J Comput Eng Technol (IJCET) 3(3)
8. Dangare CS, Apte SS (2012) Improved study of heart disease prediction system using data mining classification techniques. Int J Comput Appl 47(10):44–48
9. Das R, Turkoglu I, Sengur A (2009) Diagnosis of valvular heart disease through neural networks ensembles. Comput Methods Programs Biomed 93(2):185–191
10. Dessai ISF (2013) Intelligent heart disease prediction system using probabilistic neural network. Int J Adv Comput Theory and Eng (IJACTE) 2(3):2319–2526
11. Hartini S, Rustam Z (2020) The comparison study of kernel kc-means and support vector machines for classifying schizophrenia. Telkomnika 18(3)
12. Hussein AA (2018) Improve the performance of k-means by using genetic algorithm for classification heart attack. Int J Electr Comput Eng 8(2):1256
13. Khairuddin A, KNF KA, Kan PE (2019) A general framework for improving electrocardiography monitoring system with machine learning. Bullet Electr Eng Inf 8(1):261–268
14. Kukar M, Kononenko I, Grošelj C, Kralj K, Fettich J (1999) Analysing and improving the diagnosis of ischaemic heart disease with machine learning. Artif Intell Med 16(1):25–50
15. Kumar A, Bhatnagar R, Srivastava S (2018) Arsknn: an efficient k-nearest neighbor classification technique using mass based similarity measure. J Intell Fuzzy Syst 35(2):1633–1644
16. Latha CBC, Jeeva SC (2019) Improving the accuracy of prediction of heart disease risk based on ensemble classification techniques. Informatics in Medicine Unlocked 16
17. Malav A, Kadam K, Kamat P (2017) Prediction of heart disease using k-means and artificial neural network as hybrid approach to improve accuracy. Int J Eng Technol 9(4):3081–3085
18. Mary TS, Sebastian S (2019) Predicting heart ailment in patients with varying number of features using data mining techniques. Int J Electr Comput Eng 9(4):2675
19. Nahar J, Imam T, Tickle KS, Chen YPP (2013) Association rule mining to detect factors which contribute to heart disease in males and females. Expert Syst Appl 40(4):1086–1093
20. Nguyen TN, Nguyen TH, Ngo VT (2020) Artifact elimination in ECG signal using wavelet transform. Telkomnika 18(2):936–944
21. Olabode O, Olabode BT (2012) Cerebrovascular accident attack classification using multilayer feed forward artificial neural network with back propagation error. J Comput Sci 8(1):18–25
22. Parthiban L, Subramanian R (2008) Intelligent heart disease prediction system using canfis and genetic algorithm. Int J Biol Biomed Med Sci 3(3)
23. Pattekari SA, Parveen A (2012) Prediction system for heart disease using naïve bayes. Int J Adv Comput Math Sci 3(3):290–294
24. Peter TJ, Somasundaram K (2012) Study and development of novel feature selection framework for heart disease prediction. Int J Sci Res Publ 2(10):1–7
25. Saranya G, Pravin A (2020) A comprehensive study on disease risk predictions in machine learning. Int J Electr Comput Eng 10(4):4217

26. Shanthi D, Sahoo G, Saravanan N (2009) Evolving connection weights of artificial neural networks using genetic algorithm with application to the prediction of stroke disease. Int J Soft Comput 4(2):95–102
27. Sharma S, Srivastava S, Kumar A, Dangi A (2018) Multi-class sentiment analysis comparison using support vector machine (svm) and bagging technique-an ensemble method. In: 2018 International conference on smart computing and electronic enterprise (ICSCEE). IEEE, pp 1–6
28. Sigit R, Roji CA, Harsono T, Kuswadi S (2019) Improved echocardiography segmentation using active shape model and optical flow. Telkomnika 17(2):809–818
29. Srinivas K, Rao GR, Govardhan A (2011) Survey on prediction of heart morbidity using data mining techniques. Int J Data Mining Knowl Manag Process (IJDKP) 1:14–34
30. Sudha A, Gayathri P, Jaisankar N (2012) Effective analysis and predictive model of stroke disease using classification methods. Int J Comput Appl 43(14):26–31
31. Sultana M, Haider A, Uddin MS (2016) Analysis of data mining techniques for heart disease prediction. In: 2016 3rd international conference on electrical engineering and information communication technology (ICEEICT). IEEE, pp 1–5
32. Sundar NA, Latha PP, Chandra MR (2012) Performance analysis of classification data mining techniques over heart disease database. Int J Eng Sci Adv Technol 2(3):470–478
33. Tantimongcolwat T, Naenna T, Isarankura-Na-Ayudhya C, Embrechts MJ, Prachayasittikul V (2008) Identification of ischemic heart disease via machine learning analysis on magnetocardiograms. Comput Biol Med 38(7):817–825
34. Tiwari PK, Joshi S (2016) A review on load balancing of virtual machine resources in cloud computing. In: Proceedings of first international conference on information and communication technology for intelligent systems, vol 2. Springer, pp 369–378
35. Verma VK, Jain T (2018) Machine-learning-based image feature selection. In: Feature dimension reduction for content-based image identification. IGI Global, pp 65–73
36. Yusof AM, Ghani NAM, Ghani KAM, Ghani KIM (2019) A predictive model for prediction of heart surgery procedure. Indones J Electr Eng Comput Sci 15(3)
37. Zhang S, Li X, Zong M, Zhu X, Wang R (2017) Efficient kNN classification with different numbers of nearest neighbors. IEEE Trans Neural Netw Learn Syst 29(5):1774–1785

2D Image to Standard Triangle Language (STL) 3D Image Conversion

Manoj K. Sharma and Ashish Malik ⓘ

Abstract Nowadays, the use of 3D printing has extensively increased, which support STereoLithography or Standard Tessellation Language (STL) 3D image for printing. STL file originated from the 3D systems. STL file format works over the triangles and estimates the surfaces of a solid model without any texture and color. It specifies both binary and ASCII formats. In this paper, a method has presented to convert 2D image to 3D printer acceptable STL file and STL file to 3D object generation.

Keywords STereoLithography · STL · 2D image · 3D image conversion · Triangulation · Mesh

1 Introduction

Nowadays, the use of 3D printing has substantially increased and is being used in different applications of medical equipment's, medicine, manufacturing, art and design, etc. [1] It specifies both binary and ASCII formats. Computer-aided design and manufacturing (CAD & CAM) are the tools mainly used in 3D modeling of the machines using Selective Laser Sintering, Stereolithorgraphy, 3D printing, or Laminating Object Manufacturing algorithms [2]. In STL file format, image surface represents with the triangles which represents the coordinates of three ordered coordinate points and finally this triangular meshes known as geometrical representation of the real 3D object. It describes amorphous triangulated surface by triangles vertices and unit normal using Cartesian coordinate system with both negative and positive coordinates. STL image format does not contain scaler data, and units are arbitrary [3]. However, STL files are remarkable uses in 3D printing, but they have some limitation like they represent only the shape and outer surface of the 3D model, and second it is hard to draw the overlaps and gaps of the outer surface, which knows non-manifold edges of the model, and lastly, STL files do not contain texture, color, attributes, and internal structure of the model.

M. K. Sharma (✉) · A. Malik
Manipal University Jaipur, Dehmi-Kalan, Jaipur, Rajasthan 303007, India

© The Author(s), under exclusive license to Springer Nature Singapore Pte Ltd. 2022   391
P. Nanda et al. (eds.), *Data Engineering for Smart Systems*, Lecture Notes in Networks and Systems 238, https://doi.org/10.1007/978-981-16-2641-8_37

2 STL File Structure

As it has mentioned above, STL file format is a surface model for 3D objects, see Fig. 1. The surface space is approximated with small triangles along with their records which consists of triangle normal vector and vertex coordinates [4].

Following are certain conditions which the STL file model mind out to avoid the errors:

(1) **Right Hand Rule**: right hand rule is used to understand the axes orientation in 3D space. The right thumb points along with the Z-axis are coordinated by right handed coordinates in positive direction and motion of X-axis to Y-axis is represented by the figures curl. Handedness reverse by interchanging the labels axes and the same effect comes when the axis direction is reversed [5]

(2) **Share Vertex Rule**: Adjacent triangles are associated only two vertices.

(3) **Share Edge Line**: Two triangles can share one edge line.

(4) **Adjacent Facts**: With a triangle, there are three adjacent triangles connected.

As mentioned earlier, two formats of STL file are: (1) binary structure (2) ASCII structure [6].

2.1 STL-ASCII Format

STL-ASCII format is simple plain text format which represent the triangular mesh of an object surface [7]. However, it has large size file and because of that only it is not in much use. The syntax of the ASCII STL file is given in Fig. 2.

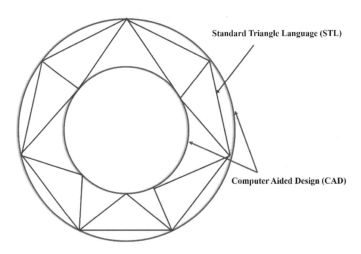

Standard Triangle Language (STL)

Computer Aided Design (CAD)

Fig. 1 CAD representation and STL approximation (3D Systems, 1987)

```
solid_object
    primary_loop nᵢ nⱼ nₖ
        secondary_loop
            grid v1ₓ v1ᵧ v1_z
            grid v2ₓ v2ᵧ v2_z
            grid v3ₓ v3ᵧ v3_z
        end secondary_loop
    end primary_loop
end solid_object
```

Fig. 2 Syntax of STL-ASCII file [7]

2.2 *STL Binary Format*

STL Binary uses IEEE floating point and integer representation [8]. The syntax of STL Binary file is given in Fig. 3.

Bytes	Data type	Description
80	ASCII	Header. No data significance.
4	unsigned long integer	Number of facets in file
4	float	i for normal
4	float	j
4	float	k
4	float	x for vertex 1
4	float	y
4	float	z
4	float	x for vertex 2
4	float	y
4	float	z
4	float	x for vertex 3
4	float	y
4	float	z
2	unsigned integer	Attribute byte count

Fig. 3 Syntax of STL binary file [8]

3 STL File Surface Representation

A tessellation pipeline is used to generate a triangular mesh for the 3D object boundary surface [3]. However, mesh also have the data to formulate the side of triangle which contain interior of 3D object. The STL file format is triangle representation which uses its vertices to represent the triangles which further use for 3D model creation. Such information is about the triangles orientation and can be represented with numerical values.

$$n_x^l \quad n_y^l \quad n_z^l$$

$$v_{1x}^l \quad v_{1y}^l \quad v_{1z}^l$$

$$v_{2x}^l \quad v_{2y}^l \quad v_{2z}^l$$

$$v_{3x}^l \quad v_{3y}^l \quad v_{3z}^l$$

A outward normal coordinates are represented by $\left(n_x^l, n_y^l, n_z^l\right)$ for the lth facet and jth vertex coordinates from lth triangle are $\left(v_{jx}^l, n_{jy}^l, n_{jz}^l\right)$ where is $l = 1, 2, \ldots, N$.

4 STL Mesh Construction

Polygonal mesh models are frequently used in 3D modeling using triangulating point. These models are widely used for rendering and design. In 3D, STL file visualization quality of the image depends on both quality and quantity of polygons [9]. Different approaches of mesh simplification have been surveyed by Talton [10], Cignoni et al. [11], and Luebke [12]. A mesh simplification approaches can be classes in three broad categories as.

4.1 Vertex Clustering

Vertex clustering [13, 16] used the principal that by mapping the points on same pixel means at that pixel in the image, only one point will be visualized and remaining points will be covered by hidden surface removal. In this way while rendering n number of points, it wastes such points, and we cannot get the desired image. This wastage can be minimized by extracting points falling on same pixel and create a new point for the representing and rendering. The vertex clustering algorithm works over the same principal. It tries to find the closest vertices and represent them with new vertex like a point.

Suppose, $G = (V, E)$ has vertices (V), edges (E). An edge is e_{kl} connected with vertex's v_k and v_l, respectively.

The connected neighbor

$$N_k = \{v_l : e_{kl} \in E e_{jk} \in E\} \tag{1}$$

4.2 Edge Contraction

Edge contraction [12, 13] is based on the concept of collapsing an edges in a vertex. Edge contraction frequently is in simplification algorithms. In this concept, two end point vertices are merged their edges to get one edge only. In this two vertices, faces incident on the edge merged in single new edge without affecting local neighborhood.

4.3 Vertex Decimation

The objective of vertex dissemination [12–15] is to minimize the triangles in mesh without affecting the original topology and geometry. For Average plane vertex v_l normal of neighboring triangles n_l centers c_l and area is A_l.

The weighted average is

$$n = \frac{\sum_i A_i n_i}{\sum_i A_i} \tag{2}$$

$$\hat{n} = \frac{n}{|n|} \tag{3}$$

$$n = \frac{\sum_i A_i y_i}{\sum_i A_i} \tag{4}$$

Distance is

$$|\hat{n} = (v_j - y) \tag{5}$$

5 Experimental Process

3D printing has substantially increased and is being used in different applications of medical equipment's, medicine, manufacturing, art and design, etc. [1] STL file requires 2D contours with triangular matrix. STL file creates series of polygons corresponding to z values. STL image surface must be connected. A detailed algorithm proposed in this work is given in Fig. 4.

In the experimental setup, we have taken one input image as shown in Fig. 5. Then after, the shape width of the image is taken for the current image it was selected 156 pixels, and color detection tolerance value is also defined. Color detection tolerance is useful for the images with slight color tone differences which is allowed for a sample image. For the present image, it has set to different color tolerance.

Figure 6 shows the converted image over different color tolerances, and difference is clearly visible. Table 1 shows the parameters defined the present 2D image to convert in 3D printer compatible STL image.

After reading image file, image width in pixels, and image color threshold for the color tolerance, it calculates the image dimension become which is 155*155, dominant color for RGB ([15.842275670675301, 12.503171666446413, 26.4560371789789]) and dominant color rate (0.90359431596545) and finally figure out the shape of an image as given in Fig. 6. In the second phase of the experiments, we determine image threshold (155 is given) and STL file base Z-input (16) and STL shape Z-input (31). Finally, STL file is created with respect to Z-dimension as given in Fig. 7.

Step 1: Select an image
Step 2: Define image pixels and color tolerance value
Step 3: Parse the image and calculate
new_image_pixels
image_dominantColors
dominantColorRate
Step 4: Emit the dominantColor, dominantColorRate to input image
Step 5: Calculate the shape of new image
Step 6: Define Pixel size in mm, STL file base size, STL file shape height
Step 7: Fuse the new parameters and write the STL file
Step 8: End

Fig. 4 Algorithm for STL creation

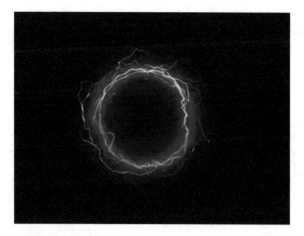

Fig. 5 Original 2D input image [17]

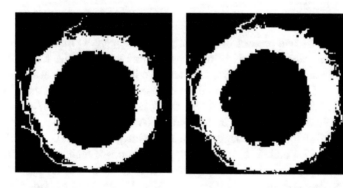

(a) Image with color tolerance 100 (a) Image with color tolerance 70

Fig. 6 Calculated shape

Table 1 Parameters

S. no	Name of parameter	Input value	Calculated value
1	Image width in pixels	155	–
2	Image color threshold	85	–
3	Image dimensions	–	155*155
4	dominantColor	–	[15.842275670675301, 12.503171666446413, 26.4560371789789]
5	dominantColorRate	–	0.90359431596545
6	threshold_input	155	
7	Unit_Input	4.16	
8	Base_Z_Input	16	
9	Shape_Z_Input	31	

Fig. 7. 3D-STL output from the input image

6 Conclusion

The workflow has two phases; in the first phase, we read 2D image file and define color tolerance value along with image width and calculate the black and white shape of an input image. During experiment, we have reduced the triangles size to decrease size of an image. Same time we have calculated dominant color for RGB ([15.842275670675301, 12.503171666446413, 26.4560371789789]) and dominant color rate (0.90359431596545). In the second phase of STL creation, we have taken input for the Z-shape of an image, z-base of an image along STL threshold value, which were 16, 31, and 4.16, respectively. The proposed technique is performing well as compared to [3–5] and can generate 3D STL file in 8 μ-second only. Since STL file does not contain color information of an image, so in our future work, we will try to create a colored STL file.

References

1. Mohammad T et al (2014) Role of build orientation in layered manufacturing: a review. Int J Manuf Technol Manage 27:47–73
2. Yan X et al (1996) A review of rapid prototyping technologies and systems. Comp Aid Des 28:307–318
3. Szilvsi-Nagy M et al (2003) Analysis of STL files. Math Comput Model 38(7–9):945–960
4. Zhou H-F et al (2016) Data generation of layer slice in FDM manufacturing. Adv Engg Res (105):419–424
5. Right Hand Rule for Cross Products. http://www.physics.udel.edu/~watson/phys345/Fall1998/class/1-right-hand-rule.html. Accessed 4 Jan 2021
6. STL (STereoLithography) File Format Family. https://www.loc.gov/preservation/digital/formats/fdd/fdd000504.shtml. Accessed 4 Jan 2021
7. Fileformat homepage. https://docs.fileformat.com/cad/stl/. Accessed 4 Jan 2021
8. Standard data formats for Fabbers. https://www.fabbers.com/tech/STL_Format. Accessed 4 Jan 2021

9. Guha S et al (2015) Mesh simplification via a volume cost measure. Int J Comp Graph Anim 5(2):53–64
10. Talton JO A short survey of mesh simplification algorithms. https://cg.informatik.uni-freiburg. de/intern/seminar/meshSimplification_2004_Talton.pdf
11. Cignoni P et al (1998) A comparison of mesh simplification algorithms. J Comp Graph 22(1):37–54
12. Luebke DP (2001) A developer's survey of polygonal simplification algorithms. J IEEE Comp Graph Appl 21(3):24–35
13. Wang D et al A vertex-clustering algorithm based on the cluster-clique. In: Sun X et al (eds) Algorithms and Architectures for Parallel Processing. ICA3PP 2014. LNCS, vol 8631. Springer
14. Martinsen K et al (2020) Tolerancing from STL data: a legacy challenge. In: 16th CIRP conference on computer aided tolerancing (CIRP CAT 2020), Procedia CIRP 92. Elsevier, pp 218–223
15. Martinsen K et al (2020) Extracting shape features from a surface mesh using geometric reasoning, In: 53rd CIRP conference on manufacturing system, Procedia CIRP 93. Elsevier, pp 544–549
16. Low KL et al Model simplification using vertex-clustering, Symposium on integrative 3D graphics, Providence RI USA. ACM, pp75–81
17. Plasma alpha ring. https://www.shutterstock.com/video/clip-28977250-fire-ice-ring-magic-ani mation-plasma-alpha. Accessed 4 Jan 2021

Improving Recommendation for Video Content Using Hyperparameter Tuning in Sparse Data Environment

Rohit Kumar Gupta, Vivek Kumar Verma, Ankit Mundra, Rohan Kapoor, and Shekhar Mishra

Abstract As we are familiar with the increasing number and demand for online videos on the internet has led to difficulty in the extraction of required data. The recommendation system is used to filter or to collect information according to the user's preference. Nowadays, the use of recommendation systems has become important so that we can find more relevant data from huge sources. Based on the attributes such as genres, we can recommend important videos to different users. There are different video recommendation problems such as cold start, data sparsity, etc. Here in this paper, we are solving the data sparsity problem which is considered as the biggest problem as it can give poor results. There are different recommendation techniques that are used to solve various recommendation problems such as content-based filtering, real-time recommendation system, hybrid recommender system, single network-based recommendation system, collaborative filtering. We are using a single-network recommendation system to enhance the quality of recommendation.

Keywords Recommender system · Collaborative filtering · Matrix factorization · Singular value decomposition (SVD) · Feature scaling

1 Introduction

As the number of online videos is increasing day-by-day many of the researchers are working hard to improve the quality of recommendation and to get users' personal preferences. Recommender systems are used as filtering mechanisms in different areas such as music and video services as playlist generators in YouTube, Spotify, and Netflix, content recommender in social media like Twitter, Facebook, and Instagram, services like product recommender in Flipkart, Amazon, Snapdeal, etc. to find the

R. K. Gupta (✉) · V. K. Verma · A. Mundra · R. Kapoor · S. Mishra
Manipal University Jaipur, Jaipur, Rajasthan, India

V. K. Verma
e-mail: vivekkumar.verma@jaipur.manipal.edu

A. Mundra
e-mail: ankit.mundra@jaipur.manipal.edu

recommended item based on user ratings and users preferences. The recommender system often suffers from data sparsity and cold start problems. Here in this paper, we are solving the data sparsity problem to get better accuracy. Data sparsity problem arises because not all the people give ratings to all the items and because of this some items are not recommended to different users because the user-item matrix becomes very sparse. Now filling these values with default data or with the video popularity will give rise to popular videos and thus degrades the quality of recommendation.

Collaborative filtering is one of the widely used techniques in machine learning which is used to retrieve important information to build personalized recommendations on the internet. Collaborative filtering algorithms are basically used to make recommendations about the user's interests by integrating preferences based on ratings and feedback which different users give to individual items. The recommender system will recommend such videos to the users which they do not rated in their past, but highly rated by their similar users called neighborhoods. For example, an e-commerce site like Flipkart or Amazon may recommend product "C" to the customers by comparing their past preferences of those who have purchased the same product. Memory-based, model-based, and hybrid approaches are used for collaborative filtering. We also walk through the singular value decomposition (SVD) a machine learning technique that is used to decompose the original matrix into different matrices, using the interesting properties of the original matrix. SVD [1] helps us to better understand the relationship between users and items as they become comparable by mapping each user and each item into a latent space.

2   Related Work

Many recommender systems [2] have been introduced and influencing a great amount of work in many different systems to improve the quality of recommendation, to bring better accuracy and to make the system efficient. We here grouped various approaches into several categories based on data pre-processing and optimization technique. Recommendation system used to retrieve necessary information to find user preferences and user ratings that different users give it to item such as books, music, videos, product from different platform [3]. Collaborative filtering based on online rating consists of array with item and rating pairs from individual user. The response to the above statement is an array with item and rating pairs for those items that the user has not rated ever [4]. To improve the quality of recommendation and personalized recommendation results, there is a need to link social network information among users. Also, trust-aware recommender system provides similar taste to the user with other users they trust [5]. Dynamic recurrent neural network to model users' dynamic interests over time in a unified framework for personalized video recommendation. The challenge is that it suffers from data sparseness problem [6]. Data sparsity problem by using watch time and defined an attribute called popularity preference. The challenge is that it cannot solve the cold start problem [7]. Cold start and sparsity problem using users' auxiliary information on other social networks is specified in [8].

Recommendation based on low-level visual features provides almost 10 times better accuracy in comparison to genre-based recommendations. The biggest challenge in this method is that they require somewhat more information which are non-scalable and very much expensive [9]. Singular value decomposition is a method which is known to be very good after the Netflix prize. This proposes a modified version of PureSVD which aims to be more efficient and accurate [1]. Personalized video recommendation service requires abundant data for understanding what type of items or videos user wants [10]. Singular value decomposition compute the unknown prediction score function of set of tests and also improves the scalability, sparsity, and efficiency of the system [11]. SVD in order to improve the quality of recommendation system produces high-quality recommendations, but the major challenge about SVD is that it undergoes with expensive matrix calculations [12]. We have used a matrix factorization a collaborative based filtering method for mapping of users and items to factor vectors by decomposing the original matrix into two matrices [13]. To reduce the dimensionality in recommender system for generating better predictions, we use SVD [14]. Enhanced content-based algorithm is proposed to solve cold start problem and to improve the quality of prediction accuracy of the recommendation system [15]. Data from Wikipedia is used to compute item-based similarities. Based on these similarities, we can compute user rating over a small set of items to other similar items thus providing better recommendation to the user profile [16].

3 System Overview

There are many ways to find similar users based on their taste to create a list of recommendations. Recommending similar items or contents to the users from the large group of data helps them to gain interest and to watch more and more relevant items. We have used a collaborative filtering technique that helps to filter out those items that users like based on their similar user. The collaborative filtering technique works by searching the most relevant data from the large group of data based on rating and feedback which different users give to individual data thus help in recommending these data to the user based on the user with similar taste. Next, we move onto the feature scaling that helps to remove outliers and extreme bias when users give extreme leniency while rating their movies, i.e., it helps in removing noise in our dataset and helps in bringing better prediction. We also perform SVD using hyper parameter tuning that helps in choosing the optimal parameter to get better accuracy. More the users' rate their item, the better the recommendation they will get. Hence, the recommendation system will work better if the number of ratings and feedback given to different items by different users' is more. We used these techniques so that users can get a better recommendation to serve up something they might like in the system.

3.1 Methodology

Since our concern in this research is to deal with the problems of data sparsity and cold start, we also decided to calculate the sparsity of the user to movie matrices for both datasets. It is given by:

$$\text{sparsity} = \frac{\left[\text{No. of ratings}\right]}{\left[\text{Total no. of elements}\right]} \tag{1}$$

On calculation, we found that the sparsity for the 100 K and 1 M datasets was approximately 94% and 96%, respectively. These values suggest that our dataset is sparse and any model performing well on these datasets can deal well with data sparsity and cold start problems.

3.2 Approach

Step 1:

Some people have an extreme bias when rating movies. While some people rate all movies that they watch with extreme leniency, i.e., 4 or 5 stars, others are extremely strict and rate all movies only 1 or 2 stars. This causes unnecessary noise in our dataset and can affect the accuracy of the predictions negatively. To solve this problem, we employ a normalization procedure which is given by:

$$b_{ui} = mu + b_i + b_u \tag{2}$$

Here, b_{ui} represents the user-item rating bias, mu represents the global average, b_i represents the movie's average rating and b_u represents the user's average rating. Using this formula, we can solve the above-described problem and move on to the next step.

Step 2:

Now, we will perform hyper parameter tuning. This helps us in fixing the optimal hyper parameters before we begin the actual learning process to maximize our accuracy. There are two main methods to do this namely grid search and random search. We found through our research that random search is outperformed by a model without any hyper parameter tuning and hence we have selected grid search for our purposes.

Step 3:

In our research we found that SVD or singular value decomposition is by far the best approach to deal with many issues that the collaborative filtering algorithm faces today. It is a method through which we can implement matrix factorization. So, we picked matrix factorization as our desired model and singular value decomposition as our method to implement it. SVD is given by:

$$A_{[m \times n]} = U_{[m \times r]} \sum \left(V_{[n \times r]} \right)^T \qquad (3)$$

Here, A is the input data matrix. Now this matrix A is to be factorized into 3 different matrices. It is factorized into U which consists of the left singular vectors, V which consists of the right singular vectors and $\sum$ which is a diagonal matrix that consists of singular values. On looking at Fig. 1, we can see how the SVD model predicts ratings for all users.

3.3 Matrix Diagram

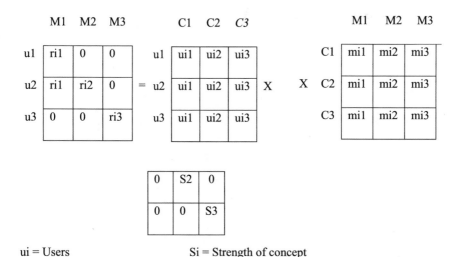

ui = Users   Si = Strength of concept
Mi = Movies   rij = rating of movie j by user i
Ci = Concept   uij = Similarly user i with concept j
Mij = Similarly of movie j with concept i.

Fig. 1 Matrix factorization using singular value decomposition

3.4 Baseline Methods

We evaluate SVD together with some baseline methods based on MAE (Mean Absolute Error) and RMSE (Root Mean Square Error). Both these metrics are negatively oriented scores which means that the lower values are better. MAE is given by,

$$MAE = \frac{1}{n}\sum\nolimits_{j=1}^{n}|y_j - \widehat{y_j}| \tag{4}$$

In this n = total number of ratings,
y_j = actual rating and $\widehat{y_j}$ = predicted rating.
RMSE is given by,

$$RMSE = \sqrt{\frac{1}{n}\sum\nolimits_{j=1}^{n}\left(y_j - \widehat{y_j}\right)^2} \tag{5}$$

In this as well the parameters are the same as MAE. We will be evaluating our approach along with the KNN-Baseline and standard SVD approaches.

4 Implementation and Results

Fig. 2 RMSE values comparison

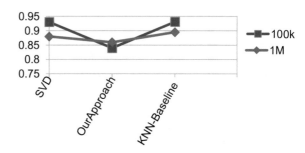

Fig. 3 MAE values comparison

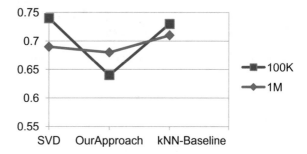

4.1 Datasets

For this research we used the MovieLens-100 K and MovieLens-1 M datasets. The MovieLens-100 k dataset consists of 1,00,000 ratings given by 1000 users on 1700 movies. The MovieLens-1 M dataset consists of 10,00,000 ratings given by 6000 users on 4000 movies [17]. The results given below show how all the components more and less contributes to the recommendation system.

Here are two table showing the performance of each of the considered algorithms on MovieLens-100 K and MovieLens-1 M datasets (Tables 1 and 2).

4.2 Results

The graph shows RMSE values for each method.

Table 1 Results for 100 K dataset

METHOD	RMSE	MAE
SVD	0.88	0.69
Our approach	0.86	0.68
kNN-Baseline	0.895	0.706

Table 2 Results for 1 M dataset

METHOD	RMSE	MAE
SVD	0.93	0.74
Our approach	0.84	0.64
kNN-Baseline	0.93	0.73

The below chart shows the MAE values for each of 1 M and 100 K datasets (Fig. 2).

As we can observe from both the tables, our approach outperforms the basic SVD in case of both the datasets and greatly outperforms KNN-Baseline algorithm in both cases (Fig. 3).

5 Conclusion

In this paper, we have introduced technique to improve the quality of recommendation which can recommend contents to users without depending on high-level semantic features which are considered costly and these features require high-level expert knowledge. We have used collaborative filtering recommendations to solve the data sparsity problem by utilizing the aggregated user information preference on video popularity to improve the inaccuracy and to bring the accuracy at optimum level by solving the insufficiency of the training data. We have used collaborative filtering to make the system effective and efficient. We have worked on the large scale dataset and thus we have proved that the sparsity problem can be solved by using the proposed singular value decomposition (SVD) algorithm and hence this algorithm can improve both the recommendation coverage and accuracy. Although we have applied the same SVD algorithm, it still outperforms the existing SVD algorithm due to feature scaling and hyper parameter tuning. This goes to show the importance of pre-processing of data. Although, this collaborative filtering algorithm has performed decently on this algorithm. It is possible to make it better by employing a hybrid approach.

Hybrid recommendation systems have become very popular recently and with good reason too. Content based and collaborative filtering algorithms together can solve the problem that they might face if applied alone. We deal with the sparsity problem by using singular value decomposition (SVD) which is efficient, simple, and practical. Also, use of SVD is beneficial to meet the personal interest of several users.

References

1. Frolov E, Oseledets I (2019) Hybrid SVD: when collaborative information is not enough. ACM Recsys
2. Ricci F, Rokach L, Shapira B (2015) Recommender systems: introduction and challenges. Recommender systems handbook, pp 1–34
3. Tanaji WT (2015) Review on recommendation systems using social networking sites, vol 3, no 4. IJEDR
4. Lemire D, Maclachlan A (2012) Slope one predictors for online rating-based collaborative filtering
5. Ma H, Zhou D, Liu C, Lyu MR, King I (2011) Recommender systems with social regularization. In: ACM internation conference
6. Gao J, Zhang T, Xu C (2017) A unified personalized video recommendation via dynamic recurrent neural networks. In: ACM
7. Tan X, Guo Y, Chen Y, Zhu W (2018) Improving recommendation via inference of user popularity preference in sparse data environment. IEICE Trans Inf Syst E101–D(4)
8. Yan M, Sang J, Xu C (2015) Unified youtube video recommendation via cross-network collaboration. In: ACM ICMR
9. Deldjoo Y, Elahi M, Cremonesi P (2016) Using visual features and latent factors for movie recommendation. CBRecSys
10. Deng Z, Sang J, Xu C (2013) Personalized video recommendation based on cross-platform user modelling. In: IEEE international conference
11. Yan M, Shang W, Li Z (2016) Application of SVD technology in video recommendation system. In: IEEE
12. OSMAN NURĠ OSMANLI (2010) A singular value decomposition approach for recommendation system
13. Koren Y, Bell R, Volinsky C (2009) Matrix factorization techniques for recommender systems. Computer 42(8):30–37. https://doi.org/10.1109/MC.2009.263
14. Sarwar B, Karypis G, Konstan J, Riedl J (2000) Application of dimensionality reduction in recommender system--a case study. WebKDD-2000 Workshop

15. Mohd Kasirun Z, Kumar S, Shamshirband S (2014) An effective recommender algorithm for cold-start problem in academic social networks. Article ID 123726, p 11
16. Katz G, Shani G, Shapira B, Rokach L (2012) Using wikipedia to boost SVD recommender system.19:03:39 UTC
17. Movielens Dataset (2019) https://grouplens.org/datasets/movielens

Importance and Uses of Telemedicine in Physiotherapeutic Healthcare System: A Scoping Systemic Review

Saurabh Kumar, Ankush Sharma, and Priyanka Rishi

Abstract Telemedicine is an emerging technology in the healthcare system, by which patients get benefits like no need for traveling, flexible treatment hours, and the possibility to better integrate skills into daily life. However, the effects of physiotherapy with telemedicine compared with usual care in patients are still inconclusive. To study the effectiveness, importance, and uses of telemedicine in the physiotherapy healthcare system. Relevant databases were obtained from Science Direct, Springer Link, PubMed, Google Scholar, MEDLINE, and Cochrane. Studies that included telemedicine and physiotherapy treatment were included in the analysis. Randomized controlled trials, controlled clinical trials, quasi-randomized studies, and quasi-experimental studies with comparative controls published in the English language were included with no restrictions in terms of date of publication. Thirty records were included for synthesis. Physiotherapy with telemedicine is equally effective and useful in patients who need rehabilitation compared with usual care; this may be an important reason to choose physiotherapy with telemedicine instead of usual care. Recently due to the COVID-19 pandemic, the healthcare system rapidly adjusts in the way in which individuals can easily and effectively access healthcare facilities. In social distancing policies related to pandemic, telemedicine is a good alternative for the conventional healthcare system, which has led to the widespread adoption of telemedicine.

Keywords Telemedicine · Physiotherapy · Importance · Uses · COVID pandemic

S. Kumar (✉) · P. Rishi
Faculty of Physiotherapy, SGT University, Gurugram, Haryana, India

A. Sharma
School of Physiotherapy, BUEST, Baddi, HP, India

© The Author(s), under exclusive license to Springer Nature Singapore Pte Ltd. 2022
P. Nanda et al. (eds.), *Data Engineering for Smart Systems*, Lecture Notes in Networks and Systems 238, https://doi.org/10.1007/978-981-16-2641-8_39

411

1 Introduction

Over the last few years, advancement in telecommunication technology has significantly increased. The medical field has been not left behind too. With the advent of technology like telemedicine and e-Medicine in the healthcare system has considerably increased choice for citizens in terms of remote access to healthcare requirements across the world. The term telemedicine is originated from the Greek word which denotes 'distance' and mederi from Latin word denotes 'heal' [1, 2]. The word telemedicine was first defined by Thomas Bud, and afterward, it was described by Reidas the telecommunication technologies used to exchange health information where geographical, social, time, and cultural barriers present [3]. Telemedicine can be utilized by various fields like patient care, education, research, administration, and public health [1]. According to World Health Organization, telemedicine defined as the use of electronic communications as well as technologies to provide clinical health services when individuals are at different locations [4]. Many people worldwide living in rustic and isolated areas struggle to timely access the good quality medical care by specialists, primarily because specialists are more likely to be located in an urban population. Also in India vast majority of the residents live in rustic areas, whereas 75% of qualified specialized health professionals practice in the urban area, 23% in semi-town areas, and just 2% like to work in rustic areas. Telemedicine has a good alternative to overcome these barriers and has the potential to bridge this distance, facilitate health care in these remote areas, and enhance patient satisfaction [1]. However, these access obstacles are not simply due to geographical remoteness, but in the current situation being the assured social distancing policies begin in response to the COVID-19 pandemic period. This has mandatory the health system to second thought regarding how traditional healthcare armed forces can be distributed to patients in rural as well as urban populations. This pandemic made the traditional healthcare system to rethink and take action. We firmly supposed that telehealth was going to become a benchmark element of care during this pandemic and in the near future [5].

Rehabilitation is a previous branch of medicine, but recent time's novel telecommunication and technology-based rehearsal have been spreader all over the globe. These kind of healthcare approaches in the branch of physiotherapy and rehabilitation are usually described as telerehabilitation, which should be included as a subfield of telemedicine to provide rehabilitation at a distance [6]. Physiotherapy rehabilitation usually starts in the hospital and continues even after discharge. Demand of physiotherapy increasing specially for home care and outpatient clinics and is difficult to meet, especially in rustic areas, and now it also becomes problematical in urban areas due to the COVID pandemic. Telerehabilitation is a fairly new emerging branch in telemedicine and is another to conventional face-to-face therapy and to facilitate the delivery of rehabilitation services to people who cannot access them [7]. The telerehabilitation aim is to provide easy access to rehabilitation services to individuals. This process may engage diverse forms of technology specially telephone, virtual reality programs, Internet-based communication, or a combination of other forms of computer systems, and technologies. These technologies are utilize for prevention, cure, rehabilitation, outcome monitoring and to educate or train the patient and caregiver. Telerehabilitation also decreases the cost of wellness care facilities in a novel and effective way [2].

There are numerous advantages and disadvantages reported for telerehabilitation. Advantages consist of that the difficulty of remoteness can be minimized, makes rehabilitative services or expert more fairly accessible, acceptable and affordable, ensure continuity of rehabilitation even after discharge, flexible exercise hours, decrease traveling cost, less time consuming, and more convenient. Disadvantage of telerehabilitation like difficulties come across by the user as well as with the equipment [8]. Despite a lot of advantages in India, the number of patients receiving the services through telemedicine remains low, and when it comes to physiotherapy services, this number still lower down. On the other hand, the number of patients requiring physiotherapeutic rehabilitation services is gradually increasing day by day. Therefore, this systematic review aimed to study the importance and uses of a telemedicine system in physiotherapy.

To review the literature that expressed the importance and uses of telemedicine in physiotherapy, relevant articles in English were retrieved through a search of Science Direct, Springer Link, Medline, PubMed, Google Scholar, MEDLINE, and Cochrane. The Cochrane Handbook for systematic reviews of interventions was followed for eligible studies. The following keywords and medical subject headings were used: telemedicine use in physiotherapy, physical therapy exercises and telerehabilitation, physiotherapy modalities, and telemedicine, and a variety of combinations of these words were entered. The references of included studies were checked for other relevant publications.

Randomized controlled trials, controlled clinical trials, feasibility studies, quasi-randomized studies, and quasi-experimental studies with comparative controls were included with no restrictions in terms of year of publication. Studies on telemedicine in any physiotherapy treated area are included if the treatment contained any physiotherapeutic modality, exercise, or combination of exercise with health education or objective to alter health-related behavior. Studies were excluded if the intervention did not contain physiotherapy treatment and telemedicine. This review did not rank or rate the quality of the studies reviewed as this was not the purpose of the review but was to summarize the studies and comment on telemedicine uses and their impotence in physiotherapy.

Based on the selection criteria, titles, and abstracts of identified studies were screened for possible inclusion. If there was non-sufficient information for consideration, the full-text article was obtained and studied. Then the study data were extracted independently and recorded according to the guidelines of the Cochrane Collaboration.

The search strategy yielded 131 results. After removing data on the bases of exclusion criteria, 96 records remained and were initially screened. 54 records were found to be eligible for full screening of which, and 30 records were included for qualitative synthesis. The full details of these studies are given in Table 1.

Table 1. Summary of studies and their findings on the effectiveness and use of telemedicine in physiotherapy

AUTHOR	TITLE	DESIGN	PARTICIPANTS	CHARACTERISTIC	TYPE OF TECHNOLOGY	CONCLUSION
Capri 2020 [9]	A Holistic Approach for Telerehabilitation in Rett Syndrome	The TCTRS Project	Not specified	Patients with Rett Syndrome	Web application	From the results it is concluded that telerehabilitation is a comprehensive, affordable, easy-to-use system in rett syndrome without burdening on families, caregivers, or rehabilitation centres.
Bennell 2019 [10]	Does a Web-Based Exercise Programming System Improve Home Exercise Adherence for People With Musculoskeletal Conditions?: A Randomized Controlled Trial	RCT	n= 262	patients with a musculoskeletal condition	Internet	A web-based exercise programming system improved home exercise adherence and confidence in patients with musculoskeletal problems.

(continued)

Table 1. (continued)

AUTHOR	TITLE	DESIGN	PARTICIPANTS	CHARACTERISTIC	TYPE OF TECHNOLOGY	CONCLUSION
Kline 2019 [11]	Improving Physical Activity Through Adjunct Telerehabilitation Following Total Knee Arthroplasty: RandomizedbControlled Trial Protocol	RCT	n= 100	Patients after TKA	Smart device application and Wearable activity sensor.	The study concluded that physical activity behavior could change by telerehabilitation. It also augments current practice and resolve poor physical activity outcomes, long-term health problems, and high costs following TKA.
Chumbler 2012 [12]	Effects of Telerehabilitation on Physical Function and Disability for Stroke Patients	Multisite RCT	n= 52	Stroke patients	telephone	The stroke telerehabilitation treatment significantly improved physical function in stroke patients.
Nield 2012 [13]	Real-Time Telehealth for COPD Self-Management Using Skype™	RCT	n= 22	Patients with COPD	videoconferencing	This study concluded that telehealth is an innovative approach for pursed-lips breathing instruction in the home and feasible and effective to reduce dysnea in COPD patients.

(continued)

Table 1. (continued)

AUTHOR	TITLE	DESIGN	PARTICIPANTS	CHARACTERISTIC	TYPE OF TECHNOLOGY	CONCLUSION
Bernard 2009 [14]	Videoconference-Based Physiotherapy and Tele-Assessment for Homebound Older Adults: A Pilot Study	A Pilot Study	n= 17	Homebound older adults	videoconferencing	Videoconference-Based Physiotherapy is feasibility to population of homebound older adults.
Eriksson 2009 [15]	Physiotherapy at a distance: a controlled study of rehabilitation at home after a shoulder joint operation.	Quasiexperimental designs with comparative controls	n= 22	Patients after shoulder hemiarthroplasty replacements	Videoconferencing	The telemedicine significantly improves pain, mobility, function and quality of life in patients after shoulder join operation.
Huijgen 2008 [16]	Feasibility of a home-based telerehabilitation system compared to usual care: arm/hand function in patients with stroke, traumatic brain injury and multiple sclerosis	Multicentre RCT	n= 70	Stroke, traumatic brain injury and multiple sclerosis patients	Home Care Activity Desk (HCAD) system application	The HCAD training was found to be as satisfied and feasible as usual care in terms of clinical outcomes for both therapists and patients. It also increases the efficiency of care.
Holden 2007 [17]	Telerehabilitation Using a Virtual Environment Improves Upper Extremity Function in Patients with Stroke	RCT	n= 11	Stroke patients	Virtual environment-based (VE) system	This study found that VE treatment sessions showed significant improvements in upper extremity functions in patients with stroke.
Multani 2006 [3]	Effectiveness of Telemedicine Services Integrated Into Physiotherapeutic Health Care System	Prospective controlled trial	n= 30	Patients of back pain	Videoconferencing, telephone, e-mail, videotapes and CDs	Telemedicine is effective in patients with back pain; provide instant care and saves time of patient as well as

Table 1. (continued)

AUTHOR	TITLE	DESIGN	PARTICIPANTS	CHARACTERISTIC	TYPE OF TECHNOLOGY	CONCLUSION
Wiborg, 2003 [18]	Teleneurology to Improve Stroke Care in Rural Areas	Feasibility study	n= 153	Stroke patients	Videoconferencing	Telecommunication using a videoconference is feasible and promising method to improve stroke care in rural areas.

Among the 30 included studies, five studies reported on total knee arthroplasty (n = 5) [7, 9, 10, 11, 19], four studies reported on stroke patients (n = 4) [20, 21, 22, 23], four on cardiac patients (n = 4) [24, 25, 26, 27], three on COPD (n = 3) [12, 28, 29], two on breast cancer (n = 2) [13, 30], one on OA knee (n = 1) [31], one on rett syndrome (n = 1) [32], one on back pain (n = 1) [3], 1 on breast, colon or rectal cancer (n = 1) [14], 1 on homebound older adults patients (n = 1) [15], 1 on colorectal cancer (n = 1) [33], 1 on hip fracture patients (n = 1) [34], 1 on shoulder hemi-arthroplasty replacement (n = 1) [35], 1 on total hip replacement surgery (n = 1) [16], 1 on musculoskeletal condition patients (n = 1) [17], 1 on ACS and undergone PCI patients (n = 1) [36] and 1 on stroke, traumatic brain injury (TBI) and multiple sclerosis (MS) (n = 1) [37] (Fig. 1).

1.1 Discussion

This systemic review illustrate that telemedicine is feasible and useful for the phys-iotherapeutic profession. The majority of included studies in present review investi-gated the effectiveness of telemedicine for rehabilitation in different disease, and they reported a significant positive effect. Telemedicine is very useful where the distance is barrier and population who reside in rural communities. However, this technology is also effective in many other barriers like work commitment, costs related, less transport facility, etc. COVID-19 pandemic is now another additional example in which traditional health care disturbed due to social distancing policies. In all these barriers, telemedicine may have a possible alternative to provide healthcare facilities effectively. Literature explored the utility of telemedicine for the rehabilitation of many conditions keep on to grow. Numerous systematic review studies have demon-strated that telemedicine can provide progress in pain, physical function, disability, and quality of life that are like to that of usual care for persons with musculoskeletal, cardiopulmonary, neurological conditions, and in pre- and post-surgery. The validity and reliability of undertaking a physiotherapy assessment via telemedicine have also been investigated. A systematic review found that assessments confirmed via telecommunication show good concurrent validity for pain, swelling, ROM, muscle strength, balance, gait, and functional assessment [38].

Multani et al. found that by telemedicine, there was significant enhancement found in pain, muscle strength, and functional capacity in the patients of low back pain. Nevertheless, use of effective results by telemedicine depends upon the efficiency of a physiotherapist for the effective explanation of the method for assessment as well as a therapeutic intervention [3]. Tousignant et al. in his study found that home telerehabilitation is as effective as usual care in reducing disability (range of motion, balance, and muscle strength) and recovering function (knee function, walking, and autonomy) in patients undergoing knee arthroplasty after two months of treatment [7, 19] Kairy et al. in his study found that most of the clinical outcomes better after telerehabilitation, but also stated that there is need of additional methodologically sound research to confirm its effectiveness [39].

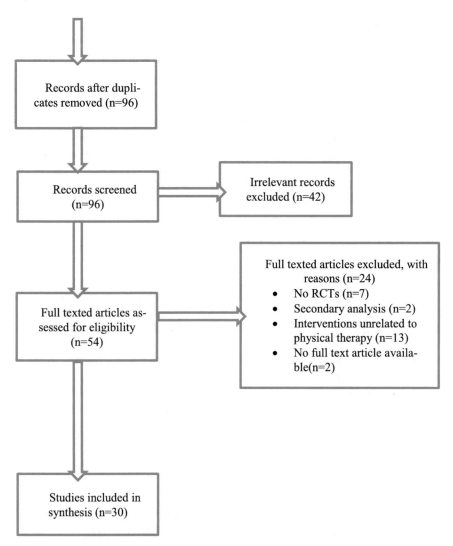

Fig. 1 Flow diagram of literature search according to PRISMA preferred reporting items for systemic review [18]

Egmond et al. in his systemic review and meta-analysis reported that physiotherapy with telerehabilitation is possible and improves the quality of life in post-surgical patients. Telerehabilitation is a valuable tool to monitor outcomes and functional progress in patients. The type of this rehabilitation helps patients to carry out their exercises frequently without any time limit or extra face-to-face visits. Telerehabilitation at home instead of hospital visits increases efficiency and perception of improved quality of life, whereas Kairy et al. evaluated that patients with physical disabilities got rehabilitation via telecommunication showed related clinical outcomes compared with common care. This is in line with the significant outcomes of patient satisfaction with telerehabilitation illustrated in studies by Beaver et al. and Cleeland et al., where telerehabilitation was reported as extra helpful to fulfill patient's requirements [18, 34]. Telemedicine is a 'work-in-progress' field where there is the opportunity to develop a lot of interesting rehabilitation paradigms according to the demand. Nowadays, with the digital revolution, telehealth is evolving from clinics to the home. The newly discovered Coronavirus (COVID-19) and social distancing have put telehealth on the front lines. This technology is relevant to treat many disabilities and diseases without frequently accessing to the hospitals. But still, there are a lot of challenges to accept it completely and need further researches to identify them and options to deal with them, although these are not determined in this paper need further researches.

2 Conclusion

In this systematic review, it was found that physiotherapy and telemedicine are equally effective for patients who need rehabilitation compared with usual care; this may be a key reason to choose physiotherapy with telemedicine instead of usual care. Recently due to COVID-19 pandemic, the healthcare system rapidly trying to adjust in the way in which individuals can access fundamental healthcare facilities. In social distancing policies related to this pandemic, telemedicine is a good alternative for the conventional healthcare system, which has led to the widespread acceptance of telemedicine. This tool is also an efficient substitute for individuals who are not getting access to specialist healthcare services due to distance.

References

1. Ramakrishnan KS, Guruprasad S (2015) Application of telemedicine & e medicine in physiotherapy-emerging trends. https://doi.org/10.13140/rg.2.1.4313.2329
2. Shenoy PM, Shenoy PD (2018) Identifying the Challenges and Cost-effectiveness of Telerehabilitation: A Narrative Review. J Clin Diagn Res 12(12):1–4. https://doi.org/10.7860/JCDR/2018/36811.12311
3. Multani NK, Singh B, Garg S (2006) Effectiveness of telemedicine services integrated into physiotherapeutic health care system. J Exercise Sci Physiother 2:87–91
4. Alan L, Davenport TE, Randall K (2018) Telehealth Physical Therapy in Musculoskeletal Practice. Orthop Sports Phys Ther 48(10):736–739
5. Cottrell MA, Russell TG Telehealth for musculoskeletal physiotherapy. Musculoskelet Sci Pract 48:102193. https://doi.org/10.1016/j.msksp.2020.102193
6. Peretti A, Amenta F, Tayebati SK, Nittari G, Mahdi SS (2017) Telerehabilitation: review of the state-of-the-art and areas of application. JMIR Rehabil Assist Technol 4(2):e7. https://doi.org/10.2196/rehab.7511
7. Tousignant M, Moffet H, Boissy P, Corriveau H, Cabana F, Marquis F (2011) A randomized controlled trial of home telerehabilitation for post-knee arthroplasty. J Telemed Telecare 17(4):195–198

8. Sarsak HI (2020) Telerehabilitation services: a successful paradigm for occupational therapy clinical services? Int Phys Med Rehabil J 5(2):93–98

9. Li LL, Gan YY, Zhang LN, Wang YB, Zhang F, Qi JM (2014) The effect of post-discharge telephone intervention on rehabilitation following total hip replacement surgery. Int J Nurs Sci. https://doi.org/1.10.1016/j.ijnss.2014.05.005

10. Eakin EG, Lawler SP, Winkler EA, Hayes SC (2012) A randomized trial of a telephone-delivered exercise intervention for non-urban dwelling women newly diagnosed with breast cancer: exercise for health. AnnBehav Med 43(2):229–238

11. Eriksson L, Lindstrom B, Gard G, Lysholm J (2009) Physiotherapy at a dis-tance: a controlled study of rehabilitation at home after a shoulder joint operation. J Telemed Telecare 15:215–220

12. Russell TG, Buttrum P, Wootton R, Jull GA (2011) Internet-based outpatient telerehabilitation for patients following total knee arthroplasty: a randomized controlled trial. J Bone Joint Surg Am 93(2):113–120

13. Nield M, Hoo GW (2012) Real-time telehealth for COPD self-management using Skype™. COPD. 9(6): 611–619

14. Dalleck LC, Schmidt LK, Lueker R (2011) Cardiac rehabilitation outcomes in a conventional versus telemedicine-based programme.J Telemed Telecare 17(5):217–221

15. Nguyen HQ, Gill DP, Wolpin S, Steele BG, Benditt JO (2009) Pilot study of a cell phone-based exercise persistence intervention post-rehabilitation for COPD.Int J Chron Obstruct Pulmon Dis 4:301–313

16. Lee YH, Hur SH, Sohn J, Lee HM, Park NH, Cho YK, et al (2013) Impact of home-based exercise training with wireless monitoring on patients with acute coronary syndrome undergoing percutaneous coronary intervention. J Korean Med Sci 28(4):564–568

17. Bennell KL, Marshall CJ, Dobson F, Kasza J, Lonsdale C, Hinman RS (2019) Does a web-based exercise programming system improve home exercise adherence for people with musculoskeletal conditions? A randomized controlled trial. Am J Phys Med Rehabil 98(10):850–858

18. Moher D, Liberati A, Tetzlaff J, Altman DG (2009) Preferred reporting items for systematic reviews and meta-analyses: The PRISMA statement. Ann Intern Med 151:264–269

19. Kline PW, Melanson EL, Sullivan WJ, Blatchford PJ, Miller MJ, Lapsley JES et al Improving physical activity through adjunct telerehabilitation following total

22 Ligibel JA, Meyerhardt J, Pierce JP, Najita J, Shockro L, Campbell N, et al (2012) Impact of a telephone-based physical activity intervention upon exercise behaviors and fitness in cancer survivors enrolled in a cooperative group setting. Breast Cancer Res Treat 132(1):205–13

21. Huijgen BC, Vollenbroek-Hutten MM, Zampolini M, Opisso E, Bernabeu M, Van Nieuwen-hoven J et al. (2008) Feasibility of a home-based telerehabilitation system compared to usual care: arm/hand function in patients with stroke, traumatic brain injury and multiple sclerosis. J Telemed Telecare 14(5):249–256

22. Arthur HM, Smith KM, Kodis J, McKelvie R (2002) A controlled trial of hospital versus home-based exercise in cardiac patients. Med Sci Sports Exerc 34:1544–50

23. Kairy D, Lehoux P, Vincent C, Visintin M (2009) A systematic review of clinical outcomes, clin-ical process, healthcare utilization and costs associated with telerehabilitation. Disabil Rehabil 31:427–47

24. Bernard MM, Janson F, Flora PK, Faulkner GEJ, Norman LM, Fruhwirth M (2009) Videoconference-based physiotherapy and tele-assessment for homebound older adults: a pilot study'. Activities, Adaptation & Aging 33(1):39–48

25. Holden MK, Dyar TA, Cimadoro LD (2007) Telerehabilitation using a virtual environment improves upper extremity function in patients with stroke. IEEE Trans Neural Syst Rehabil Eng 15:36–42

26. Wiborg A, Widder B (2003) Telemedicine in stroke in swabia project. Teleneurology to improve stroke care in rural areas: The Telemedicine in Stroke in Swabia (TESS) Project. Stroke 34(12):2951–2956

27. Mani S, Sharma S, Omar B, Paungmali A, Joseph L (2017) Validity and reliability of Internet-based physiotherapy assessment for musculoskeletal disorders: a systematic review. J Telemed Telecare 23(3):379–391

28. Latham NK, Harris BA, Bean JF, Heeren T, Goodyear C, Zawacki S, et al. (2014) Effect of a home-based exercise program on functional recovery following rehabilitation after hip fracture: a randomized clinical trial. JAMA 311(7):700–708

29. Scalvini S, Zanelli E, Comini L, Tomba MD, Troise G, Giordano A (2009) Home-based exercise rehabilitation with telemedicine following cardiac surgery. J Telemed Telecare 15(6):297–301

30. Piqueras M, Marco E, Coll M, Escalada F, Ballester A, Cinca C, et al.(2013) Effectiveness of an interactive virtual telerehabilitation system in patients after total knee arthoplasty: a randomized controlled trial. J Rehabil Med 45(4):392–6

31. Pinto BM, Papandonatos GD, Goldstein MG (2013) A randomized trial to promote physical activity among breast cancer patients. Health Psychol 32(6):616–626

32. Capri T, Fabio RA, Iannizzotto G, Nucita A (2020) The TCTRS project: a holistic approach for telerehabilitation in rett syndrome. Electronics 9(3):49. https://doi.org/10.3390/electroni cs9030491

33. Chumbler NR, Quigley P, Li X, Morey M, Rose D, Sanford J, et al (2012) Effects of telerehabilitation on physical function and disability for stroke patients: a randomized, controlled trial. Stroke 43(8):2168–2174

34. Odole AC, Ojo OD (2014) Is telephysiotherapy an option for improved quality of life in patients with osteoarthritis of the knee? Int J Telemedicine and Appl: 1–9. https://doi.org/10.1155/2014/903816

35. Piron L, Turolla A, Agostini M, Zucconi C, Cortese F, Zampolini M, et al.(2009) Exercises for paretic upper limb after stroke: A combined virtual-reality and telemedicine approach. J Rehabil Med 41(12):1016–1020

36. Pinto BM, Papandonatos GD, Goldstein MG, Marcus BH, Farrell N (2013) Home-based physical activity intervention for colorectal cancer survivors. Psychoncology 22:54–64

37. Kortke H, Zittermann A, El-Arousy M, Zimmermann E, Wienecke E, Korfer R (2006) New East-Westfalian postoperative therapy concept: a telemedicine guide for the study of ambulatory rehabilitation of patients after cardiac surgery. Telemed J E-health: the Official Journal of the American Telemedicine Association 12(4):475–483

38. Beaver K, Tysver-Robinson D, Campbell M, Twomey M, WilliamsonS, Hindley A, et al. (2009) Comparing hospital and telephone follow-upafter treatment for breast cancer: randomised equivalence trial. BMJ 338:a3147

39. Cleeland CS, Wang XS, Shi Q, Mendoza TR, Wright SL, Berry MD,et al. (2011) Automated symptom alerts reduce postoperative symptom severity after cancer surgery: a randomized controlled clinical trial. J ClinOncol 29(8):994–1000

Feature Exratction of PTTS System and Its Evaluation by Standard Statistical Method Mean Opinion Score

Sunil Nimbhore, Suhas Mache, and Sidhharth Mache

Abstract This paper we targeted the testing and evaluation of the PALI-TTS system (PTTS) for performance and accuracy using the statistical approach of mean opinion score (MOS). The text-to-speech (TTS) system is to convert an arbitrary given text into a corresponding spoken waveform or artificial production of human speech. In this PTTS system recorded speech of digits, vowels, consonants and isolated words which are used for feature extraction. The PTTS system generated voice quality was calculated by standard statistical measurement MOS. The overall accuracy of the PTTS system depends on the intelligibility of synthesized speech. The overall accuracy and evaluation parameters which was used for this PTTS system gives the good voice quality of the sentences, and listers' result was 83% and 3.92, respectively.

Keywords PTTS · Intangibility · COV · Mean opinion score (MOS)

1 Introduction

The text-to-speech system (TTS) converts text into voice using a speech synthesizer it is the artificial production of human speech [1]. In recent years, a lot of research is going on speech synthesis. Speech plays important role in day-to-day life of communication. Speech synthesis it is the process of converting the written text into machine-generated synthetic speech [2]. We have selected concatenative unit selection method to develop text-to-speech (TTS) synthesis for Pali language. Concatenative speech synthesis systems read a text and render into speech by concatenating pre-recorded speech units [3]. Corpus-based strategies (unit choice) utilize an enormous stock to choose the units and link. We have designed and developed an intelligible and natural sounding corpus-based concatenative speech

S. Nimbhore (✉) · S. Mache
Dr. Babasaheb Ambedkar Marathwada University, Aurangabad, MH, India

S. Mache
R. B. Attal Arts, Science and Commerce College Georai, Beed, MH, India

© The Author(s), under exclusive license to Springer Nature Singapore Pte Ltd. 2022   423
P. Nanda et al. (eds.), *Data Engineering for Smart Systems*, Lecture Notes
in Networks and Systems 238, https://doi.org/10.1007/978-981-16-2641-8_40

synthesis system for Pali language. The implemented system is processing the inputted text for analyze, normalized and transcribed into a phonetic representation. The unit selection algorithm is based on the best network path of the units [4].

In this work, the different unit of sizes such as vowels, consonants, syllables, digits, and words have experimented. In unit selection-based concatenative speech synthesis, joint cost also known as concatenative cost, which measures how well two units can be joined together [5, 6].

The paper is organized as follows: Section 1 presents the introduction part of speech synthesis which is summarized. In Sect. 2, the methodology is depicted in three categories such as test data, accuracy test, and subjective evaluation metrics. The Sect. 3 describes the actual implementation part of PTTS system which represents the feature extraction techniques of Pali digit, vowels, consonants and isolated Pali words, etc. Sect. 4 describes the result discussion with accuracy test, overall performance of PTTS system, and mean opinion score [MOS]. The last section of paper is end with conclusion and acknowledgement.

2 Methodology

The aim of testing and evaluation of the TTS system is to determine the system performance and accuracy. It also used to judge the speech quality in terms of its similarity to the human voice and its ability using MOS.

2.1 Test Data

Test information assumes an urgent part in testing; it can impact the entire testing measure and can influence the test results harshly. Test data is sufficient enough to cover all the functionalities of the system under test. Different functionalities of the TTS system can be evaluated by examining overall output speech. It should be designed in such a way that it covers all possible variations including numerals, vowels, consonants, words and connected words. The text-to-speech system is evaluated by three different methods, i.e., objective test, subjective test, and acoustic measurements of speech.

The objective test contains accuracy test and subjective test intelligibility test by mean opinion score [7]. Digits, vowels, consonants and words of Pali language in Devanagri script with its pronunciation are given in Table 1.

Table 1 Pronunciation structure of Pali language with Devanagri scripts

अंक (Digits)	०- सुय्य१ ,-एक२ ,- द्वे३ ,- तयो४ ,- चतु५ ,- पञ्च६ ,- छ७ ,- सत्त८ ,- अठ९ ,- नव१० ,-दस					
स्वर (Vowels)	अ	आ	इ	ई	उ	
	ऊ	ए	ओ			
व्यंजन (Consonants)	क	ख	ग	घ	ड. च	छ
	ज	झ	त्र ट	ठ	ड	ढ
	ण					
	त	थ	द	ध	न	
	प	फ	ब	भ	म	
	य	र	ल	व		
	स	ह	ळ			
शब्द (Words)	अनुली ,उय्यान ,चन्द्वारो ,कस्सको ,वज्जो ,बलीवद्द , मयूरो ,ससुरो ,कट्टक ,कुसलस्स ,सब्ब					

2.2 Accuracy Test

In accuracy calculation, proper selection of test data is important. All such data whose expected output is well defined can be considered for accuracy test [8]. For accuracy measure, it just checks the pronunciation of total correct data such as numerals (digits), vowels, consonant, and words with a total number of the text of the above input. The formula is;

$$\text{Accuracy} = \frac{\text{No. of Correct Pronounced Speech Data}}{\text{Total No. of Text Data}} \times 100$$

2.3 Subjective Evaluation Metrics

Intelligibility testing is done by mean opinion score (MOS). The effective performance of a text-to-speech synthesis system can be properly measured by conducting subjective listening tests [8]. A mean opinion score (MOS) test was conducted. MOS is the arithmetic mean of all the individual scores, and it gives the numerical indication of the perceived speech quality. To check the intelligibility of synthesized speech, as the part of this evaluation, we selected ten (ten) sentences and ten listeners. The listeners were asked to give a rating from 1 to 5 to each utterance. The definition of the rating is given in Table 2.

Table 2 Mean opinion score (MOS)

Mean opinion score (MOS)	Quality rating
5	Excellent
4	Good
3	Fair
2	Poor
1	Bad

3 Implementation

To evaluate the system, we have tested all possible variations including numerals, vowels, consonants, words, and connected words. However, these results in discussion of Pali vowels, consonants, digits, and words. The mean and standard deviation of pitch frequency of speech signals have been calculated and correlated using the coefficient variance formula. The overall mean and standard deviation (SD) for finding the closeness of pitch values on the basis of these pitch values are closely related to the overall average of the PALI-TTS system using MOS (Table 3, 4, 5 and 6).

4 Results Discussion

To assess the framework, we have tried all potential varieties including numerals, vowels, consonants, and associated words of PTTS system which was tested. The overall system associated words are evaluated and calculated by the statistical approach mean opinion score (MOS formula was given in Table 7.

Table 3 Calculated mean, standard deviation, and variance of Pali digit

Pali Digit	Pitch of Uttr_1	Pitch of Uttr_2	Pitch of Uttr_3	Pitch of Uttr_4	Pitch of Uttr_5	Mean	COV	SD
/१/	151.7	163.3	141.9	153.2	158.7	153.8	6.1	8.0
/२/	139.9	156.0	144.8	149.6	149.3	147.9	12.0	6.0
/३/	144.4	151.3	149.4	146.9	146.4	147.7	5.3	2.6
/४/	149.0	161.5	156.2	159.4	162.7	157.8	10.9	5.4
/५/	149.8	157.9	156.5	156.8	157.0	155.6	6.5	3.2
/६/	145.5	154.1	151.2	149.0	151.6	150.3	6.4	3.2
/७/	153.6	162.2	156.2	163.7	161.1	159.4	8.4	4.2
/८/	147.5	153.3	158.2	148.6	150.3	151.6	8.6	4.3
/९/	146.8	151.9	151.2	151.3	148.3	150.0	4.5	2.2
/१०/	146.4	141.9	137.1	144.0	143.0	142.6	6.5	3.2

Table 4 Calculated mean, standard deviation, and variance of Pali vowel

Pali Vowels	Pitch of Uttr_1	Pitch of Uttr_2	Pitch of Uttr_3	Pitch of Uttr_4	Pitch of Uttr_5	Mean	COV	SD
/अ/	163.1	158.7	156.1	151.4	153.1	156.5	9.3	4.6
/आ/	148.3	158.8	151.6	150.7	149.5	151.8	8.2	4.1
/इ/	155.8	174.4	153.5	152.1	150.6	157.3	19.5	9.7
/ई/	148.4	162.9	150.9	151.0	155.2	153.7	11.4	5.7
/उ/	158.2	158.1	151.5	151.2	149.2	153.7	8.4	4.2
/ऊ/	153.2	163.5	161.8	153.9	152.8	157.0	10.3	5.1
/ए/	151.4	162.3	148.0	155.3	149.3	153.3	11.4	5.7
/ओ/	138.7	146.6	140.4	150.8	130.4	141.4	15.6	7.8

Table 5 Calculated mean, standard deviation, and variance of Pali consonants

Pali Vowels	Pitch of Uttr_1	Pitch of Uttr_2	Pitch of Uttr_3	Pitch of Uttr_4	Pitch of Uttr_5	Mean	COV	SD
/क/	155.7	158.3	154.4	160.5	153.4	156.5	5.8	2.9
/ख/	152.2	153.4	146.7	164.0	151.7	153.6	12.7	6.3
/ग/	149.2	156.3	150.3	150.3	153.7	151.9	5.9	2.9
/घ/	143.6	146.4	143.0	157.7	148.0	147.8	11.8	5.9
/च/	157.6	159.1	147.8	149.6	155.4	153.9	9.9	4.9
/छ/	150.9	157.8	156.4	151.8	148.6	153.1	7.7	3.8
/ज/	151.5	154.2	153.3	153.5	155.9	153.7	3.1	1.5
/झ/	143.7	153.5	152.8	148.3	146.7	149.0	8.3	4.1
/ट/	146.2	162.1	154.1	154.3	157.8	154.9	11.7	5.8
/ठ/	135.6	143.4	137.9	139.2	138.0	138.8	5.7	2.8

Table 6 Calculated mean, standard deviation, and variance of Pali consonants

Words	Pitch of Uttr_1	Pitch of Uttr_2	Pitch of Uttr_3	Pitch of Uttr_4	Pitch of Uttr_5	Mean	COV	SD
/वानरीदो/	150.3	149.1	147.8	147.1	148.1	148.5	2.5	1.2
/मक्कुट/	159.5	157.5	156.7	153.5	150.9	155.6	6.8	3.4
/मयुरो/	150.1	147.3	149.1	143.6	150.0	148.0	5.4	2.7
/गरुड/	134.6	136.7	144.3	139.9	134.0	137.9	8.5	4.2
/सुवो/	157.7	157.1	155.4	158.2	148.4	155.4	8.0	4.0
/सारिका/	153.0	151.6	152.8	150.6	151.7	152.0	1.9	0.9
/कोकिलो/	162.6	150.8	157.7	156.9	149.4	155.5	10.7	5.0
/कपिलो/	145.2	165.9	167.6	161.0	155.1	148.5	18.2	4.1
/काक/	150.4	149.3	151.9	149.2	147.4	154.8	11.2	1.6
/कुकुटो/	165.1	159.4	161.2	161.1	158.8	161.1	4.9	2.4

Table 7 Calculated system accuracy mean, standard deviation, and variance of Pali consonants

Digits or numerals	$\text{Accuracy} = \frac{100}{100} \times 100 = 100\%$ (1)
Vowels	$\text{Accuracy} = \frac{8}{8} \times 100 = 100\%$ (2)
Consonants	$\text{Accuracy} = \frac{32}{32} \times 100 = 100\%$ (3)
Syllables	$\text{Accuracy} = \frac{341}{341} \times 100 = 100\%$ (4)
Words	$\text{Accuracy} = \frac{71}{100} \times 100 = 71\%$ (5)
Connected words	$\text{Accuracy} = \frac{42}{100} \times 100 = 42\%$ (6)
ANN words	$\text{Accuracy} = \frac{68}{100} \times 100 = 68\%$ (6)

Table 8 Overall accuracy of PTTS system

Test	Type of data	Accuracy (%)
1	Vowels	100
2	Consonants	100
3	Syllables	100
4	Digits (1–100)	100
5	Short words	71
6	Connected words	42
7	ANN trained words	68
Average		**83. 00**

4.1 Overall Performance of the System

The general presentation of the framework is registered by ascertaining the level of right phonemes (i.e., consonants and vowels), 1–100 digits, short words, connected words, and ANN trained connected words calculated on 100 words of Pali language. Every one of these tests show that the precision of the created PTTS framework was **83.00%** (Table 8).

4.2 Subjective Evaluation (Listing Tests MOS)

An ordinary improvement technique for an open-area PTTS content includes gathering an enormous delegate reference text and choosing a subset (target text) to cover important phonetic and prosodic settings, explicitly: phonetic units (e.g., diaphones), sentence types (e.g., questions and articulations), and expression lengths [9]. While targeting the measures that are valuable in contrasting point by point framework qualities, the successful exhibition of a text-to-speech blend framework can be appropriately estimated by leading abstract listening tests [9]. This test finding the connections among understandability and fathom ability in

Table 9 Mean opinion score of PTTS system

Sentences	L1	L2	L3	L4	L5	L6	L7	L8	L9	L10	MOS
S_1	4	3	4	4	5	4	5	5	4	4	4.2
S_2	3	4	4	4	3	4	4	4	4	4	3.8
S_3	4	4	4	3	4	5	4	4	3	5	4.0
S_4	4	4	4	4	4	4	3	4	4	5	4.0
S_5	4	3	3	5	4	4	5	3	4	5	4.0
S_6	5	4	5	4	5	5	4	5	5	5	4.7
S_7	4	3	4	4	4	4	3	3	4	4	3.7
S_8	3	4	3	4	3	4	3	4	3	4	3.5
S_9	4	3	4	3	4	4	4	3	4	4	3.7
S_10	4	4	3	4	3	3	3	4	4	4	3.6
Avarage											**3.92**

discourse synthesizers and attempts to plan a proper appreciation task for assessing the discourse synthesizers' conceivability [10].

The mean assessment score (MOS) test was directed. MOS is the number-crunching mean of the apparent multitude of individual scores and it gives the mathematical sign of the apparent discourse quality. To check the immaterialness of incorporated discourse, as the piece of this assessment, we chose ten sentences and ten audience members. The crowd was drawn nearer to give a rating from 1 to 5 to each verbalization. The meaning of rating of the voice scaling intangibility was given in Table 3. 1-tBad, 2-poor, 3-Fair or reasonable, 4-Good and 5-Exellent or outstanding. The tested scaling of mean opinion score results are given in Table 9.

The Table 9 consist of S_1 to S_10 which are the ten different sentences and L1 to L10 are the ten different listeners. The mean opinion score (MOS) of each sentences and listeners calculated average result was 3.92. This average result show good intangibility synthesized speech near by the listener's average approximately.

5 Conclusion

The assessment strategies were intended to check the framework precision and discourse nature of the combined discourse by the audience members. The assessment was finished by the few MOS levels, for example, digits, vowels, consonants, and words level. The general exactness of the PTTS framework was 83%. We have directed the emotional posting test for estimating of the coherent integrated discourse utilizing the mean opinion score (MOS) test. The mean opinion score (MOS) test was 3.92; it is very acceptable and close by audience voice. The mathematical sign shows the apparent discourse quality of the speech which marked decent reach.

Acknowledgements The authors gratefully acknowledge full research cultivated support of the Head, Department of Computer Science & IT, Dr. Babasaheb Ambedkar da University, Aurangabad (MH) India, for providing the infrastructure for this research work.

References

1. Mache SR, Baheti MR, Mahender CN (2015) Review on text-to-speech synthesizer. Int J Adv Res Comput Commun Eng 4.8:540
2. Nimbhore SS, Ramteke GD, Ramteke RJ (2015) Implementation of english-text to marathi-speech (ETMS) synthesizer. Int Organ Sci J Comput Eng (IOSR-JCE) 17(1): 34–43. eISSN: 2278–0661, p-ISSN: 2278–8727. Ver. VI
3. More SS, Borde PL, Nimbhore SS (2018) Isolated Pali Word (IPW) feature extraction using MFCC & KNN based on ASR. IOSR J Comput Eng (IOSR-JCE) 20(6): 69–74. e-ISSN: 2278–0661,p-ISSN: 2278–8727, Ver. II
4. Dutoit T (1997) An introduction to text-to-speech synthesis. vol 3. Springer Science & Business Media
5. Tatham M, Morton K (2005) Developments in speech synthesis. John Wiley & Sons
6. Black AW, Campbell N (1995) Optimising selection of units from speech databases for concatenative synthesis
7. Vepa J, King S (2006) Subjective evaluation of join cost and smoothing methods for unit selection speech synthesis. IEEE Trans Audio Speech Lang Process 14(5):1763–1771
8. Vepa J, King S (2004) Subjective evaluation of join cost and smoothing methods
9. Klatt DH (1987) Review of text-to-speech conversion for English. J Acoust Soc Am 82.3:737–793
10. TDIL (2014) Text to speech testing strategy Version 2.1, pp 1–46
11. Rosenberg A, Ramabhadran B (2017) Bias and statistical significance in evaluating speech synthesis with mean opinion scores. In: Proceedings of the 18th annual conference of the international speech communication association (Interspeech 2017). Stockholm, Sweden, pp 20–24
12. Chang Y-Y (2011) Evaluation of TTS systems in intelligibility and comprehension tasks. In: Proceedings of the 23rd conference on computational linguistics and speech processing. Association for Computational Linguistics

Computational Analysis of a Human–Robot Working Alliance Trust in Robot-Based Therapy

Azizi Ab Aziz and Wadhah A. Abdulhussain

Abstract This paper presents a computational analysis of human–robot working alliance trust with behaviour change intervention domain. The model can be used as a trust evaluation module that supports robot-based therapy in maintaining trust human trust towards robots. Simulation experiments under various parameter settings indicated that the model could generate reasonable behaviour of distinct types of chosen cases. Moreover, by equilibria analysis, the model's stability state has been established, and by automated checking, the model's fundamental empirical-based properties have been confirmed.

Keywords Working alliance · Robot-based therapy · Trust in digital therapy · Computational cognitive modelling

1 Introduction

In psychology, trust can be represented as a set of choices of unknown people in situations where individuals passively witness their behaviour, with consequences on their own decisions. Within human–robot interaction, this concept plays an essential role to conserve the trustworthiness, a requirement for such robots to be trusted by humans to perform the assigned tasks. This requirement provides a factor of its motivational activities for humans to engage actively for collaborative tasks. For example, when the individuals are working closely with a therapeutic robot to reach the intended goal, trust plays a vital component in improving belief in following the robot's suggestions. While humans may trust a robot to vacuum dirty floors, there are still some concerns for certain personal aspects due to possible distrust in specific contexts. For example, through a vacuum robot's malfunctions (e.g. partial cleaned room) may not have an enormous impact in daily living. However, having a similar malfunction in high-risk tasks (e.g. autonomous vehicles or surgery robots) can create catastrophic consequences. Regardless of the risk, robotic technology and other autonomous systems

A. A. Aziz (✉) · W. A. Abdulhussain
Relational Machines Group, Human-Centred Computing Lab, School of Computing,
Universiti Utara Malaysia, 06010 Sintok, Kedah, Malaysia
e-mail: aziziaziz@uum.edu.my

© The Author(s), under exclusive license to Springer Nature Singapore Pte Ltd. 2022
P. Nanda et al. (eds.), *Data Engineering for Smart Systems*, Lecture Notes in Networks and Systems 238, https://doi.org/10.1007/978-981-16-2641-8_41

offer possible advantages through supporting individuals in achieving their intended missions. Within cognitive and psychological perspectives, behavioural change and cognitive therapy are among essential areas to be investigated [1–3]. In general, trust has been explored in various domains (including human factors, social psychology, and behavioural organization) in identifying possible connections between humans and machine [5]. The wide range of contexts within which trust has been researched leads to several representations and perceptions of trust. In this paper, our fundamental purpose is to demonstrate the high-level concept of human–robot working alliance trust modelling via examples/analyses in the literature. This paper is structured as follows. In Sect. 2, the theoretical constructs of trust in human–robot working alliance factors are presented, and Sect. 3 describes the details of the model. The simulations traces and results of the model are explained in Sect. 4. The model has been confirmed by mathematical analysis and automated logical checking properties of simulation traces (Sect. 5). Lastly, Sect. 6 summarizes this paper.

2 Trust and Human–Robot Working Alliances

Within conventional therapy environment, the therapeutic alliance is abstracted as an accord between the client and the therapist on the objectives of therapy and the therapeutic responsibilities required to accomplish those objectives. Also, it offers the social bond between client and therapist. In the past, human–robot trust was dedicated primarily to collaboration/teamwork within groups of autonomous agents, machines or robots. However, the latest trend shows there is an increasing interest in leveraging human involvement effectively by enhancing trust in the human–robot working alliance [1, 4, 7]. Contrary to the autonomous system, which is devised mainly to take humans out of the loop, the therapeutic robot requires people and robots to work together in constant and comparatively long-term interaction. In the area of behavioural change, for robots to participate as a working alliance in persuading people to carry out recommended activities, they must have some of the human-like (e.g. counsellor, coach) capabilities that enable fluid and effective teamwork/collaboration with humans [6, 6]. Thus, the therapeutic alliance is the emergent property that stems from this teamwork/collaboration. Trust is also an important factor to consider as changes of trust may significantly affect the outcomes of the teamwork, consequently affecting behavioural change programme outcomes [9, 11].

The attention on trust in human–robot working alliance emerges because several previous works show that humans tend to trust robots similarly to how they trust other humans [7, 10]. Therefore, it is a major concern that people may misunderstand the possible risk associated with handing over decisions to a robot. Depending on the types of assigned tasks, the definition of trust is particularly appealing as different tasks deal with various levels of risk. An example of this risk is a robot in the life-threatening condition which is not attempting to reward humans for their conformity but rather to mitigate risk to save human life [6, 12]. When it comes to the behaviour change intervention, a robot needs real-time feedbacks to maintain the trust that leads to positive outcomes in intervention; otherwise, the main risk will be potential rejections from users. Several components are imperative to regulate trust in the human–robot working alliance. First, human factors such as personality (based on personality traits), openness for interaction, perceived controllability, and appraisal about progress are vital components to allow trust in the robot. A considerable amount of literature has been published on the components, as mentioned earlier. These studies are [6, 11, 13, 14]. In general, openness and agreeableness are often associated with successfulness in human–robot interaction. Perceived controllability shows conditions where the robot is under the user's control. It means robots' actions are predictable as to when performed by humans [7, 13]. Thus, appraisal about the progress enables users to find their expressions and take rights in the therapeutic activity. It has been confirmed that feedback informed therapy decreases dropout rates, enhances client experiences and results and improves the user's perception of positive well-being.

Other important robot-based concepts, such as automation level, physical embodiment and social behaviours, provide a realistic idea of how the robot-based therapy will work. Automation level refers to the degree of an automated task (from Level 1 to 10). These levels are a set of range from complete human control to complete computer/ robot control. For instance, Level 1 indicates to the user, does the task and turns it over to the robot/computer to execute the rest. In contrast, Level 10 demonstrates the robot /computer and does the action if it determines it should be done. Also, the automation level of a system changes an agent's capacity to make some critical decisions based on evidence/knowledge on its own without any other external control requests [5, 11]. Thus, the robot's automation level modifies the robot's ecosystem, whether humans can intervene in the robot's control loop or even depending on the task. The robot/computer tells the human operator only if it decides the operator should be told. When it comes to human–robot interaction, social behaviours capabilities allow humans to use sophisticated social cues and interaction. That concept includes human-centred multi-modal communication and teamwork [8]. This factor is a unifying feature because social robots communicate and synchronize their behaviour through verbal, non-verbal or affective-based sensory systems. Other components like transparency, reliable behaviours, perceived competency and perception about robot drive positive perception from user's perspectives about trust-related perception with human–robot working alliances [3, 9, 11, 16].

First, transparency operationalized as the user's understanding of why a robot behaved expectedly. It allows some technical understanding to the level to which the technology's operating rules and logic are apparent to users. This suggests that transparency will increase in importance as a system's autonomous capabilities increase as it allows users to calibrate their trust of the systems during information uncertainty [3]. Reliable behaviours relate to whether the technology exhibits the same expected behaviour over time (through the understanding of technological constructs and perceived level of automation). When it comes to perceived competency, it states how competence (in terms of human-like subjects) judge the behaviours or possible outcomes of the robot. Through this idea, the robot should be able to convey social behaviours (e.g. prosody, tone, turn-taking and emotional expressions) to be considered as "trustworthy" [9]. Perception about robots deals with how humans behave towards robots as collaborators/teams and, hence, correspondingly decide robots as "equal" co-workers. This notion is related to the perceived intelligence that asserts how intelligent and human-like subjects pass judgment on the robot's behaviours or performances. Still, humans' pre-conceived perceptions about robots can be manipulated by science fictions concepts that may lead to highly overstated beliefs. Therefore, the human should have some awareness of the robot's abilities and what signifies suitable interaction. Other concepts, such as perceived risk and distrust, deal with the destruction of trust [15]. Perceived risk helps explain why users often do not move from the desired stage to the action stage, trusting the decision made by the robot. This could be viewed as either the robot is highly adaptive but tends to be risky or not very adaptive and conservative [15, 16]. Therefore, to avoid this, the robot should be attentive to every detail the user wants and reassure them by answering all their questions. Another concept that eliminates trust is distrust. Distrust in robot could be the most prominent dividing force in the implementation of robot-based therapy. In general, several findings suggests that users might trust robot more if they had more experience with it, less sceptical, and control over how it is used rather than being told to follow orders from a mysterious robotic system [5, 7, 17].

3   Computational Modelling

This section illuminates the execution of *network-oriented modelling* method (based on *temporal causal network*) to design and implement a computational model of human–robot working alliance trust. Consequently, all the identified factors from related literature were utilized to conceptualize the model, as described in Fig. 1.

To construct a quantifiable model of trust, formal specifications trust measurement within human-to-robot perspective must be designed.

Transparency, Reliable Behaviours, Perception about Robot

First, transparency (Tr) is related to the weighted contributions of physical embodiments (Pe), social behaviours (Sb), perceived controllability (Cl) and reliable behaviours (Rb). This concept explains that higher transparency (through the perception of embodiment, visible automation and behaviours) may mitigate the "cry wolf" effect, a phenomenon normally detected in high-risk decision-making where the threshold to activate an "alarm" is habitually set very low to consider all conditions as being "critical".

$$Tr(t) = \alpha_{Tr}(w_{r1} \cdot Pe(t) + w_{r2} \cdot Sb(t)) + (1 - \alpha_{Tr}) \cdot (w_{r3}.Cl(t) + w_{r4} \cdot Rb(t)) \tag{1}$$

$$Rb(t) = \beta_{Rb} \cdot Lp(t) + (1 - \beta_{Rb}) \cdot (w_{b1} \cdot Pc(t) + w_{b2} \cdot Al(t)) \tag{2}$$

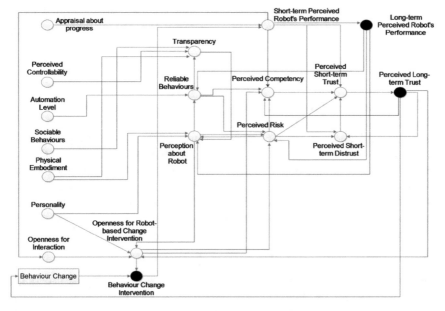

Fig. 1 Conceptual model of human–robot working alliance trust

$$Pc(t) = (1 - \Pr(t)) \cdot$$
$$\left(\lambda_{Pc} \cdot \left(w_{p1} \cdot Lp(t) + w_{p2} \cdot Ps(t) \right) + (1 - \lambda_{Pc}) \cdot \left(w_{p3} \cdot Pe(t) + w_{p4} \cdot Tr(t) \right) \right) \quad (3)$$

Next, reliability (Rb) was shown to affect trust and participants' self-assessment of performance. It relates to the combination of a robot's performances (Lp), perceived controllability and perceived automation level (Al). Perception about a robot (Pc) is calculated using the positive contribution of current long-term performance, personality (Ps) (openness and agreeableness), physical embodiment and transparency. However, perceived risk (Pr) reduces positive perception about the robot.

Openness for Robot-based Change Intervention, Openness for Interaction, Perceived Competency, and Risk

Openness to intervention is one of the vital core components of the successfulness of any therapy. It denotes receptivity *to* new ideas and new experiences. For the openness on robot-based change intervention (Oc), this component can be measured by assessing the openness in the intervention (Oi), perception about a robot, individual's personality and accumulated (long-term) trust (Ls).

$$Oc(t) = \gamma_{Oc} \cdot (w_{o1} \cdot Oi(t) + w_{o2} \cdot Pc(t) + w_{o3} \cdot Ps(t)) + (1 - \gamma_{Oc}) \cdot Ls(t) \quad (4)$$

$$Oi(t) = \beta_{Oi} \cdot Oi_{norm}(t) + (1 - \beta_{Oi}) \cdot Sp(t) \quad (5)$$

$$Cy(t) = \left(\lambda_{cy} \cdot Ls(t) + \left(1 - \lambda_{cy}\right) \cdot \left(w_{y1} \cdot Rb(t) + w_{y2} \cdot Oc(t) + w_{y3} \cdot Sp(t)\right) \cdot (1 - \mathrm{Pr}(t))\right) \tag{6}$$

$$\mathrm{Pr}(t) = (1 - (\alpha_{\mathrm{Pr}} \cdot Lp(t) + (1 - \alpha_{\mathrm{Pr}}).(w_{d1} \cdot Rb(t) + w_{d2} \cdot Pc(t) + w_{d3} \cdot Oc(t)))) \tag{7}$$

The openness towards interaction (Oi) is influenced by perception about short-term robot's performance (Sp) and the individual's openness norm (Oi_{norm}) when the robot has particular skills to support intervention processes, the competency (Cy) levels are crucial as it will improve trust towards a human–robot alliance. Components like long-term trust, reliable behaviours, openness for change intervention and short-term performance increase perceived competency while perceived risk negates the level. The effect of perceived risk (Pr) is determined through the contrast sum contribution of performance, reliable behaviour, perception about robot and openness on robot-based intervention.

Perceived Short-term Trust, Distrust and Performance

Perceived short-term trust (Ss) is essential in shaping *human* interactions with one another and with *robots*. Within human–robot working alliances, the combination of perceived risk and distrust (Sd) reduce short-term trust while perceived performance and competency always provide positive feedback towards trust.

$$Ss(t) = (1 - (\mathrm{Pr}(t) \cdot Sd(t))) \cdot (\gamma_{Ss} \cdot Sp(t) + (1 - \gamma_{Ss}) \cdot Cy(t)) \tag{8}$$

$$Sd(t) = (1 - (\beta_{Sd} \cdot Sp(t) + (1 - \beta_{Sd}) \cdot Ls(t))) \cdot \mathrm{Pr}(t) \tag{9}$$

$$Sp(t) = \beta_{sp} \cdot Ap(t) + \left(1 - \beta_{Sp}\right) \cdot Bi(t) \tag{10}$$

Distrust (Sd) creates negative feedback towards many aspects of human–robot working alliances. The sum contribution of short-term performance (Sp), long-term trust (Ls) and competency mitigate the formation of human–robot working alliance distrust. Short-term performance (Sp) is generated when an individual's appraised the robot's performance and accumulated effects of change intervention (Bi) are positive.

Behaviour Change Intervention, Long-term Perceived Robot's Performance and Perceived Long-term Trust

Here, behaviour change intervention (Bi) builds or reduces over time. When the weightage combination (w_{bi}) between behaviour change component (Bc) and openness on robot-based intervention (Oc) is higher than the previSous behaviour change intervention and decay (λ_{decay}) in belief in change intervention multiplied with the contribution factor, λ_{Bi}, then the behaviour change belief increases. Otherwise, it declines based on its preceding level and influencing component. This circumstance also can be applied to explain the related phenomenon for all temporal relations (e.g.

long-term performance (*Lp*) and long-term trust (*Ls*)), based on their corresponding parameters and attributes.

$$Bi(t + \Delta t) = Bi(t) + \lambda_{Bi}\big(((w_{bi1} \cdot Bc(t) + w_{bi2} \cdot Oc(t)) - Bi(t)) - \lambda_{\text{decay}}\big). \\ Bi(t) \cdot (1 - Bi(t)) \cdot \Delta t$$

(11)

$$Lp(t + \Delta t) = Lp(t) + \lambda_{Lp.}(Sp(t) - Lp(t)) \cdot (1 - Lp(t)) \cdot \Delta t \qquad (12)$$

$$Ls(t + \Delta t) = Ls(t) + \lambda_{Ls.}(Ss(t) - Ls(t)) \cdot (1 - Ls(t)) \cdot \Delta t \qquad (13)$$

It is also essential to address that the change process is assessed in a time interval between *t* and *t* + Δ*t*.

4 Simulation Experiments

In this study, the proposed model was executed using a numerical programming platform. Several simulations based on selected cases have been performed using the developed formal specifications to determine the human–robot working alliance trust. Several unique patterns of trust have been discovered. These different types are accomplished by setting the levels to range from 0 to 1 (as can been seen in Table 1). These weights, contribution and regulation rates can also be adapted to simulate unique individual traits.

This paper shows only the simulation runs for three fictional individuals with a different personality profile due to the extreme amount of possible sequences. Figure 2 displays the variables concerning the short-term effects: (i) perceived risk, (ii) perceived distrust, (iii) reliable behaviour and (iv) competency. The simulation

Table. 1 Settings of the simulations

Factors/Cases	#1	#2	#3
Ap	0.8	0.5	0.4
Cl	0.7	0.6	0.3
Al	0.8	0.5	0.4
Sb	0.9	0.6	0.5
Pe	0.7	0.5	0.2
Ps	0.7	0.4	0.1
Oi	0.8	0.5	0.3
Bc	0.7	0.5	0.1
Oi$_{norm}$	0.9	0.5	0.1
Bc$_{norm}$	0.9	0.5	0.1

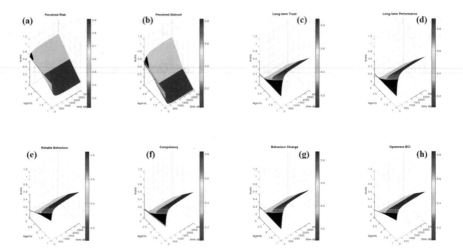

Fig. 2 Simulation results for **a** perceived risk, **b** perceived distrust, **c** long-term trust, **d** long-term performance, **e** reliable behaviours, **f** competency, **g** behaviour change intervention and (**d**) openness towards BCI

results (Fig. 2) show that a person in Case #1 has a lower perceived risk than other persons in Case #2 and #3. Thus, it mitigates possible negative impacts in perceived distrust and hence improves both perceived competency and reliable behaviours.

From this graph (Fig. 2a, b), it can be seen that the initial perceived risk and distrust levels are slightly high and later decrease as the trust and openness towards robot-based intervention improve. This finding is consistent with [7, 9, 15]. It is also important to support the successfulness of following the advice and tasks assigned by the therapeutic robot.

In this experiment, an improvement in the competency and reliable levels (and other concepts) will influence to the development of long-term trust, performance and perception in behaviour change intervention (as in Fig. 2c, d, g). These simulation results reflect those of [2, 5, 9, 13, 17] who also find similar behaviours in their empirical study. The next section, therefore, moves on to discuss the correctness of the model.

5 Evaluation

In this section, the stationary points and equilibria occurring in the model have been analysed. A stationary point of a state Z at time t can be derived when $dZ(t)/dt = 0$. Therefore, it can be concluded that the computational model Z is in equilibrium at t when all states have a stationary point at t. this condition can be obtained by rewriting all mathematical specifications in a continuous form of differential equations. Based on this assumption, we will have an equilibrium stage when;

$$dBi/dt = 0, dLp/dt = 0, dLs/dt = 0$$

Thus the following equations are found:

$$\lambda_{Bi} \cdot \left(\left((w_{bi1} \cdot Bc + w_{bi2} \cdot Oc) - Bi\right) - \lambda_{decay}\right) \cdot Bi \cdot (l - Bi) = 0 \tag{14}$$

$$\lambda_{Lp} \cdot (Sp - Lp).(1 - Lp) = 0 \tag{15}$$

$$\lambda_{Ls} \cdot (Ss - Ls) \cdot (1 - Ls) = 0 \tag{16}$$

Theoretically, it can be presumed that the equilibrium phase exists when the difference between the current accumulated impact for all temporal specifications and short-term instantaneous specifications is equal to zero. Thus, the equilibria points of these specifications can be made by distinguishing cases (1–10). In this paper, only three equilibria points are considered; $Bi = 0$, $Lp = 1$, and $Ls = 0$. For these cases, it can be obtained that the subsequent cases can determine the values for the equilibria:

Case #1: $Bi= 0$.

For this case, by (10), it follows that

$Sp = \beta_{Sp}.Ap$, assuming $\beta_{Sp} \neq 0$.

Case #2: $Lp= 1$.

From Eqs. (2), (3), and (7), it follows that

$$Rb = \beta_{Rb} + (1 - \beta_{Rb}) \cdot (w_{b1} \cdot Pc + w_{b2} \cdot Al)$$
$$Pc = (1 - Pr) \cdot \left(\lambda_{Pc}\left(w_{p1} + w_{p2} Ps\right) + (1 - \lambda_{Pc}).\left(w_{p3} \cdot Pe + w_{p4} \cdot Tr\right)\right)$$

By Eq. (7), this is equivalent to

$$Pr = (1 - (\alpha_{Pr} + (1 - \alpha_{Pr}) \cdot (w_{d1} \cdot Rb + w_{d2} \cdot Pc + w_{d3}.Oc)))$$

Case #3: $Ls= 0$.

For this case, by Eq. (4), it follows that

$$Oc = \gamma_{Oc} \cdot (w_{o1} \cdot Oi + w_{o2} \cdot Pc + w_{o3} \cdot Ps)$$

Moreover, from (9), it follows that, $Sd = 1-\beta_{Sd}.Sp$.

Next, the temporal trace language (TTL) representations are used by specifying a set of dynamic statements that are (or are not) expected to hold. These expressions will be automatically used to verify these statements based on generated temporal traces. The implementation of TTL supports formal specification and assessment of dynamic properties, encompassing both qualitative and quantitative features. This type of verification intends to examine whether the simulation model performs as it should. A typical case of a property that may be verified is whether no unanticipated situations arise such as a variable running out of its boundaries.

VP1 = Physical embodiment and social behaviours of a relational agent improves trust in human–robot alliances

VP1 $\equiv \forall \gamma$: TRACE, $\forall t1, t2$:TIME, $\forall R1,R2, D1,D2$:REAL
[state$(\gamma,t1)$|= has_value(physical_embodiment, R1) &
state$(\gamma,t1)$|= has_value(sociable_behaviours, R2)
state$(\gamma,t1)$|= has_value(LT_trust, D1) &
state$(\gamma,t2)$|= has_value(LT_trust, D2) &
$t2 > t1 + d$ & $R1 > 0.8$ & $R2 > 0.7$ & $D1 > 0.6]$ $\Rightarrow$ $D2 \geq D1$

VP2 = Robot's performance increases humans' perception over robot ability to help them

VP2$\equiv \forall \gamma$:TRACE, $\forall t1, t2$:TIME, $\forall M1, M2, H1, H2$:REAL
[state$(\gamma, t1)$|= has_value(performance, M1) &
state$(\gamma, t1)$|= has_value(perception_robot, H1) &
state$(\gamma, t2)$|= has_value(performance, M2) &
state$(\gamma, t2)$|= has_value(perception_robot, H2) &
$M1 \geq 0.5$ & $t2 > t1$ & $H1 > 0.5]$ $\Rightarrow$ $H2 > H1$.

VP3: Stability of Variable x

VP3$\equiv \forall \gamma$: TRACE, $\forall t1, t2$: TIME, tb, te:TIME, $\forall J1,J2$:REAL
[state$(\gamma,t1)$|= has_value(x, J1) &
state$(\gamma,t2)$ |= has_value(x, J2) &
$tb < t1 < te$ & $tb < t2 < te]$ $\Rightarrow$ $J1 - \alpha \leq J2 \leq J1 + \alpha$.

This property can be used to verify in which situations a certain variable does not fluctuate or change after a series of time-steps (stable point).

VP4: Perceived risk results in the low perceived performance

VP4 $\equiv \forall \gamma$: TRACE, $\forall t1, t2$:TIME, $\forall J1,J2, L1,L2$:REAL
[state$(\gamma,t1)$|= has_value(perceived_risk, J1) &
state$(\gamma,t2)$|= has_value(perceived_risk, J2)
state$(\gamma,t1)$|= has_value(performance, L1) &
state$(\gamma,t2)$|= has_value(performance, L2) &
$t2 > t1 + d$ & $J1 > 0.5]$ $\Rightarrow$ $L1 \geq L2$.

6 Conclusion

In this paper, the formal representation of human–robot working alliance trust model was introduced. Using a numerical programming environment, several simulation experiments under distinctive parameter settings have been executed. Although an extensive empirical validation is left for upcoming work, these experimental results have pointed out that the model can generate several trust circumstances when humans use a robot-based therapy platform. Our method has resulted in two interesting simulation results. First, the developer of a computational trust model can evaluate possible trustworthiness from user's perspectives in the first place by incorporating correct considerations (e.g. personalized parameters) into the computational human–robot working alliance trust function. Furthermore, using this model, we can analyse possible scenarios and make it available to improve interaction design and support repertoire. Besides, we conducted a mathematical analysis to determine possible equilibria points. Also, several expected simulation traces of the model have been verified. For example, these traces indicated that all variables remained within their boundaries, and the deliberated equilibria points are confirmed. The automated temporal trace language was also used to verify against simulation traces that ensued from the chosen scenario. Based on several parameter settings, these logical properties turn out to succeed, which gives formal evidence that the model performs as anticipated. Besides, it allows us to get more computational and theoretical insights into the temporal dynamics of trust formation processes. As our concluding remark, we would like to highlight that our human–robot working alliance trust modelling method can be expanded to other formal modelling approaches where various integration and parameter settings are involved.

References

1. Aziz AA, Ghanimi MHA (2020) Reading with robots: a personalized robot-based learning companion for solving cognitively demanding tasks. Int J Adv Sci Eng Inform Technol 10(4):1489–1496
2. Baxter P, Belpaeme T, Canamero L, Cosi P, Demiris Y, Enescu V, Nalin M (2011) Long-term human-robot interaction with young users. In: IEEE/ACM human-robot interaction 2011 conference (robots with children workshop), pp 1–4
3. Bodenhagen L, Fischer K, Weigelin HM (2017) The influence of transparency and adaptability on trust in human-robot medical interactions. In: 2nd workshop on behavior adaptation, interaction and learning for assistive robotics
4. Chen JYC, Lakhmani SG, Stowers K, Selkowitz AR, Wright JL, Barnes M (2018) Situation awareness-based agent transparency and human-autonomy teaming effectiveness. Theor Issues Ergon Sci 19(3):259–282
5. Hoff KA, Bashir M (2015) Trust in automation: integrating empirical evidence on factors that influence trust. Hum Factors 57(3):407–434
6. Kwiatkowska M, Lahijanian M (2016) Social trust: a major challenge for the future of autonomous systems
7. Langer A, Feingold-Polak R, Mueller O, Kellmeyer P, Levy-Tzedek S (2019) Trust in socially assistive robots: considerations for use in rehabilitation. Neurosci Biobehav Rev104:231–239

8. Leite I, Martinho C, Paiva A (2013) Social robots for long-term interaction: a survey. Int J Soc Robot 5(2):291–308
9. Lewis M, Sycara K, Walker P (2018) The role of trust in human-robot interaction. In: Foundations of trusted autonomy. Springer, Cham, pp 135–159
10. Martelaro N, Nneji VC, Ju W, Hinds P (2016) Tell me more designing HRI to encourage more trust, disclosure, and companionship. In: 2016 11th ACM/IEEE international conference on human-robot interaction (HRI), pp 181–188
11. Oleson KE, Billings DR, Kocsis V, Chen JYC, Hancock PA (2011) Antecedents of trust in human-robot collaborations. In: 2011 IEEE international multi-disciplinary conference on cognitive methods in situation awareness and decision support (CogSIMA). IEEE, pp 175–178
12. Onyeulo EB, Gandhi V (2020) What makes a social robot good at interacting with humans? Information 11(1):43
13. Ososky S, Sanders T, Jentsch F, Hancock P, Chen JYC (2014) Determinants of system transparency and its influence on trust in and reliance on unmanned robotic systems. In: Unmanned systems technology XVI (Vol. 9084, p 90840E). International Society for Optics and Photonics
14. Pickering JB, Engen V, Walland P (2017) The interplay between human and machine agency. In: International conference on human-computer interaction. Springer, pp 47–59
15. Salem M, Lakatos G, Amirabdollahian F, Dautenhahn K (2015) Would you trust a (faulty) robot? Effects of error, task type and personality on human-robot cooperation and trust. In: 2015 10th ACM/IEEE international conference on human-robot interaction (HRI), pp 1–8
16. Schaefer KE, Chen JYC, Szalma JL, Hancock PA (2016) A meta-analysis of factors influencing the development of trust in automation: implications for understanding autonomy in future systems. Hum Factors 58(3):377–400
17. Ullman D, Malle B (2016) The effect of perceived involvement on trust in human-robot interaction. In: 2016 11th ACM/IEEE international conference on human-robot interaction (HRI), pp 641–642

Novel Intrusion Prevention and Detection Model in Wireless Sensor Network

Neha Singh and Deepali Virmani

Abstract WSN have gotten progressively one of the most sultry exploration areas in the field of computer science because of their broad scope of uses, including military, transportation, and detection applications. To guarantee the security and reliability of WSN, an intrusion detection system ought to be set up. This IDS must be viable with the attributes of WSNs and fit for recognizing the most significant conceivable number of security dangers. This paper proposes an intrusion prevention and detection system in wireless sensor networks. The proposed model is a combination of FzMAI and Support Vector Machine. This technique performs in two phases: The first phase prevents intrusions from passing a wireless sensor network, and in the second phase, malicious nodes that joined the system are detected using Support Vector Machine. The proposed model is verified on WSN-DS dataset and yields an accuracy of 99.97%. The proposed model is compared with five existing algorithms: SVM, Naïve Bayes', Random Forest, KNN, and Decision Tree and outperforms all five algorithms.

Keywords Wireless sensor networks · Intrusion detection · Intrusion prevention · FzMAI

1 Introduction

Wireless sensor networks (WSNs) comprises sensor hubs sent in a way to gather data about general conditions [4]. It is the technology comprised of hubs of sensors deployed over the region to recover data about the climate. It is made out of numerous sensor hubs with different capacities like detecting, handling, and transferring [12]. The WSN is adaptable and simple to implement [14]. WSN has various applications

N. Singh (✉)
University School of Information, Communication and Technology, Guru Gobind Singh Indraprastha University, Dwarka, Delhi 110078, India

D. Virmani
Department of Computer Science Engineering, Bhagwan Parshuram Institute of Technology, Rohini, Delhi 110089, India

like in the military, medical care to accumulate information and information transmission, transportation services, etc. [1]. Applications of wireless sensor networks are expanding day-by-day. This expansion is making many lives simpler and better. Because of security issues and restricted asset vitality, they are powerless against security attacks. This sort of attack targets delivering an organization unequipped for offering ordinary assistance by focusing on either the organization's data transfer capacity or its network. These attacks accomplish their objective by sending at a casualty a flood of packets that overwhelms his organization or preparing limit denying admittance to his clients [3]. For any good application of wireless sensor networks, there is an increase in the threat of attacks. These attacks lead to a severe loss of efficiencies, data breach, and monitory loss. Intrusion detection is the process of detecting activity malicious in nature. These malignant exercises or interruptions are intriguing from a PC security point of view. Intrusion detection frameworks are one of the significant pieces of PC security [2]. An intrusion detection system (IDS) is a framework security innovation for identifying weakness abuses against a PC framework that examines network/framework capacities [2]. IDS are categorized based on data collection which analyzes host data files that execute on the particular host. IDS are categorized on detection techniques such as signature-based detection which analyzes the malicious activity using the stored information of previous attacks and anomaly-based intrusion detection analyze the malicious activity by checking the abnormal behavior of the system [2]. This leads to a big research gap for researchers to work on. Many researchers are working on making the best possible models to prevent and detect intrusions in sensor networks. In this paper, a model for the prevention and detection of attacks in wireless sensor networks is proposed—the proposed model in the integration of two models. First model, FzMAI, uses fuzzy rules to block the attacks from invading the system. Those attacks that still manages to enter the network are detected by using the Support Vector Machine algorithm. Integrating both the techniques to make one model has many advantages. Firstly, the model provides a prevention model that will prevent attacks from entering the system. Secondly, even if a few attacks manage to enter the system, the model will detect these attacks in the network. These advantages of the proposed model will help increase the efficiency of the network, reduce the risk of a data breach, and prevent monitory loss.

Our Contribution:

1. Designed a model that prevents the attack from entering the wireless sensor network.
2. Designed a model that detects the attacks that have entered the model.
3. This model is tested on WSN-DS dataset.
4. The proposed method is compared with five existing with five existing algorithms: SVM, Naïve Bayes', Random Forest, KNN, and Decision Tree.

2 Related Work

In [2], author provides a review that attempts to give an organized and thorough diagram of the survey on the identification of malicious activities. From the current anomaly detection techniques, every procedure has relative qualities and shortcomings. This study gives an investigation of existing abnormality recognition procedures, and how the strategies utilized in one territory can be applied in another application space. In [15], author proposed to examine and assess NGIPS from shielding the network. Examination technique incorporates situation and geography entrance. In paper [7], the author has proposed an Intrusion Prevention System by using a Neural Network alongside SVM. The proposed framework gives a security arrangement that is competent enough. In [6], author has presented the comparison of intrusion detection. The author has presented the difficulties while forming an IDS followed by the principal necessities of a decent competitor detection schemes. The investigation of each plan in these classes is introduced, demonstrating the downsides. Before ending every class, the overall favorable circumstances and deficiencies of every classification is expressed. This sort of preventive instruments shaped the principal protection line for WSNs. Nonetheless, a few attacks like wormholes, sinkhole, could not be distinguished utilizing this sort of preventive techniques. In [9], the author has acquainted a framework with look at changed intrusion in system. The investigation of the framework is done in ns2. A point-by-point investigation of different parameters is made and calculated. From this examination and in like manner reason, it is inferred that the framework is generally proper for shortcoming location in WSN. Jose et al. [5] has used three assorted datasets for finding the precision of five major algorithms. The exploration is to check whether another dataset WSN-DS gives a prevalent precision when stood out from existing datasets on similar algorithms. In [4], wireless sensor networks comprise of sensor hubs sent in a manner to accumulate information about the overall condition. The author presented open exploration issues identified with WSN security. In [3], author explains WSN applications information transmission through remote organizations. Because of the shortcomings in the WSN, the sensor hubs are helpless against the majority threats. Rassam et al. [10] presents an examination of different malicious detection methods over different datasets and delineates how these calculations have developed with time. The investigation depends on different boundaries, for example, different sorts of innovations used to propose new frameworks and number of distributions over some stretch of time, the exactness of intrusion detection rate. The examination infers that there is a noteworthy increment in new interruption discovery frameworks after some time. The investigation additionally infers that scientists are slowly moving from old informational collections to the new informational index to approve their frameworks. In [13], author presents one of the strong rule-based techniques known as fuzzy logic in which the knowledge is the combination of rules depicted as IF-THEN structure. It gives a blend of the boundary parameters. Fuzzy reasoning can be adapted with system in various perspectives. In [11], new dataset is made from Network Simulator 2 (NS-2) using LEACH. The new dataset is called WSN-DS. The author in [8] suggested a system known as advanced hybrid intrusion detection system (AHIDS) that uses a clustered framework that support limiting energy utilization.

3 Proposed Model

Proposed model works in two phases. First phase consists of prevention of attacks through FzMAI, and second phase consists of detection of attacks through SVM. Both the phases are explained below.

3.1 FzMAI

FzMAI is a fuzzy-rule based system used for prevention of attacks in wireless sensor networks. FzMAI has three steps:

- Feature selection.
- Membership function computation.
- Fuzzified rules.

The initial step is to examine WSN-DS dataset. Five boundaries are chosen out of 18, and one new boundary is proposed utilizing existing boundaries. These boundaries are chosen utilizing feature selection. The subsequent advance is to characterize membership functions. FzMAI utilizes six participation capacities as input [2]. The estimations of every enrollment work for these boundaries are characterized in the model [2]. These membership functions are utilized in fuzzy rules. The third step is to characterize fuzzy rules for preventing intruders from invading the system as shown in Fig. 1 [2].

3.2 SVM

• SVM is one of the most powerful supervised machine learning algorithms. • This classification models are highly effective when dealing with both linearly separable as well as nonlinearly separable data. • Training of Support Vector Machine (SVM) model requires performing kernel trick and finding the hyperplane, which becomes more and more complex as the size of the training dataset increases. Predicting new datapoints involves calculations with various hyperplanes so as the number of classes increases the process becomes more complex.

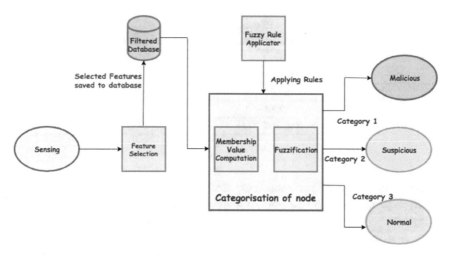

Fig. 1 FzMAI for intrusion prevention in WSN

Algorithm:

• Dataset U with total of n tuples and m independent attributes and a total of p number of classes.

$U = \text{list}(a_1, a_2, \ldots, a_m, class_k)$

$U \rightarrow Training\ dataset$

$(a_1, a_2, \ldots, a_m) \rightarrow m\ independent\ attributes$

$class_k \rightarrow Depenedent\ (Target)\ Categorical\ Variable\ out\ of\ m\ total\ classes : 1 <= k <= p$

• Training Support Vector Classification Machine on dataset U

$Model\ (SVM) \rightarrow f(U, K(V, W), C, Gamma)$

$U \rightarrow Dataset$

$K(V,W) \rightarrow Kernel\ Function$

$C \rightarrow Complexity\ Parameter$

$Gamma \rightarrow Gamma\ Parameter$

Table 1 Overall statistics of proposed algorithm

	MALICIOUS	NORMAL
Sensitivity	0.9997	0.9998
Specificity	0.9998	0.9998
Pos-Pred value	0.9983	0.9999
Neg-Pred value	0.9999	0.9987
Balanced accuracy	0.9997	0.9998

• Making prediction for a new datapoint V involves the checking which part of the hyperplane does the new point lies and the same class is assigned to the new data-point.

Prediction(Model(SVM),V) $\rightarrow$ $class(Hyperplane)$

4 Result Analysis

The proposed model in executed in R-Studio. The dataset used for testing the validity of the proposed model is WSN-DS. This dataset is developed using Leach protocol, and method has been defined in NS2 for data collection named as WSN-DS dataset. Five classes are defined in this dataset—blackhole attack, greyhole attack, flooding attack, scheduling attack and normal based on 18 attributes. The proposed model is compared with five existing algorithms: SVM, Naïve Bayes', Random Forest, KNN, and Decision Tree.

Table 1 gives the overall statistics of the proposed algorithm with respect to all above-defined parameters. The sensitivity of proposed model is 0.9997 and 0.9998 for malicious and normal node, respectively. The specificity is coming out to be 0.9998 for both malicious and normal nodes. The Pos-Pred value is 0.9983 and 0.9999 for malicious and normal nodes, respectively. The balanced accuracy of the proposed model is coming out to be 0.9997 for malicious and 0.9998 for normal nodes.

Figure 2 shows the accuracy of the proposed model with different parameters. The sensitivity of proposed model for malicious nodes is 99.97% and for normal node is 99.98%. The specificity is coming out to be 99.98% for both malicious and normal nodes. The Pos-Pred value is 99.83% and 99.99% for malicious and normal nodes, respectively. The balanced accuracy of the proposed model is coming out to be 99.97% for malicious and 99.98% for normal nodes.

Figure 3 provides the accuracy analysis of the proposed model. As is can be seen, the proposed algorithm has the highest accuracy of 99.97%, followed by Random Forest with 99.94% accuracy and SVM with 99.64% accuracy. The least accuracy is of Naïve Bayes' with 94.46% accuracy.

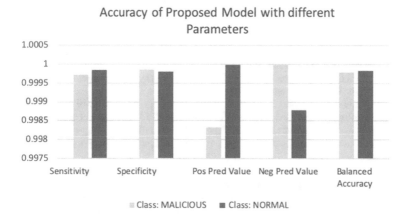

Fig. 2 Accuracy of proposed model with different parameters

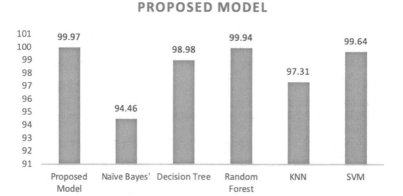

Fig. 3 Accuracy analysis of proposed model

5 Conclusion

The scope of wireless sensor networks has grown exponentially over the past few years. This expansion is making many lives simpler and better. For any good application of wireless sensor network, there is an increase of threat of attacks. These attacks lead to severe loss of efficiencies, data breach, and monitory loss. This paper proposes an intrusion prevention and detection system in WSN. The proposed model is a combination of FzMAI and Support Vector Machine. The model is tested on WSN-DS dataset and yields an overall efficiency of 99.97%. The proposed model is compared with five existing algorithms: SVM, Naïve Bayes', Random Forest, KNN, and Decision Tree and outperforms all five algorithms. The analysis concludes that the proposed model is yielding better accuracy as compared to all five existing techniques.

References

1. Almomani I, Al-Kasasbeh B, Al-Akhras M (2016) WSN-DS: a dataset for intrusion detection systems in wireless sensor networks. J Sensors
2. Alrajeh NA, Khan S, Shams B (2013) Intrusion detection systems in wireless sensor networks: a review. Int J Distrib Sensor Netw 9(5)
3. Constantinides C, Shiaeles S, Ghita B, Kolokotronis N (2019) A novel online incremental learning intrusion prevention system. In: 2019 10th IFIP international conference on new technologies, mobility and security (NTMS). IEEE, pp 1–6
4. Duppa GIP, Surantha N (2019) Evaluation of network security based on next generation intrusion prevention system. TELKOMNIKA 17(1):39–48
5. Jose S, Malathi D, Reddy B, Jayaseeli D (2018) A survey on anomaly based host intrusion detection system. J Phys Conf Ser 1000(1):012049. IOP Publishing
6. Kashyap R (2020) Applications of wireless sensor networks in healthcare. In: IoT and WSN applications for modern agricultural advancements: emerging research and opportunities. IGI Global, pp 8–40
7. Kuila P, Jana PK (2020) Evolutionary computing approaches for clustering and routing in wireless sensor networks. In: Sensor technology: concepts, methodologies, tools, and applications. IGI Global, pp 125–146
8. Maksimović M, Vujović V, Milošević V (2014) Fuzzy logic and wireless sensor networks—a survey. J Intell Fuzzy Syst 27(2):877–890
9. Patil S, Chaudhari S (2016) DoS attack prevention technique in wireless sensor networks. Proc Comput Sci 79:715–721
10. Rassam MA, Maarof MA, Zainal A (2012) A survey of intrusion detection schemes in wireless sensor networks. Am J Appl Sci 9(10):1636
11. Singh N, Virmani D (2018) Adequacy scrutiny of intrusion detection techniques over discrete datasets. Int J Eng Technol 626–630
12. Singh N, Virmani D (2020) Competence computation of attacks in wireless sensor network. J Statis Manage Syst 1–13
13. Singh N, Virmani D (2020) Computational method to prove efficacy of datasets. J Inform Optim Sci 1–23
14. Singh R, Singh J, Singh R (2017) Fuzzy based advanced hybrid intrusion detection system to detect malicious nodes in wireless sensor networks. Wireless Commun Mobile Comput
15. Singh N, Virmani D, Gao XZ (2020) A fuzzy logic-based method to avert intrusions in wireless sensor networks using WSN-DS dataset. Int J Comput Intell Appl 2050018

Shadow Detection from Real Images and Removal Using Image Processing

Sumaya Akter Usrika and Abdus Sattar

Abstract Automatic shadow detection and removal have been used in many image processing systems such as video surveillance, scene interpretations, and object recognition. Ignoring the presence of shadows in images can cause serious problems such as object merging, object loss, misinterpretation, and alteration makeup in visual processing applications such as segment, group analysis, and follow-up. Many algorithms had it proposed to books, related to the acquisition and removal of images and videos. Comparative testing and capacity building of existing methods in the video has already been reported, but we do not have the same in case the images are still standing. This paper provides the complete existing dignity detection survey and removal technique reported in the current situation image. The test metrics involved in strategies for finding and removing strategies are also discussed with the inefficiencies of common metrics such as the accuracy of the pixel, precision, recall, and F-score in the acquisition phase which is also checked. Plenty and quantity of the selected methods are also tested. Ku to our knowledge all of this is a special first article that discusses ways to detect and remove shadows from real photos.

Keywords Image processing · Shadow removal · Object detection · Object recognition

1 Introduction

We live in the age of modern technology. Here, almost every things are dependent on technology. People are getting used to technology to make their life easy and more comfortable. Modern technology was just an advancement of old technology,

S. A. Usrika · A. Sattar (✉)
Daffodil International University, 102, Mirpur Road Dhanmondi, Dhaka 1207, Shukrabad, Bangladesh
e-mail: abdus.cse@diu.edu.bd

S. A. Usrika
e-mail: sumaya25-811@diu.edu.bd

© The Author(s), under exclusive license to Springer Nature Singapore Pte Ltd. 2022 451
P. Nanda et al. (eds.), *Data Engineering for Smart Systems*, Lecture Notes in Networks and Systems 238, https://doi.org/10.1007/978-981-16-2641-8_43

the effects of technology in modern life were immeasurable, we use technology differently sometimes the ways we use different technologies end up damaging our lives or the society we leave behind. Modern technology by technology is not so new in the most important cases. For example, communication technologies have evolved over the years, these days we use an email with images that have been an advancement of fax.

The way it works and the purpose of degrading it is still photographs and videos are not the same. For videos, details from previous frames are available with dignity, and in an image, we should rely on geometrical as well the statistical feature of the shadow in one image are directly the shadow of the part. To achieve a better performance, shadow detection was widely used in the video surveillance system as object tracking [1] and automatic driving [2]. The image depends on the type of input images such as internal, external, or satellite image acquisition and deletion finds application in object recognition [3] and description of the scenes [4]. Such as pictures it is often analyzing, including the geometry of antiquities. Dignity, to get a 3D analysis of the material to be release object geometry [4] or obtaining direction of light sources [5]. Other important applications include enhancing the localization of objects and measurements, especially in the aerial imagery of saw buildings [3, 6, 7] to rediscover the 3D scene [8] or to detect cloud and their shadows [9]. Comparative experiments with the formation of existing shadow energy. The video acquisition method had already been reported by [10–12]. Shadow algorithms are divided into two-layer tablets, mainly mathematical and determined by Andrea et al. [10] and four representatives of the algorithms described in detail. The article [11] distinguishes methods based on object/location and starting point. Another paper [12] divides reporting activities to date into content-based taxonomy consisting of four categories: chromaticity, physical, geometry, and composition. The authors of [10–12] provided an evaluation of the measurements and quantity of activities reported in time using a standalone collection of available videos. Another interesting review of Dee and Santos [13] is a multi-facet field a conversation about a shadow that sheds light on the elements of belief in the formation of dignity. This update discusses, as well as the view of shadows in humans and the perspective of the machine, and the ways in which the human cognitive system uses information from the shadow. However, the above review videos are also specified; other similar images had not been reported yet. Such a review could help researchers, who are planning to do so work locally.

Nowadays, we see here that almost every things are dependent on technology. People are getting used to technology to make their life easy and comfortable. Modern technologies are just the advancement of old technology, the effects of technology on modern life are immeasurable, we used technology in a variety of ways, and sometimes the way we use different technologies ends up damaging our health or the society that we leave behind photograph or image. When we take pictures, we see a picture with shadows, because we cannot understand about this image or this image to make the system a competent recommendation must be used to have the power to decide for itself. Decision making should require the ability to dig into the data. This makes us eager to make this kind of foundation workable. Our work is closely related to image processing techniques. The goal of this project was to design a system to detect shadow removal when capturing images. This system will detect the suspicious removal photograph. We will detect the sentiment between the test conversations. Here, we are proposing the method of designing a system that can automatically analyze the conversation and give feedback with the image. The most important common approach to be image analysis consists of detecting the occurrence of features (image) of known value. There are some works on analyzing image data. Some of these tried to analyze the large data of the image. They use sentimental analysis to detect positive–negative sentiment. The following questions were raised to guide the research.

- Can we collect row data of the image?
- Can I preprocess the row data to use for the machine learning approaches?
- Can we machine the learning process correctly detect or identify the category of the shadow remover?

2 Literature Review

In the last few work decades, lots of effective ways of achieving dignity in the fields of photography or video sequencing had been introduced. In this video sequence, shadows acquisition be used to improved target acquisition performances, which lies to be the basis for the position. In the last year, with the continued enlargement of the social frugality and the advancement of information technology, it will become a program which has increased dramatically. As a major problem for intellectuals to monitor programs, in the field of computer vision research, target acquisition for moving purposes has been one of the hottest problems. Therefore, the removal of shadows was an important and challenging issue in the directional images [3].

It is divided into at least two categories in the existing methods: the model-based approach and the feature-based approach. On the other hand, the model-based approaches need to be establishing shadows to be a statistical model based on shadow attributes and then evaluator whether every pixel was a shadow space or not based on mathematical models. Cucchiara et al. [13] suggest the basis for a method in the parameter model. An unused model-based approach is suggested in [14]. Feature-making techniques often use them to be image-related elements [15], such as brightness, color, complementary information, and other details in order to be a judge. The paper [6, 8, 14] reviewed the dignity model and extended their improvement on it. Recently, many researchers had been proposed a series of innovative ways to obtain shadows to use the convolution neural network (CNN) model. Khan et al. [14] firstly apply in-depth study method of obtaining dignity by training the two networks to find the location of the shadows and the edge of the shadow, respectively. Predict posteriors based on excerpted elements were provided with being a random field model to produce a smooth shading path. Vincent et al. [15] used two ad networks to obtain dignity. The first network is used to extract the original markers. Along with the original graph, these mark to apply to the second network to get adjusted to image results to mark. In [14], relying on color and texture markers, the SVM

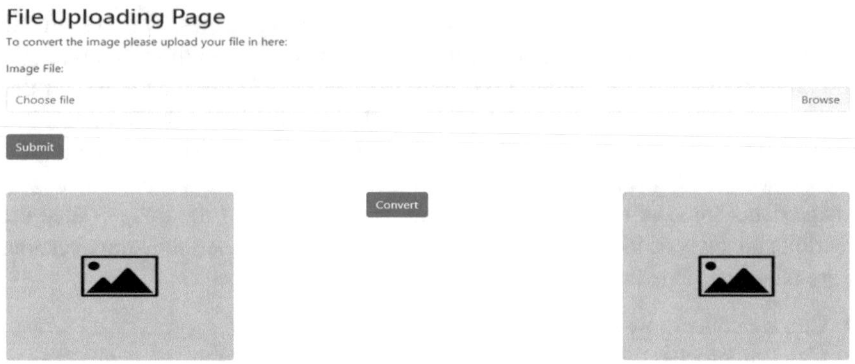

File Uploading Page
To convert the image please upload your file in here:

Image File:

Choose file   Browse

Submit

Convert

Fig. 1 Enter interface of shadow removal

separator was used, obtained a pre-shadow map, combined with the real image, and embedded in a pre-made CNN network to extract being the image effects. Nguyen et al. [15] introduced conditional opposition networks (cGAN). The effect of the generator was a shadow and discriminatory marks separating a true and false mark. The friction between the generator and the generator enables the generator to locate the shadow space. This approach has been a major improvement compared to the method [13]. Similarly, Le et al. [14] used GAN to improve the network's ability to distinguish a shadow area. Existing methods for obtaining CNN shadows often to being use cascade networks [12, 13] or GAN [14, 15], stated that difficulty in the real-time acquisition and also during the training model (Fig. 1).

It is noted that digital image processing technology is continuously developing and the use of image recognition was also growing and growing in Web services. Common areas had a shadow app especially for the following locations: video surveillance, security, and HCI field. It has been also using in the medical and health section, public safety application, agro field section, industry labor field, and also military defense and aerospace-related field. Since a large amount of data viewing required large resources to be bandwidth, storage, and transfer, and to store or process high-definition green video (HD) [5], the cost of transferring was much needed. Therefore, how we can integrate visual analysis with Web network technology has been very valuable, especially in remote areas of the Web.

The paper [14, 15] stated that we need to study very lost cost hardware devices because of the limited hardware and software technology used in image processing technology. Therefore, it is found that HTML5 technology developed which is easier and more efficient. On the other hand, this HTML5 technology helps significantly reduce the cost, once the content of Web application has been transforming. It also supports live update to the user and gets real-time feedback from the user applications contents. Dignity itself is part of dignity when the light was completely absent. The shadows are created by the object behind the event which was called being cast shadows. There are two regions divided for the shadow cast to continue lightly, the first one is umbra, and the second one is the penumbra region. The umbra region is

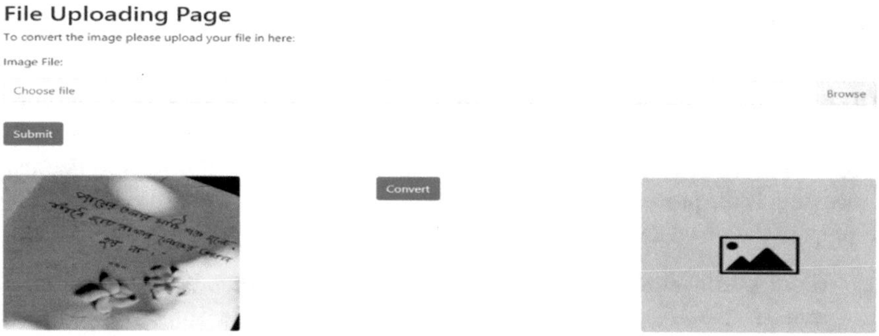

File Uploading Page

To convert the image please upload your file in here:

Image File:

Choose file   Browse

Submit

Convert

Fig. 2 Uploading shadow image

working for the dark central part where there is no way to receive the light. If we see, Fig. 2 shows that the shadows are simple and others called penumbras. On the other hand, we can get the dignity that we use both combinations of umbrella and the penumbra region [14].

3 Research Methodology

3.1 Research Subject and Instrumentation

The subject of the study was that the area of study was reading and research to clarify understanding. Not only a clear understanding, but also the research studies were responsible for providing relevant information on the various parameters of the study. Instrumentation, on the other hand, refers to the equipment or tools that investigators should have used.

3.2 Data Collection

To research these specific fields, the fastest and most prominent was data. The data were, in fact, regarded as the heart of machine learning as a process. And in our study, there was no other data. Therefore, it has been our most challenging tasks for us researched. I build up our data sets by analyzing being lots of journal.

3.3 Data Processing

The data processing used the data mining process, and it is involved to train the raw and real data in the comprehensible format. Normally, we find the real data is incomplete, with many errors and defects, inconsistency, etc. It is proved that data processing is the processing part to solve the kinds of issues. The bellow shows how to prepare the raw data for further processing and shows the step by step.

(a) Data purification is used to plug in lost values, smooth data pointing or removing outliers, and resolve inconsistencies.
(b) Data integration is using large amounts of data, data cubes, or files.
(c) Data modification is using generalization and integration.
(d) Data reduction is using the volume reduction but produces the same or similar analysis result.
(e) Data optional is the part of the data is reduced, replacing the de-recognition value.

3.3.1 Algorithm

What we see mainly is that the text tends to have a background with a consistent color everywhere, so the unused effect had that place as well. I suggest adding a factor, αi, as determined by our integrated shadows map for each pixel in the inputs image to match the local domain color intensity with earth index color. Specifically,

$$Include := c_i = c_i\, \alpha i \tag{1}$$

where c_i ne $-c_i$ is the RGB color intensity of the dignified input with extensions not included in pixels i, respectively. We found the background and background size of the text in the text and customize the background size with the earth index to produce shadow map for each RGB pixel, α. Inserting this shadow map in the file, the insert image produces that final result.

3.3.2 Local and Global Background Color

To get background colors of the area, we start by splitting the installation images into small scattered blocks. In each of those blocks, we combine pixels sizes into two groups ourselves. Labels as a paper or text background. Both local and global data, we used to be Gaussian hybrid model (GMM) equivalent to be expectation–maximization (EM) and start with k-means clustering. Typically, documents usually had a brightly colored text with a bright background. According to this, assign a high-definition collection center like to be your local RGB color, I, when I define a pixel in the center of the current blocks. In the program block cases with fixed colors (e.g., all in the background), the collections were almost identical to the same geography so choosing the higher one stills applies. Next, we found universal background colors. We take the intensity of the pixels in all inputs, rather than the local regions, and combine the two categories of paper and texts, as before. Also, the collection means you had the maximum value which was listed as the background strength. Finally, we search all the details for the power of the original inputs and provide the closest to the background collection as the ultimate global RGB references, g. Note that it was useless. The definition of a collection can be used in these steps instead of the closest force, and it has been found conceptually that this method improves the results slightly.

3.3.3 Computing Shadow Map

The background colors to be the same real background colors. Background color deviate from this due to lighting effects such as shadows and blurring. To remove the influences of lighting, we calculate the ratio of background and local color to produce a shadows map such as: $\alpha i = ig$, (2) where i was the localization intensity in pixels i and g was the global background reference for all pixels. Additionally, ai maps each pixel inputs file to the background color of the reference and, when applied to the input images (Eq. 1), produces the unused end results.

3.4 Statistical Analysis

The above system, configuration reflects the internal and external configuration of the system modules integrated into a single package for the first program. The following sections provided were a brief overview of the tool used and the details of the implementation of the various modules of the advanced program that have been going from start to finish. The entire system is made up of Windows operating system, PhpStorm IDE, and PyCharm IDE platform.

4 Result Analysis

Shadow remover detection framework had to be to develop on the machine have the Windows 10, core i5 processor with 8 GB RAM. The system had been developed in Python and Php in the backend, and Javascript was used to be in the front end. MySQL was used for storing was related data in this framework.

For coding in Python, we had used the latest version of PyCharm which was 2018.2.4 with the Python version 3.6. For coding in PHP, we had used to be the latest version of PhpStorm which was 2018.2.2 with PHP version 7. System configuration follows showing the internal and external configuration of the system modules that are integrated together into a single package to become the first system. The following

Fig. 3 Remove shadow image

sections are provided with a brief overview of the existing tools and implementation details of the various modules of the program developed from start to finish. The whole system is built into the Windows Operating System again using PyCharm and PhpStorm IDE.

4.1 Experiments Results

In this section, discuss about a shadows removing systems. Here, build a system that can show us an interface. In this system, build the process to shadow remove from any type of image. In this system, anyone can remove shadow by image. This system can be used from anywhere of this world using Internet. This system is hosted with a Public IP. Thus, anyone in anywhere can remove shadow by image of this system.

Figure 1 shows a snapshot of the systems of remove shadow by the image. In here, we will give the snapshot of the enter page of the interface.

Now we will give the shadow image in the interface of our system that will show the shadow image remove the shadow. This page will be seen after uploading shadow image.

Here, we will give the necessary interface of the remove shadow of the system (Fig. 3).

4.2 Descriptive Analysis

In this section, we will discuss about the interface of the detection model. Here, the system management can have the image with shadow and got the image without shadow in real time. If anyone upload image, still the system can monitor the image in real time. It stores the image with shadow after analyzing in an image without

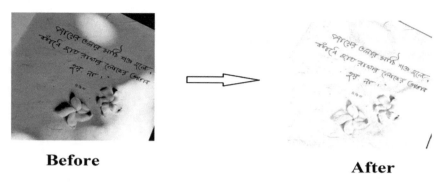

Before   **After**

Fig. 4 Remove shadow image before to after

shadow proof of the image. In this every section, we see the result when uploading image with shadow in this system after removing shadow of the image (Fig. 4).

After obtaining that accuracy, the highest results come from the shadows as to why, once satisfied, if we are trying to increase our accuracy, we should be well prepared for the database. All images must be the same number. At the time, to increase the level of accuracy, data purification was no exception. When more details are processed, more accurate predictions will be made by the divider.

5   Conclusion and Future Work

It had no doubt that there are a lot of researches works on image processing especially in the English Language. When the outcome of such kind of works is taking a revolutionary change in our computing life, recently, such kind of research is being increased this time. It gets some outstanding real-life applications on the blessing of such kind of research works. But it is a matter of great regrets that there has no such research work on shadow remover. But it is the hope for us that many researchers from various countries have started to do research on this field. This research work did some approaches of shadow remover to classify its category. The primary aim was to develop a system that can automatically detect the shadow images. We have introduced a method of removing shadow from text image (e.g. document, menu, receipt) by making shadows maps, or pixel measurements, corresponding to the background colors in the global references. My approach work power, as indicated by quality and proportion, in many types of examples contains a large number of solid shadows in both controlled and real-world settings. In this paper, we had provided an in-depth study of existing methods of detecting and removing shadows organized into the most useful category. Test metric involved in strategies for finding and removing dignity was also being reviewed. Ways such as recovery methods and time-consuming learning methods. The image in the painting method works well in the region of small parks, but the expansion of in painting is a large clip that the hole brings a difficult calculation. Also, dark things exist it is usually wrong as shadows. Many options are available to video sequence which was a specific application and will not be used for still images. Dignity acquisition and removal from single images with different geometries features and textures that reflected different display parameters remain a major challenge.

References

1. Forman G (2003) An extensive empirical study of feature selection metrics for text classification. J Mach Learn Res 1289–1305
2. Pang B, Lee L (2005) Seeing stars: exploiting class relationships for sentiment categorization with respect to rating scales. ACL 115–1243
3. Tumey P, Littman ML (2003) Measuring praise and criticism: inference of semantic orientation from association. ACM Trans Inform Syst 315–346
4. Liu S, Lee I (2015) A hybrid sentiment analysis framework for large email data. In: 2015 10th international conference on intelligent systems and knowledge engineering (ISKE). IEEE, pp 324–330
5. Feng S, Wang D, Yu G, Yang C, Yang N (2009) Sentiment clustering: a novel method to explore in the blogosphere. Springer
6. Li N, Wu DD (2010) Using text mining and sentiment analysis for online forums hotspot detection and forecast. Decis Support Syst 48(2):354–368
7. Balasubramanyan R, Routledge BR, Smith NA (2010) From tweets to polls: linking textsentiment to public opinion time series
8. Klimt B, Yang Y (2004) The enron corpus: a new dataset for Email classification research. Springer
9. Sharma AK, Sahni S (2011) A comparative study of classification algorithms for spam Emaildata analysis. Int J Comput Sci Eng 3(5):1890–1895
10. Sahami M, Dumais S, Heckerman D, Horvitz E (1998) A Bayesian approach to filteringjunk e-mail
11. Mohammad SM, Yang TW (2011) Tracking sentiment in mail: how genders differ onemotional axes
12. Hangal S, Lam MS, Heer J (2011) Muse: reviving memories using Email archives. ACM
13. van Prooijen J-W, van Vugt M Conspiracy theories: evolved functions and psychological mechanisms. Perspect Psychol Sci 0(0):1745691618774270, PMID: 30231213.
14. Karen M (2017) Douglas and Ana Caroline Leite, suspicion in the workplace: organizational conspiracy theories and work-related outcomes. Br J Psychol 108(3):486–506
15. Al-Amrani Y, Lazaar Md, Eddine Elkadiri K (2017) Sentiment analysis using supervised classification algorithms. In: Proceedings of the 2nd international conference on big data, cloud and applications, ACM, p 61

A Study on Pulmonary Image Screening for the Detection of COVID-19 Using Convolutional Neural Networks

Shreyas Thakur, Yash Kasliwal, Taikhum Kothambawala, and Rahul Katarya

Abstract The coronavirus, also known as COVID-19, has now spread to almost all parts of the world causing widespread illness and deaths. Not only control but also testing individuals for this virus has become a challenge in a variety of sections of the society. The rising number of cases, and the shortage of testing kits for the virus has motivated us to explore other methods of testing for the virus. This study has tried to explore various research papers and studies, to come up with an efficient method for the detection of the virus, using datasets based on the chest X-ray and computed tomography (CT) scans of individuals. This study has found that visual inspection of the CT scan and X-rays of the chest of the individuals has led to a time-consuming process, but using convolutional neural network models, there exists a strong possibility of coming up with a data-driven deep learning model, which would act as a classifier between infected and healthy individual. This study has showed that convolutional neural network-based models have a capability of providing an alternative to the current methods of testing for the virus.

Keywords Classifier · Convolutional neural networks · COVID-19 · CT scans · Deep learning · X-ray

1 Introduction

In December 2019, several individuals were found exposed to unidentified cases of pneumonia, which was later found to be caused by a virus called 2019-nCoV.

It was found to have originated from the wet markets of Wuhan, China. This virus has been found to spread from human-to-human transmission and hence has spread to almost all the parts of the world. This has caused wide spread illness and deaths and also has disrupted economic activities around the world [1, 2].

As a consequence, the need for testing COVID-19 suspects to curb the spread has skyrocketed. The gold standard of detection of COVID-19 virus is by making individuals undergo a reverse-transcription polymerase chain reaction (RT-PCR) test, in

S. Thakur (✉) · Y. Kasliwal · T. Kothambawala · R. Katarya
Department of Computer Science, Delhi Technological University, New Delhi, India

© The Author(s), under exclusive license to Springer Nature Singapore Pte Ltd. 2022
P. Nanda et al. (eds.), *Data Engineering for Smart Systems*, Lecture Notes in Networks and Systems 238, https://doi.org/10.1007/978-981-16-2641-8_44

which the presence of viral nuclei acid in the blood sample of the individual confirms the disease. This testing process has not only been found to be time consuming, but also has shown poor sensitivity in detecting the virus in its early stages. The examination of CT scan X-ray images of suspects has been found to be a non-evasive method of detecting the diseases, as it confirms the disease by the affirming the presence of certain characteristic manifestations in the lungs. This testing process has also proven to have a better sensitivity in the early stages of the disease [3, 4].

To improve upon the RT-PCR test's fallacies, an alternative procedure has been proposed. Artificial intelligence and deep learning methods have found to be very efficient in predicting the presence of the virus, by scanning the CT X- ray images of the chest of individuals suspected of having the disease. Computerized CNN models have been developed, which have not only helped radiologists and medical professionals help in the confirmation and quantification of the disease, but have also streamlined the work process by minimizing the physical contact between the patient and medical professionals working in the testing facilities [3, 5].

Due to the emergence of computerization in the detection of COVID-19, this study has tried to study various ways that the data driven methods have worked in the confirmation and quantification of COVID-19 and tried to improve the testing process with the use of convolutional neural networks and data science.

In this study, Sect. 2 has defined the various terminologies used by the study along with the performance matrix used to compare the performance of various studies. Section 3 consists of the literature review of various studies this study has reviewed. Section 4 consists of the comparison of the performance of various studies, along with the results and the scope for future work in this field. This is followed by the references for the source of literature which were read and reviewed for this study.

2   Terminologies Used

2.1   Pulmonary Imaging

Alterations or changes in the lung function can be attributed to the changing lung structure, which can be identified on the basis of destruction or remodeling of the airway wall and lung parenchyma present inside the lungs. Several imaging techniques like the CT scan or MRI scan have been historically practiced to monitor the changes in lung structure [11]. Taking CT scans of the chest have proven to be an efficient and reliable way of pulmonary imaging by the medical professional community. Through chest CT scanning, the presence or absence of disease in each pulmonary segment can be clearly recorded [12, 13] (Figs. 1 and 2).

Fig. 1 Chest CT scan for a
healthy person [21]

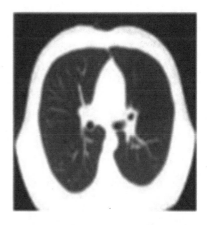

Fig. 2 Chest X-ray for a
healthy person [22]

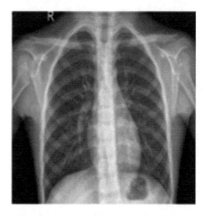

2.2 Convolutional Neural Networks

Human beings possess a strong ability to recognize patterns in images, but this generally takes a long duration of time. Computerizing the process of image recognition
delivers a quick way to scan through large volumes of data for image recognition
[6, 7]. Convolutional neural networks have shown great results in the recent years in
large-scale image recognition, as computers have made great progress in interpreting
unstructured data in the recent years. Convolutional neural networks mainly use the
concepts of image recognition and object detection to classify images into different
classes, along with the identification and quantification of the feature on which the
classification is based [8, 9]. Even though image recognition using convolutional
neural networks has seen big success in the recent years, it has its limitations. CNN
consists of simple neural networks, and hence, it requires multiple iterations on a
large training dataset. It is because the neural network employs a descending gradient
process, which causes the CNN iterate slowly towards the optimization point [10]
(FIg. 3).

Fig. 3 Schematic
representation of convolution
neural network
implementation

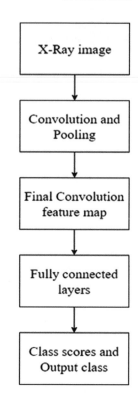

3   Literature Review

This study has studied and compared various testing methods currently employed to test individuals for this virus, and has looked into their efficiency and performance.

- Our first study had used the data of 131 patients infected with COVID-19, from three different hospitals in China. Fused lesions, new lesions of increased lung density were identified to be the signs of progression of disease. This study did not use any computerized data-driven technique to compute the progression of the disease, as all the CT scan images were screened individually by two radiologists. But this study did help us know the features to look for to affirm the progression of the disease [14].
- As the first study lacked computerization, we moved on to the second study which had used a deep convolutional neural network to detect COVID-19 automatically, and separate it from other forms of pneumonia, from the chest X-Ray images of people suspected of being infected with the disease. The dataset used consisted of 1241 images, consisting of healthy, COVID-19, pneumonia bacterial- and pneumonia viral-infected pulmonary X-ray scans of infected people. Deep learning was once again used, in the form of the CoroNet model, to detect various kinds of pneumonia or COVID-19, and was able to produce results with overall accuracy

of 89.6% [15]. The overall accuracy achieved by the model in this study could be improved.

- For our third study, to detect COVID-19 from chest X-ray of patients, deep learning was implemented with the help of PyTorch1.1.0, DenseNet (name given to the model). Image preprocessing was employed by this study, as target axial slides had to be selected from 300–500 axial images per X-ray. All the statistical analysis was performed with the help of Python 3.7.6 and sklearn 0.22.1. DenseNet used a dataset of 295 patients, with 149 healthy and 146 infected patients, which were divided into three sets, training, validation, and testing. It was able to deliver accurate results with accuracy of 99% in the training and 98% in the testing datasets [16].

- The fourth study explored many different ways to classify patients with confirmed COVID-19, bacterial pneumonia, and healthy individuals, but the best results were observed using deep learning to make convolutional neural network models. This was due to the high accuracy performance of the method in image recognition and object detection. This study separated the development of the deep learning model into two steps, feature extraction, and fine tuning. Data augmentation, with several preprocessing steps were also performed. Dataset of 112 COVID-19 infected, pneumonia bacterial, and healthy individuals' chest X-ray images were used by this study. This study developed a model which was able to detect COVID-19 with an accuracy of 95% [17].

- The final study used an AI-based, deep learning model to detect the presence of COVID-19, as well as predicted a quantification of the disease burden. This study used a dataset consisting of 56 COVID-19 infected individuals and 51 healthy individuals' chest X-ray images [18, 19] (Table 1).

As seen from the table above, CNNs have very high accuracy for pulmonary image classification and outperform visual inspection. But the performance of CNNs differs with model and nature of sample.

4 Discussion

From our study, we can infer that CNNs show much better performance than simple visual inspection by experts. But the degree of improvement also depends on the model implemented. It was also observed that preprocessing of images played a crucial role in the overall accuracy. The models also showed a loss of accuracy when the number of classes increased. While study 3 does have excellent accuracy, it also has the fewest number of samples, which is in complete juxtaposition to study 2. Therefore, the goal is to create a model which maintains high performance accuracy that does not deteriorate with sample size. Two-set classifiers also are easier to execute and provide with decisive results, while increasing the number of classes increases the diagnostic capabilities of the model.

Table 1 Representation of various implementations of pulmonary image screening for COVID-19

Study	Dataset size	Techniques used	Inference
1	131 samples of pulmonary X-ray images of individuals infected by COVID-19	Visual inspection	• Visual inspection by two radiologists to find common patterns in CT scan X-rays of infected individuals • Fused lesions and new lesions found as common traits in infected individuals • Did not use any data-driven automated technique
2	1242 chest X-ray images of 4 different data sets; healthy, COVID-19, pneumonia viral, and pneumonia bacterial-infected individuals	Convolutional neural networks	• Used image recognition and CNN to classify images into infections • Used preprocessing techniques for more efficient use of data • Delivered overall accuracy of 89.6
3	295 chest X-ray images; 149 healthy and 146 COVID-19 infected individuals, divided into three sets, training, validation, and testing	Convolutional neural networks	• 2-set classifier • Data was used in an efficient and systematic way with several preprocessing steps to reduce the working load on the classifier • Delivered accuracy of 99% in training and 98% in the testing datasets
4	336 chest X-ray images, 112 each of three data sets; pneumonia bacterial, COVID-19 infected, and healthy individuals	Convolutional neural networks Visual inspection	• Compared different CNN models to find out the one with the best accuracy • 3-set classifier • Broke the process into two major steps; image recognition, and object detection • Accuracy of 95% was achieved with the best model
5	107 pulmonary X-ray images, consisting of healthy, and 56 COVID-19 infected individuals	Artificial intelligence Deep learning	• Used deep learning to develop a 2-set classifier • AI-based approach for image recognition

5   Conclusion and Future Work

In this study, the problem of testing COVID-19 suspects was introduced. The short-comings of the RT-PCR test were inspected, and it was concluded that computerization can aid in this predicament. Then, basic terminologies which were going to be mentioned in this paper were briefly explained. Subsequently, the study reviewed papers which dealt with pulmonary image screening using CNNs. Considering the review, it was ultimately inferred that models' performance was higher for preprocessed images, smaller sample size, and fewer classes. From our study, all authors have reached the unanimous conclusion that developing a CNN model attains an effective image classifier. This conclusion helps us in detecting COVID-19 with the use of chest CT scans and X-rays. Consequently, we propose to develop a CNN model which can provide effective results for larger sample size and compare its performance with existing models.

References

1. Novel CPERE (2020) The epidemiological characteristics of an outbreak of 2019 novel coronavirus diseases (COVID-19) in China. Zhonghua liu xing bing xue za zhi= Zhonghua liuxingbingxue zazhi 41(2):145
2. Perlman S (2020) Another decade, another coronavirus, 760–762
3. Li L, Qin L, Xu Z, Yin Y, Wang X, Kong B, Bai J, Lu Y, Fang Z, Song Q, Cao K (2020) Artificial intelligence distinguishes COVID-19 from community acquired pneumonia on chest CT. Radiology
4. Fang Y, Zhang H, Xie J, Lin M, Ying L, Pang P, Ji W (2020) Sensitivity of chest CT for COVID-19: comparison to RT-PCR. Radiology, 200432
5. Fang Y, Zhang H, Xie J, Lin M, Ying L, Pang P, Ji W (2020) Sensitivity of chest CT for COVID-19: comparison to RT-PCR. Radiology
6. Koch G, Zemel R, Salakhutdinov R (2015) Siamese neural networks for one-shot image recognition. In: ICML deep learning workshop, vol 2
7. Pinheiro P, Collobert R (2014) Recurrent convolutional neural networks for scene labeling. In: International conference on machine learning, pp 82–90
8. Sumahasan S, Addanki UK, Irlapati N, Jonnala A Object detection using deep learning algorithm CNN
9. Perwej Y, Chaturvedi A (2012) Neural networks for handwritten English alphabet recognition. arXiv preprint arXiv:1205.3966.
10. Zhang J, Shao K, Luo X (2018) Small sample image recognition using improved convolutional neural network. J Vis Commun Image Represent 55:640–647
11. Washko GR, Parraga G, Coxson HO (2012) Quantitative pulmonary imaging using computed tomography and magnetic resonance imaging. Respirology 17(3):432–444
12. Abbo L, Quartin A, Morris MI, Saigal G, Ariza-Heredia E, Mariani P, Rodriguez O, Munoz-Price LS, Ferrada M, Ramee E, Rosas MI (2010) Pulmonary imaging of pandemic influenza H1N1 infection: relationship between clinical presentation and disease burden on chest radiography and CT. Br J Radiol 83(992):645–651
13. Mayo JR, Remy-Jardin M, Müller NL, Remy J, Worsley DF, Hossein-Foucher C, Kwong JS, Brown MJ (1997) Pulmonary embolism: prospective comparison of spiral CT with ventilation-perfusion scintigraphy. Radiology 205(2):447–452
14. Li X, Zeng W, Li X, Chen H, Shi L, Li X, Xiang H, Cao Y, Chen H, Liu C, Wang J (2020) CT imaging changes of corona virus disease 2019 (COVID-19): a multi-center study in Southwest China. J Transl Med 18:1–8
15. Khan AI, Shah JL, Bhat MM (2020) Coronet: a deep neural network for detection and diagnosis of COVID-19 from chest x-ray images. Comput Methods Programs Biomed, p 105581
16. Yang S, Jiang L, Cao Z, Wang L, Cao J, Feng R, Zhang Z, Xue X, Shi Y, Shan F (2020) Deep learning for detecting corona virus disease 2019 (COVID-19) on high-resolution computed tomography: a pilot study. Annals Transl Med 8(7)

17. Makris A, Kontopoulos I, Tserpes K (2020) COVID-19 detection from chest X-Ray images using deep learning and convolutional neural networks. medRxiv
18. Gozes O, Frid-Adar M, Greenspan H, Browning PD, Zhang H, Ji W, Bernheim A, Siegel E (2020) Rapid ai development cycle for the coronavirus (covid-19) pandemic: initial results for automated detection & patient monitoring using deep learning ct image analysis. arXiv preprint arXiv:2003.05037.
19. Bashivan P, Rish I, Yeasin M, Codella N (2015) Learning representations from EEG with deep recurrent-convolutional neural networks. arXiv preprint arXiv:1511.06448.
20. Shi H, Han X, Jiang N, Cao Y, Alwalid O, Gu J, Fan Y, Zheng C (2020) Radiological findings from 81 patients with COVID-19 pneumonia in Wuhan, China: a descriptive study. Lancet Infect Diseases
21. Sharma S (2020) Drawing insights from COVID-19-infected patients using CT scan images and machine learning techniques: a study on 200 patients. Environ Sci Pollut Res, 1–9
22. Tuncer T, Dogan S, Ozyurt F (2020) An automated residual exemplar local binary pattern and iterative relieff based corona detection method using lung x-ray image. Chemomet Intell Lab Syst 203:104054

Analyzing Effects of Temperature, Humidity, and Urban Population in the Initial Outbreak of COVID19 Pandemic in India

Amit Pandey, Tucha Kedir, Rajesh Kumar, and Deepak Sinwar

Abstract On January 2020, India reported its first case of Coronavirus disease 2019 (COVID19), caused because of acute respiratory syndrome coronavirus 2 (SARS-CoV-2). As of November 18, 2020, total of 8.91 million cases have been reported in the various parts of India, causing 131,000 deaths. In India, the disease has moved to its next stage, and now it is spreading through community level transmissions. It has now become extremely important to perform researches, exposing the factors affecting the transmission rate of SARS-CoV-2 in the community. The current study includes the ecological factors, like overall population and population distribution in various parts of India, together with other meteorological factors such as temperature and humidity to study the outbreak of COVID19 pandemic in India. This study states that there is a high correlation between the (Temperature / Humidity) ratio, overall population, urban population, and the COVID19 pandemic cases in India. The derived values of Spearman's rank correlation coefficient (φ) for the associations are φ [N_Population][N_Cases_Fst_25Dys] = 0.772932, φ [N_Population][N_Cases_Lst_2Wek] = 0.704289, φ [N_Urban_Pop][N_Cases_Fst_25Dys] = 0.812030, φ [N_Urban_Pop][N_Cases_Lst_2Wek] = 0.792325, φ [N_(Tem/Hum)_Fst_25][N_Cases_Fst_25Dys] = 0.540255, and φ [N_(Tem/Hum)_Lst_2Wek][N_Cases_Lst_2Wek] = 0.67697. The outbreak of SARS-CoV-2 virus varies depending on the factors population, urban population, and (Temperature / Humidity) ratio in different states of India and have significant influence on the COVID19 pandemic outbreak.

Keywords Coronavirus · COVID19 · Population · Temperature · Humidity · Pandemic

A. Pandey (✉) · T. Kedir · R. Kumar
College of Informatics, BuleHora University, Bule Hora, Ethiopia
e-mail: amit.pandey@live.com

D. Sinwar
Department of Computer and Communication Engineering, Manipal University, Jaipur, India

© The Author(s), under exclusive license to Springer Nature Singapore Pte Ltd. 2022   469
P. Nanda et al. (eds.), *Data Engineering for Smart Systems*, Lecture Notes in Networks and Systems 238, https://doi.org/10.1007/978-981-16-2641-8_45

1   Introduction

The initial outbreak of COVID19 was reported in December 2019, in the Wuhan city, situated in the Hubei Province of China [1–3]. In a period of more than a month, on January 30, 2020, the first case of Coronavirus disease 2019 (COVID19), caused by the acute respiratory syndrome coronavirus 2 (SARS-CoV-2), was detected in Kerala, state of India [4, 5]. The initial cases of COVID19 outbreak in India consisted of those patients, who were associated with the recent history of foreign travel. Later, this COVID19 outbreak has shifted to next stage and started spreading through community-level transmissions. As of November 18, 2020, total of 8.91 million cases have been reported throughout India, causing 131,000 deaths [4, 5].

Earlier researches have been carried out to study the effects of meteorological conditions, such as temperature and humidity and ecological factors, such as gender and age to understand the COVID19 outbreak patterns. Menebo et al. analyzed the effects of the climatological factors in the transmission of COVID19 in Oslo city, the capital of Norway [6]. They have used Spearman's correlation analysis over various temperature inputs, wind speed and precipitation data to establish the fact that there is a negative correlation between the precipitation and the COVID19 cases. However, the temperature has a positive correlation with the number of COVID19 cases. Similar studies have been carried out by Ma et al. in Wuhan, China, using the general additive model [7] and Pani et al. in Singapore [8], establishing the fact that the temperature is positively associated with the number of COVID19 cases.

In country like India, with world's second-largest population and where a large fraction of population stays in urban areas, it becomes utterly significant to study the transmission rates of SARS-CoV-2 considering the factors of overall population and population distribution together with other meteorological factors to understand the outbreak pattern of the pandemic. In this study, such meteorological factors, together with some ecological factors, have been considered to understand the outbreak of the COVID19 pandemic in India. Using the data accumulated from the Ministry of Health and Family Welfare Web site and other social and news media, this study demonstrates the existing correlation between the meteorological and ecological factors with the outbreak of COVID19 pandemic in India. Further, this information is useful to take the necessary steps for the containment of the COVID19 pandemic outbreak. As many states in India are going through the regular seasonal shift, it will cause changes in their environmental conditions, such as temperature and humidity.

2   Duration and Data Repositories

2.1   Time Frame

Two time frames have been considered for the analysis. This study considers the initial months of COVID19 transmission for the analysis. Initially, the pandemic was not spreading through the community-level transmissions, and hence, effects of the factors like temperature and humidity can be clearly seen in the pandemic transmission in that duration. First time frame considered for the analysis is of 25 days, which starts from the day when the first COVID19 case was recorded in a particular state. The second time frame considered is of two weeks, i.e., from April 01, 2020, to April 14, 2020. The period of first 25 days shows the correlation of COVID19 cases with the (Temperature/Humidity) ratio and urban population in the initial stage of the COVID19 pandemic, and the second time span shows the association of COVID19 cases with the (Temperature/Humidity) ratio and urban population in various states of India during the second phase of COVID19 transmission, when it started spreading through community level transmissions.

2.2 Epidemiological Data

The epidemiological data is retrieved from Ministry of Health and Family Welfare, Government of India Web site [8], further information from social media, and news media has been used to summarize the data. Initially, the epidemiological data was categorized by the names of the states in India. Further, the study has considered the data from only those states where the COVID19 transmission was not influenced by any external factors, so that the association of the COVID19 cases with the (Temperature/Humidity) ratio, and urban population can be studied without the influence of any external influence. Information gathered from social and news media was used to identify these external influences, such as the gathering of a particular community in Delhi region during the COVID19 spread, mass return of Indian migrant workers to few particular states and invasion of COVID19 in Dharavi region of Maharashtra state, the most densely populated slums in Asia [9–16].

2.3 Weather Data

The weather data is retrieved from the timeanddate.com. It includes the temperature and humidity data in various states of India [17]. The study considers the average of the temperature and the humidity values in different states of India during the two considered time frames.

2.4 Population Data

The data related to overall population and rural population is taken from the Web site of planning commission, body of Government of India [18].

3   Data Analysis Techniques

3.1   Spearman's Rank Correlation

Initially, the Spearman's rank correlation is used in this study (See Eq. 1), to find
the association between the attributes like overall population, urban population,
(temperature / humidity) ratio with the COVID19 pandemic cases in various states
of India.

$$\varphi = 1 - \frac{6 \sum \delta^2}{n\left(n^2 - 1\right)} \tag{1}$$

φ is the Spearman's rank correlation coefficient.

δ is the difference in the rank, given to the two variables.

n is the number of observations taken.
    Here,

$$\varphi = \begin{cases} +1 & \text{shows a possitive correlation} \\ 0 & \text{shows there is No correlation} \\ -1 & \text{shows a negative correlation} \end{cases}$$

    A positive value of φ shows a positive correlation between the attributes. It means,
with the rise in the value of one attribute the value of other attribute will also increase,
and closer the value of φ is to $+1$, stronger is the association between the attributes.
In similar fashion, a negative value of φ shows a negative correlation between the
attributes, it means if the value of one attribute will increase, then the value of
other attribute will decrease. The closer the value of φ gets to -1, the stronger is
the negative correlation among the attributes. Although, the value of $\varphi = 0$ only
reflects that the attributes are completely independent of each other, and there is no
correlation between them. Also, the nonzero values of φ closer to $+1$ or -1 only
reflect that there is a strong correlation between the attributes, which may or may
not be of linear order.

3.2   Regression Analysis

Further, we have used regression analysis to plot the regression graphs (See Eq. 2),
showing the associations between the COVID19 cases with overall population, urban
population, and (Temperature/Humidity) ratio. Here, the number of COVID19 cases
is the response variable, and the overall population, urban population, and (Temper-
ature/Humidity) ratio are the predictor variables for the three different associations,

respectively. Linear regression was used on the studied data to plot the associations among the considered factors and the number of COVID19 cases. In general, linear regression can be represented as,

$$y_i = \beta_0 + \beta_1 x_i + \varepsilon_i \qquad (2)$$

Here,

y_i is the response variable.

β_0 is the intercept value.

β_1 is the slope of the line.

x_i is the predictor variable.

ε_i is the error in estimation of the response variable.

4   Result and Discussion

Table 1 shows the Spearman's rank correlation values for various associations between overall population, urban population, (Temperature/Humidity) ratio for both the time periods, and number of COVID19 cases in various states of India.

4.1   Correlation Between COVID19 Pandemic Cases and Overall Population

The study considered epidemiological data in the period of two time frames, and there is a positive correlation found between the COVID19 pandemic cases and the population of the various states in India, for both the time frames (See Table 1). The value of Spearman's rank correlation coefficient for the associations are,

$$\varphi[\text{N\_Population}][\text{N\_Cases\_Fst\_25Dys}] = 0.772932$$
$$\varphi[\text{N\_Population}][\text{N\_Cases\_Lst\_2Wek}] = 0.704289$$

These positive values clearly states that there is strong positive correlation between COVID19 pandemic cases and population in various states of India. It also reflects that there will be higher number of COVID19 pandemic cases in the states with higher population. Figure 1a, b shows the regression plots for the associations between the population of the various states with the COVID19 cases in first 25 days and the COVID19 cases in other two weeks, respectively. In both the cases, the positive slope of the regression line reflects that the states with higher population have more number of COVID19 cases.

Table 1 Spearman's rank correlation analysis between overall population, urban population, (Temperature/Humidity) ratio, and number of COVID19 cases in two time frames

Index	N_Population	N_Urban_Pop	N_(Tem/Hum)_Fst_25	N_Tem/Hum_Lst_2Wek	N_Cases_Fst_25Dys	N_Cases_Lst_2Wek
N_Population	1	0.899248	0.562077	0.772129	0.772932	0.704289
N_Urban_Pop	0.899248	1	0.65839	0.755557	0.81203	0.792325
N_(Tem/Hum)_Fst_25	0.562077	0.65839	1	0.76668	0.540256	0.624247
N_(Tem/Hum)_Lst_2Wek	0.772129	0.755557	0.76668	1	0.485877	0.67697
N_Cases_Fst_25Dys	0.772932	0.81203	0.540256	0.485877	1	0.740407
N_Cases_Lst_2Wek	0.704289	0.792325	0.624247	0.67697	0.740407	1

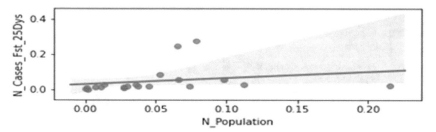

(a) Regression plot between population of states and COVID19 cases in first 25 days.

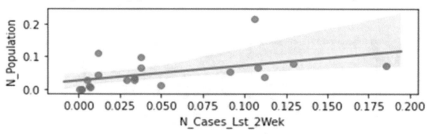

(b) Regression plot between population of states and COVID19 cases in last two weeks.

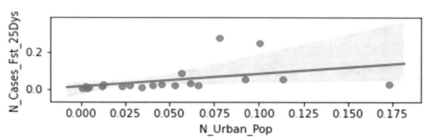

(c) Regression plot between urban population of states and COVID19 cases in first 25 days.

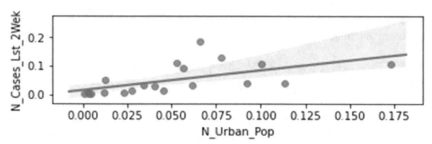

(d) Regression plot between urban population of states and COVID19 cases in last two weeks.

Fig. 1 Regression plots showing associations between overall population, urban population, (Temperature/Humidity) ratio, and COVID19 cases in the two different time frames

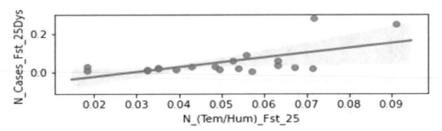

(e) Regression plot between (Temperature/Humidity) ratio for first 25 days in various states of India and COVID19 cases in first 25 days.

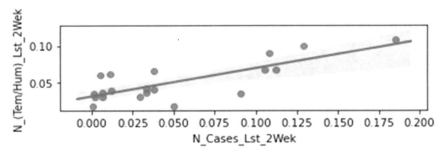

(f) Regression plot between (Temperature/Humidity) ratio for last two weeks in various states of India and COVID19 cases in last two weeks.

Fig. 1 (continued)

4.2 Correlation Between COVID19 Pandemic Cases and Urban Population

There is a positive correlation between the COVID19 pandemic cases and the urban population in the various states of India. (See Table 1) The value of Spearman's rank correlation coefficient for the associations are,

$$\varphi[N\_Urban\,Pop\,][N\_Cases\_Fst_2 5Dys] = 0.812030$$
$$\varphi[N\_Urban\_Pop][N\_Cases\_Lst_2 Wek] = 0.792325$$

These positive values clearly states that there is a strong positive association between COVID19 pandemic cases and urban population (Fig. 1). It also reflects that there will be higher number of COVID19 pandemic cases in the states with higher urban population. Figure 1c shows the regression plot for the association between the urban population in the states and the COVID19 cases in first 25 days, whereas Fig. 1d shows the regression plot for the association between the urban population in the states and the COVID19 cases in other two weeks. In both the plots, the positive slope of the regression line reflects that the states with higher urban population have more number of COVID19 cases.

4.3 Correlation Between COVID19 Pandemic Cases and (Temperature/Humidity) Ratio

There is a positive correlation between the (Temperature/Humidity) ratio and the COVID19 pandemic cases, recorded during the first 25 days, in various states of India (See Table 1). The value of Spearman's rank correlation coefficient for the association is,

$$\varphi[N\_(Tem/Hum)\_Fst\_25][N\_Cases\_Fst\_25Dys] = 0.540255$$

Further, this value of the Spearman's rank correlation coefficient has increased for the association of the (Temperature/Humidity) ratio and the COVID19 pandemic cases, recorded during the second time frame of two weeks. The value of Spearman's rank correlation coefficient for the association is,

$$\varphi[N\_(Tem/Hum)\_Lst\_2Wek][N\_Cases\_Lst\_2Wek] = 0.67697$$

This positive value clearly states that there is strong positive correlation between COVID19 pandemic cases and (Temperature / Humidity) ratio for both the time frames. It also reflects that there will be higher number of COVID19 pandemic cases in the states with higher (Temperature/Humidity) ratio. Figure 1e, f represents the regression plots for the associations of (Temperature/Humidity) ratio with COVID19 cases in first 25 days and COVID19 cases in other two weeks, respectively, and their positive slopes indicates that the states with higher (Temperature/Humidity) ratio have higher number of COVID19 cases. The (Temperature/Humidity) ratio for both the above-mentioned associations is calculated separately, as they are associated with two different time frames.

5 Conclusion

The study based on the records of temperature and humidity in various states of India together with the overall population and urban population factors states that there is significantly high positive correlation between the COVID19 pandemic cases and the considered meteorological and ecological factors like (Temperature / Humidity) ratio, overall population, and urban population. India is the country that embraces lots of geographical and environmental diversities, and the outcomes of this study are beneficial in taking the essential steps for the containment of the COVID19 pandemic outbreak in various states of India.

References

1. Zhu N et al (2020) A novel coronavirus from patients with pneumonia in China. New England J Med
2. Zhou P, Yang X, Wang X et al (2020) A pneumonia outbreak associated with a new coronavirus of probable bat origin. Nature 579:270–273. https://doi.org/10.1038/s41586-020-2012-7
3. Ghinai I et al (2020) First known person-to-person transmission of severe acute respiratory syndrome coronavirus 2 (SARS-CoV-2) in the USA. The Lancet. https://doi.org/10.1016/S0140-6736(20)30607-3
4. World Health Organization. Novel coronavirus (2019-nCoV). Available at: https://www.who.int/emergencies/diseases/novel-coronavirus-2019.
5. 2020 coronavirus pandemic in India. Available at: https://en.wikipedia.org/wiki/2020_coronavirus_pandemic_in_India 2020 coronavirus pandemic in India
6. Menebo MM (2020) Temperature and precipitation associate with COVID-19 new daily cases: a correlation study between weather and COVID-19 pandemic in Oslo, Norway. Sci Total Environ 2020:737. https://doi.org/10.1016/j.scitotenv.2020.139659
7. Ma Y, Zhao Y, Liu J, He X, Wang B, Fu S et al Effects of temperature variation and humidity on the death of COVID-19 in Wuhan, China. Sci Total Environ 2020:724. https://doi.org/10.1016/j.scitotenv.2020.138226
8. Pani SK, Lin NH, RavindraBabu S (2020) Association of COVID-19 pandemic with meteorological parameters over Singapore. Sci Total Environ 2020:740. https://doi.org/10.1016/j.scitotenv.2020.140112
9. Update on COVID-19. Available at: https://www.mofw.gov.in/#site-advisories
10. Coronavirus: About 9,000 Tablighi Jamaat members, primary contacts quarantined in country, MHA says. Available at: https://timesofindia.indiatimes.com/india/coronavirus-about-9000-tablighi-jamaat-members-primary-contacts-quarantined-in-country-mha-says/articleshow/74948832.cms
11. How Nizamuddin markaz became Covid-19 hotspot; more than 8,000 attendees identified. Available at: https://www.hindustantimes.com/india-news/how-nizamuddin-markaz-became-a-hotspot-for-covid-19-8-000-people-identified/story-zedKlaDNTPpbDLdMsgExAM.html
12. Indonesians among foreigners from 40 countries attended Tablighi Jamaat gathering: Sources. Available at: https://www.aninews.in/news/national/general-news/379-indonesians-among-foreigners-from-40-countries-attended-tablighi-jamaat-gathering-sources20200403152720/
13. 1445 out of 4067 Covid-19 cases linked to Tablighi Jamaat: Health Ministry. Available at: https://www.hindustantimes.com/india-news/1-445-out-of-4-067-covid-19-cases-linked-to-tablighi-jamaat-health-ministry/story-eK8oimpTN6qCZcnUAYlrDN.html
14. Fighting Covid-19: After the long walk, jobless migrants head home by bus. Available at: https://www.business-standard.com/article/current-affairs/fighting-covid-19-after-the-long-walk-jobless-migrants-head-home-by-bus-120032900041_1.html
15. Coronavirus | Migrant workers to be stopped, quarantined at borders, says Centre. Available at: https://www.thehindu.com/news/national/coronavirus-centre-warns-lockdown-violators-of-14-day-quarantine/article31198038.ece
16. Coronavirus: Maharashtra orders closure of malls, cinema halls in five cities until 31 March, Available at: https://www.livemint.com/news/india/coronavirus-maharashtra-orders-closure-of-malls-cinema-halls-in-five-cities-until-31-march-11584102050561.html
17. Coronavirus: Cases in Maharashtra surge, states shut down public places | Developments, Available at: https://www.indiatoday.in/india/story/coronavirus-cases-in-maharashtra-surge-states-shut-down-public-places-developments-1655635-2020-03-15, Past weather Information. Available at: https://www.timanddate.com/weather/?type=historic&query=India
18. Census 2011 (Final Data)—Demographic details, Literate Population (Total, Rural & Urban). Available at: https://web.archive.org/web/20180127163347, http://planningcommission.gov.in/data/datatable/data_2312/DatabookDec2014%20307.pdf

FOFS: Firefly Optimization for Feature Selection to Predict Fault-Prone Software Modules

Somya Goyal

Abstract Early prediction of fault-prone modules in software development is desirable to ensure the high-quality end product. It highlights the modules which are susceptible to faults and incur high development cost. Curse of dimensionality hinders the performance of classifiers and threatens the accuracy of classifiers to predict fault-prone software modules. This study proposes firefly optimization to select most informative features from high-dimensional defect dataset. It works on the swarm intelligence of fireflies and searches the high dimensional feature space using multi-objective optimization criteria. It looks for a minimal feature subset giving maximum accuracy of classifier. The experimental study utilizes the publicly available NASA corpus. The study compares the proposed model with traditional models statistically. It is inferred from the experiments that the proposed model performs better than the competing techniques.

Keywords Software defect prediction (SDP) · Metaheuristics · Dimensionality reduction · Firefly · Feature selection (FS)

1 Introduction

In software development, early fault prediction is a vital research area. It improves the chances of detecting and fixing the bugs at minimal cost. It ensures the delivery of error-free products in time. It allows to focus testing on these buggy modules, and the chances to debug them increase. The early detection and correction of defects cause less cost than later stages. Software failure causes huge money loss; for instance, the failure of NASA spacecraft caused the loss of $125 million. It was the result of a small data conversion bug [1]. Early software defect prediction using machine learning has always attracted the attention of researchers [2–6]. The defect dataset may contain irrelevant, redundant and insignificant features that threaten the accuracy of classifiers to predict the fault-prone modules. Feature selection improves the performance of

S. Goyal (✉)
Department of Computer and Communication Engineering, SCIT Manipal University Jaipur, Jaipur 303007, Rajasthan, India

© The Author(s), under exclusive license to Springer Nature Singapore Pte Ltd. 2022
P. Nanda et al. (eds.), *Data Engineering for Smart Systems*, Lecture Notes in Networks and Systems 238, https://doi.org/10.1007/978-981-16-2641-8_46

the classifier by selecting the most significant feature subset from the large space of features. Few researchers follow statistical methods to obtain minimal feature subset [7] and others follow the metaheuristics for feature selection [8, 9].

Metaheuristics are optimization techniques which ensure to reach the global optimum. These are of three types—(1) Nature-inspired evolutionary algorithms, for example, genetic algorithms (GA), (2) Based on theories of physics, for example, big-bang-big-crunch (BBBC) and (3) Population-based swarm intelligence (SI) methods, for example, particle swam optimization (PSO) [10]. Firefly algorithm was introduced in 2009 by Yang [11]. It is based on swarm intelligence (SI) of fireflies.

1.1 Motivation

Many researchers have contributed into the field of SDP. They have prominently advocated the application of machine learning (ML) techniques for developing SDP classifiers with pretty good accuracy [2–6]. These studies have not considered the issue of 'curse of dimensionality'. Further, a few researchers deployed statistical solution [7] to FS. The studies [8, 9, 12–17] favoured metaheuristics to tackle the FS issue. Neither their results are generalized nor did they consider the multi-objective optimization. Here lies the motivation behind this study; this raises a question whether FOFS—a multi-objective optimization using firefly algorithm performs better than the traditional methods of SDP.

1.2 Objectives of the Study

The motive is to demonstrate the positive impact of FOFS on the accuracy and the prediction power of SDP classifier by selecting the most significant features effectively. For this purpose, two research questions are formulated which are stated as below:

(1) Does the proposed FOFS performs better than the traditional classifiers and results in better accuracy?
(2) Is the answer to above-stated research question justified in the statistical aspects?

1.3 Contribution

This paper contributes a novel feature selection technique—FOFS and improves the performance of SDP classifiers. This paper demonstrates the statistical proof that the proposed FOFS improves the prediction power of SDP classifiers.

1.4 Organization

The paper is organized as follows. The related work is highlighted in Sect. 2. The research methodology and experimental set-up is discussed in detail in Sect. 3. The experimental results are reported and investigated to answer the research questions in Sect. 4. The conclusions are drawn in Sect. 5 and future scope of the study is also discussed.

2 Literature Review

This section highlights the related literature work. Goyal et al. [8] deployed GA to tackle high-dimensional feature space of dataset and demonstrated the improved performance of SDP classifiers. Cai et al. [9] used cuckoo search for selecting the features then, filtered dataset is fed to the SDP classifier. Overall contribution is to improve the performance of the classifier by negating the impact of irrelevant and redundant features. Moussa et al. [14] deployed PSO-GA fusion to improve the performance of classifiers by removing redundant features. Anbu et al. [15] proposed traditional firefly algorithm (FA) to select the significant features for defect datasets. Rong et al. [16] highlighted the bat algorithm for software defect prediction and demonstrated the metaheuristics to work well for SDP. Abdi et al. [17] presented the fusion of association rules and swarm intelligence to predict software quality. However, all the studies claim to achieve better performance accuracy of SDP classifier, but none customized the traditional metaheuristics algorithm completely and autonomously to select features from a large space of feature set in a generalized way. So, it counts for a gap to be fulfilled in this proposed work.

3 Research Methodology

This section explains the experimental set-up, the proposed model, description of datasets and evaluation metrics utilized for this experimental study.

This work proposes customized firefly optimization technique for feature selection (FOFS). The defect dataset is pre-processed, and then, it is fed to the artificial neural network (ANN)-based classifier. The choice of classifier is made from the popularity and the wide acceptance for SDP from the literature surveys [2–8]. The data for conducting experiments are taken from PROMISE corpus [17, 18]. The performance of the model is measured in terms of AUC, ROC and accuracy. The model is depicted as Fig. 1.

The experiment is designed using CM1, KC1, KC2, PC1 and JM1—the SDP datasets available publicly from PROMISE corpus [19]. The performance evaluation

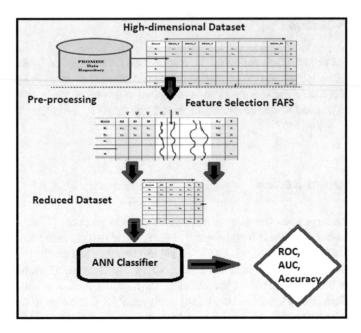

Fig. 1 Proposed FOFS SDP model

criteria are receiver operating curve (ROC), area under the ROC curve (AUC) and accuracy [2, 5, 6, 9, 14–17].

4 Analysis of Experimental Results

This section reports the experimental results and answers the research questions.

4.1 Performance of Proposed FOFS ANN Classifier Over Traditional Classifiers

Using ROC, AUC and accuracy, the performance of the proposed classifier is compared with traditional classifiers. The most popular classifiers from the studies are artificial neural network (ANN), decision tree (DT), naïve Bayes (NB), K-nearest neighbour (KNN) and support vector machine (SVM). The AUC values for all the traditional classifiers and the proposed FOFS model are recorded as in Table 1. It is clear from the results that for all datasets, the proposed FOFS–ANN model performs better than all the traditional classifiers.

Table 1 Comparison over AUC

Dataset	ANN	NB	TREE	KNN	SVM	Proposed FOFS-based ANN
CM1	0.73	0.66	0.52	0.57	0.67	0.89
JM1	0.70	0.49	0.50	0.47	0.90	0.91
KC1	0.75	0.49	0.70	0.41	0.70	0.87
KC2	0.83	0.51	0.66	0.69	0.75	0.89
PC1	0.70	0.49	0.67	0.61	0.62	0.88

Further, the performance is evaluated over accuracy measure, and all the experimental results over accuracy metrics are reported in Table 2. Over accuracy, the proposed model performs better than the traditional classifiers.

The ROC curve for all the traditional classifiers and FOFS-based SDP model are plotted and reported as Fig. 2. The best curve passing through the left top-most corner is the curve for the proposed FOFS model. It is the measure of both sensitivity and

Table 2 Comparison over accuracy

Dataset	ANN	NB	TREE	KNN	SVM	Proposed FOFS based ANN
CM1	0.83	0.84	0.83	0.84	0.83	**0.92**
JM1	0.81	0.74	0.73	0.42	0.86	**0.93**
KC1	0.82	0.80	0.82	0.28	0.84	**0.90**
KC2	0.83	0.71	0.75	0.80	0.76	**0.93**
PC1	0.92	0.90	0.91	0.78	0.93	**0.94**

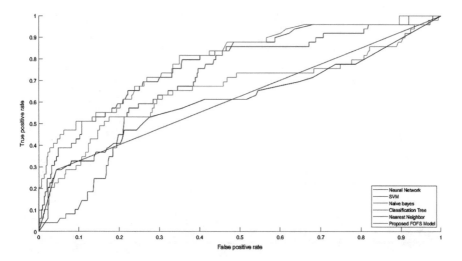

Fig. 2 ROC curve for traditional classifiers and the proposed FOFS model

specificity. It shows that the proposed model has the potential to perform best among the competing classifiers.

The experimental results are displayed as box-plots for accuracy measure in Fig. 3. From the figures, we find out the technique with high median and lesser outliners. Figure 4 reports the box-plots for AUC measure for all six classifiers.

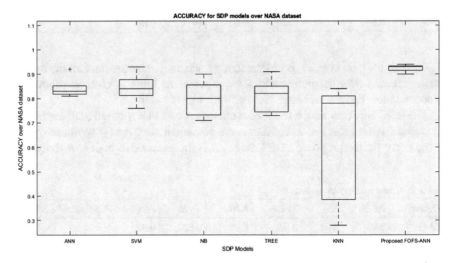

Fig. 3 Plots of ACCURACY measure

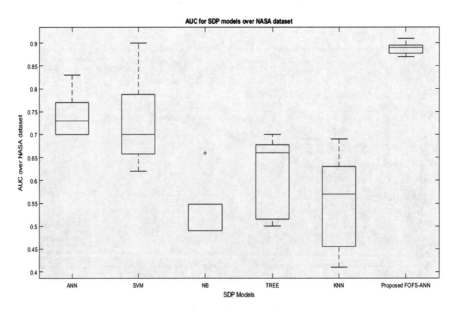

Fig. 4 Plots of AUC measure

Friedman's ANOVA Table

Source	SS	df	MS	Chi-sq	Prob>Chi-sq
Columns	51.6667	4	12.9167	20.67	0.0004
Error	8.3333	20	0.4167		
Total	60	29			

Fig. 5 Result friedman test with p-statistic of 0.0004

From all the above results, it is evident that the proposed model performs better than artificial neural network (ANN), decision tree (DT), naïve Bayes (NB), K-nearest neighbour (KNN) and support vector machine (SVM) over ROC, AUC and accuracy.

4.2 Statistical Analysis of the Experimental Study

It is essential to statistically validate the comparison made between the performance of proposed model and traditional models. Friedman's test is used for statistical proof [20]. It is assumed that (H_0) "The performance reported by proposed FOFS-ANN and the performance reported by other traditional classifiers are not different". Figure 5 shows the result of the test at 95% confidence level, which is very small (less than 0.05). Therefore, (H0) can not be accepted and it can be inferred that the proposed FOFS-based SDP model is better than the traditional classifiers.

5 Conclusion

An accurate prediction model is proposed to effectively predict whether the module is faulty with the high-dimensional dataset provided. Software defect prediction is important to perform targeted testing and to reduce the testing efforts. In case of a large number of redundant and irrelevant features, the prediction power of a classifier gets negatively affected as the majority class dominates the minority. In literature, various predictors have been proposed to handle SDP using traditional ML techniques. But they do not work well with redundant features. Some researchers contributed methods to select features, but still, the prediction power needs to be improved. In this paper, FOFS-based classifier is proposed with metaheuristic algorithm to select features. The performances of all the classifiers using accuracy, ROC and AUC are compared statistically. The experiments are conducted in MATLAB and the findings are as follows—from the experiments, it can be inferred that the performance of the proposed FOFS-based ANN classifier is the best with the highest value for AUC measure with the best ROC curve. For future work, the study is proposed to be extended for other datasets. Deep learning structures can also be deployed for future work.

References

1. NASA (2015) https://www.nasa.gov/sites/default/files/files/Space_Math_VI_2015.pdf
2. Goyal S, Bhatia PK (2021) Empirical software measurements with machine learning. Computational intelligence techniques and their applications to software engineering problems. CRC Press, Boca Raton, pp 49–64. https://doi.org/10.1201/9781003079996
3. Goyal S, Parashar A (2018) Machine learning application to improve COCOMO model using neural networks. Int J Inform Technol Comput Sci (IJITCS) 10(3):35–51. https://doi.org/10.5815/ijitcs.2018.03.05
4. Goyal S, Bhatia PK (2019) A non-linear technique for effective software effort estimation using multi-layer perceptrons, In: Proceedings of international conference on machine learning, big data, cloud and parallel computing (COMITCon). IEEE, pp 1–4. https://doi.org/10.1109/COMITCon.2019.8862256
5. Goyal S, Bhatia P (2020) Comparison of machine learning techniques for software quality prediction. Int J Knowl Syst Sci (IJKSS) 11(2):21–40. https://doi.org/10.4018/IJKSS.2020040102
6. Goyal S, Bhatia PK (2021) Software quality prediction using machine learning techniques. In: Sharma MK, Dhaka VS, Perumal T, Dey N, Tavares JMRS (eds) Innovations in computational intelligence and computer vision. Advances in intelligent systems and computing, vol 1189. Springer, Singapore, pp 551–560. https://doi.org/10.1007/978-981-15-6067-5_62
7. Goyal S, Bhatia PK (2020) Feature selection technique for effective software effort estimation using multi-layer perceptrons. In: Proceedings of ICETIT 2019. Lecture Notes in Electrical Engineering, vol 605. Springer, Cham, pp 183–194. https://doi.org/10.1007/978-3-030-30577-2_15
8. Goyal S, Bhatia PK (2019) GA based dimensionality reduction for effective software effort estimation using ANN. Adv Appl Math Sci 18(8):637–649
9. Cai Y, Niu S, Geng J, Zhang Z, Cui J, Li J, Chen J (2020) An undersampled software defect prediction method based on hybrid multi objective cuckoo search. Concurrency Comput Pract Exper 32(5):e5478
10. Darwish A (2018) Bio-inspired computing: algorithms review, deep analysis, and the scope of applications. Future Comput Inform J 3(2):231–246, ISSN 2314–7288. https://doi.org/10.1016/j.fcij.2018.06.001
11. Yang XS (ed) (2017) Nature-inspired algorithms and applied optimization, vol 744. Springer
12. Zhang L, Mistry K, Lim CP, Neoh SC (2018) Feature selection using firefly optimization for classification and regression models. Decision Support Syst 106:64–85, ISSN 0167–9236. https://doi.org/10.1016/j.dss.2017.12.001
13. Maza S, Zouache D (2019) Binary firefly algorithm for feature selection in classification. In: 2019 international conference on theoretical and applicative aspects of computer science (ICTAACS), Skikda, Algeria, pp 1–6. https://doi.org/10.1109/ICTAACS48474.2019.8988137
14. Moussa R, Azar D (2017) A PSO-GA approach targeting fault-prone software modules. J Syst Softw 132:41–49
15. Anbu M, Anandha Mala GS (2019) Feature selection using firefly algorithm in software defect prediction. Cluster Comput 22:10925–10934. https://doi.org/10.1007/s10586-017-1235-3
16. Rong X, Li F, Cui Z (2016) A model for software defect prediction using support vector machine based on CBA. Int J Intell Syst Technol Appl 15(1):19–34
17. Abdi Y, Parsa S, Seyfari Y (2015) A hybrid one-class rule learning approach based on swarm intelligence for software fault prediction. Innov Syst Softw Eng 11(4):289–301

18. Sayyad S, Menzies T (2005) The PROMISE repository of software engineering databases. University of Ottawa, Canada. http://promise.site.uottawa.ca/SERepository
19. (PROMISE) http://promise.site.uottawa.ca/SERepository.
20. Lehmann EL, Romano JP (2008) Testing statistical hypothesis: Springer texts in statistics. Springer, New York

Deceptive Reviews Detection in E-Commerce Websites Using Machine Learning

Sparsh Kotriwal, Jaya Krishna Raguru, Siddhartha Saxena, and Devi Prasad Sharma

Abstract Online reviews influence the buying decision of customers. Thus, false reviews can hamper sales of a product because a potential buyer would not buy it upon seeing negative reviews. Detection of these spam or fake reviews are necessary for the good of the users. A number of methods were used in the past for spam review detection; some used methods like manual labelling of a small dataset and then using it as base to check authenticity of other reviews, but this is not considered a viable option considering the amount of manual work required. This problem occurs due to the scarcity of clearly labelled datasets for spam/not spam reviews. The problem seems synonymous to the problem of detecting fake news on social media sites but it is not because there is no ground truth for fact checking. A very important fact about spam reviews is that they bear similarity and are full of adjectives because a spammer writes reviews mainly with the intent of either defaming a product or rating it higher than the actuality. Review rating and verified purchase are features that play a vital impact in detection of the fakeness of a review. So here we try to categorize customer reviews into fake and not fake on the basis of review-centric feature so that a reviewer may know whether a review they are reading is a potential spam. Various review-centric features include the rating of the review, product with which the review is related and the authenticity of the review.

Keywords E-commerce · Spam detection · Support vector machine · Logistic regression · Naïve Bayes · Random forest · NLTK

1 Introduction

The vast increase in the availability of internet has impacted all of our way of living and day-to-day activities. Along with the growth of internet, there is a direct proportional increase in the field of online shopping as well. Through online shopping or e-commerce, people can buy and sell products through digital markets. People

S. Kotriwal · J. K. Raguru (✉) · S. Saxena · D. Prasad Sharma
School of Computing and Information Technology, Manipal University Jaipur, Jaipur, India
e-mail: jaya.krishna@jaipur.manipal.edu

© The Author(s), under exclusive license to Springer Nature Singapore Pte Ltd. 2022
P. Nanda et al. (eds.), *Data Engineering for Smart Systems*, Lecture Notes in Networks and Systems 238, https://doi.org/10.1007/978-981-16-2641-8_47

even have provision to review or comment on the product they have bought or not bought as well. Reviews plays a vital role in online shopping platform as they may impact many other people who are interested in buying similar type of products, and reviews given for that product may influence buyers about their decisions. Even the manufacturers also use these reviews as a valid feedback and update their products design and service in their later versions. Using these reviews, e-commerce platform gets feedback about quality of service like speed of delivery, quality of product, etc. Unfortunately due to spammers, there are some fake reviews also getting posted in e-commerce either to fame or defame a particular product. Spammers aim to either create a false popularity of the product to attract more sales or negative popularity about the product. It is a general tendency of an individual to know in advance about the product in which they are interested in before buying. E-commerce has become a significant means of shopping in everyone's lifestyle. The motives for posting fake reviews are (1) it will reach to many people in a very small amount of time, and it is very low of cost for spammer to post their review. (2) Information diffusion in e-commerce-like platform will reach to maximum people, and it is easy to post without any validations or verifications. Posting deceptive reviews in e-commerce is more vast and fast in volume and may have high impact on individual's decision. Fake reviews can alter the way people react to genuine review. The impact of fake reviews can also disrupt trustworthiness about the product and entire ecosystem. Fake reviews in e-commerce pose a lot of challenges. Like, deceptive reviews are written intentionally to misguide buyers as well as sellers, which makes it difficult for buyers based on the content. A high percentage of a person's decision to buy a product is decided by the fact that "What other people think about it", people are highly influenced by the reviews and ratings given by other users who have a hands-on experience of device or a product. But what happens if the review turns out to be skewed or in other words, the characteristics of a product differs from the actuality, the buyer is deceived. Companies adopt various techniques to increase the ratings of the product by adding reviews of their own. Some of these techniques are:

Sock-puppetting—a single user creates various accounts and adds reviews to the product. This means that the reviews may seem different and by different ids in the review section but are added by the same person.

Astroturfing—this is a method in which companies recruit large number of individuals to review a certain product and make it look authentic. These people usually have no other reviews other than that on the product.

Review Brushing—a fake purchase for a product is made and is delivered to an anonymous person. The buyer then after a few days of the purchase writes a review for the same. Now websites like amazon treat them as "verified purchase" and the review is considered a true one. The Amazon's review weighting algorithms are not able to sort out the difference here in this case.

So, here we aim to resolve this problem as this would help bigger platforms like Amazon and Flipkart to just identify spam reviews which cause them a lot of problems in selling their products.

2 Related Work

Sentiment analysis approach for spam review detection has been used several number of times. Jindal and Liu [1] did the pioneering work of review spam detection which aimed at duplicate detection in online reviews also taking the sentiment of the review into account. Saumya and Singh [2, 3] applied the sentiment analysis approach on an unbalanced and used self-labelling using an algorithm which took sentiment score of review and its comments, due to the imbalance in the dataset, synthetic reviews were added using SMOTE and ADASYN. F1-Score of 0.91 was obtained by using random forest with ADASYN. Review-centric and reviewer-centric features are also widely used approaches for detecting spam. Review-centric spams can be untruthful, and reviews may be of certain brands only or are simply non-review texts. Crowford et al. [2, 4] used review-centric and reviewer-centric features (like bursty reviews, Amazon verified purchases, Cosine Similarity) separately and achieved a maximum F-Score of 0.89 and 0.75, respectively, for each. Soni and Prabhakar [4, 5] used a deepwalk approach, in which only information about reviewers were used like timing of reviews, number of reviews available. Support vector machine gave a maximum accuracy of 0.87. Li et al. [5, 6] classified reviews based on their helpfulness and comment evaluation. They labelled 20% reviews into top-helpful, mid-helpful and low-helpful, and then, human labelling was done into spam and non-spam. Finally, labelling through naïve Bayes was done to achieve an accuracy of 0.58. Ahmed et al. [6, 7] applied normal pre-processing on the review dataset using count vectorizer, and feature extraction was then done through TF and TD_IDF. The application of N-grams with Lagrange's SVM yielded an accuracy of 92%. In [8], authors for fake review detection in electronics area of consumers proposed a feature framework-based fake review detection. In this model, they have framed this in four sections, the initial dataset is classified based on reviews of four cities, then devising the definition of a fake review detection-based feature framework, then classification of fake reviews, and finally, evaluation of fake review under each city considered for evaluation. In this work, authors claimed that AdaBoost classifier is able to produce better results in comparison with other classifiers. In [9], authors have analysed fake news postings, sources of this news and subjects of this news in online social networks. In this paper, they have also proposed a new method based on automatic deep diffusive neural network-based fake news detection model. In this, model is basically a composition of feature learning representation and credibility label inference. Through the proposed model using information from textual latent features, they were able to study the depictions of sources of news articles and subjects instantaneously. In this paper, using latent feature representation and explicit feature learning from hybrid feature learning unit (HFLU) studied the news articles and information fusion is performed with the use of gated diffusive unit in deep diffusive network. In Table 1, accuracy or F1-score of various methods used in the literature with the proposed method is shown.

Table 1 Comparison of various methods used in the literature

Title	Method used	Accuracy or F1-score
Detection of spam review: a sentiment analysis approach	Random forest with ADASYN (best result achieved in this case)	F1-Score = 0.91
Survey of review spam detection using machine learning approach	Review-centric + bigram + SVM + LIWC	F1-Score = 0.89
Effective machine learning approach to detect group of fake reviewers	Word2Vec + SVM	F1-Score = 0.87
Learning to identify review spam	Naïve Bayes + co-training algorithm + semi-supervised learning	F1-Score = 0.637
Detecting opinion spams and fake news using text classification	LSVM + Tfidf + unigram	92% accuracy

3 Methodology

There are three basic steps involved in predicting the success or failure of a movie using sentiment analysis with the help of machine and learning. (1) Analysing and collecting reviews. (2) Pre-processing and feature extract. (3) Training and testing.

(1) Analysing and Collecting Reviews: Amazon sells nearly 4000 products per minute and a lot of them get reviews from maybe a certified user or a general reviewer. These reviews influence highly the decision of the next lot of people who are potential buyers of a product. The review dataset has 21,000 reviews that are labelled in spam and non-spam. Many exploratory data analyses were performed on the review dataset, and the general characteristics of a SPAM review and a REAL review were compared, with some parameters being the length of the review, the stop words count, caps count, or the correct name of the product being written in the review, and how ratings are related to review.

(2) Pre-processing and Feature Extraction: In most of the cases, data consists of many features that can be unrelated, redundant and deceptive in nature which actually increase difficulty in data processing. In machine learning-based algorithms, dimensionality reduction is a basic step for pre-processing through which redundant and unrelated data is removed. Through this, learning accuracy can also be improved. In [10], authors have given a brief survey about feature selection and feature extraction techniques in machine learning. Text pre-processing plays a very important role by converting the text into a simpler form so that machine learning can be performed efficiently.

Tokenization—Tokenization is breaking the sentence into individual words. This helps the algorithm to understand and relate each and every word present in the sentence.

Stop Words Removal—Stop words are words like 'a', 'an', 'the' which are present in the sentence. These words does not mean much to the classifier, and thus, we first tokenize the sentence and then remove these words. NLTK has stop words library to remove these words.

Lemmatization—Lemmatization is reducing the word to its base form. Large number of words like playing, plays, played are different but have the same root form which is play. To make sure that the classifier treats these words the same and ensure that they have the same meaning, lemmatization is done.

N-Grams—N-grams (bigrams were used) is clubbing n-words together so that the classifier can club to words together to predict the words that come next. This is useful because fake reviews usually contains the same text and classifier can know whether the same clubbing was present in different reviews and predict them.

(3) Training and testing: Training and testing is done using different classifiers and different ways.

Method 1: Training and testing without pre-processing and only using labels and unclean text.

Method 2: Training and testing after cleaning, tokenization, lemmatization and n-grams (bigrams).

Method 3: Training and testing after including extra features like rating, verified_purchase, product _category, etc.

Method 4: Final testing using both step2 and step3.

Overall steps involved in processing reviews

1. Dataset collection.
2. Analysing collected reviews.
3. Stop words removal from reviews.
4. Tokenization of reviews.
5. Lemmatization of reviews.
6. Applying N-Grams on review.
7. After Step 6 without extra features.

   7.1 Train the model using classifiers.
   7.2 Testing for data accuracy.

8. After Step 6 with extra features like verified purchase is considered.

   8.1 Train the model using classifiers.
   8.2 Testing for data accuracy.

Dataset is split into two sets—training (80%) and testing (20%) and further used tenfold cross validation.

We used four machine learning models—support vector machine (SVM), logistic regression, naïve Bayes and random forest classifier to learn the behaviour of different reviews using the training set and then used the testing set to test the trained models. We obtained the best accuracy from support vector machine (F1-score: 0.80).

Table 2 Comparison of various methods used in literature

Machine learning model	Accuracy without cleaning text	Accuracy after stop words removal, lemmatization and N-Grams	Accuracy after including extra features	Final accuracy
Naïve Bayes	**64.9**	67.5	70.3	69.6
SVM	61.3	**69.2**	**81.9**	**80.92**
Random forests	63.6	66.9	77.4	69.04
Logistic regression	64.8	68.4	82.06	79.67

Table 3 Accuracy, precision, recall and f1-score

Model	Accuracy	Precision	Recall	F1-Score
Logistic regression	0.797	0.799	0.797	0.796
Naïve Bayes	0.646	0.650	0.646	0.645
Support vector classifier	0.804	0.808	0.804	0.803
Random forest	0.704	0.704	0.704	0.704

4 Experimental Results

For uncleansed text, naïve Bayes and logistic regression gave almost equal and best accuracies, with SVM giving the least accuracy. After data pre-processing, that is, after removing stop words, doing lemmatization and applying bigrams, the best accuracy was given by our support vector classifier followed by logistic regression. After taking into account other characteristics such as confirmed purchase and review rating, the greatest accuracy attained was in the case of logistic regression, closely followed by support vector classifier or SVM, both of which provided about 82 percent accuracy. For the final accuracy, best results were obtained for SVM and logistic regression, respectively, both giving about 80% accuracy in predicting spam reviews. Support vector machine had not performed well in case of uncleansed text but performed well after the inclusion of extra features. In Table 2, all the various machine learning models and their accuracies with various methods are shown. Various machine learning-based models used in this study are logistic regression, naïve Bayes, SVM and random forest. After applying various models on the training classifier, accuracy, precision, recall and f1-score got after the processing are shown in Table 3.

5 Conclusion and Future Scope

Reviews play a crucial role for both the reviewer and the seller. Majority of the users who do online shopping rely on online reviews. They also allow the product owners to identify the flaws in their product and help them in taking decisions regarding the same. But in the current scenario, it is a difficult task for the consumer to distinguish between real and fake reviews. Sales of large number of products are affected because of the same. Large number of reviews are written every day, and it is very difficult for a common user to identify its authenticity. Spammers write reviews in such a way that seem authentic, but they differ from real reviews in various ways like the words they use related to product or their review length. This helps us in identifying and removing these reviews which may help both the user and the seller. Various e-commerce companies like Amazon are also working towards the same as they get a huge amount of reviews every day and handpicking and cleaning them is not possible, so taking the help of a ML algorithm goes a long way in achieving this goal. We achieved a F-1 Score of 0.8 in case of support vector classifier and logistic regression which is better than a few models studied above.

In future, we would like to improve the accuracy of the system by introducing some reviewer-centric features. We would further strive to introduce timestamp of comments to find the frequency of the spam reviews coming from a spammer. Deep learning algorithms like Word2Vec and GloVe can be introduced for enhancing the accuracy. For taking extra features into account, TF-IDF vectorizer [6] can be used. Data from websites other than Amazon can be collected for enhancing the training ability of the model.

References

1. Jindal N, Liu B (2007) Review spam detection. In: Proceedings of the 16th international conference on world wide web. ACM, pp 1189–1190
2. Saumya S, Singh JP (2018) Detection of spam reviews: a sentiment analysis approach. Csi Trans ICT 6(2):137–148
3. Jindal N, Liu B (2008) Opinion spam and analysis. In: Proceedings of the 2008 international conference on web search and data mining. ACM, pp 219–230
4. Crawford M, Khoshgoftaar TM, Prusa JD, Richter AN, Al Najada H (2015) Survey of review spam detection using machine learning techniques. J Big Data 2(1):23
5. Soni J, Prabakar N (2018) Effective machine learning approach to detect groups of fake reviewers. In: Proceedings of the 14th international conference on data science (ICDATA'18), Las Vegas, NV, pp 3–9
6. Li FH, Huang M, Yang Y, Zhu X (2011) Learning to identify review spam. In: Twenty-second international joint conference on artificial intelligence
7. Ahmed H, Traore I, Saad S (2018) Detecting opinion spams and fake news using text classification. Secur Privacy 1(1):e9
8. Barbado R, Araque O, Iglesias CA (2019) A framework for fake review detection in online consumer electronics retailers. Inf Process Manage 56(4):1234–1244
9. Zhang J, Dong B, Yu PS (2019) Deep diffusive neural network based fake news detection from heterogeneous social networks. In: 2019 IEEE international conference on big data (Big Data), Los Angeles, CA, USA, pp 1259–1266. https://doi.org/10.1109/BigData47090.2019.9005556
10. Khalid S, Khalil T, Nasreen S (2014) A survey of feature selection and feature extraction techniques in machine learning. In: 2014 Science and information conference, London, pp 372–378. https://doi.org/10.1109/SAI.2014.6918213

A Study on Buying Attitude on Facebook in the Digital Transformation Era: A Machine Learning Application

Bui Thanh Khoa, Ho Nhat Anh, Nguyen Minh Ly, and Nguyen Xuan Truong

Abstract The attitude will have a massive impact on the purchase, and the customer will have a very different attitude when they shop on social networks, i.e., Facebook. Because of the above relationship, the purpose of this research is to find out the key factors affecting the buying attitude of young people on Facebook, especially during the digital transformation era. The research combined qualitative and quantitative research methods. Moreover, the development of computing strongly impacts on the research result. This study's result based on the supervised learning of machine learning pointed out that informativeness, entertainment, interactivity, credibility, and personalization positively affected the buying attitude; irritation and negative politeness negatively impacted on the buying attitude of young people on Facebook in the digital transformation era. Some managerial implications were also proposed.

Keywords Buying attitude · Social network · Negative politeness · Informativeness · Entertainment · Interactivity · Credibility · Personalization · Irritation · Machine learning

1 Introduction

There are over 2.7 billion monthly active users as of the second quarter of 2020 [8]. Facebook is the biggest social network worldwide. In the third quarter of 2012, the number of active Facebook users surpassed one billion, making it the first social network ever. Facebook is a social network for communicating with friends and the fertile online marketplace behind e-commerce platforms [21]. Due to the impact of digital transformation, individuals can choose Facebook as a place of sale or choose Facebook to shop if they cannot shop directly. Individual to business sellers who operate on Facebook focus on customer attitude, which is the most important factor

B. T. Khoa (✉) · H. N. Anh · N. M. Ly · N. X. Truong
Industrial University of Ho Chi Minh City, Ho Chi Minh City, Vietnam
e-mail: khoadhcn@gmail.com; buithanhkhoa@iuh.edu.vn

[11, 18]. If the customer has a positive attitude, attracting customers to shop online on Facebook will increase in the present and the future.

However, the above studies are mainly focused on the media like SMS advertising [23], or online advertising [3], or the electronic commerce site [12]. Studies of Facebook's direct influence on user attitude on the current Facebook e-marketplace are still limited. With the advantage of being a source of impact, do social media's above characteristics affect consumers when buying goods on Facebook e-marketplace? That content is part of the research question of this paper.

Besides, social interaction has been viewed by many researchers as a factor leading to customer buying behavior. Stemming from the concept of subjective norms in the theory reasonable action (TRA) [1], the customer's buying behavior is easily affected by the behavior of the social networking community such as pressing the "Like" button, "Love" button, or share a post, or comment on a post [25]. However, these effects are not always positive, significantly when the evidence may be interfered with by technology, or a relationship, or from a shopper's lack of awareness on duty. Brown and Levinson [4] mentioned negative politeness in communication when communicators mainly perform gossip acts without paying attention to content. Therefore, this research and inheriting previous research also add an appropriate factor for the Facebook environment: negative politeness through communication affecting shoppers' attitude on Facebook.

The econometrics is an increasingly popular and essential tool in analyzing data, making predictions about data trends and behavior [27]. It is used in economic research and in data science to build predictive operations models, customer behavior, financial trend forecast, and risk management [22]. Machine learning has caused a technology fever worldwide in the past few years. In the academic world, there are thousands of scientific articles on the subject each year. A wide range of applications using machine learning has emerged in all life areas, from computer science to less relevant disciplines such as physics, chemistry, medicine, and politics [19].

This paper aimed to base on the supervised learning of machine learning for finding out the relationship between the informativeness, entertainment, interactivity, credibility, personalization, irritation, negative politeness, and the buying attitude on Facebook. Besides the Introduction part, this paper was organized with the Literature Review, Research Method, Research Result, Discussion, and Conclusion.

2 Literature Review

Attitude can be defined as feeling favorable or unfavorable to an object, influencing a person to act or behave predictably toward products or services [1]. A consumer's buying attitude is understood and described excellent or bad ratings based on a person's perceptions, feelings, and executive tendencies about an object or idea [16]. Technology development has created many advantages for social networks such as informativeness, entertainment, credibility, interactivity, personalization, and irritation for users [3, 14]. The previous studies showed that informativeness, entertain-

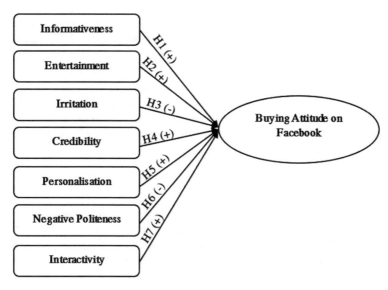

Fig. 1 Theoretical model

ment, credibility, interactivity positively affect consumer attitude, and irritation negatively impacts consumer attitude [12, 23]. Furthermore, negative politeness through comments or reviews usually does not affect consumer attitude or negatively impact [13]. The theoretical model is shown in Fig. 1.

One of the most effective and popular products in the world today is social media. Besides, the Internet has led to the proliferation of online social networks or community Web sites. The nature of social networks is connecting and sharing. Social media users' psychology is to want to express themselves, and their ego and social media create an ideal environment to show that, and above all, attract attention. That is the peak of sharing is the most significant reason users turn to social networks. Besides, social media also develops because of entertainment. People increasingly need to play, and social media is one of the best recreational facilities available to bring people comfort and relaxation.

Informativeness shows the ability of the advertisements to inform consumers about products and services [2]. Informativeness is the ability of advertising to provide relevant information effectively [26]. It was also stated that the effectiveness of information delivery would allow the consumers to search for information without limitations in terms of time and place. This effect is very critical to consumers on the Internet [5]. Information is considered a valuable motivation that leads recipients to react very positively to advertisements [3]. Hence, the research proposed the hypothesis:

H1: Informativeness has a positive effect on the buying attitude of young customers as they shop on Facebook.

The degree of entertaining information on Facebook is not only crucial to the success of advertising delivered on traditional media, but it is also an essential factor in order to succeed in the Internet context [9]. It was found that entertainment referred to the ability of the advertising to fulfill an audience's needs for aesthetic enjoyment, fun diversion, or emotional pleasure. In other words, entertainment is perceived to be consumers' opinions regarding how enjoyable, pleasing, exciting, and fun to see that they find advertisements to be [24]. Entertainment aims to enhance the viewers' excitement on Facebook, attracting more hits, and the consumer's buying attitude is very high [28]. Hence, the research proposed the hypothesis:

H2: Entertainment has a positive effect on buying attitude to young customers as they shop on Facebook.

Irritation refers to any offending effects that may go against what a user values [26]. In the context of advertising, irritation has been defined as employing tactics in advertising that is perceived to be annoying, offensive, insulting, deceptive, or overly manipulative [32]. When competing for consumers' attention, advertisers' intrusive tactics can be annoying to the audiences [23]. Hence, the research proposed the hypothesis:

H3: Irritation has a negative effect on buying attitude to young customers as they shop on Facebook.

Cheung et al. [7] defined the credibility in advertising as a trustworthy and recognizable statement in the listener's mind. The advertising credibility is influenced by various factors, especially its reputation and the messenger's reputation [10]. The perceived advertisement credibility was among the first constructs that were empirically examined and significantly influenced consumers' attitude toward advertising [2, 23]. Therefore, the H4 was proposed:

H4: Credibility has a positive effect on buying attitude to young customers as they shop on Facebook.

Personalization is the process of matching and tailoring advertising content to individual consumers' characteristics and preferences on a particular Web site in which each customer is treated uniquely [24]. Personalization aims to improve the user's experience of a service [34]. Thus, personalization can ensure that visitors to Web sites can be exposed to the most appropriate and appealing advertising messages [31] and have positive benefits ranging from improved advertising and the Web site [6]. Therefore, it would be plausible to assume that Facebook advertising's personalization would be considered an essential factor was influencing consumers' attitude toward Facebook advertising. Hence, the research proposed that:

H5: Personalization has a positive effect on buying attitude to young customers as they shop on Facebook.

Brown and Levinson [4], according to their theory, are the heart of deference behavior. However, as the above example shows, deference in Japanese culture focuses on the hierarchical social structure between addresser and addressee [20] rather than on people's desire to be free of imposition as suggested by Brown and Levinson. When the addressee assumes a higher or lower social status, the addresser is expected to acknowledge this social relationship, and show his or her (even ostensible) dependence (amae) on it, by making deferent impositions (if inferior), or by displaying his or her disposition to take care of the addressee (if superior). Today, most people interact in a virtual world where they can see each other through a computer screen or communicate by text. This context has prompted many studies to shift their focus to computer communication. Virtual interaction has entailed social networks, where people communicate with each other through the Internet. Facebook is one of the most popular social networks. With 1.1 billion users worldwide, Facebook's use as a means of communication is beyond doubt. Therefore, being negative politeness can affect consumers' buying attitude [33], the ability to return buying is lower than before if negative politeness overcomes. Hence, the H6 was proposed:

H6: Negative politeness has a negative effect on buying attitude to young customers as they shop on Facebook.

Interactivity is the degree of interaction between users with different types of ads. Sukpanich and Chen [30] made assertions about Interactivity in three groups: "person-person," "person-message," and "person-computer." For online advertising, the Facebook social network is a potential environment for advertising because this is the environment where the interaction between people and people is at the highest level, especially Facebook social network. Hence, the buying attitude could be impacted by interactivity.

H7: Interactivity has a positive effect on buying attitude to young customers as they shop on Facebook.

3 Research Methodology

3.1 Research Method

The mixed methods were qualitative research and quantitative research to achieve research objectives. Qualitative research is done through focus groups. Discuss ongoing research issues to gather the group members' opinions. The focus group survey consisted of 10 members with experience buying online on Facebook. Focus groups are conducted under the author's chair with a discussion guideline.

The observed variables are adjusted to bring high survey value through quantitative research on focus groups research. After completing the survey, the next step is to conduct qualitative research. Preliminary research is one of the necessary methods in qualitative research. Preliminary research gives an overview of the survey, completes new issues, identifies risks, and is a stepping stone before it is put into the official study. The preliminary survey was conducted by direct interviews with Facebook users in Ho Chi Minh City. In one day, 50 samples were collected. The data are analyzed in the test to give an overview of the data and complete the survey.

The observed items were measured on the 5-level Likert scale, which includes 1 = strongly disagree; 2 = disagree; 3 = neutral; 4 = agree; 5 = strongly agree. Specifically, (1) informativeness includes 4 observed items, (2) entertainment and (3) personalization including 3 observed items, (4) credibility and (5) interactivity including 4 observed items, (6) irritation includes 5 items, (7) negative politeness includes 4 observed items, and finally 5 observed items for the buying attitude variable.

After obtaining the official scale, the study was conducted with an estimated sample size of 210 customers shopping from Facebook; a survey by sharing questionnaires online with people using Facebook on any social media. As a result, there are 325 data collected, all saying they are using Facebook and of which 290 people have made a purchase on Facebook. Of which, 46.2% are male and 53.8% are female. Respondents are young people, aged from 15 to 21 years old, accounting for 92.6%. The collected data will be processed via SPSS software and Python. Once encrypted and cleaned, the data will be processed and analyzed to describe young people's buying attitude toward Facebook.

3.2 Machine Learning in Multiple Regression

Machine learning (ML) is the study of computer algorithms that automatically improve through experience [19]. Machine learning's critical goal is to make computers smarter, automatically learn and form knowledge from experience, and be more useful in communicating with people. Today, machine learning is overgrowing, proving that this is the introduction of products based on artificial intelligence such as Alexa, robot Sophia, and Google Assistant [19].

Linear regression is a famous model applied with far-reaching applications in economics and many different social fields. It assumes a significant role in machine learning, one of the fundamental supervised machine learning approaches due to its comparative simplicity.

Multiple regression is an extension of simple linear regression. It is used to predict a variable's value based on the value of two or more other variables. The variable, which was predicted, is called the dependent variable (or sometimes the outcome, goal, or criteria variable) [29]. Regression usually targets a predicted value (Y) based on other independent variables (X). This paper uses the ordinary least squares (OLS) method in the linear regression to find the optimal equation that shows the optical correlation between X and Y. Python language and the machine learning algorithm are used to build an algorithm to find the regression equation based on survey data [29].

4 Result

The mean of the items in Table 1 pointed out customers' assessment of the research factors. Mean results indicated that customers consider Facebook sales advertising annoying and unreliable due to direct outside influences.

This study tested the reliability and convergent validity through the Cronbach's alpha and exploratory factor analysis (EFA) using the principal component analysis method with varimax rotation to reduce the observed variables into a set of different constructs. All the Cronbach's alpha must be greater than 0.7 to ensure the reliability of the measurement scale. Table 2 showed that all constructs' scale was more significant than 0.7; hence, all scales were reliable. Moreover, the KMO coefficient = 0.913, which was more than 0.5; therefore, the EFA was appropriate. In Bartlett's test of sphericity, the Sig. = 0.00, which is less than 0.05, pointed out that all observed variables were correlated. Furthermore, the factor loading of all items was more significant than 0.5. All the scales in this study got the convergent validity.

The study divided the dataset into two datasets including the training set, accounting for 70% of the data (N = 203/290) to build regression function based on the initialized operator for linear regression function y (linear regression), and the testing set, accounting for 30% of the data (N = 87/290) to check the correctness of the built regression function. Running the regression model on the training set using Python, the result pointed out irritation(IRR) and negative politeness (IMP) have a negative impact on buying attitude (ATT). Moreover, credibility (CRE), informativeness (INF), interactivity (INT), personalization (PER), and entertainment (ENT) have the positive impact on buying attitude (ATT).

This study applied the Sklearn Linear Model computing in Python to build the linear regression based on the Training set. In the next step, the study predicted the test results based on the regression function. Moreover, the research also compared the prediction results and the testing set, obtained the following results as Fig. 2. The variable data Y test or Y forecast based on the regression function we see the projections similarity through scatter distribution:

The study determined the correlation between the independent variables and the dependent variable. In statistics, correlation is the degree to which a pair of variables is linearly related between two random variables or two-variable data. All significant values were equal to 0.00, which is less than 0.05; hence, the independent variables were correlated with youth customers' buying attitude. The multivariate linear regression results showed that seven independent variables explain 94.5% of the change in the young customers' buying attitude (adjusted R square = 0.945). Durbinal "Watson coefficient = 1.943, ranging from 1 to 3; therefore, there is no autocorrelation in the regression estimation residuals in this study (Table 3).

Table 1 Descriptive statistics

Scale	Code	Mean	Skewness	Kurtosis
Facebook provides information about products/services	INF1	3.87	−1.275	2.423
Facebook provides useful information	INF2	3.84	−0.692	0.14
Facebook is continually updating information	INF3	3.96	−1.116	1.659
I learned a lot from being with Facebook	INF4	3.83	−0.954	1.406
The content on Facebook is enjoyable	ENT1	3.89	−1.109	1.984
The content on Facebook is interesting	ENT2	3.89	−0.771	0.545
Products and services on Facebook are very eye-catching	ENT3	3.74	−0.738	0.469
It does not annoy me to see sales live-stream appearing on my Facebook page	IRR1	1.87	−0.943	0.272
The advertising breaks me off as using Facebook	IRR2	2.06	−1.198	0.852
I am comfortable when I see Facebook ads	IRR3	1.81	−0.723	0.09
Facebook ads do not distract me from other content	IRR4	2.7	−0.495	−0.358
I feel satisfied as objectionable content appears	IRR5	2.96	−1.024	1.124
I use Facebook as a reference to buy products or services.	CRE1	3.41	−0.488	−0.495
Content about products/services on Facebook is quite practical, partly comes from people who already know	CRE2	3.62	−0.336	−0.586
Facebook products/services are trusted	CRE3	3.06	0.126	−0.6
The content on Facebook is convincing	CRE4	3.35	−0.269	0.078
Facebook recommendations that match my interests	PER1	3.69	−0.973	1.212
The content on Facebook is of close interest to me	PER2	3.78	−0.859	0.824
The content on Facebook is what I expected	PER3	3.4	−0.238	−0.451
Products/services that are rated well on Facebook are right	IMP1	2.58	0.418	−0.734
The content, likes, shares, comments of Facebook posts are realistic	IMP2	2.39	0.704	−0.55
Commenting on products/services on Facebook is natural	IMP3	2.67	0.377	−1.16
Facebook comments/reviews are not affected by objective factors (by a majority, by trend)	IMP4	2.53	0.492	−0.863
Facebook offers a high level of product/service awareness	INT1	3.66	−0.586	0.727
Facebook makes it easier to access products/services	INT2	3.88	−0.608	0.303
I often click on the content of articles, advertisements to understand the product/service better	INT3	3.69	−0.857	0.762

(continued)

Table 1 (continued)

Scale	Code	Mean	Skewness	Kurtosis
I communicate with the seller through private messages	INT4	3.87	−0.889	1.383
I love reading about products/translations on Facebook	ATT1	3.63	−0.497	0.352
I see products/services on Facebook (before going to the store)	ATT2	3.91	−0.802	1.059
I enjoy the community values that Facebook brings	ATT3	3.8	−0.461	0.105
I consider Facebook the right choice for buying online	ATT4	3.36	−0.33	−0.397
I react positively to the content that Facebook products/services suggest to me	ATT5	3.53	−0.376	0.166

Table 2 Cronbach's alpha and exploratory factor analysis (EFA)

Construct	Cronbach's alpha	Loading factor
Irritation (IRR)	0.884	[0.841–0.877]
Credibility (CRE)	0.875	[0.825–0.852]
Informativeness (INF)	0.876	[0.822–0.866]
Interactivity (INT)	0.876	[0.820–0.865]
Negative politeness (IMP)	0.869	[0.813–0.862]
Personalization (PER)	0.938	[0.880–0.942]
Entertainment (ENT)	0.88	[0.802–0.855]
Buying attitude (ATT)	0.803	[0.733–0.822]
Kaiser-Meyer-Olkin = 0.913		
Bartlett's test of sphericity, sig. = 0.00		

The study result showed that the all significant values are less than 0.05, so all independent variables significantly impact youth's buying attitude with 99% confidence. Therefore, the linear regression model was formed as follows:

$$ATT = -0.1759 * IRR + 0.2042 * CRE + 0.0945 * PER - 0.2020 * IMP$$
$$+ 0.1455 * INT + 0.0670 * ENT + 0.1331 * INF + 0, 19451 \quad (1)$$

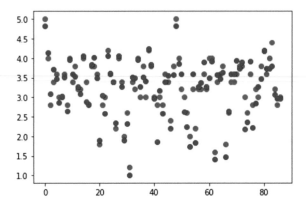

Fig. 2 Compare between the prediction results and the testing set

Table 3 OLS regression results

```
                              OLS Regression Results
===============================================================================
Dep. Variable:                   ATT   R-squared:                       0.945
Model:                           OLS   Adj. R-squared:                  0.943
Method:                Least Squares   F-statistic:                     475.6
Date:               Mon, 30 Nov 2020   Prob (F-statistic):          7.08e-119
Time:                       20:54:26   Log-Likelihood:                 92.917
No. Observations:                203   AIC:                            -169.8
Df Residuals:                    195   BIC:                            -143.3
Df Model:                          7
Covariance Type:             nonrobust
===============================================================================
                 coef    std err          t      P>|t|      [0.025      0.975]
-------------------------------------------------------------------------------
const          1.9451      0.107     18.160      0.000       1.734       2.156
IRR           -0.1759      0.015    -11.735      0.000      -0.205      -0.146
CRE            0.2042      0.018     11.415      0.000       0.169       0.240
PER            0.0945      0.016      5.765      0.000       0.062       0.127
INF            0.1331      0.021      6.356      0.000       0.092       0.174
IMP           -0.2020      0.016    -12.367      0.000      -0.234      -0.170
INT            0.1455      0.021      7.096      0.000       0.105       0.186
ENT            0.0670      0.019      3.515      0.001       0.029       0.105
===============================================================================
Omnibus:                       3.671   Durbin-Watson:                   1.943
Prob(Omnibus):                 0.160   Jarque-Bera (JB):                3.482
Skew:                         -0.320   Prob(JB):                        0.175
Kurtosis:                      3.035   Cond. No.                         87.6
===============================================================================
```

5 Discussion

This study aimed to identify factors of Facebook that affect consumers' buying attitude, especially young people. Results of qualitative research and quantitative surveys with a sample size of 290 participants to test the linear structure model, of which, purchasing attitude is affected by such factors as informativeness, irritation, credibility, personalization, negative politeness, interactivity, entertainment. Some theoretical and practical contributions were mentioned as follows:

One is that informativeness of Facebook always has a positive impact and has a considerable influence (Beta = 0.1331) on the purchase attitude. Therefore, the provision of information regularly is the main advantage of attracting potential customers and nurturing the business's customer base [2, 3]. The trust of consumers can be obtained by personal experience or being introduced by acquaintances. Therefore, retail businesses need to regularly communicate, creating closeness and trust with their products and brands [5, 26].

Secondly, irritation has impacted negatively on shopping attitude, with Beta = − 0.1759. Therefore, when participating in the business on Facebook, businesses need to pay attention to the nature of the content; content needs to be in the right case and increase efficiency [23]. Therefore, the more consumers feel annoyed by the content that the advertisements through social media bring, the more they will no longer intend to buy the advertised product, reducing advertisement effectiveness on Facebook [32].

Thirdly, credibility is also one of the positive factors influencing the buying attitude (Beta = 0.2042), through which businesses need to build trust with customers, reputation will bring long-term benefits for business [15]. Advertising and selling on Facebook are honest because it partly comes from people who already know about the product/service/brand information; hence, they will believe easily [16].

Fourthly, personalization has a positive impact on the Facebook shopping attitude (Beta = 0.0945). Personalization will create a specific and different choice or programs for each set of customers, which will help reduce costs but bring high efficiency [24, 34]. Understanding each customer segment is important because when there is a gathering of each group of objects, businesses can make a more accurate and appropriate call to action [6]. If the business captures each audience segment's information through online behaviors, the sites' behavior helps refine the strategy and optimize its target audience. From there, businesses can create unique, differentiated experiences between an individual and a business brand.

Fifthly, negative politeness harmed the buying attitude in Facebook shopping (Beta = − 0.2020); according to the survey, the result pointed out that negative politeness has a negative impact on customers' buying attitude. Even though a picture or status on Facebook is posted on one's page with highly interactive, the spread can reach millions of other Facebook users [22]. Therefore, the content posted on Facebook of current users is no longer wrapped around personal life stories, but sometimes they also share many other things, maybe a favorite article, a story [33]. Humanities call for widespread negative information such as shocking images, false information, and even offensive statements to other individuals and organizations. In recent years, content on Facebook has significantly affected real social life, confusing public opinion or causing consumers to misunderstand and buy behaviors that are not in line with demand.

Sixthly, interactivity (Beta = 0.1455) is attracted and appreciated by many different factors, including interaction such as share and comment. This result proves that consumers will have higher buying intent if the ads via social networks give them more interaction and sociality [30]. Social interaction is demonstrated by the fact that users can use various text, images, videos, and links to follow and share new products with other users. Simultaneously, highly interactive messages to social communities and influential individuals will substantially impact consumers' attitudes. The businesses need to take advantage of this to increase billions conversion rate.

Lastly, entertainment positively impacts on the buying attitude (Beta = 0.0670). Facebook brings many hedonic value for users and customers. Businesses can take advantage of this for a wiser approach strategy. Accordingly, if an advertisement through social media brings entertainment to consumers as much as possible, their shopping intentions will increase [28]. It can be seen that the primary and mandatory requirement for advertising programs and activities is the ability to provide information and entertain viewers [24]. Only when consumers perceive these two values, they create a positive attitude toward advertising. The entertainment of Facebook is measured by the activities that bring pleasure and comfort to the consumer and shown in the content that the ad is conveying, the simplicity, ease of understanding and unique is also a way to help users not feel bored when receiving advertising information on social networking sites in order to influence buying attitude.

6 Conclusion

The best e-commerce sites in the world have already done integration with social networks. Social media channels are like arms reaching out from that facility to increase interactivity like regional sales reps in a virtual store environment [17]. Shopping on Facebook is becoming a new trend, called f-commerce; customers feel convenient with all the shopping process done in one application. They are interested in products through sharing from friends or advertising on Facebook, then continue the shopping process without switching apps; when logging in, they often face problems and experience password reset process. The flexibility of customer support is another reason why they like to shop on Facebook. Consumers are very concerned with the process and the creation of requests to check order status, change the time or place of delivery. This advantage makes it easy to shop on Facebook as the store's customer support is available to respond immediately to address their request. Research has shown a negative relationship between irritation and negative politeness, and a positive relationship between informativeness, credibility, personalization, interactivity, and entertainment. Simultaneously, through the multivariate regression algorithm that is considered a part of machine learning, the research has partly approached the analysis of survey data based on machine learning algorithms, creating a premise for further studies.

Research is done as scientifically and entirely as possible; however, it still has some limitations. First, the research has only conducted general experiments on young people's reality surveys on the social network Facebook, although Facebook is a relatively complete social networking site with many kinds of information from education and health. Second, on the sampling method, due to the time and data frame, the author can only choose random samples, which reduces the reliability of the research. Third, the research focuses only on the influence of Facebook on young people's buying attitude during digital transformation. Moreover, this study calculated data based on the primary machine learning technique as the regression model.

Subsequent studies may overcome some of this research's limitations by focusing on one factor that explicitly affects the specificity of that social network or other social media platforms. Besides, it is possible to build a sampling and sampling framework for investigation according to the probabilistic method to create higher research reliability. Several research variables can be enhanced to create extensive research, especially research variables about the factors that influence purchasing decisions on social media, which influence buying intent, repeat purchasing, or the factor affecting customer loyalty through repeat advertising of social networks. Finally, hypothesis function, cost function, and gradient descent can be applied to make the result more precise.

References

1. Ajzen I, Fishbein M (1975) Belief, attitude, intention and behavior: an introduction to theory and research. Addison-Wesley, Reading, MA
2. Al Khasawneh M, Shuhaiber A (2013) A comprehensive model of factors influencing consumer attitude towards and acceptance of SMS advertising: an empirical investigation in Jordan. Int J Sales Market Manage Res Dev 3:1–22
3. Ariffin SK, Aun TL, Salamzadeh Y (2018) How personal beliefs influence consumer attitude towards online advertising in malaysia: to trust or not to trust? Glob Bus Manage Res 10
4. Brown P, Levinson SC (1987) Politeness: some universals in language usage. Cambridge University Press, Cambridge
5. Chen L-D, Nath R (2004) A framework for mobile business applications. Int J Mobile Commun 2:368–381
6. Chen P-T, Hu H-H (2010) The effect of relational benefits on perceived value in relation to customer loyalty: an empirical study in the Australian coffee outlets industry. Int J Hosp Manage 29:405–412
7. Cheung MY, Luo C, Sia CL, Chen H (2014) Credibility of electronic word-of-mouth: informational and normative determinants of on-line consumer recommendations. Int J Electron Commerce 13:9–38. https://doi.org/10.2753/jec1086-4415130402
8. Clement J (2020) Facebook: number of monthly active users worldwide 2008–2020. https://www.statista.com/statistics/264810/
9. Ducoffe RH (1995) How consumers assess the value of advertising. J Curr Issues Res Adver 17:1–18
10. Fang Y-H (2014) Beyond the credibility of electronic word of mouth: exploring eWOM adoption on social networking sites from affective and curiosity perspectives. Int J Electron Commerce 18:67–102. https://doi.org/10.2753/jec1086-4415180303
11. Fredricks AJ, Dossett DL (1983) Attitude-behavior relations: a comparison of the Fishbein-Ajzen and the Bentler-Speckart models. J Personal Soc Psychol 45:501
12. Gao Y, Wu X (2010) A cognitive model of trust in e-commerce: evidence from a field study in China. J Appl Bus Res (JABR) 26
13. Hu Y, Tafti A, Gal D (2019) Read this, Please? The role of politeness in customer service engagement on social media
14. Kharajo VE, Kharajo VE (2020) Investigating the impact of social media interactivy on buying behaviour of consumer. Int J Manage Reflect 1:14–20

15. Khoa BT (2020) Electronic loyalty in the relationship between consumer habits, groupon website reputation, and online trust: a case of the group on transaction. J Theor Appl Inform Technol 98:3947–3960
16. Khoa BT (2020) The impact of the personal data disclosure's trade-off on the trust and attitude loyalty in mobile banking services. J Promot Manage Ahead-of-print. https://doi.org/10.1080/10496491.2020.1838028
17. Khoa BT (2020) The role of mobile skillfulness and user innovation toward electronic wallet acceptance in the digital transformation era. In: 2020 international conference on information technology systems and innovation (ICITSI), pp 30–37. IEEE. https://doi.org/10.1109/ICITSI50517.2020.9264967
18. Khoa BT, Nguyen HM (2020) Electronic loyalty in social commerce: scale development and validation. Gadjah Mada Int J Bus 22:275–299. https://doi.org/10.22146/gamaijb.50683
19. Ma L, Sun B (2020) Machine learning and AI in marketing-connecting computing power to human insights. Int J Res Market 37:481–504
20. Matsumoto Y (1988) Reexamination of the universality of face: politeness phenomena in Japanese. J Pragmat 12:403–426
21. Mosquera R, Odunowo M, McNamara T, Guo X, Petrie R (2020) The economic effects of Facebook. Exp Econ 23:575–602
22. Mullainathan S, Spiess J (2017) Machine learning: an applied econometric approach. J Econ Perspect 31:87–106
23. Najiba NMN, Kasumab J, Bibic ZBH (2016) Relationship and effect of entertainment, informativeness, credibility, personalization and irritation of generation Y's attitudes towards SMS advertising. In: Proceedings of the 3rd international conference on business and economics, pp 213–224, Shah Alam, Malaysia. https://doi.org/10.15405/epsbs.2016.11.02.20
24. Nguyen HM, Khoa BT (2019) Perceived mental benefit in electronic commerce: development and validation. Sustainability 11:6587–6608. https://doi.org/10.3390/su11236587
25. Nguyen MH, Khoa BT (2019) Customer electronic loyalty towards online business: the role of online trust, perceived mental benefits and hedonic value. J Distrib Sci 17:81–93. https://doi.org/10.15722/jds.17.12.201912.81
26. Oh L-B, Xu H (2003) Effects of multimedia on mobile consumer behavior: an empirical study of location-aware advertising. In: ICiS 2003 proceedings 56
27. Pokharel S (2007) An econometric analysis of energy consumption in Nepal. Energy Policy 35:350–361
28. Raney AA, Bryant J (2019) Entertainment and enjoyment as media effect. In: Oliver MB, Raney AA, Bryant J (eds) Media effects: advances in theory and research. Routledge, Taylor & Francis Group, New York
29. Raschka S, Mirjalili V (2017) Python machine learning. Packt Publishing Ltd
30. Sukpanich N, Chen L-D (2000) Interactivity as the driving force behind E-commerce. AMCIS 2000 Proceedings 244
31. Tynan D (2018) Personalization is a priority for retailers, but can online vendors deliver? https://www.adweek.com/digital/personalization-is-a-priority-for-retailers-online-and-off-but-its-harder-than-it-looks-in-an-off-the-shelf-world/
32. Van der Waldt DR, Rebello T, Brown W (2009) Attitudes of young consumers towards SMS advertising. Afr J Bus Manage 3:444–452
33. Watts RJ (2003) Politeness. Cambridge University Press, New York
34. Zo H (2003) Personalization versus customization: which is more effective in E-services? In: AMCIS 2003 proceedings 32

Performance Evaluation of Speaker Identification in Language and Emotion Mismatch Conditions on Eastern and North Eastern Low Resource Languages of India

Joyanta Basu⬤, Tapan Kumar Basu⬤, and Swanirbhar Majumder⬤

Abstract This paper describes the impact of spoken language and emotional variation in a multilingual speaker identification (SID) system. The development of speech technology applications in low resource languages (LRL) is challenging due to the unavailability of proper speech corpus. This paper illustrates performance analysis of SID in six Eastern and North Eastern (E&NE) Indian languages and an emotional corpus of six basic emotions. For this purpose, six experimentations are carried out using the collected LRL of E&NE data to build speaker identification models. Speaker-specific acoustic characteristics are extracted from the speech segments in terms of short-term spectral features, i.e., shifted delta cepstral (SDC) and partial correlation (PARCOR) coefficients. Gaussian mixture model (GMM) and support vector machine (SVM)-based models are developed to represent the speaker-specific information captured through the spectral features. Apart from that, to build the modern SID i-vectors, time delay neural networks (TDNN) and recurrent neural network with long short-term memory (LSTM-RNN) have been considered. For the evaluation, equal error rate (EER) has been used as a performance matrix of the SID system. Performances of the developed systems are analyzed with different emotional native and non-native language corpus in terms of speaker identification (SID) accuracy in six different experiments.

Keywords Low resource language (LRL) · Speaker identification (SID) · Shifted delta cepstral (SDC) · Partial correlation (PARCOR) coefficients · i-vectors · Linear discriminant analysis (LDA) · Probabilistic linear discriminant analysis (PLDA) · Deep neural network (DNN) · Time delay neural networks (TDNN) · Recurrent neural network (RNN) · Long short-term memory (LSTM)

J. Basu (✉)
CDAC Kolkata, Salt Lake, Sector-V, Kolkata 700091, India
e-mail: joyanta.basu@cdac.in

T. K. Basu
Department of Electrical Engineering, IIT, Kharagpur, West Bengal 721302, India

S. Majumder
Department of Information Technology, Tripura University, Tripura 799022, India
e-mail: swanirbhar@ieee.org

1 Introduction

Speaker identification (SID) is a machine-learning problem based on artificial intelligence and pattern matching approaches. To represent the speaker's identities, an enrollment stage is required to create the speaker's models. Speaker's voice is dependent on different acoustic properties like recording channel conditions, recording environment, speaker traits, and spoken language. The reason for the degradation of SID performance is a mismatch between enrollment and evaluation or testing speakers' data. It may occur due to any of the above said acoustic dimensions for the SID system. Former studies on SID focus majorly on monolingual applications. In this kind of application, for both enrollment and testing speakers are spoken in the same language. Most of the studies focused on mismatch of environment/recording/channel [1], not much on spoken language [2, 3] and emotion mismatch condition [4–6].

For the past several years, speech technology research in India has focused on some of the major Indian languages, and the expertise itself is now developed for real-world applications. Although existing systems for language identification (LID) [7], automatic speech recognition (ASR) [8], text-to-speech synthesis (TTS) [9], speaker recognition (SR) [10], etc., are available for some majorly spoken world languages, commercially feasible speech technology solutions for Indian Languages are still not available. The main obstacle in customizing this technology for various Indian languages is the lack of appropriately transcribed speech corpus in these languages. Some indigenous efforts have been reported in prior works [9, 11, 12] on the development of Indian language corpus for ASR and TTS-based domain-specific applications. However, the same for speaker and language identification problem and in E&NE Indian languages is very rare.

Speaker discriminative models like the i-vector model [13, 14] are used for modern SID systems. More recently, DNN [15] is being proposed, like the x-vector model [16]. However, a large amount of data is required for all these models. The training data is often not sufficient for the low resource scenario, for such data-driven models. SID problem is dependent on speakers' speech data in the different forms of spoken variants within a language like read speech, free speech, query-response, and conversational speaking modes. However, MFCC [17], LPC [18], and PLP [7] are recognized as the most useful features that have stood the test of time. Further, the field of speaker recognition was dominated by probabilistic methods such as hidden Markov models (HMMs) [19], Gaussian mixture models (GMMs) [20], support vector machines (SVMs) [21], and artificial neural networks (ANNs) [22]. Indigenous efforts with search space optimization for SID applications based on popular modeling techniques have been reported in [23] and in [24].

With the above motivation, here we have collected the speakers' data in six native languages from the East and North Eastern regions of India. All these languages are low-resourced because they do not have available digital language data and resources (like prior literature, language grammar and dictionary, scripts, etc.). The languages that are considered for this work are namely Assamese (AS), Bangla (BN),

Hrangkhawl (HK), Mizo (MI), Nagamese (NG), and Santali (SA) as referred native languages (NL) in this paper. Apart from this, we have collected other non-native languages (NNL) like Indian English (IE) and Hindi (HN) from the same set of speakers, as these are frequently spoken languages in these regions. Authors in this work investigated speaker identification from emotional audio data also (like anger, fear, happy, sad, surprise, and neutral) and studied in detail to find out the SID performance using different state-of-the-art techniques. Major contributions of this present work are as follows:

- Collection of NL and NNL as well as emotional corpus from the speakers.
- Based on the collected data, we have designed several experiments and built a text-independent baseline SID system using different state-of-the-art features and classifiers. Finally, we have reported the performance of the system on the developed corpus for language and emotion mismatch conditions.

2 About Speech Database

Tables 1 and 2 show the native (six different LRL) and non-native (Hindi and Indian English) language and emotion wise speaker distribution, respectively. A total of 85 speakers' data is collected for the experiments in different E&NE languages of around 16.13 h for native language and 12.67 h in the non-native language including emotional speech corpus of 29 and 14 speakers, respectively. The age group of the speakers is between 15 and 55 years. The same set of speakers participated in both the native and non-native languages data collection process. For the emotional speech

Table 1 Native language (NL) wise speaker distribution

Sl	Language name	# Speakers	# Emotional speakers	Recording in hours
1	Assamese (AS)	15	5	2.75
2	Bengali (BN)	15	9	3.24
3	Hrangkhawl (HK)	10	2	1.50
4	Mizo (MI)	15	4	2.87
5	Nagamese (NG)	15	2	2.62
6	Santali (SA)	15	7	3.15
	Total	**85**	29	**16.13**

Table 2 Non-native language (NNL) wise speaker distribution

Sl	Language name	# Speakers	# Emotional speakers	Recording in hours
1	Hindi (HN)	50	8	8.35
2	Indian English (IE)	35	6	4.32
	Total	**85**	**14**	**12.67**

corpus, we have created 139 sentences in six different emotions. Speech corpus are recorded in a home, office, and studio environment with a 22.05 kHz sampling rate and using 16-bit digitization format.

3 Experimental Setup

Kaldi, an open-source toolkit [25], and python [26] are used to extract features and build classification models for this research work.

3.1 Experimental Data

For experimental purposes, we have distributed the corpus as shown in Table 3 to analyze the combined effect of emotion and language mismatch condition. Four different experiments were carried out during the SID performance evaluation of language and emotion mismatch conditions. On the other hand, another two experiments were carried out to analyze emotion and language mismatch conditions separately as shown in Table 4.

Table 3 Speech corpus distribution to analyze the combined effect of emotion and language mismatch condition

Purpose	Experiment 1	Experiment 2	Experiment 3	Experiment 4
Train	NNL_Readout	NL_Readout	NNL_Readout NNL_Emotion	NL_Readout NL_Emotion
Test	NNL_Emotion NL_Readout NL_Emotion	NL_Emotion NNL_Readout NNL_Emotion	NL_Readout NL_Emotion	NNL_Readout NNL_Emotion

Table 4 Speech corpus distribution to analyze emotion and language mismatch separately

Purpose	Experiment 5	Experiment 6
Train	NL_Readout	NNL_Readout NNL_Emotion
Test	NL_Emotion_Neutral NL_Emotion_Anger NL_Emotion_Fear NL_Emotion_Happy NL_Emotion_Sad NL_Emotion_Surprise	NL_Readout_Emotion_AS NL_Readout_Emotion_BN NL_Readout_Emotion_HK NL_Readout_Emotion_MI NL_Readout_Emotion_NG NL_Readout_Emotion_SA

3.2 SID System and Evaluation Metrics

A front-end signal-processing module is required to extract features from speech. Front-end processing generally consists of voice activity detection (VAD) and feature extraction block. VAD is performed to remove non-speech portions from the signal. On the other hand, the feature extraction block extracts suitable features from the speech. In the experiment, shifted delta cepstral (SDC) features are extracted from MFCC (13 vectors) feature. Also, we have extracted PARCOR coefficients [27] of 13 vectors. MFCC features are derived from a speech frame of 25 ms with an overlap of 10 ms from 24 filter bands placed as per the mel-scale. Shifted delta cepstral coefficients are based on four variables. In this study, the SDC parameter specification is 7–1–3–7. For each frame, with seven direct MFCC coefficients, 49 SDC coefficients are appended, so 56 coefficients are used for this study.

In this study, the GMM [28] and SVM [29] were used to build up traditional speaker models and evaluation of test utterances. In GMM, 1024 mixtures have been used to build models. Radial basis function (RBF) kernel is used for SVM to build models. 400-dimensional i-vectors [14, 30] using 2048 Gaussian mixtures are extracted. Transforms like LDA [31] and PLDA [32] are applied, where the dimensions of the i-vectors is 150 after using LDA reduction. As a baseline DNN system, two different approaches were used, namely TDNN [33] and LSTM-RNN [34]. 56-dimension feature vectors of SDC are used as input to both TDNN and LSTM-RNN. For TDNN, six hidden layers are used. Second, as well as for the last hidden layers, no splicing is used. It is decided empirically. Each layer contains 650 units and the activation function is a rectified linear unit (ReLU). Similar to the first layer of TDNN architecture, the affine transformation layer is used in the LSTM network as the first layer with a sequential context. 512 cells are used after that stacked Long Short-Term Memory layer.

Evaluation metrics for the performance of the SID system are equal error rate (EER). False acceptance rate (FAR) and false rejection rate (FRR) are used to predetermine the threshold values of EER. The equal error rate is determined when FAR and FRR are equal.

4   Results and Discussion

The system performance of the SID system for experiment 1 to experiment 4 is shown in Fig. 1. In these experiments, we have analyzed the performance of language and emotion mismatch conditions together for native and non-native language with six different emotions. In these four experiments, we have used SDC and PARCOR coefficients as a feature of the system and GMM, SVM, i-vector, i-vector with LDA, PLDA, TDNN, and LSTM-RNN as classifiers. The performance of SID in terms of EER is varying from 22.36 to 9.73% using different feature types and classifier combinations for these experiments. It has been observed that the SDC feature set

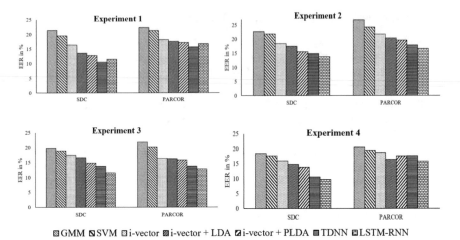

Fig. 1 Performance of SID system to analyze the combined effect of emotion and language mismatch condition

using the LSTM-RNN classifier performs better than other features and classifiers for the SID task with the corpus except experiment 1. In this experiment, training and testing languages are different. We trained the speakers' model using native language (NL) and test using non-native language (NNL) like Indian English or Hindi spoken by the speaker and vice versa. So due to the language mismatch as well as emotion mismatch condition, we have tested the SID performance. We received the best EER (9.73%) for experiment 4 using SDC and LSTM-RNN combination. In Fig. 2, we have shown two different experiments (5 and 6) the performance of the SID system to analyze emotion and language mismatch separately. In these experiments, we have used only SDC as features because it has been observed from the above experiments that SDC shows the best results than PARCOR coefficients. It has been observed that emotion plays an important role in SID performance. We received the best result for neutral emotions and achieved 7.18% EER using TDNN. Other emotions

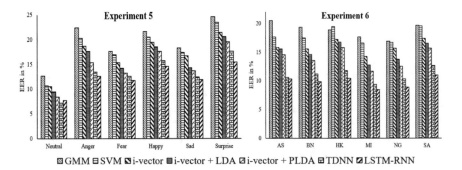

Fig. 2 Performance of SID system to analyze emotion and language mismatch separately

are shown the impact on SID performance and performance is deteriorated due to various prosody parameters. Similarly, experiments 6 analyzed the impact language mismatch condition only. In this experiment also using LST-RNN, we achieved the best results. It has been observed that speakers of MI and NG languages achieved the best result 8.45% and 8.86%, respectively. This is due to the reason that in Mizoram people use Mizo and English both as the official language. Moreover, Nagaland English is the state language. Due to this reason, we achieved good EER in these two languages for SID performance. To reduce the mismatch condition, we need to explore other features. However, there are options to include prosodic features and modify the feature set and classifiers specific parameters or architecture to perform the SID experiments for better accuracy.

5 Conclusion

This research study describes experimental findings on the SID system on language and emotion mismatch conditions. The LRL speech corpus consists of six different E&NE Indian languages as well as common languages Hindi and English. Voice samples include of total 85 native speakers including an emotional corpus of 29 speakers. Same 85 speakers participated in non-native language recording including emotional corpus from 14 NNL speakers. Experiments on speaker identification are carried out using this speech corpus, with different feature sets and state-of-the-art classifiers in language and emotion mismatch condition. Performances of SID systems have been assessed using different feature-classifier combinations, and six different experiments were carried out. We received the best performance with EER is 9.73% (using SDC feature and LSTM-RNN classifier) for experiment number 4. In experiment 5, it has been observed that emotions play an important role in SID. We received an EER of 7.18% for neutral emotion; but for other emotions like surprise, happy, and anger, EER is on the higher side. However, there are opportunities to achieve higher accuracy using traditional x-vector or other DNN architecture. In addition to spectral features, researchers can also explore and include prosodic and source features for identifying speakers.

Acknowledgements For supporting speech data collection from native speakers of North-East India, authors would like to acknowledge the North Eastern Regional Institute of Science and Technology (NERIST), Arunachal Pradesh, India. The authors thank all native speakers of E&NE Indian states for their contribution. The authors are also grateful to the Centre for Development of Advanced Computing (CDAC), Kolkata, India for the necessary support to carry out the research activities.

References

1. Reynolds D (1997) Comparison of background normalization methods for text-independent speaker verification
2. Akbacak M, Hansen JHL (2007) Language normalization for Bilingual speaker recognition systems. In: 2007 IEEE international conference on acoustics, speech and signal processing—ICASSP '07, vol 4, pp IV-257-IV-260. https://doi.org/10.1109/ICASSP.2007.366898
3. Misra A, Hansen JHL (2014) Spoken language mismatch in speaker verification: an investigation with NIST-SRE and CRSS Bi-Ling corpora. In: 2014 IEEE spoken language technology workshop (SLT), pp 372-377. https://doi.org/10.1109/SLT.2014.7078603
4. Koolagudi SG, Fatima SE, Rao KS (2012) Speaker recognition in the case of emotional environment using transformation of speech features. In: Proceedings of the CUBE international information technology conference on - CUBE '12, p 118. https://doi.org/10.1145/2381716.2381739
5. Jawarkar NP, Holambe RS, Basu TK (2012) Text-independent speaker identification in emotional environments: a classifier fusion approach. In: Sambath S (ed) Advances in intelligent and soft computing. Springer, Berlin, Heidelberg, pp 569-576
6. Meftah AH, Mathkour H, Kerrache S, Alotaibi YA (2020) Speaker identification in different emotional states in Arabic and English. IEEE Access 8:60070-60083. https://doi.org/10.1109/ACCESS.2020.2983029
7. Hegde RM, Murthy HA (2005) Automatic language identification and discrimination using the modified group delay feature. In: Proceedings of 2005 international conference on intelligent sensing and information processing, pp 395-399. https://doi.org/10.1109/ICISIP.2005.1529484
8. Besacier L, Barnard E, Karpov A, Schultz T (2014) Automatic speech recognition for under-resourced languages: a survey. Speech Commun 56:85-100. https://doi.org/10.1016/j.specom.2013.07.008
9. Baby A, Thomas A, Consortium TTS (2016) Resources for Indian languages. CBBLR – Community Based Building of Language Resources, Brno, Czech Republic: Tribun EU, pp 37-43
10. Campbell JP (1997) Speaker recognition: a tutorial. Proc IEEE 85(9):1437-1462. https://doi.org/10.1109/5.628714
11. Aanchan Mohan SU, Rose R, Ghalehjegh SH (2014) Acoustic modelling for speech recognition in Indian languages in an agricultural com-modities task domain. Speech Commun 56:167-180
12. Basu J, Khan S, Roy R, Bepari MS (2013) Commodity price retrieval system in Bangla. In: Proceedings of the 11th Asia Pacific conference on computer human interaction - APCHI '13, pp 406-415. https://doi.org/10.1145/2525194.2525310
13. Tang Z, Wang D, Chen Y, Li L, Abel A (2018) Phonetic temporal neural model for language identification. In: IEEE/ACM Trans Audio Speech Lang Process 26(1):134-144. https://doi.org/10.1109/TASLP.2017.2764271
14. Dehak N, Kenny PJ, Dehak R, Dumouchel P, Ouellet P (2011) Front-end factor analysis for speaker verification. IEEE Trans Audio Speech Lang Process 19(4):788-798. https://doi.org/10.1109/TASL.2010.2064307
15. Richardson F, Reynolds D, Dehak N (2015) Deep neural network approaches to speaker and language recognition. IEEE Signal Process Lett 22(10):1671-1675. https://doi.org/10.1109/LSP.2015.2420092
16. Snyder D, Garcia-Romero D, McCree A, Sell G, Povey D, Khudanpur S (2018) Spoken language recognition using x-vectors. In: Odyssey, pp 105-111
17. Manchala V, Prasad S, Janaki VK (2014) GMM based language identification system using robust features. Int J Speech Technol 17(2):99-105. https://doi.org/10.1007/s10772-013-9209-1
18. Mansour E, Sayed MS, Moselhy AM, Abdelnaiem AA (2013) LPC and MFCC performance evaluation with artificial neural network for spoken language identification. Int J Signal Process Image Process Patt Recognit 6(3):55-66. 10.1.1.360.2692

19. Rabiner LR (1989) A tutorial on hidden Markov models and selected applications in speech recognition. Proc IEEE 77(2):257–286. https://doi.org/10.1109/5.18626
20. Reynolds D (2015) Gaussian mixture models. In: Li SZ, Jain AK (ed) Encyclopedia of biometrics. Springer, Boston, pp 827–832
21. Campbell WM, Sturim D, Reynolds DA (2006) Support vector machines using GMM super-vectors for speaker verification. IEEE Signal Process Lett 13(5):308–311. https://doi.org/10.1109/LSP.2006.870086
22. Dey NS, Mohanty R, Chugh KL (2012) Speech and speaker recognition system using artificial neural networks and hidden Markov model. In: 2012 international conference on communication systems and network technologies, pp 311–315. https://doi.org/10.1109/CSNT.2012.221
23. Khan S, Basu J, Bepari MS (2012) Performance evaluation of PBDP based real-time speaker identification system with normal MFCC vs MFCC of LP residual features. In: Lecture Notes in Computer Science (including subseries Lecture Notes in Artificial Intelligence and Lecture Notes in Bioinformatics), vol 7143 LNCS, pp 358–366
24. Khan S, Basu J, Bepari MS, Roy R (2012) Pitch based selection of optimal search space at runtime: speaker recognition perspective. In: 2012 4th international conference on intelligent human computer interaction (IHCI), pp 1–6. https://doi.org/10.1109/IHCI.2012.6481822
25. Povey D et al (2011) The Kaldi speech recognition Toolkit
26. Python. https://www.python.org/
27. Soni T, Zeidler JR, Ku WH (1995) Behavior of the partial correlation coefficients of a least squares lattice filter in the presence of a nonstationary chirp input. IEEE Trans Signal Process 43(4):852–863. https://doi.org/10.1109/78.376838
28. Reynolds DA Speaker identification and verification using Gaussian mixture speaker models. Speech Commun 17(1–2):91–108
29. Vapnik VN (2000) The nature of statistical learning theory. Springer, New York
30. Dehak N, Dehak R, Kenny P, Brümmer N, Ouellet P, Dumouchel P (2009) Support vector machines versus fast scoring in the low-dimensional total variability space for speaker verification
31. Balakrishnama S, Ganapathiraju A (1998) Linear discriminant analysis-a brief tutorial. Inst Signal Inf Process 18:1–8
32. Prince SJD, Elder JH (2007) Probabilistic linear discriminant analysis for inferences about identity. In: 2007 IEEE 11th international conference on computer vision, pp 1–8. https://doi.org/10.1109/ICCV.2007.4409052
33. Lang KJ, Waibel AH, Hinton GE (1990) A time-delay neural network architecture for isolated word recognition. Neural Netw 3(1):23–43. https://doi.org/10.1016/0893-6080(90)90044-L
34. Hochreiter S, Schmidhuber J (1997) Long short-term memory. Neural Comput 9(8):1735–1780. https://doi.org/10.1162/neco.1997.9.8.1735

Autonomous Wheelchair for Physically Challenged

B. Kavyashree⑩, B. S. Aishwarya, Mahima Manohar Varkhedi, S. Niharika, R. Amulya, and A. P. Kavya⑩

Abstract An area of bodily challenged people makes it very hard to use traditional wheelchairs. To move patients or physically disabled people from one place to another either in hospital or in day to day life, wheelchairs are used. Wheel chairs are designed such that they could be used either with the help of attendee or it can be self-propelling. To lessen the complexities for people who do not have power to move the wheelchair, it is automated. Accelerometer is used to control the wheelchair using head movements. Ultrasonic sensors are used to discover obstacles, joystick is used to provide input through hand, and Bluetooth module is used to interface smartphone to give the voice command. The special modes can be chosen in keeping with the requirement the use of a switch. It is a low cost machine which has unique methods incorporated to manipulate the wheelchair.

Keywords Wheelchair · Accelerometer · Bluetooth module · Joystick · Ultrasonic sensor

1 Introduction

As per statistical report by World Health Organization (WHO) and World Bank, 15% of total population experience some form of disability [4], day by day the percentile is increasing due to road accidents and disease like paralysis. About 1.85% of population need wheelchair for their daily life [5]; in most cases where the disability is low, manual or electrical wheelchair is enough, but in severe cases where the patients are solely dependent on external help find it difficult to maneuver. To overcome such problems, several companies have introduced "Smart Wheel Chairs." Smart wheel chair can be defined as modified electrical wheelchair which is equipped with various sensors and control system which makes it easier for disabled person to maneuver without the help of relatives or nurses, giving full control to the user, and eliminating

B. Kavyashree (✉) · B. S. Aishwarya · M. M. Varkhedi · S. Niharika · R. Amulya · A. P. Kavya
Vidyavardhaka College of Engineering, Mysuru, Karnataka, India

A. P. Kavya
e-mail: kavya.ap@vvce.ac.in

© The Author(s), under exclusive license to Springer Nature Singapore Pte Ltd. 2022   521
P. Nanda et al. (eds.), *Data Engineering for Smart Systems*, Lecture Notes in Networks and Systems 238, https://doi.org/10.1007/978-981-16-2641-8_50

the user's responsibility in moving the wheel chair manually. Gyro sensor is used as input devices. The gyro sensor will detect the movements made by user's hand in X and Y axis and sends it to a Atmega328p-based Arduino board. The Arduino then sends corresponding signals to motor controller which drives the motor in respective direction.

2 Literature Survey

Bhagat et al. [4] Microcontroller is used to obtain the preferred operation; robot is a tool that can carry out tasks routinely given the commands. Computational intelligence entails the programming instructions. Robotic car directs it each time an impediment is detected. Ultrasonic sensors are used to detect the object with the help of the commands given the wheelchair movements forward, backward, right or left.

Sharath Babu Rao and Anusha [5] Depending on the human instructions in accordance of audio and head or hand gesture, the automation is done. These gestures are used as inputs. Head gesture is captured by the usage of accelerometer. The wheelchair proposed here has multiple benefits which include reduced complexity, clean controlling, low fee, and incredible reliability as compared to other already existing wheelchairs.

Nirmal [6] The wheelchair can be controlled using voice control, joystick. Patient health condition will be informed to the doctor via text message. Servo motor and voice recognition are used for the movement of the wheelchair. It helps people who are physically disabled without relying on other people.

Nipanikar et al. [7] Wheeled robots are constructed with the help of motors. DC motors are used for the purpose of movement. It mainly consists of control system which includes H-bridge motor circuit. Implementation of these H-bridge motor circuit will assist to make the wheeledrobots versatile and could work in any conditions.

Gupta [8] It describes the development of a wheelchair with voice recognition with the use of embedded system. Android software is developed on android smartphone. Arduino is used to execute all the commands. Arduino Uno, HC05 Bluetooth module are used. This wheelchair is a mixture of mechanical, electrical, and communication system.

3 Methodology

Figure 1 shows the block diagram of autonomous wheelchair for physically challenged people. The battery, which is the source, supplies electrical power to the wheelchair. The battery output is connected to the regulated power supply (RPS).

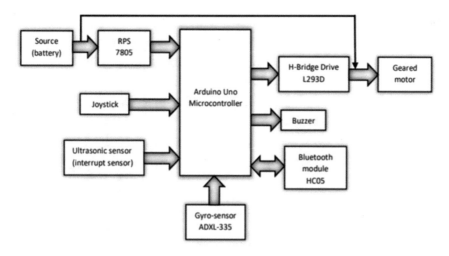

Fig. 1 Block diagram of autonomous wheelchair for physically challenged

A regulated power supply has a range of 6.8–42 V. The output of the RPS, which provides the working voltage, is connected to the Arduino microcontroller.

Joystick is used to control the movement of wheelchair manually. It is used as the input device to send command to the processing unit. Arduino board with controller ATMega328p is used as the processing unit. After receiving input, the processing unit processes the input and sends the digitized signal to the motor driving IC, which in turn controls the movement of the wheelchair.

In case of any presence of objects in the movement path of wheelchair, the ultrasonic sensor senses it and sends the input to the microcontroller which in turn stops the movement of wheelchair, thereby avoiding collision with the object. Here, buzzer produces a beep sound as soon as it encounter any obstacle.

ADXL-335 is a MEMS device which converts the gesture motions made by the user into electrical data which is sent to ATMega328-based Arduino board. The microcontroller then process the input received and drives the motors in corresponding direction; to move forward, both motors rotate in forward direction, and for reverse direction, the motors will rotate in backward direction.

Bluetooth module HC05 acts as serial communication between the user and the vehicle. This recognizes the voice of the user and acts according to it. A switch is provided to select either of the methods that are one method at a time. Then, the output of the Arduino microcontroller is connected to the H-bridge drive which is used to amplify the current. DC geared motor requires a huge amount of torque to move the vehicle easily. Based on the logic that is analyzed and produced from the microcontroller the vehicle moves.

The Fig. 2 represents the flowchart of the autonomous wheelchair. The algorithm is as follows,

Fig. 2 Flowchart of the
autonomous wheelchair for
physically challenged

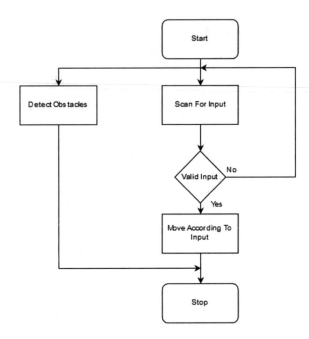

- Start the vehicle.
- Scan for the sensor inputs given by the user according to the person's desire.
- The system then checks if the given inputs is valid or not. If the given input is valid, it goes to the next step. But if the given input is invalid, it scans for sensor input again.
- Once the input is said to be valid, the system moves according to the type of the sensor input given by the user.
- The system is then stopped as desired by the user.
- If the systems encounter any obstacle at any point of time, the system suddenly stops.

4   Hardware Description

4.1   Arduino Uno

It is an Atmega328p-based microcontroller board which consists of 14 digital I/O pins, among them 6 pins configured to work as PWM outputs, 6 has analog input pins, 16 MHz crystal, a USB type A connector, and a power jack. In this paper, the Arduino is used to process the signals received from gyro sensor and ultrasonic sensor to drive the motor.

Fig. 3 Working of ultrasonic sensor

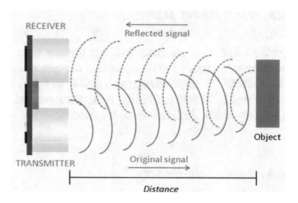

4.2 Adxl-335

It is 3 axis acceleration measurement systems. It consists of polysilicon surface-micro machined sensor which is built on top of a silicon wafer. It produces analog voltage as output which in turn depends on acceleration. In this paper, the ADXL-335 is used as gyro sensor which is mounted on the user's hand to detect the gesture made in X and Y axis to drive the motor in that particular direction.

4.3 Ultrasonic Sensor

It is a sensor which is used to measure distance between the object and itself. It consists of two transducers which are transmitter and receiver, the transmitter sends ultrasonic waves which are reflected back to the receiver by the object, the time taken by waves to bounce back is proportional to the distance between the object and the sensor. The working of the ultrasonic sensor is shown in Fig. 3. In this paper, the sensor is used to detect any obstacle in the path of the sensor to avoid collision.

4.4 BLDC Driver

It is high voltage and current brushless motor driver; it operates on voltage range from 12-36v and has maximum current rating of 15A. It consists of MOSFETs which switch accordingly to drive the BLDC motor.

4.5 Hub Wheel Motor

As the name says it is motor which is directly built into the hub of the wheel, this eliminates the need of drive trains and the mechanical system behind it. These are brushless DC motor which operates on 24–48 V DC. In this paper, these are used for movement of wheelchair in left, right, forward, and reverse direction. The motors are controlled by Arduino through the driver.

4.6 Lead-Acid Battery

This is used to supply power to the BLDC motor drivers which drives the hub wheel motors, Arduino, gyro sensor, and ultrasonic sensor operate on 5 V which is provided with the help of LM7805 voltage regulator IC by converting 12 V into 5 V.

4.7 HC-05 Bluetooth Module

Bluetooth module is interfaced with microcontroller for the purpose of serial communication. Bluetooth SPP (Serial Port Protocol) module is used in this paper.

5 Results and Discussions

Figure 4a, b, c shows the implementation of autonomous wheel chair. Two ultrasonic sensors are used, one in front and the other is in back to detect the obstacles. Ultrasonic sensor is placed to detect any obstacles in the path. If the ultrasonic sensor detects the distance less than 30 cm, the wheelchair will automatically stop. Then the user has to use the backward command to move away from obstacles; as soon as the wheelchair moves away from the obstacles, the distance detected by the ultrasonic sensor will be more than 30 cm, now the wheelchair can move in any direction depending on the user command. For the purpose of gesture control, accelerometer sensor is used.

By measuring the amount of static acceleration due to gravity, it is possible to find the angle, the tilt of the device with respect to earth, and thereby sensing the dynamic acceleration. Using this, the direction of movement can be analyzed and signals are sent to the driver circuit to rotate the motor in that respective direction. Joystick command can also be used to control the direction of the wheelchair, by switching the toggle switch present on the arm rest the joystick command can be used instead of gesture command. A Bluetooth module is used to control the wheelchair through the voice command. The wheelchair designed here can withstand a weight of 100 kg. The speed of the wheelchair depends on the weight; for a weight of 60–80 kg, the

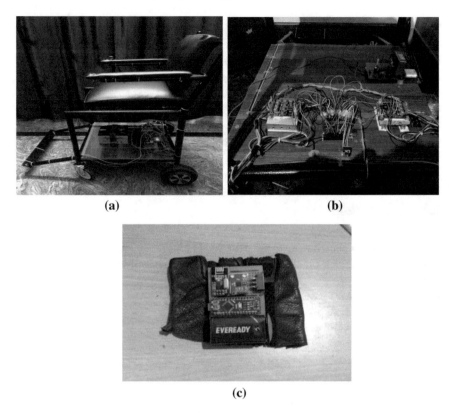

(a)   (b)

(c)

Fig. 4 **a**, **b**, **c** Implementation of autonomous wheel chair

wheelchair can move at a speed of 6–7 km/hr, and for a weight of 80 to 100 kg, it can move at a speed of 4–5 km/hr.

6   Advantages and Applications

Autonomous wheelchair helps the physically challenged to travel freely, hence avoiding accidents and thefts making them independent. The autonomous wheelchair can be easily installed, operated with ease, and it comes with good reliability. Physically challenged people who can operate joystick or press the touch screen of the control device can use this autonomous wheelchair.

6.1 Advantages

- Can be adopted for existing wheelchairs thus reducing the cost.
- Helpful for paralyzed people who are unable to use their hands to move the wheelchair.
- Reduces the physical strain exerted on the body.
- The system cost is economical.
- Gives sense of freedom for physically challenged people.
- Increases mobility.

6.2 Applications

- Used in hospitals for handicapped patients. The wheelchair comes with flexibility to move in different directions which helps the physically challenged people to move freely.
- Used in healthcare centers.
- Physically handicapped individuals in industries.
- Used in schools and colleges.

7 Conclusion

Physically challenged people lack adequate physical or mental strength which prevents them from living a normal life. This paper establishes a prototype which helps such kind of people by providing autonomous vehicles for transportation. The above-mentioned reason is the motivation for this work. The objective of this work is to create an autonomous vehicle model which includes sensors, joystick, voice recognitions, and gesture movements to help specific group of physically challenged people (e.g., Blind, handicapped, deaf and dumb, etc.) to move easily. This will greatly help in reducing the human dependency and ease the life of physically challenged people.

References

1. Alam Md. E, Kader Md. A, Hany U, Arjuman R, Siddika A, Islam Md. Z (2019) A multi-controlled semi-autonomous wheelchair for old and physically challenged people. In: 1st international conference on advances in science, engineering and robotics technology (ICASERT2019) DOI:10.1109/ ICASERT.2019.8934663
2. Hartman A, Nandikolla VK (2019) Human-machine interface for a smart wheelchair. J Robot, vol 2019, Article ID 4837058

3. Priyanayana S, Buddhika AG, Jayasekara P (2018) Developing a voice controlled wheelchair with enhanced safety through multimodal approach. IEEE region 10 humanitarian technology conference
4. Bhagat K, Deshmukh S, Dhonde S, Ghag S, Waghmare V (2016) Obstacle avoidance robot. Int J Sci Eng Technol Res (IJSETR) 5(2):439
5. Sharath Babu Rao D, Anusha T (2015) Gesture controlled wheelchair. Int J Sci Eng Technol Res (IJSETR) 4(8):2838
6. Nirmal TM (2014) Wheelchair for physically and mentally disabled persons. Int J Electric Electron Res 2(2):112–118
7. Nipanikar RS, Gaikwad V, Choudhari C, Gosavi R, Harne V (2013) Automatic wheelchair for physically disabled persons. Int J Adv Res Electron Commun Eng (IJARECE) 2(4):466
8. Gupta V (2010) Working and analysis of the H-bridge motor driver circuit designed for wheeled mobile robots. IEEE Xplore, p 441

Watermarking of Digital Image Based on Complex Number Theory

Nadia Afrin Ritu, Ahsin Abid, Al Amin Biswas, and M. Imdadul Islam

Abstract The objective of this paperwork is to implement a watermarking algorithm using complex number theory. Here, each pixel of the original image is considered as the real part of each complex number, and the imaginary part corresponds to the pixel of the watermark. The matrix consists of the absolute value of the complex number treated as the watermarked image. The real and imaginary components of each complex number cannot be retrieved without the phase angle, which is treated as the key to a secured watermarked image. The phase angle of each complex number transmitted through a secured channel. At the receiving end, when both the magnitude and phase angle of each pixel estimated properly, then the real and imaginary components of each complex number corresponding to the pixel of original and watermarked images are determined with some accuracy.

Keywords Digital watermarking · Cross-correlation · Signal to noise ratio · RSA · Big modular arithmetic

1 Introduction

The approach of concealing digital information that provides data protection and data security is called a digital watermark. It is inserted bits into a digital image, audio, or video. The copyright information distinguishes them. Different printed watermarks that are deliberate to be somewhat noticeable and fully invisible, digital watermarks are designed to be inaudible, or in the case of audio recordings.

Patil and Rindhe [1] proposed a new embedding algorithm (NEA) of digital watermarking. Here, the authors used nonsubsampled contourlet transform (NSCT) and discrete cosine transform (DCT) watermarking technique called a hybrid watermarking. The obtained results show the comparison between Cox's additive embedding and the NEA algorithm. To increase the extraction efficiency, Kim et al. [2]

N. A. Ritu (✉) · A. Abid · A. A. Biswas · M. Imdadul Islam
Department of CSE, Jahangirnagar University, Dhaka, Bangladesh

M. Imdadul Islam
e-mail: imdad@juniv.edu

© The Author(s), under exclusive license to Springer Nature Singapore Pte Ltd. 2022
P. Nanda et al. (eds.), *Data Engineering for Smart Systems*, Lecture Notes in Networks and Systems 238, https://doi.org/10.1007/978-981-16-2641-8_51

proposed a robust digital watermarking technique based on the method of combined transformation for image content. A combination of discrete wavelet transform (DWT) and alpha blending formula was found in [3] to extract the watermark image. Watermarking is a component of steganography that is used for message security has introduced by Bhatt et al. [4]. In this case, the secured image is embedded in a base image so that only the base image is visualized. Some private key is incurred with the watermarked image to retrieve the watermark image. Jaju and Chowhan [5] introduced a difference between RSA and modified RSA algorithms along with time and security. In [6], the authors focused on the digital watermarking technique that provides data security. The different watermarking technique provides some security mechanisms such as frequency domain watermarking, spread spectrum, and spatial domain techniques. Here, we have used the spatial domain LSB method that is easy and simple to implement for the security of images. In [7], the authors mainly indicated the difference between the digital watermarking technique and the reversible image watermarking technique.

The digital watermarking technique is used to mask information that cannot be easily retrieved, but reversible image watermarking can use the interpolation process to restore the actual image without any misinterpretation.

The rest of the paper is organized as Sect. 2 provides the literature review. Section 3 deals with the proposed algorithm of a digital watermark. Section 4 provides results based on the analysis of Sect. 3. Section 5 compares this work with existing methods. Lastly, Sect. 6 concludes the entire analysis.

2   Literature Review

Due to various issues, the aspects of this type of research problem are prominently complicated so that an intense investigation is essential to deal with this. In this section, we have observed various recent related research articles to uncover the gap between the existing works. This section helps to give an advanced solution in this context. The summary of the details of our findings is presented below.

Jiang et al. [8] introduced a new zero-watermarking algorithm based on tensor mode expansion for the color image. The results showed its robustness to the noise attack and other well-known image-processing attacks. Begum and Uddin [9] reviewed the most imperative watermarking techniques with the conclusion that DWT is high-quality and hardy technique in this regard because of its multi-resolution characteristics. Additionally, they identified essential qualifications for the watermarking system as robustness, capacity, and imperceptibility. For color remote sensing images, Li et al. [10] introduced a nonblind digital watermarking method. To work on it, the authors applied the quaternion wavelet transform (QWT) and tensor decomposition. The test results showed that their technique is more robust than usual QWT and DWT. Their approach balances the trade-off between robustness and imperceptibility. Ahmaderaghi et al. [11] introduced a novel blind watermarking framework. To work on it, the authors employed the discrete shearlet transform

(DST). The test results showed the enhanced performance of the working approach. Using dual-tree complex wavelet transform (DTCWT) and the L1-norm function, a quantization-based watermarking method has introduced by Liu et al. [12]. The test results showed that the introduced technique is robust enough toward the noise and gain attack with good obscurity, JPEG compression, filtering, and amplitude scaling. In [13], the authors mainly focused on the significance of SVD in watermarking and recognized the false positive problems (FPPs) of many SVD-based watermarking schemes. Here, to demonstrate how FPPs and attacks could grievously influence the considered watermarking approach, the author performed a reliability test. To check FPPs, the proposed method employed the three potent attacks. These FPP analyses can assist in knowing the security matters associated with SVD-based watermarking techniques in general. Wang et al. [14] proposed an algorithm that has extensive theoretical significance with practical usefulness in copyright protection and has solid robustness to many symmetric and asymmetric attacks confronted with other zero-watermarking algorithms. Research work conducted by Zebbiche et al. [15] introduced a blind additive image watermarking system in the DTCWT domain with taking benefit of a perceptual masking model that exploits the human visual system (HVS) characteristics at the embedding stage. The obtained test results showed that the introduced approach outperforms the related intricate watermarking schemes in terms of robustness and imperceptibility. Jaiswal and Ravi [16] presented many features such as overview, framework, techniques, applications, challenges, and limitations for digital watermarking. A comparative study between DWT and DCT watermarking techniques with their benefits and drawbacks has been presented here. They also worked to classify digital watermarking in all the known features such as robustness, perceptivity, and many others. From the result, they concluded that the proposed approaches are much suitable in terms of PSNR value. Wang et al. [17] proposed robust zero-watermarking algorithms. To work on it, the authors used logistic mapping and polar complex exponential transform. The obtained test results illustrate that the applied approach can counter geometric attacks with other well-known related attacks and has notable benefits over the present zero-watermarking techniques.

3 Methodology

3.1 Proposed Digital Watermarking Algorithm

The detailed processing of the watermark algorithm can be represented as:

a. Watermark and the original image will be an input of our methodology where input watermark image will be hidden in the original image finally.
b. As it is an RGB image, we will calculate the remainder r and quotient q for every channel at the watermark image.

c. Then we apply the RSA algorithm to create it more securely and use a big modular algorithm to handle large prime in RSA.

The following Fig. 1 will describe more clearly.

3.2 Algorithm of Digital Watermark

a. Read the original image $\mathbf{O}$ and watermark image $\mathbf{W}$.
b. Determine magnitude $M(i, j) = |O(i, j) + j\sqrt{\alpha} \cdot W(i, j)| = \sqrt{O^2(i, j) + \alpha W^2(i, j)}$ and phase angle $\phi(i, j) = \tan^{-1}\left(\frac{\alpha W(i,j)}{O(i,j)}\right)$ at (i, j), where $O(i, j)$ and $W(i, j)$ are the pixels at (i, j) of original and watermark image, respectively, and weighting factor $0 < \alpha < 1$.
c. Apply RSA algorithm on both $M(i, j)$ and $\phi(i, j)$ like: $M_{RSA}(i, j) = E_{RSA}\{M(i, j)\}$ and $\phi_{RSA}(i, j) = E_{RSA}\{\phi(i, j)\}$
d. Transmit encrypted $M(i, j)$, i.e., $M_{RSA}(i, j)$ as the watermarked image to the receiver through the public channel and $\phi_{RSA}(i, j)$ as the private (both phase angle itself and RSA parameters) key through the secured channel.
e. At the receiving end decrypts the magnitude and phase angle at (i, j) as: $\hat{M}(i, j) = D_{RSA}\{M_{RSA}(i, j) + n(t)\} = M(i, j) + e(i, j)$ and $\hat{\phi}(i, j) = D_{RSA}\{\phi_{RSA}(i, j) + n'(t)\} = \phi(i, j) + e'(i, j)$, where $n(t)$ and $n'(t)$ are the noise of public and secured channels, respectively, at the instant of arrival of the pixel. Here $e(i, j)$ and $e'(i, j)$ are the error of the decrypted data at (i, j).
f. Recover the pixels of original and watermark image as:

$$\hat{O}(i, j) = \text{Re}\left[\hat{M}(i, j)\right] = \hat{M}(i, j) \cos\phi(i, j) \text{ and } \hat{W}(i, j) = \text{Im}\left[\hat{M}(i, j)\right] = \hat{M}(i, j) \sin\phi(i, j),$$ where $\hat{O}(i, j)$ and $\hat{W}(i, j)$ are the estimated value of pixels original and watermark image because of the noisy channel.

3.3 Big Modular Arithmetic

To implement the RSA algorithm, we have to compute modular arithmetic. If we take the small number as the public or private key, it will not be secure. So, to make it more secure, we have to take a large number as a key. But if we take a large number, overflow occurs. Here, we use big modular arithmetic. Now, we used several steps to describe the big modular arithmetic as shown in Fig. 2.

a. Try to divide the power by two. If the power is an odd number, subtract one from the power. If the power is even, direct divide it by two.
b. Do step 1 until power does not convert into zero.
c. Multiply the result according to the operation occurs.
d. Calculating all the results, return the final result.

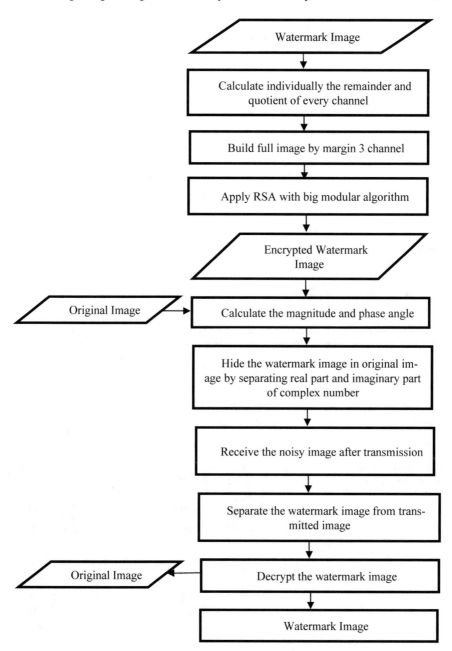

Fig. 1 Processing of original image

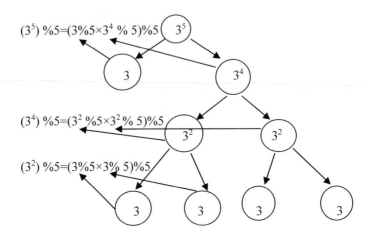

Fig. 2 Example of big modular arithmetic

4 Result and Discussion

The original image, watermark image, encrypted watermark, transmitted image, received, and recovered image are shown in Fig. 3 for -25 dB additive white Gaussian noise (AWGN) channel taking $\alpha = 0.1$. The quality of the images has deteriorated with the reduction in SNR. The recovered image resembles the original one when SNR is greater than -20 dB.

We hide a black and white watermark in a colored image in Figs. 4 and 5. Then notice the effect of results for different SNR. The impact of α is shown in Figs. 6

Fig. 3 Recovery of RGB watermark under -25 dB channel

Original Image

Watermark Image

Encrypted Watermark

Transmitted Image

Received Image

Recovered Image

Fig. 4 Recovery of grayscale watermark under Fig. 25 dB channel

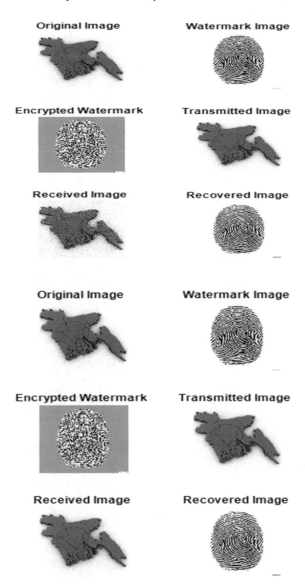

Fig. 5 Recovery of grayscale watermark under 25 dB channel

and 7; here the watermark becomes visualized ($\alpha > 0.5$) on the combined image for a larger value of α; hence, the security of the watermark is destroyed. Recovery of the watermark is improved with the larger value of α shown in Fig. 8 where the cross-correlation coefficient (between original recovered watermark) is plotted against SNR of the channel taking α as a parameter. There is a trade-off between security and recovery of watermark images. Here, we found a clear recovered watermark image even though at high SNR. Though SNR is 40 dB, we successfully

Fig. 6 Recovery of RGB watermark by considering α = 0.7

Original Image

Watermark Image

Encrypted Watermark

Transmitted Image

Received Image

Recovered Image

Fig. 7 Recovery of RGB watermark by considering α = 0.1

Original Image

Watermark Image

Encrypted Watermark

Transmitted Image

Received Image

Recovered image

recovered the watermark image from the encrypted image. Now, we recover the image for different α values.

The watermark is less visualized at a glance for the larger value of α ($0 < \alpha < 0.3$) observed from Fig. 7. The impact of α on the cross-correlation coefficient is more

Fig. 8 Variation of cross-correlation coefficients against SNR on RGB image by considering α as a parameter

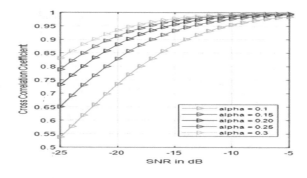

Fig. 9 Comparison between our proposed algorithm and previous methodologies

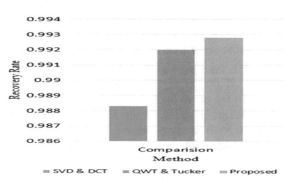

prominent on the RGB image than the grayscale image found in Fig. 8. The variation of cross-correlation coefficients against SNR on RGB image is shown in Table 1. Here, we have compared our research with two other works that have a notable accuracy. One of these methodologies [8] has been applied to SVD and DCT methods [10]. And another one has been applied QWT and Tucker decomposition method. In both research works, Gaussian noise has been used. The comparative result between our proposed algorithm and previous methodologies has been presented in Fig. 9.

Table 1 Variation of cross-correlation coefficient

SNR in dB	Taking different α for RGB image			
	0.1	0.3	0.5	0.7
−60	0.0088	0.0269	0.0447	0.0557
−40	0.1145	0.2657	0.3885	0.4897
−20	0.7360	0.9345	0.9715	0.9840
0	0.9951	0.9993	0.9997	0.9998
20	1.0000	1.0000	1.0000	1.0000
40	1.0000	1.0000	1.0000	1.0000
60	1.0000	1.0000	1.0000	1.0000

5 Comparisons with Existing Methods

Here, we examine the performance of proposed digital watermarking algorithms compared to other watermarking algorithms [8, 10, 17]. Observe the following instances: Most of the existing digital algorithms for watermarking are worked on grayscale images, and almost zero algorithms for watermarking are available. So for finding some related algorithms, the comparison is very difficult. Two evaluation metrics are widely used in this article for comparisons such as SNR dB and cross-correlation coefficient. Estimating the dissimilarity between the watermark and original images is called SNR. As the value of SNR increases, then the encrypted watermark is mostly invisible. On the other hand, the cross-correlation coefficient is to calculate the similarity measurement within the encrypted watermark and the original watermark. Three algorithms are selected in this study with better quality for experiments of contrast. In contrast, therefore with the algorithm in [8], there are more advantages to the proposed algorithm which is that it uses color images for both the original image and watermark image but existing zero-watermarking algorithms uses color images as the original image only and also compared with the correlation coefficient in [8, 10, 17], the proposed algorithm is better to work with different SNR values on RGB image taking α as a parameter rather than existing zero watermark uses NC values under different attacks. However, in most of the cases, the correlation values of our proposed method are higher than existing methods. Both of the above cases show that the approach proposed achieves exceptional robustness and better performance than existing algorithms.

6 Conclusion

In this work, the digital watermarking technique is implemented with a simple algorithm. The result reveals the recovery of base and watermarking images effectively. Here the recovery of the image is verified with cross-correlation coefficients but still, we have the scope to verify it with other statistical parameters and features of an image. We can apply other encrypted algorithms, for example, elliptic curve cryptography, triple DES, blowfish, two fish, AES, etc. instead of RSA. We can compare the model with other watermarking techniques like discrete cosine transform (DCT) and DWT in the context of accuracy and process time. If we use channel coding (e.g., convolutional coding), it will give more immunity to additive white noise. The trade-off between the degree of security watermark and the complexity of the encryption-decryption algorithm (especially process time) can be made to reach an optimum point.

References

1. Patil HM, Rindhe BU (2016) Study and overview of combined NSCT-DCT digital image watermarking. In: 2016 international conference on global trends in signal processing, information computing and communication (ICGTSPICC), pp 302–307
2. Kim M, Li D, Hong S (2014) A robust digital watermarking technique for image contents based on DWT-DFRNT multiple transform method. Int J Multimedia Ubiquit Eng 9(1):369–378
3. Narang M, Vashisth S (2013) Digital watermarking using discrete wavelet transform. Int J Comput Appl 74(20)
4. Bhatt S, Ray A, Ghosh A, Ray A (2015) Image steganography and visible watermarking using LSB extraction technique. In: 2015 IEEE 9th international conference on intelligent systems and control (ISCO), pp 1–6
5. Jaju SA, Chowhan SS (2015) A modified RSA algorithm to enhance security for digital signature. In: 2015 international conference and workshop on computing and communication (IEMCON), pp 1–5
6. Kaur G, Kaur K (2013) Image watermarking Using LSB (least significant bit). Int J Adv Res Comput Sci Softw Eng 3(4)
7. Luo L, Chen Z, Chen M, Zeng X, Xiong Z (2010) Reversible image watermarking using interpolation technique. IEEE Trans Inform Forensics Secur 5(1):187–193
8. Jiang F, Gao T, Li D (2020) A robust zero-watermarking algorithm for color image based on tensor mode expansion. Multimed Tools Appl 79:7599–7614
9. Begum M, Uddin MS (2020) Digital image watermarking techniques: a review. Information 11(2):110
10. Li D, Che X, Luo W, Hu Y, Wang Y, Yu Z, Yuan L (2019) Digital watermarking scheme for color remote sensing image based on quaternion wavelet transform and tensor decomposition. Math Methods Appl Sci 42(14):4664–4678
11. Ahmaderaghi B, Kurugollu F, Rincon J, Bouridane A (2018) Blind image watermark detection algorithm based on discrete shearlet transform using statistical decision theory. IEEE Trans Comput Imaging 1(4):46–59
12. Liu J, Xu Y, Wang S et al (2018) Complex wavelet-domain image watermarking algorithm using L1-norm function-based quantization. Circuits Syst Signal Process 37:1268–1286
13. Makbol NM, Khoo BE, Rassem TH (2018) Security analyses of false positive problem for the SVD-based hybrid digital image watermarking techniques in the wavelet transform domain. Multimed Tools Appl 77:26845–26879
14. Wang C, Wang X, Xia Z, Zhang C (2019) Ternary radial harmonic Fourier moments based robust stereo image zero-watermarking algorithm. Inf Sci 470:109–120
15. Zebbiche K, Khelifi F, Loukhaoukha K (2018) Robust additive watermarking in the DTCWT domain based on perceptual masking. Multimed Tools Appl 77:21281–21304
16. Jaiswal KR, Ravi S (2018) Robust imperceptible digital image watermarking based on discrete wavelet & cosine transforms. Int J Adv Res Comput Eng Technol (IJARCET) 7(2)
17. Wang C, Wang X, Chen X et al (2017) Robust zero-watermarking algorithm based on polar complex exponential transform and logistic mapping. Multimed Tools Appl 76:26355–26376

A Computer Vision Approach for Automated Cucumber Disease Recognition

Md. Abu Ishak Mahy, Salowa Binte Sohel, Joyanta Basak, Md. Jueal Mia, and Sourov Mazumder

Abstract Agriculture is the greatest labor agency in Bangladesh. As our country is an agricultural country, most of the people of Bangladesh depend on agricultural products for their livelihood. Lack of opportunities and facilities, natural disasters and diseases of crops are main obstacles in growth of agricultural sector in our country. Diseases in plants are significant hitch constraints the quality of crops. As important agricultural crops are under threat due to various plant diseases and pests, the quality of the crop can be affected by various diseases. So, it is very necessary to detect the disease at the proper period to the cultivator. When crop disease could be easily track out, it can be effective for monitoring and restrain disease for agriculture and food safety. We have worked with cucumber disease detection in our project because cucumber is a much needed grain in our country. If the cucumber is affected, it will cause a huge damage to the economy of the country. Since cucumber has many qualities, if it is transited, it will have a huge detrimental effect on nutrition. A system called computer vision and machine learning is used to detect crop diseases promptly and accurately which we used to detect cucumber diseases. Diagnosis of plant disease through this technology is very profitable and easy. This is because it reduces huge workload of crop monitoring and it can detect the symptoms of all diseases at a very early stage. This system includes pattern recognition and creating database and classification, approximation, optimization and data clustering. We are going to implement this goal by thinking of the people and helping the farmers to cultivate properly.

Keywords Agricultural crops · Computer vision · Cucumber diseases · Pattern recognition · Classification

Md. Abu Ishak Mahy (✉) · S. B. Sohel · J. Basak · Md. Jueal Mia · S. Mazumder
Department of CSE, Daffodil International University, Dhaka, Bangladesh

© The Author(s), under exclusive license to Springer Nature Singapore Pte Ltd. 2022
P. Nanda et al. (eds.), *Data Engineering for Smart Systems*, Lecture Notes in Networks and Systems 238, https://doi.org/10.1007/978-981-16-2641-8_52

1 Introduction

Everyone directly or indirectly depends on agriculture. According to Food and Agriculture Organization (FAO), about 60% people around the world depends on agricultural sector for their existence. In Bangladesh economy, agriculture is the most significant department [1]. This department provides employment to more than 63% population of this country. Agriculture contributes 19.6% to the nationwide GDP [2]. Cucumber, *Cucumis sativus*, is also a part of this department. It is a herbaceous warm-season crop in the family Cucurbitaceous. It is widely cultivated in our country. In this country, about twenty-two thousand two hundred and eighty-six acres of land is used for cucumber farming. Every year, farmers produce approximately 54,854 metric tons cucumbers all over the country [2]. This cucumber is a great source of vitamin B and good for skin and hair. It also helps to fight and prevent cancer. Cucumber also plays a significant role in weight loss which reduces cholesterol and blood pressure. But disease-infected cucumber causes harmful effects. Cucumber disease have features of severe disease casualty, numerous and fast transit. Agricultural productivity is fully dominated by Bangladeshi economy. Hence, disease detection of plants can play a significant role in the sector of agriculture. This is a special and favorable procedure to spy on an herbaceous disease in the early stages and as an innate disease detection strategy. According to our project, we can find if there are any mentioned diseases on cucumber; alongside, we can also find a fresh cucumber [3]. Physical approach is slow and difficult for identification of these kind of diseases. But automated system method is quite easy and accurate. So, we work with automated system to help the farmers by identifying the diseases before their cultivation and it also saves money.

There have been some related works on disease detection of two cucumber diseases using artificial neural network. In this system, the input will capture the image of the infected crop and trained by neural network. Then find out the features that would be appropriate to diagnose crop diseases. Another approach is machine vision-based papaya disease recognition. They proposed an agro-medical system which can recognize diseases from the image of papaya. This proposed system is almost in real-time system. The disease classification has been sourced with support vector machine. Their project attained 90.15% accuracy. Computer vision-based local fruit recognition, which is an expert system based on computer vision that is working on captured pictures and recognize what fruits this is. The accuracy of their system is 94.61% which is quite well. The other approach was cucumber disease recognition which is based on (SVM) support vector machine and image processing system. The system of cucumber disease detection based on color feature is more accurate and faster.

In our research paper, the computer vision system is raised. It works on capturing images through mobile device and can be able to identify which disease the cucumber is affected with. We worked on four types of cucumber diseases as well as fresh cucumber, those diseases are Belly rot, Pythium fruit rot, Scab cucumber and Mosaic virus. We collected images then trained them and then extracted the features from images by clustering. We have extracted 500 data features through clustering. From the previous research papers, it can be concluded that our project will be very up to date and, it has an excellent accuracy level which is the most needed in case of making a successful research paper.

2 Literature Review

A system is proposed by Pawar et al. [4] for the identification of two cucumber diseases using artificial neural network. Those diseases are: downy mildew and powdery mildew. They insert an input to this system, and this input will be used to capture the image of the infected crop. Those images need to be trained through neural network. Then they find out the features that would be appropriate to diagnose crop diseases. Image features usually include color, size and texture in image processing method. They have used ANN which will give symptoms of disease for diagnosis and treatment of cucumber. Another approach is machine vision based by Habib et al. [5]; papaya disease recognition where they proposed an agro-medical system which can recognize diseases from the image of papaya. They set a feature for diseases recognition. To substance the characteristics of images, here image processing method is used. They used support vector machine (SVM) for the classification of diseases. This proposed system is almost in real-time system. The disease classification has been sourced with support vector machine. Their project attained 90.15% accuracy, which is excellent as well as pledge, where there remains a dynamic future conduct with a huge data set of pictures with wide range of this papaya disease. Another approach is computer vision-based local fruit recognition by Mia et al. [6] which is an expert system based on computer vision that is working on captured pictures and recognizes what fruits this is. In the midst of endemic fruits, they choose six fruits called amla, sugar-apple, bilombo, elephant apple, orboroi and sapota for experiment. They perform a preprocessing method on those images, and they use segmentation for extracting the features of images. After that, support vector machine (SVMs) algorithm is used for classification. An expert system is called machine vision for recognition of fruit. They used two feature set and that consists of ten features. To extract the features, image processing system is used. Their system accuracy level is 94.61% which is good enough. Another is an in-depth exploration of automated jackfruit disease recognition abstract by Habib et al. [7] where they used an agro-medical expert system, which results with digital images that recognize diseases. At first, they select a feature. A segmentation is k-means clustering is used to detect the disease affected area of the images of jackfruit and extracting nine classifiers separately in the features, and image processing technique is used to do this. For classification of the diseases, the classifiers of the index are seven eminent metrics. Random forest is got of all other classifiers which attains 90% accuracy. They attempt to detect the jackfruit disease with strong features into classify jackfruit diseases. In order to simplify future research, an in-depth classification of jackfruit disease recognition is used. Their approached project is very competitive. The other approach by Youwen et al. [8] was cucumber disease recognition which is based on (SVM) support vector machine and image processing system. At first, they used vector median filter to remove cucumber disease leaf color images. Then the pattern of statistics and the morphology of mathematics were introduced in the images of segment. Finally, spot on leaves and the color properties were extracted. The system of detecting cucumber disease is based on color feature is more accurate and faster. Another related work by Sharmin et al. [9] is a local fish recognition. It is a machine vision-based project. Then they converted images in gray-scale and they formed gray-scale histogram. Then the image segmentation took place where they used histogram-based method. Then after calculation, SVM gives them the accuracy of 94.2%. Another one is registration status prediction by Mia et al. [10] using machine learning for students of private university. They have done this for saving private university from direct marketing strategy in terms of budget and sustainability. They applied a total of seven classifiers. They have used a data set of a thousand private university students in Bangladesh. They have got a top accuracy of 85.76%.

3   System Architecture

The system architecture of a computer vision-based expert system for detecting cucumber disease is shown in Fig. 1. For example, a farmer or a gardener is deduced to capture an image of an apprehensively disease-attacked cucumber with a cellphone or other handheld device, in which the mobile application of the expert system has

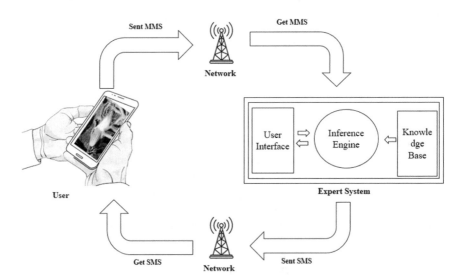

Fig. 1 The architecture of computer vision based system for cucumber disease recognition

already been installed. The entire expert system consists of some integrated components that are related among themselves. They are an inference engine, a knowledge base acquirement facility and an explanation facility. The inference engine handles the entire workflow inside the expert system. It is the part of the expert system which delivers predictions, suggestions and answers in a way similar to a human expert. Here is our system architecture shown in Fig. 1. Through our project, we have created a model that can detect and identify four types of cucumber disease and also fresh cucumber and extract features from the pictures through clustering. To implement this project, some steps need to be taken such as image collection, preprocessing, clustering data, extraction of features classification, confusion matrix and diagnosis.

4 Research Methodology

4.1 Approach Followed

We have taken 500 color image data set. This data set consists of diseased cucumber and fresh cucumber images. So, transformation into a fixed-size image from an arbitrary-sized image takes place. Then, the contrast in image is increased by using an image processing technique. Then, the histogram-equalized image or processed image goes through a transformation of RGB color space into L*a*b* color space. Then, this transformed image is segmented out with k-means clustering [6]. We select the first feature as first-order statistical process to detect diseases or freshness in cucumber. It predicts features of separate pixel values by waving the interaction between picture pixels. The main characteristics of first-order statistical process are mean, standard deviation, kurtosis, skewness and variance. The GLCM is used for extracting second-order texture data from given images. It is a well-set device. It is a kind of matrix in which the number of columns and rows of the matrix are equal to the pixel values of given images of the texture. This matrix narrates the frequency of relationship between two gray levels within the area of supervision [6]. In GLCM, there are 14 characteristics, and among them, the most effective characteristics are contrast, correlation, energy, homogeneity and entropy. Here the flowchart of our approach is given in Fig. 2.

Confusion matrix is a method of calculating functionality for the classification of machine learning. It helps to know the functionality of the given classification model on a set of given test data. As we have five types of data sets for our project, we need to take dimensions of $(n * n)$ where $n = 5$, which means we take five class problem. So, for a confusion.

Matrix of five-class problem summarizes the number true positives (TP), true negatives (TN), false positives (FP) and false negatives (FN). [6] Now these values can be calculated as shown in Eqs. (1)–(4).

$$TP = a_{ii} \tag{1}$$

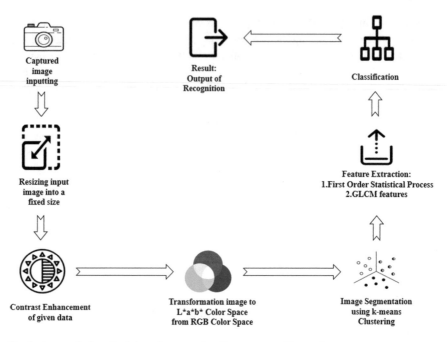

Fig. 2 Approached methodology for cucumber disease recognition system

$$FP_i = \sum_{j=1, j \neq i}^{n} a_{ji} \tag{2}$$

$$FN_i = \sum_{j=1, j \neq i}^{n} a_{ji} \tag{3}$$

$$TN_i = \sum_{j=1, j \neq i}^{n} \sum_{k=1, k \neq i}^{n} a_{jk} \tag{4}$$

After calculating these values, we need to calculate the accuracy, precision, specificity, sensitivity, false negative rate (FNR) and false positive rate (FPR) of the system by using those values of confusion matrix. We divided our data into two individual data sets. One of them is a training data set and another one is a testing data set. We have kept the highest amount of data in our training data set. We can measure the performance of the system in phrase of these given metrics with the help of those test data sets [5]. The rules of calculating the percentage values of accuracy, specificity, sensitivity, precision, FNR and FPR are given in Eqs. (5)–(10).

$$\text{Accuracy} = \frac{TP + TN}{TP + TN + FP + FN} \times 100\% \tag{5}$$

$$\text{Precision} = \frac{TP}{TP + FP} \times 100\% \tag{6}$$

$$\text{Specificity} = \frac{TN}{FP + TN} \times 100\% \tag{7}$$

$$\text{Sensitivity} = \frac{TP}{TP + FN} \times 100\% \tag{8}$$

$$FNR = \frac{FN}{FN + TP} \times 100\% \tag{9}$$

$$FPR = \frac{FP}{FP + TN} \times 100\% \tag{10}$$

We calculate the total performance of classifiers with the help of given equation of evaluation matrix are shown in Eqs. (11)–(20). We have taken eight classifiers and calculated the average values of those equations for our all five-class problems.

4.2 Description of Cucumber Disease

Mosaic Virus. Due to the cucumber mosaic virus (CMV), cucumber mosaic happens. CMV is the most destructive virus which is widely contagious in temperate as well as in tropical areas. Symptoms come into sight 7–14 days after tainting at the temperatures of 79–89°F. Light- to dark-green spots appear first. Infected cucumbers are smaller and formless shows mosaic patterns on aril with light as well as dark green color [11].

Belly Rot. Due to the soil borne fungus *Rhizoctonia solani*, belly rot is produced. The disease is deadly to cucumber since the germs can attack all the cucurbits. Rhizoctonia solani is a very common soil inhabitant which invades cucumbers in contact with soil. This disease can ensue over (46–95 °F) temperatures, but greater damaging portion is near 80 °F. Belly rot symptoms ensue on underside and blossom end which produces water-drabble, tan to brown bruise [12].

Scab. *Cladosporium cucumerinum* fungus is the reason for the scab disease of cucumber. Cucumber spots maybe white to gray and may have a yellowish tinge. As the age of infection, leaf spot falls out the center. The scab travels as a seed transmitted over infected seed or moist air motion. Disease spreads in cold (around 70 °F) and wet climates [13].

Pythium Fruit Rot. Pythium fruit rot is the result of a few fungal organisms. Pythium spp. are found in most, if not all, soils of the world. When cucumber contacts with wet ground, those infection occurs. The Pythium fruit rot pathogens are rapidly propagated among the fields in watering. The symptoms of the disease begin as tawny, water-soaked blowing that speedily become massive, watery and limp and putrefied. This disease makes the fruit as unsalable [14].

(a)   (b)

(c)   (d)

Fig. 3 Cucumber crop disease. **a** Mosaic virus. **b** Belly rot. **c** Scab disease. **d** Pythium fruit rot (Cottony leak)

All of these diseased cucumbers are shown in Fig. 3.

4.3 Features

In our project, we select the first feature as a first-order statistical process to detect diseases or freshness in cucumber. It predicts features by separate pixel values by waving the interaction between picture pixels. The main characteristics of first-order statistical process are mean, standard deviation, kurtosis, skewness and variance. Their equations are given below [4]:

Mean. Mean is the average equilibrium of redundancy of the image or texture. Mean as defined in Eq. (11) as:

$$\mu = \frac{1}{n} \sum_{I=1}^{n} x_i \tag{11}$$

Standard Deviation. This statistic is used to determine variability. It shows how much variation or scattering exists from the average. It indicates a low deviation that points to a high standard deviation that tends to close to the mean and those data is scattered over the points. Standard deviation is defined as in Eq. (12) as:

$$\sigma = \sqrt{\frac{1}{n}\sum\nolimits_{i=1}^{n}(x_i - u)^2} \tag{12}$$

Kurtosis. Kurtosis is the rule of "tailed ness" of the chance of ordering random value from a real value. Kurtosis is defined as in Eq. (13) as:

$$K = E\left[\left(\frac{x - \mu}{\sigma}\right)^4\right] \tag{13}$$

Skewness. Skewness is used to measure a data set's symmetry. Here, normal distribution is 0 and symmetric which have skewness as zero. Skewness is defined as in Eq. (14) as:

$$\gamma = E\left[\left(\frac{x - \mu}{\sigma}\right)^6\right] \tag{14}$$

Variance. In statistical process, variance measures how far the set values are spread out and their average values. Variance is defined as in Eq. (15) as:

$$\sigma^2 = \frac{\sum_{i=1}^{N}(GS_i - \mu)^2}{N} \tag{15}$$

The GLCM is used for extracting second-order texture data from given images. It is a well-set device. It is a kind of matrix in which the number of columns and rows of the matrix are equal to the pixel values of given images of the texture. This matrix narrates the frequency of relationship between two gray levels within the area of supervision [6]. In GLCM, there are fourteen characteristics, and among them, the most effective characteristics are contrast, correlation, energy, homogeneity and entropy. Their equations are given below:

Contrast. Contrast measures the intensity of a pixel of the image. It is fixed with the difference in the color and luster of an object. Contrast is defined as in Eq. (16) as:

$$C = \sum\nolimits_{i=0}^{G-1}\sum\nolimits_{j=0}^{G-1}(i - j)^2[p(i, j)] \tag{16}$$

Correlation. It is the method of measuring the gray level's linear dependency from its neighboring pixels. Correlation is defined as in Eq. (17) as:

$$\rho = \sum_{i=0}^{G-1} \sum_{j=0}^{G-1} \frac{ij\,P(j, j) - \mu_{XYy}}{\sigma_x \sigma_y} \qquad (17)$$

Energy. It measures the limitation of repetitions of pixel pairs. It also measures the uniformness of an image. Hence, pixels are similar and energy value would be large. Energy is defined as in Eq. (18) as:

$$E = \sum_{i=0}^{G-1} \sum_{j=0}^{G-1} [P(i, j)]^2 \qquad (18)$$

Homogeneity. IDM is homogeneity. IDM feature is obtained to measure the local homogeneity and measures the closeness distribution of GLCM elements to GLCM diagonal. It has a large range of values to determine whether the picture is textured or not textured. Homogeneity is defined as in Eq. (19) as:

$$H = \sum_{i=0}^{G-1} \sum_{j=0}^{G-1} \frac{P(i, j)}{1 + (i - j)^2} \qquad (19)$$

Entropy. Entropy characterizes the texture of an input image. When the elements of the matrix are same, entropy's value will be maximum. Entropy is defined as in Eq. (20) as:

$$S = -\sum_{i=0}^{G-1} \sum_{j=0}^{G-1} P(i, j) \log[p(i, j)] \qquad (20)$$

5 Experimental Evaluation

Recognition of cucumber diseases using computer vision system is shown in Fig. 4. Firstly, we have grabbed 500 color pictures of four diseased cucumber which are captured with different view of points and size. These pictures have been sent to a proposed system. These data sets are made to train with machine learning; MATLAB is used for machine learning toolbox. The size of the images has been changed into 350 × 300 pixels and completed with contrast enhancement. Then, k-means clustering is used to extract all the features from these images. In this case, the images are in the form of three cluster. Then, we select the first best and the other best cluster from the three cluster of images. And through this cluster, there are thirteen features of cucumber diseases that came out. We have taken 500 data features through clustering. We extract the features to detect the diseases of cucumber.

In our project, we extract the feature from two statistical feature process: first-order statistical feature and second-order GLCM feature. The main characteristics of first-order statistical features are mean, standard deviation, kurtosis, skewness and

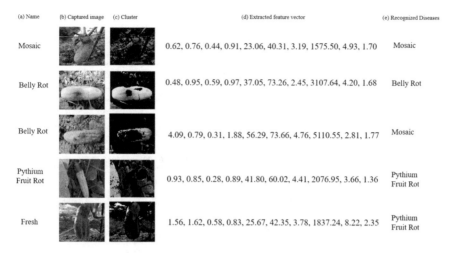

(a) Name	(b) Captured image	(c) Cluster	(d) Extracted feature vector	(e) Recognized Diseases
Mosaic			0.62, 0.76, 0.44, 0.91, 23.06, 40.31, 3.19, 1575.50, 4.93, 1.70	Mosaic
Belly Rot			0.48, 0.95, 0.59, 0.97, 37.05, 73.26, 2.45, 3107.64, 4.20, 1.68	Belly Rot
Belly Rot			4.09, 0.79, 0.31, 1.88, 56.29, 73.66, 4.76, 5110.55, 2.81, 1.77	Mosaic
Pythium Fruit Rot			0.93, 0.85, 0.28, 0.89, 41.80, 60.02, 4.41, 2076.95, 3.66, 1.36	Pythium Fruit Rot
Fresh			1.56, 1.62, 0.58, 0.83, 25.67, 42.35, 3.78, 1837.24, 8.22, 2.35	Pythium Fruit Rot

Fig. 4 Experimental outcome. **a** Name. **b** Image. **c** Segmented image. **d** Extracted feature vector. **e** Recognized diseases

variance. The GLCM is used for extracting second-order texture data from given images. In GLCM, there are fourteen characteristics, and among them, the most effective characteristics are contrast, correlation, energy, homogeneity and entropy. To calculate functionality for classification of machine learning, confusion matrix is used. Then we calculate the total performance of classifiers such as accuracy, precision, specificity, sensitivity, FNR (false negative rate) and FPR (false positive rate) of the system. Our experimental output is shown in Fig. 4. Finally, we can say that, our expert system gives better outcome.

To achieve the desired result, we need these classifier values. After all the calculations, we have got our desired result. We got 82.78% accuracy which is a good percentage. As we took just 500 data sets, we got this accuracy. We will increase our data sets. Then our accuracy level will be increased. Then we have done some experimental steps to find some output. The experimental output in given Fig. 4.

We have discussed about confusion matrix and its equations above. Now we have calculated all confusion matrix values (TP, FN, FP and TN) and equations (accuracy, specificity, sensitivity, precision, FNR and FPR) for every classifier. Here is an example for confusion matrix of a classifier given Table 1. The average values for every classifier is given in Table 2.

6 Conclusion and Future Work

In our project, there are thirteen features and image processing system should be applied to extract all the features in this regard. We have got our outcome based on these features and attain a desired output. But our departure to complete our project

Table 1 Class-wise confusion matrix values of random forest

Class	Accuracy (%)	Precision (%)	Specificity (%)	Sensitivity (%)	FNR (%)	FPR (%)
Mosaic	74.85	49.82	72.34	65.39	38.61	27.66
Belly rot	78.92	49.94	85.57	55.69	48.32	19.43
Scab	89.50	52.94	94.52	40.91	59.09	5.48
Pythium fruit rot	91.20	37.04	94.33	27.78	72.22	5.67
Fresh	85.14	57.56	91.11	45.46	54.55	8.89

Table 2 Average values (accuracy, specificity, sensitivity, precision, FNR and FPR) for every classifier

Classifier name	Accuracy (%)	Precision (%)	Specificity (%)	Sensitivity (%)	FNR (%)	FPR (%)
Random forest	82.78	49.66	87.57	45.94	54.56	13.73
Random committee	80.64	48.09	86.37	45.12	54.88	13.63
IBK	78.69	43.66	86.24	43.58	56.42	13.76
K Star	78.45	43.33	86.12	43.71	56.29	13.88
Random subspace	77.26	46.92	84.38	34.66	65.34	15.62
Bagging	76.91	41.71	84.55	37.50	62.49	15.45
Random tree	76.91	39.32	85.07	39.30	60.70	14.93

was not so easy and simple as we have had to face many obstacles and even many challenges to come to this positive result. By analyzing the work of this project, we can say that our system is up to date and work well enough. The accuracy level of our project is 82.78% and which is quite well. The classification of cucumber diseases will make a huge contribution in science as well as the agricultural sector. We have seen development of computer vision nowadays and seen many problems too. Which problems cannot be solved yet to achieve higher accuracy for existing predictions. What we have achieved as the final outcome can be further improved through a lot of data collection and increasing data training. So, it can be said that this cucumber disease detection has become a much more promising future with a huge data set. Hopefully in the future, we will be able to work with a huge amount of data set about diseases of cucumber for performing the best experiment.

References

1. EXPO net. http://www.expo2015.org/magazine/en/economy/agriculture-remains-central-to-the-world-economy.html. Accessed 6 Aug 2020
2. Nations Ecyclopedia. https://www.nationsencyclopedia.com/economies/Asia-and-the-Pacific/Bangladesh-AGRICULTURE.html. Accessed 6 Aug 2020
3. PlantVillage. https://plantvillage.psu.edu/topics/cucumber/infos. Accessed 6 Aug 2020
4. Pawar P, Turkar V, Patil P (2016) Cucumber disease detection using artificial neural network. In: 2016 international conference on inventive computation technologies (ICICT), vol 3. IEEE, pp 1–5
5. Habib MT, Majumder A, Jakaria AZM, Akter M, Uddin MS, Ahmed F (2020) Machine vision-based papaya disease recognition. J King Saud Univ-Comput Inform Sci 32(3):300–309
6. Mia MR, Mia MJ, Majumder A, Supriya S, Habib MT (2019) Computer vision based local fruit recognition. Int J Eng Adv Technol 9(1):11, October 2019
7. Habib MT, Mia MJ, Uddin MS, Ahmed F (2020) An in-depth exploration of automated jackfruit disease recognition. J King Saud Univ-Comput Inform Sci
8. Youwen T, Tianlai L, Yan N (2008) The recognition of cucumber disease based on image processing and support vector machine. In: 2008 congress on image and signal processing, vol 2. IEEE, pp 262–267
9. Sharmin I, Islam NF, Jahan I, Joye TA, Rahman MR, Habib MT (2019) Machine vision based local fish recognition. SN Appl Sci 1(12): 1529
10. Mia MJ, Sattar A, Biswas AA, Habib MT (2019) Registration status prediction of students using machine learning in the context of private University of Bangladesh. Int J Innov Technol Exploring Eng (IJITEE) 9(1):November 2019. ISSN: 2278-3075
11. Mosaic Virus Diseases of Cucumber. https://www.seminis-us.com/resources/agronomic-spotlights/mosaic-virus-diseases-of-cucumber/ Accessed 22 Sept 2020
12. Rhizoctonia solani (belly rot). https://wiki.bugwood.org/Rhizoctonia_solani_(belly_rot). Accessed 25 Sept 2020
13. Scab of Cucurbits. https://extension.umn.edu/diseases/scab-cucurbits. Accessed 25 Sept 2020
14. High Plains Integrated Pest Management. https://wiki.bugwood.org/HPIPM:Pythium_Fruit_Rot. Accessed 25 Sept 2020

An Automated Visa Prediction Technique for Higher Studies Using Machine Learning in the Context of Bangladesh

Asif Ahmmed, Tipu Sultan, Sk. Hasibul Islam Shad, Md. Jueal Mia, and Sourov Mazumder

Abstract Nowadays in Bangladesh, one of the most common scenarios for our students is moving to other countries for higher studies. In order to succeed, students need to pick the right direction before applying for a higher education visa. This article aims to implement which student's visa will get approved or rejected for their higher studies abroad by using machine learning. Throughout this analysis, we predict the visa for higher studies based on student data. Then we process the data (such as cleaning, transformation, integration, standardization, and feature selection). We used multiple classification methods later on, i.e., C4.5 (j48), k-NN, naive Bayes, random forest, SVM, neural network, for classifying these models. Depending on the outcome of the study, it was observed that the accuracy, confusion matrix, and other factors of a random forest classifier are more compatible than others. GRE score and undergraduate CGPA, are two of the most significant attributes that rely on deciding the success of visa acceptance for higher studies also discovered.

Keywords Higher studies · Machine learning · Classification techniques · Confusion matrix · Random Forest

1 Introduction

At this time, getting a visa is a very tough process for a student who wants to apply for their higher studies in abroad. Many students in Bangladesh apply for a visa, but many of them get a rejection because of lacking analysis of their previous academic or non-academic work. Studies abroad are a dream for most of the students in a developing country.

For the Scholarship programs, the Bangladesh Government has already taken many steps so that students can easily apply for higher studies abroad. Bangladesh Government recently interconnected many developed countries to help Bangladeshi students get visas for higher studies. The latest report from the UNESCO Institute

A. Ahmmed (✉) · T. Sultan (✉) · Sk. Hasibul Islam Shad1 (✉) · Md. Jueal Mia1 (✉) · S. Mazumder (✉)
Department of CSE, Daffodil International University, Dhaka, Bangladesh

for Statistics indicates that in 2017, 60,390 Bangladeshis pursued higher education abroad. A total of 34,155 Bangladeshis enrolled in universities in Malaysia, 5,441 in the United States of America, 4,652 in Australia, 3,599 in the United Kingdom in 2017, Canada in 2028, Germany in 2008, India in 1099, Saudi Arabia in 870, Japan in 810, and the United Arab Emirates in 637, according to UNESCO records [1].

The goal of this work is to create a model of classification to classify which students have a chance to accept their visas and which do not. We evaluate their profiles before jumping into this program using their academic results, job experience, research papers, and IELTS–TOFEL, GRE–GMAT scores, etc. To predict their visa, it is always important to know about their previous academic result. IELTS is the internationally recognized test most generally used and approved for that [2]. To make our proposed classifier more specific in this work, in addition to an academic performance, we have collected data from different universities in Bangladesh which includes student's academic results, job experience, and research experiences they have already faced. Firstly, we preprocessed our data set in different steps, such as cleaning, transformation, integration, standardization, and feature selection [3]. Afterward, we further labeled the dataset and built a classification model and applied different types of algorithms to predict the student's visa YES or NO for higher studies.

Selecting the right study track for an abroad study is very significant for every student because this is something that determines the academic and professional achievement of a student. So, students should have clear knowledge about these things.

2 Related Works

A very few papers have been published in the recent years in which some works like ours have been performed using machine learning. But a lot of methods have been developed over the past few years using machine learning and classification.

Biswas et al. [4] proposed the enrollment and dropout of students using the machine learning classifier in the post-graduation degree. Seven classifier rules have been applied, including naive Bayes, random forest, J48, logistic, JRip, multilayer perceptron, and SVM. They considered SVM to be the highest accuracy at 85.76% and 79.65%, while random forest reached the lowest. Mia et al. [5] proposed the case of the Private University of Bangladesh, registration status prediction of students using machine learning. Seven classifiers rules have been applied, including naive Bayes, multilayer perceptron, locally weighted learning (LWL), random forest, random tree, and part are applied in this context. They considered LWL to be the highest at 86.36% accuracy, while random tree reached the lowest at 74.24%. Nilashi et al. [6] proposed a recommendation system for the tourism industry using cluster ensemble and prediction machine learning techniques. As a technique for minimizing dimensionality, they use adaptive neuro-fuzzy inference systems (ANFIS) and principal component analysis (PCA), support vector regression (SVR), and self-organizing map (SOM) and expectation–maximization (EM) as two well-known clustering methods. Their studies conclude that cluster sets can have greater predictive precision in comparison to methods that focus entirely on single clustering techniques for the proposed recommendation system. Ahmed et al. [7] proposed a recommendation system which are the factors of contributing programming skill and CGPA as a CS graduate by using mining educational data. They have applied the decision tree, support vector machine, and naive Bayes classifier algorithm on student's academic results. They considered LWL to be the highest at 86.36% accuracy, while Random Tree reached the lowest at 74.24%. Kurniadi et al. [3] proposed a recommendation system for predicting scholarship recipients using k-nearest algorithm. They apply the k-nearest neighbor (k-NN) algorithm model for forecasting. In forecasting students who are most likely to earn the scholarship, the k-NN algorithm with the highest precision score of 95.83%. Goga et al. [8] proposed a recommender for improving student's academic performance. They first gathered student enrollment records from Babcock University, Nigeria, and then developed models using multi-layer perception learning algorithms based on classification trees and WEKA. Random tree was adopted as the best algorithm in the domain of this analysis and this generalized method has represented as a building block. Yukselturk et al. [9] proposed the dropout students classify, four data mining approaches neural network, naive Bayes, k-NN, and decision tree. They used the tenfold approach of cross-validation at the time of implementation. Detection sensitivities for k-NN, NB, NN, and DT were 87%, 76.8%, 73.9%, and 79.7%, respectively.

3 Methodology

The methodology section is organized by the four subsections namely, (A) Dataset Description, (B) Classifier Description (C) Confusion Matrix and Classifier Evaluation Metrics, (D) Implementation Procedure. The details of these four subsections are presented below.

3.1 Dataset Description

Researchers from various regions find that it is very important for higher study to predict the GRE score of students or the CGPA graduation program. They considered the academic outcome of students, IELTS ranking, city, agency, manage fund, work experience, job experience, and some more data to forecast their final success and create the model classification that will help them make the decision they should follow for the visa Prediction in the higher study program [10]. Even more so than literature, with its immeasurable success and the kind of obstacles it faces, we found the only undergraduate program as our aim of the study. In fact, in higher research, anyone who is not able to take on such difficulties is sure to suffocate. It is also very important to calculate GRE and CGPA for higher analysis [11]. So, using various regression algorithms, we create a model to forecast CGPA and GRE Score to build a prediction model based on 20 attributes, including their academic achievement, work and study background, thesis release, personal experiences of current students with similar academic achievements [12].

There are a few steps in our proposed method:

1. Data collection
2. Pre-processing

   I.   Data cleaning.
   II.   Transformation.
   III. Integration.
   IV.   Standardization.
   V.   Feature selection.

3. Data mining.

3.2   Classifier Description

In this work, based on the academic and non-academic outcomes of students, we predict visas for higher studies. We process those data (like cleaning, transformation, integration, standardization, feature selection) [13]. Later we used different classification techniques, i.e., C4.5 (j48), *k-NN*, naive Bayes, random forest, SVM, neural network to classify these profiles. Providing their definition below.

Random Forest is an assembly operation. It runs with a myriad of decision trees. The decision tree algorithm and tree bagging are used but the distinction is that overlearning is used [14].

SVM is an algorithm for supervised machine learning that can be used for both classification and regression problems. It uses a technique called the kernel trick to transform your knowledge and then finds an optimal boundary between the possible outputs based on these transformations [15].

Neural Network is a sequence of algorithms that, through a mechanism that mimics the way the human brain works, aims to identify fundamental associations in a collection of data. Neural network applies to neuron structures, either biological or artificial in nature, in this context [16].

Decision Trees is an algorithm based on the classification of both numeric and nominal groups. A decision tree can be used to visually and explicitly represent decisions and decision making. In data mining, a decision tree describes data (but the resulting classification tree can be an input for decision making) [17].

K-nearest Neighbor is the shortest supervised algorithm. It is a learning algorithm that is not parametric and lazy. Data sets are typically divided into several groups, and k-NN's task is to learn from these training datasets and forecast future data [18].

Naive Bayes lies in the algorithm of a probabilistic class. Suppose, in a sample, there are 20 independent variables. Only one variable at a time is taken into account by naive Bayes. Not only is it an algorithm, but it also refers to a complete set of algorithms [19].

3.3 Confusion Matrix and Classifier Evaluation Metrics

The confusion matrix is a method used by a classification algorithm to display the number of instances correctly predicted or incorrectly predicted [20]. To evaluate the consistency of a classifier, this matrix can be used. This is where we are working on the two-class problem [6]. It shows the true positives, true negatives, false positives, and false negatives in the case of the two-class issue. The following words apply to this work:

TP = True Positive = Predicted as a positive and initially a positive class member

TN = True Negative = Predicted as a negative and initially a negative class member

FP = False Positive = Predicted as a positive but initially a positive class member

FN = False Negative = Predicted as a negative but initially a negative class member

Multiple standard terminologies for confusion matrix evaluation for two classes

$$\text{Accuracy} \quad ACC = \frac{(TF + TP)}{(TP + FN + FP + TN)} \tag{1}$$

$$\text{Recall or Sensitivity or True Positive rate} \quad TPR = \frac{TP}{(TP + FN)} \tag{2}$$

$$\text{True Negative rate or Specificity} \quad SPC = \frac{TN}{(FP + TN)} \tag{3}$$

$$\text{False Positive Rate} \quad FPR = \frac{FP}{(TN + FP)} \tag{4}$$

$$\text{False Negative Rate} \quad FNR = \frac{FN}{(FN + TP)} \tag{5}$$

$$\text{Precision} \quad PPV = \frac{TP}{(FP + TP)} \tag{6}$$

$$\text{F - Measure or F1 Score} \quad \text{F1} = \frac{2\text{TP}}{(2\text{TP} + \text{FN} + \text{FP})} \tag{7}$$

3.4 Implementation Procedure

The whole process is known as classification. At first, we have collected the data by the survey. Most of the data is collected from university students who already got the visa or were rejected. Then we have built up a database. We perform the data preprocessing on our dataset. Data preprocessing includes data selection, data cleaning, and data transformation. If there is any missing or error value in data, then we rechecked the data for better preprocessing. We labeled the data for our experiment purpose. After labeling the data, we split our data into two parts, namely the train and test data set. We have used 1215 students' data for the experiment. Train data set contains 815 students' data which is 67% of the total data. Test data set contains 400 (33%) data. We have used different types of algorithms such as random forest, SVM, k-NN, naive Bayes, decision tree, neural network for finding the accuracy, and after that, we found the best accuracy algorithm to average all algorithms. We have used confusion matrix because the performance of classification models is usually evaluated by a confusion matrix (Fig. 1).

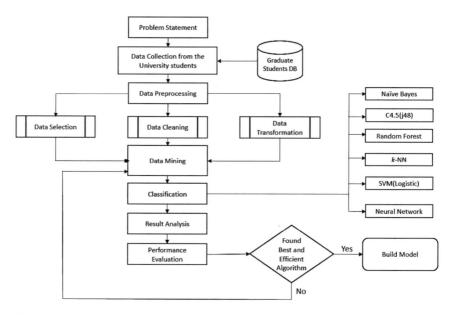

Fig. 1 Working procedure of visa prediction

These are Random forest, naive Bayes, k-NN, SVM, decision tree, and neural network and we took the best six. Of all the rules, the best accuracy and the best model is random forest [21].

4 Result and Discussion

Performance of classification models is usually evaluated by a confusion matrix. The confusion matrix contains information from a classifier based on the original and predicted classification.

Table 1 Six classifiers of the confusion matrix

Classifier name	True positive	False negative	False positive	True negative
C4.5(J48)	192	10	8	190
Naive Bayes	180	8	20	192
Neural network	188	9	12	191
Random forest	194	6	6	194
SVM (poly kernel)	187	8	13	192
k-NN	187	7	13	193

About 400 student activities are placed in the test set over the deployment period, where the substantial acceptance of 200 visas is yes or approved. Otherwise, the significant visa status of 200 students is either non-existent or rejected [22]. After implementation, for all the classifiers mentioned in Table 1, we have initiated a confusion matrix. At present, for the competent classifier and worst classifier, we will explain in depth the speculative outcome of the uncertainty matrix.

From Table 1, it is understood that the random forest classifier is precisely able to predict that among 200 students, 194 students will obtain their visa. So, the 200 applicants, the remaining 6 students are inaccurately classified as refusing their visas. In comparison, this classifier can correctly predict that among 200 students, 194 students will not get their visas [23]. So, of the 200 candidates, the remaining 6 students are incorrectly classified that they are going to register.

Table 2 Comparison of seven classifier's results based on six performance evaluation metrics

Classifier name	Accuracy (%)	Precision (%)	Recall (%)	F-measure (%)	ROC area (%)	AUC (%)
C4.5(J48)	95.50	95.50	95.50	95.50	94.70	93.50
Naive Bayes	93.00	93.20	93.00	93.00	96.50	98.20
Neural network	94.75	94.80	98.40	94.70	98.00	97.70
Random forest	97.00	97.00	97.00	97.00	99.40	98.50
k-NN	95.00	95.00	95.00	95.00	95.90	96.70
SVM (poly kernel)	94.00	94.80	94.80	94.70	94.80	98.90

From Table 2, random forest achieved greater correctness than all classifiers, and of all classifiers, naive Bayes obtained the lowest accuracy. Random forest's precision is 97%, which is decent enough. Naive Bayes then obtained the lowest precision of 93%, which is sufficiently poor for random forest assimilation. Naive Bayes 93%, which is less than random forest 97%, which is above all, neural network and SVM are the same 94.75% and C4.5 95.5% [24]. So, the best is the random forest model since it is above all.

The relationship between therapeutic sensitivity and accuracy for any potential cut-off is seen by the ROC curve. The curve of the ROC is a graph with:

$$\text{The x - axis showing } 1 - \text{Specificity} = \text{False Positive Fraction}$$

$$= \frac{FP}{FP + TN}. \qquad [13]$$

A perfectly skilled model is described by a line moving from the bottom left of the plot to the top left and then from the top right to the top right [7]. For the final model, an operator may trace the ROC curve and choose a threshold that creates a desirable ratio between false positives and false negatives [25].

From this, we can see that in every aspect, the random forest provides the best outcomes. We will now see in Fig. 2 the classification outcome and ROC curve analysis.

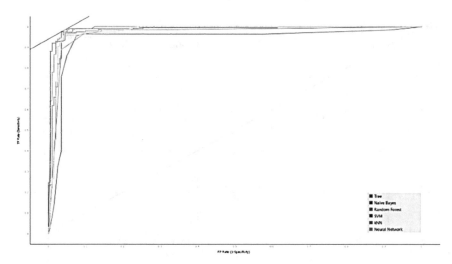

Fig. 2 ROC curve for student visa

Here we have discussed our questionnaire, sorted variable list for the transformation and preprocessing of data. We have found that attributes in the section personal experience are mostly different for all. But for separate characteristics, there is no class distinction between males and females.

From the frequency table, we came to know that the variable value is not the same. Also figured out the importance. Our prediction model can be used for Bangladeshi students who want to get a visa in scholarship. Day by day, technology is changing. Often the method of schooling. In this predictive model, a lot of development can be achieved. Data from other significant facets, such as Bachelor of Medicine, pharmacy, electrical engineering, etc., can be taken from the model.

5 Comparative Analysis

Table 3 shows that our accuracy rate is higher than the work of the others. We can see in this table that each data size is almost the same, and all the machine learning techniques used are almost the same. Our best classifiers are random forest with 97% accuracy and Mia et al. [6] and Ahmed et al. [4] are the best classifiers in the other case, SVM. This is almost a higher degree of precision.

Table 3 Results of the comparison of our work and others' works

Method	Object(s)	Data size	Technique used	Algorithm	Best classifiers	Accuracy (%)
This work	Students	400	Machine learning	SVM, random forest, neural network, etc.	Random forest	97.00
Kurniadi et al. [3]	Students	434	Machine learning	k-NN	k-NN	95.83
Biswas et al. [4]	Students	501	Machine learning	SVM, random forest, neural network, etc.	SVM	86.00
Mia et al. [5]	Students	344	Machine learning	SVM, naïve Bayes, logistic, J48, etc.	SVM	85.76
Ahmed et al. [7]	Students	455	Machine learning	SVM, random forest, neural network, etc.	SVM	89.00

6   Conclusions

It has been shown in this study that the random forest algorithm can work better than any other classifier algorithms. So random forest is the correct way to evaluate the applicants for visa approval, proven by the value of 97.00% achieving consistency based on the results that have been conducted. This initiative would assist students overseas with their higher education. In the future, we planned to add more features, with more student information than the current dataset, to our data collection. We will also try to gather the dataset from several other universities and apply the dataset with more classifier algorithms.

References

1. Foreign Scholarship Info for Bangladeshi Students. https://bit.ly/3jy8SdN. Accessed 20 Sep 2020
2. Chalmers D (2011) Progress and challenges to the recognition and reward of the scholarship of teaching in higher education. High Educ Res Dev 30(1):25–38
3. Kurniadi D, Abdurachman E, Warnars HLHS, Suparta W (2018) The prediction of scholarship recipients in higher education using k-Nearest neighbor algorithm. In: IOP conference series: materials science and engineering, vol 434, no 1, p 012039
4. Biswas AA, Majumder A, Mia MJ, Nowrin I, Ritu NA. Predicting the enrollment and dropout of students in the post-graduation degree using machine learning classifier
5. Mia MJ, Sattar A, Biswas AA, Habib MT. Registration status prediction of students using machine learning in the context of Private University of Bangladesh
6. Nilashi M, Bagherifard K, Rahmani M, Rafe V (2017) A recommender system for tourism industry using cluster ensemble and prediction machine learning techniques. Comput Ind Eng 109:357–368

7. Ahmed SA, Khan SI (2019) A machine learning approach to predict the engineering students at risk of dropout and factors behind: Bangladesh perspective. In: 2019 10th international conference on computing, communication and networking technologies (ICCCNT). IEEE, pp 1–6
8. Goga M, Kuyoro S, Goga N (2015) A recommender for improving the student academic performance. Procedia-Soc Behav Sci 180:1481–1488
9. Yukselturk E, Ozekes S, Türel YK (2014) Predicting dropout student: an application of data mining methods in an online education program. Eur J Open Distance e-Learn 17(1):118–133
10. BESSiG-Bangladeshi Expat & Student Society in Germany. https://bit.ly/34qQGwa. Accessed 01 Oct 2020
11. HigherStudyAbroad—Global Hub of Bangladeshis. https://bit.ly/33tsom8. Accessed 25 Sep 2020
12. Study in UK & Canada for Bangladeshi students. https://bit.ly/3ldvalC. Accessed 3 Oct 2020
13. Han J, Pei J, Kamber M (2011) Data mining: concepts and techniques. Elsevier
14. Random Forest. https://en.wikipedia.org/wiki/Random_forest. Accessed 21 Nov 2020
15. SVM. https://en.wikipedia.org/wiki/Support_vector_machine. Accessed 21 Nov 2020
16. Neural Network. https://en.wikipedia.org/wiki/Neural_network. Accessed 21 Nov 2020
17. Decision Tree. https://en.wikipedia.org/wiki/Neural_network. Accessed Accessed 21 Nov 2020
18. KNN. https://en.wikipedia.org/wiki/K-nearest_neighbors_algorithm. Accessed 21 Nov 2020
19. Naïve Bayes. https://en.wikipedia.org/wiki/Naive_Bayes_classifier. Accessed 21 Nov 2020
20. Apply for a U.S. VISA. https://bit.ly/2I6CsJL. Accessed 20 Sep 2020
21. Billah MA, Ahmed SA, Khan SI, Chittagong B. Factors that contribute programming skill and CGPA as a CS graduate: mining educational data. Datab Syst J Board 33
22. Approved Visa's Data from Higher Studies. https://bit.ly/2EXJxLc. Accessed 1 Aug 2020
23. Rejected Visa's Data from Higher Studies. https://bit.ly/34n6Wyz. Accessed 10 Aug 2020
24. Weka. https://en.wikipedia.org/wiki/Weka_(machine_learning). Accessed 10 May 2020
25. Orange (software). https://en.wikipedia.org/wiki/Orange_(software). Accessed 10 May 2020

COVID-19 Safe Guard: A Smart Mobile Application to Address Corona Pandemic

Soma Prathibha, K. L. Nirmal Raja, M. Shyamkumar, and M. Kirthiga

Abstract The novel coronavirus, COVID-19, has been declared as a pandemic all across the world by the World Health Organization (WHO). World lockdown and social distancing has been considered as ideal tools to prevent community transmission. That too, in a country like India which is densely populated, it is very difficult to prevent the community transmission even during lockdown. The people are forced to move out of their houses during this pandemic for emergency purposes. The people moving out during this pandemic situation for unavoidable situations, i.e., to get their groceries, for medical emergencies and so on, have the probability of getting infected by COVID-19. Also in case of some medical emergency, people may take the path which is blocked and return back and again take a different route. The people also does not know whether a particular path is blocked until they travel through the specific path. This might put the life of person in danger. The proposed mobile Android application gives the safest route and notifies the user if the path the user chose is blocked in advance. This saves a lot of time of the user by preventing the user to move, to and fro the blocked areas and then choosing a different route. This application also help those who is forced to move out during for a medical emergency. The proposed Android application helps the people to get updates regarding COVID-19 in their locality and make them feel safe during this pandemic.

Keywords COVID-19 pandemic · Safe path · Quarantine · Android application

S. Prathibha
Department of Information Technology, Sri Sairam Engineering College, West Tambaram, Chennai, Tamil Nadu, India
e-mail: prathibha.it@sairam.edu.in

K. L. Nirmal Raja · M. Shyamkumar · M. Kirthiga (✉)
Sri Sairam Engineering College, West Tambaram, Chennai, Tamil Nadu, India
e-mail: sec19ec062@sairamtap.edu.in

K. L. Nirmal Raja
e-mail: sec19ec137@sairamtap.edu.in

M. Shyamkumar
e-mail: sec19ec048@sairamtap.edu.in

1   Introduction

The novel coronavirus (COVID-19) has put a stop to the normal routines of everyday life for mankind. And while social distancing is the course of action to take until told otherwise, moves are not always something that can wait. Daily provisions/medicine may be purchased during the stipulated time (6 am–2.30 pm), but social distancing must be strictly maintained. But people are not expected to buy provisions/medicines every day. During this COVID-19 pandemic, people fear to move out of their houses as the number of positive cases is increasing day by day. Also in countries like India, they have taken measures to block the regions affected by virus to control the spread of the virus. Few houses or paths are blocked when people in a particular region is affected by COVID-19. Several thousands of city residents across various localities will have to live an isolated existence for an indefinite period inside the containment zones demarcated for localization and subduing of the COVID-19 pandemic.

As per the operational guidelines issued by the government, residents of the zones will have to confine themselves to their homes. Movement of persons within the zone is not allowed as well. Perimeter of the zone will be cordoned off with eight-foot barricades and access to all roads and thoroughfares will be closed barring one common entry and exit point. Every entry and exit will be recorded and monitored. Size of each zone will depend on the incidence of the positive cases, with an approximate radius of 100–500 m. For each positive case, 100 households will be counted in for containment, and as such, a single apartment or gated community too could be announced as the containment zone.

During this pandemic situation, people have to be tested as earlier as possible. They should know about the information related to COVID-19 to act wisely and get themselves tested in the early stages. This improves the success rate of the treatment. The people moving out during this pandemic situation for unavoidable situations, i.e., to get their groceries, for medical emergencies and so on, have the probability of getting infected by COVID-19. Also in case of some medical emergency, people may take the path which is blocked and return back and again take a different route. The people also does not know whether a particular path is blocked until they travel through the specific path. This might put the life of person in danger. Also it becomes difficult for the visually impaired to maintain social distancing when they move out. During post-COVID period, it becomes difficult for the people to identify the places where vaccines are available. In this work, a smart mobile application COVID-19 Safe Guard is proposed to provide solution to these problems [1, 2, 3].

2   Literature Survey

In the last six months, there has been a lot of mobile applications available to help people to face this pandemic. The existing COVID-19 mobile applications help in providing information related to COVID-19 pandemic, for tracking the pandemic,

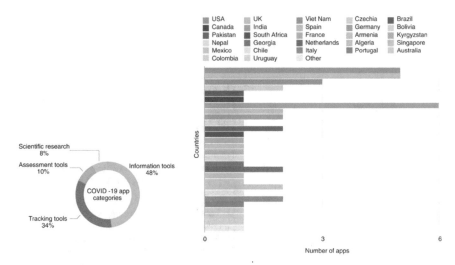

Fig. 1 Various COVID-19 application all over the world

for assessment, and for scientific research. Figure 1 shows the overview of all the applications related to COVID-19 all over the world. In [3], they have mentioned that about 114 apps were created between April 27 and May 2. Of these, 54.4% were on Android and 45.6% were on IOS.

In this section, the comprehensive list of various mobile applications [4, 5] used in different countries are summarized in listed in Table 1 [6].

In [7], authors have designed apps in such a way that the doctors themselves view the citizens who are affected by COVID-19, and the information is shared with the doctor for better treatment. In [8], authors propose an app which facilitates to find the location history of the user using JSON file which contains the location details of the user. The readability analysis to evaluate the comprehensibility of the privacy policies of seven contact-tracing apps currently in use was done in [9]. They have mentioned about the use of readability test tool, a free web-based reliability calculator to assess the contents of the privacy policies of these apps [10, 11]. In [12 and 13], Bluetooth technology is used for contract tracing to find the region of COVID clusters. In papers [14, 15], they have reviewed the protocols of the apps which has the feature of contract tracing. The data given plays a powerful role; the fear of the people is that the information like location and personal details may be misused. They have finally mentioned about the advantages and disadvantages of a few of the mobile applications. Most of the applications mentioned in Table 1 uses Bluetooth technology. This was seen as major disadvantage. Use of Bluetooth technology is the main cause of battery drain. Proposed smart mobile application does not use Bluetooth technology; hence, the users need not worry about the issues related to battery drain. Also as mentioned in Fig. 1, no specific application is designed for both COVID related updates and stress-busting activities. The proposed smart mobile application provides both of these features in combo.

Table 1 List of existing COVID-19 mobile apps used in different countries (Source-https://en.wik
ipedia.org/wiki/COVID-19_apps)

S. no.	Name	Country of origin	Features and advantages
1	Aarogya setu **Aarogya Setu** र सुरक्षित | हम सुरक्षित | भारत सुरक्षित	India	It gives information about the risk of getting COVID-19 for the users, and also it helps the users to identify COVID-19 symptoms and their risk profile. It also gives updates on local (in their area) and national COVID-19 cases. If applied for E-pass, it will be available shortly, as it is mandatory for travelling. Allows the users to assess the risk level of their Bluetooth contacts
2	COVID safe	Australia	It uses Bluetooth technology on the mobile phone to look for other devices with COVID safe installed. User's device will take a note of contact had with other users by securely logging the other user's reference code. It also gives information when the other users are tested positive
3	Stopp Corona **STOPP** **CORONA**	Austria	Helps the users to take appropriate measures to maintain social distancing, gives necessary information about self-quarantine as a precaution. User will be notified instantly and anonymously if one of the saved encounters report to have contracted the virus
4	Corona Alert Canada	Canada	It works based on Bluetooth technology. Once the user is tested positive, he or she receives a one-time key, and after which the app asks for permission to share the random code from last 14 days with a central server. If codes matches, then the other users will be notified that they are exposed
5	Smittestop	Denmark	This app helps the user to inform that he or she is tested positive in a crowded area. This can be done without revealing the identity of the user. Also the user's family members and close ones have to be informed as the virus may even spread with their contacts

(continued)

3 Proposed System

In this work, it is proposed to develop COVID-19 Safe Guard mobile application
which does not require Bluetooth technology. The features of the proposed mobile
application are discussed below.

Table 1 (continued)

S. no.	Name	Country of origin	Features and advantages
6	CareFuji careFIJI	Fuji	It uses Bluetooth technology to check if the person near the user is COVID-19 positive. This data is encrypted and stored in user's mobile phones. It is an open-source reference of Trace together mobile application

- Map module: Identifies test centers nearby-helping users to get tested earlier
- Chatbot: Gives necessary information to users whenever required
- Determination of safe path and makes people to move to destination safely and quickly
- Shows COVID count state wise alerts people whenever the count of positive cases is high and informs about the precautionary methods
- During the post-COVID-19 period, identifies the vaccines providers near them
- Few entertainment activities like games and exercise for people who are in quarantine

The application design and the detailed analysis of each module are elaborated in the next subsection.

3.1 Proposed System Modules

The application mainly consists of six main modules. First the splash screen (see Fig. 2) pops which shows the logo of the application. Next is the dashboard (see Fig. 3) with six grids each representing each module appears. The modules are chatbot, test centers, vaccine centers, news updates, safe path, and quarantine. A new screen appears for each module (see Figs. 4, 5, 6, 7, 8, 9, and 10). Soon, a new module for visually impaired will be introduced as an updated version (see Fig. 8). The features of the modules are given below in detail.

Chatbot. Gives necessary information to users whenever required (see Fig. 4). The features which the chatbot provides are

Fig. 2 Splash screen

Fig. 3 Dashboard

Fig. 4 Screenshot of chatbot
module

Fig. 5 Screenshot of test center module

Fig. 6 Screenshot of news updates screen

Fig. 7 Screenshot of
vaccine center module

Fig. 8 Screenshot of safe
path screen

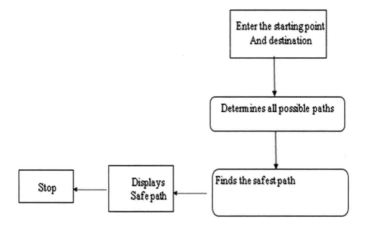

Fig. 9 Work flow of safe path module

Fig. 10 Screenshot of quarantine module

- Information regarding COVID-19 (symptoms).
- A self-assessment test (gets alerted if he has more symptoms).
- Information regarding E-pass (steps to get e-pass).

Test Centers. This module identifies test centers nearby (see Fig. 5). Our chatbot alerts the user in early stages of the disease (self-assessment test), and with the test center module, the user can find the nearby test centers and helping users to get tested earlier. As the famous problem states "Earlier the better," in case the user is COVID-19 positive, the severity of the disease is very less. Hence, the user can be treated easily, and the success rate of getting cured is really high.

News Updates. This modules shows COVID count state wise (see Fig. 6).

- Alerts people whenever the count of positive cases is high
- "Prevention is better than cure"
- Helps the users to stay safer and take the necessary precautionary steps.

Vaccine Centers. This module identifies the nearby vaccine centers (see Fig. 7).

- During the post-COVID-19 period, identifies the vaccines providers near them.

   This helps them to get vaccinated soon and stay safer.

Safe Path. This module shows the safest path to the user by determining the number of COVID positive cases (see Fig. 8). It shows the path with least COVID-19 counts.
   The criteria for choosing the safe path is given below.

- Less number of COVID positive cases
- Zero blocked paths
- Post-COVID period, Maximum number of people vaccinate.

   The module helps the user to choose the right path and helps the user to stay safe during unavoidable circumstances. The workflow of the module is shown in Fig. 9.

Quarantine. Proposed mobile application also has a few entertainment activities such as games and exercises for the people in quarantine to relieve their stress (see Fig. 10).
   The proposed mobile application is implemented using Android studio. It works based on current location of the user. It is also enabled with voice for its ease of use for the visually impaired. Application has various modules.
   The user can choose any of the modules as per their wish. For the safe path module, the user has to enter the destination. Once the information of source to destination is fed to the application, it finds all the possible paths to reach the destination. It calculates the number of positive cases in each of the different paths. Finally, the application gives the safest route, the route where there is less number of COVID-19 positive cases to the user. Therefore, this application helps the user to move to the desired location safely. In case if the user is visually impaired, the application guides

the user with voice messages. The user need not type the destination, and instead, a voice message can be sent. The message is then recognized by the application, and the application guides the user to reach the destination.

4 Conclusion and Future Work

COVID-19 Safe Guard gave us the solutions to most of the problems that people face during this pandemic. Some of them are during this pandemic, some people are not aware whether they are affected or if they are undergoing the symptoms of the disease. Most of the people get tested only when the severity of the cold or fever is high. Sometimes treatment fails as the severity of the disease is high. The people moving out during this pandemic situation for unavoidable situations, i.e., to get their groceries, for medical emergencies and so on, have the probability of getting infected by COVID-19. Also it becomes difficult for visually impaired to maintain social distancing when they move out. During post-COVID period, it becomes difficult for the people to identify the places where vaccines are available. People feel stressed when they stay in homes for a long period of time.

In the future work, maintaining social addressing for the visually impaired person will be addressed. Sensors are connected to the walking stick of the user, and when any other people is in 6 m radius, it alerts the user. This avoids them to get infected easily. Figure 11 shows the idea of the future project. Here the signals are sent out and incase of any disturbance, it returns back. The distance "d" is calculated. If

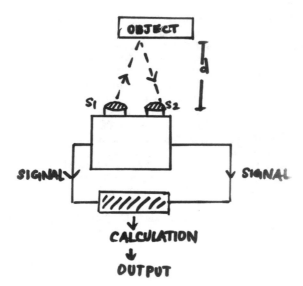

Fig. 11 Block diagram of social distancing for visually impaired person system

the distance d is <6 m, then the user gets the warning alarm. This is done with the help of arduino. Arduino was used to alert the user incase the d > 6 m. This way, the application COVID-19 Safe Guard helps the visually impaired to remain safe in their way out during any emergency cases.

$$^{235}U\ 0.72\%$$
$$^{238}U\ 99.27\%$$

References

1. https://www.moving.com/tips/moving-during-coronavirus-epidemic-heres-what-you-need-to-know/
2. https://www.thehindu.com/news/cities/Hyderabad/coronavirus-guidelines-issued-for-containment-zones/article31340772.ece
3. Collado-Borrell R, Escudero-Vilaplana V, Villanueva-Bueno C, Herranz-Alonso A, Sanjurjo-Saez M (2020) J Med Internet Res, August 25, 22(8):e20334
4. Mallik R, Hazarika AP, Dastidar SG, Sing D, Bandyopadhyay R. Received 29 April 2020. Revised 13 June 2020. Accepted 18 June 2020. Published online: 1 July 2020
5. Cho H, Ippolito D, Yu YW (2020) Contact tracing mobile apps for COVID-19: privacy considerations and related trade-offs. arXiv preprint arXiv:2003.11511
6. Wang J (2020) Mathematical models for COVID-19: Applications, limitations, and potentials. J Public Health Emergency 4
7. Abbas R, Michael K. COVID-19 contact trace app deployments: learnings from Australia and Singapore. IEEE Consum Electron Mag 9(5): 65–70, 1 September 2020. https://doi.org/10.1109/MCE.2020.3002490
8. Hang A, Dascalu M-I, Stanica I (2020) Contact tracing app for containing diseases spread. In: 2020 zooming innovation in consumer technologies conference (ZINC), Novi Sad, Serbia, pp 216–217. https://doi.org/10.1109/ZINC50678.2020.91618
9. Zhang M, Chow A, Smith H, J Med Internet Res December 3, 22(12):e21572
10. Sharma T, Bashir M (2020) Use of apps in the COVID-19 response and the loss of privacy protection. Nat Med 26:1165–1167. https://doi.org/10.1038/s41591-020-0928-y
11. https://en.wikipedia.org/wiki/COVID-19_apps
12. Ahmed N et al (2020) A survey of COVID-19 contact tracing apps. IEEE Access 8:134577–134601. https://doi.org/10.1109/ACCESS.2020.3010226
13. Berglund J. Tracking COVID-19: there's an app for that. IEEE Pulse 11(4):14–17, July–August 2020. https://doi.org/10.1109/MPULS.2020.3008356
14. Islam MN, Islam I, Munim KM, Islam AKMN (2020) A review on the mobile applications developed for COVID-19: an exploratory analysis. IEEE Access 8:145601–145610. https://doi.org/10.1109/ACCESS.2020.3015102
15. Chowdhury MJM, Ferdous MS, Biswas K, Chowdhury N, Muthukkumarasamy V. COVID-19 contact tracing: challenges and future directions. IEEE Access. https://doi.org/10.1109/ACCESS.2020.3036718

A Novel Mathematical Model to Represent the Hypothalamic Control on Water Balance

Divya Jangid and Saurabh Mukherjee

Abstract For the stability maintenance process of internal environment of human body, various parameters are responsible such as temperature, water, glucose and a number of hormones participated. Homeostasis is an important substantial process which maintains the threshold value for all these basic parameters of human body. Osmoregulation, energy balance, respiratory system and appetite control are four major components of human homeostasis process. A homeostasis process which balances the water level in human body is osmoregulation. The threshold value for the hydration level of each cell is retained by osmoregulation process. Osmoregulation controls the water level in different internal organs of human body. Some of the internal organs participated in this process such as hypothalamus, pituitary gland, kidney, etc. Osmoregulation starts from the smallest unit like cell till the respiratory system. But the domain for the introduced mathematical model is limited to the organ level only. An antidiuretic hormone (ADH), i.e., vasopressin is the main regulator for the osmoregulation process at organ level. Hypothalamic Water Balance Mathematical Model (HWBMM) is being designed and deployed to represent the basic functioning of water regulation process of human body.

Keywords Hypothalamus · Hypothalamic hunger regulation model · Water balance · Osmoregulation · Anti-diuretic hormone (ADH)

1 Introduction

The stable state of different functions of human body is maintained by the homeostasis process. The factors which participate in balancing different physiological parameters of homeostasis are hunger regulation, energy balance, osmoregulation

D. Jangid · S. Mukherjee (✉)
Faculty of Mathematics and Computing, Department of Computer Science, Banasthali Vidyapith, Banasthali, Tonk 304022, Rajasthan, India
e-mail: msaurabh@banasthali.in

D. Jangid
e-mail: jangid.divya@gmail.com

© The Author(s), under exclusive license to Springer Nature Singapore Pte Ltd. 2022
P. Nanda et al. (eds.), *Data Engineering for Smart Systems*, Lecture Notes in Networks and Systems 238, https://doi.org/10.1007/978-981-16-2641-8_55

(water balance), thermoregulation and so on [1, 2]. Mathematical modeling is an influential and impactful way to define, analyze and postulate the homeostatic and endocrine functions (systems) [3] of human body like hunger regulation [4, 5], water balance, and hypothalamus–pituitary–adrenal (HPA) system [6]. Our major concern is about the water regulation process (osmoregulation) with hypothalamic effect. It depends on many parameters like blood volume, blood pressure, osmotic pressure and so on. Osmotic pressure is the main simulator for osmoregulation, and it quantifies the tendency of water during osmosis. The extra water from human body is drained out during water osmosis. Osmotic pressure is directly proportional to the water requirement by ferent internal organs. In other, words we can say that osmotic pressure prevents the water level from too much diffusion or too much concentration [7]. If the water level is too much higher than the average threshold, the osmosis process will start, whereas thirst (Dry Mouth Theory 1779) is a significant indicator for the decrease in water level from the specific threshold value. Thirst helps in maintaining hydration level of human body and stops the osmosis process [8, 9]. The negative feedback process of osmolality instigates and regulates the feeling of thirst, and it increases the release of ADH on the circumventricular organs (hormonal neurons) [10]. The circumventricular organs influence the hunger regulation and initiate thirst signals [11]. Water balance (osmoregulation) and hunger regulation are strongly connected. Similarly, hunger and thirst are directly connected in the same way. Nowadays, this connection is effected by today's eating behavior, but the basic parameters for this connection cannot be denied [7, 12]. Both hunger and thirst are very strong sensation, but thirst is more stable and effective. Thirst instigates the process of water intake which is further integrated by different nerve, hormonal and osmotic signals [9]. The osmotic signals (osmoreceptors) and hormonal signals are received and generated by hypothalamic nuclei, respectively. Supraoptic (SON) and paraventricular (PVN) nuclei located at the anterior region of hypothalamus majorly participated in osmoregulation. The eating behavior and the fluid regulation is executed together efficiently and smoothly by the neuro-signal generated at subfornical organs [13]. During the recent studies on mice depicts the effect of sleep and active state on homeostasis level of water regulation. The vasopressin, the anti-diuretic hormone, polarizing effect increases the electrical activity of neurons at subfornical organ during active state which result into more water intake, whereas during final state of sleep, vasopressin, secreted by magnocellular neuro-secretory cells at the hypothalamic nuclei (supraoptic and PVN), increase the water reabsorption by internal organs (kidney) until recommence of water intake [14, 15]. The initial phase of Thirst detection starts with osmotic signals generated at gastrointestinal and instigate the throat osmolarity level (dryness in throat) which is balance by the hypothalamic signals by consumption of water [16]. Also the increase in the ADH hormone concentration intensifies the stimulation to the hypothalamus pituitary and adrenal axis [17, 19].

2 Objectives

The objective of our model (HWBMM) is to present a base mathematical model which describes the hypothalamic impact in maintaining the water balance at homeostatic level. It also represents the mapping between the osmotic signals and hormonal signals, the spikes generated and regulated at hypothalamus for the feeling of thirst (water requirement) and water balancing in human body. We also consider some basic parameters for the analysis report which only gives hypothetical validation.

3 Motivation

For hunger regulation, a general mathematical representation was being introduced, named hypothalamic hunger regulation model (HhRM) [4]. Our model (HWBMM), i.e., hypothalamic water balancing mathematical model is also dependent on the concept of HhRM and the optimized HhRM [4, 5]. Moreover, at physiological level osmoregulation process for human has been studied regularly with psychological and analysis perspective, but this physiological process was never described from the mathematical aspect. However, some of the researchers have represented mathematical model on endocrine system. These models concentrate on the impact and functioning of hypothalamus and pituitary gland in endocrine functioning in human body. A mathematical model based on the physiology of HPA systems is defined where the positive and negative feedbacks of homeostasis used as the key regulator [11]. A differential equation based mathematical is used to define the endocrine system where all the hormone concentration and endocrine gland represented via differential equations (linear and nonlinear) [3].

4 Hypothalamic Water Balance Mathematical Model

The methodology for the development of the mathematical model can be explained in multistep process. Each step is specific to the activity of thirst and osmosis process. The presence of the vasopressin hormone and change (increase/decrease) in its secretion start the process at physiological level. The whole process of HWBMM (water balance) is explained with block diagram in Fig. 1.

Step-1: Osmotic Signal: Increase and decrease in the water level from its homeostatic level identify by the negative feedback homeostatic level (negative feedback process for water regulation). The parameters participate in this process, osmotic pressure, updated from it stabilized value. This change perceived by osmoreceptors, i.e., homeostatic neurons. Vasopressin hormone secretion also updates from the homeostatic level. The non-uniformly generated random numbers are used for the parametric value of osmotic signals.

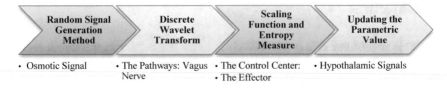

Random Signal Generation Method	Discrete Wavelet Transform	Scaling Function and Entropy Measure	Updating the Parametric Value
• Osmotic Signal	• The Pathways: Vagus Nerve	• The Control Center: • The Effector	• Hypothalamic Signals

Fig. 1 Steps for HWBMM with the mathematical terms

Step-2: Pathway for the Signal: The osmotic signals from the osmoreceptors further regulated to hypothalamic nuclei, supraoptic and paraventricular nuclei via vagus nerve. The osmotic signal travels the path of vagus nerve by blood stream which destined the osmotic signals to forebrain. The discrete wavelet transform is being used to represent the osmotic signal with fractal constant.

Step-3: Control Center and Effectors: The hypothalamic signals generated at anterior part, i.e., supraoptic and paraventricular nuclei which further communicate with the posterior pituitary gland to release vasopressin, the antidiuretic hormone into the blood stream. The scaling function with entropy measure is applied for the activation of the hypothalamic signal. The variance property of the scaled signals is also used to optimize the parametric value of the signals.

Step-4: Updating the Parametric Value: The vasopressin hormonal signals are transferred from the internal organs (adrenal gland (kidney), lungs) to hypothalamic nuclei (PVN and SON). It decreases the rate of loss of water from the and vice versa. The physiological effect of ADH on the internal organs is either water absorption or water secretion. Increase in ADH is responsible for decrease in loss of water, and decrease in ADH is responsible for increase in loss of water from different internal organs which participated in osmoregulation process (Fig. 2).

HWBMM's assumptions and functions: The process starts with the assumption that vasopressin hormone releases positively. Osmoregulation can be divided into two subprocesses hyperosmolality and hypoosmolality. A binary function $f(o)$ is used to start the process after vasopressin release and identification between hyperosmolality (increase in secretion of vasopressin concentration (VC)) and hypoosmolality (suppress the secretion of vasopressin concentration):

$$f(o)=\begin{cases} 1 \text{ for ST} <= \text{VC} <= \text{HT(HyperOsmolality(Thirst))} \\ 0 \text{ for LT} <= \text{VC} <= \text{HT(HyperOsmolality)} \end{cases}$$

where ST = Stable Threshold Value for VC, LT = Lowest Threshold Value for VC, HT = Highest Threshold Value for VC according to homeostasis process, i.e., osmoregulation. O(s) is osmotic signals generated from osmoreceptors and projected to hypothalamic nuclei. AD(s) represent the antidiuretic hormonal signals. We also use Daubechies wavelet function D4 (AD(s)) and scaling function Sf (h) for hormonal signals. Constant fractals function frac (h) used with wavelet transformation pattern

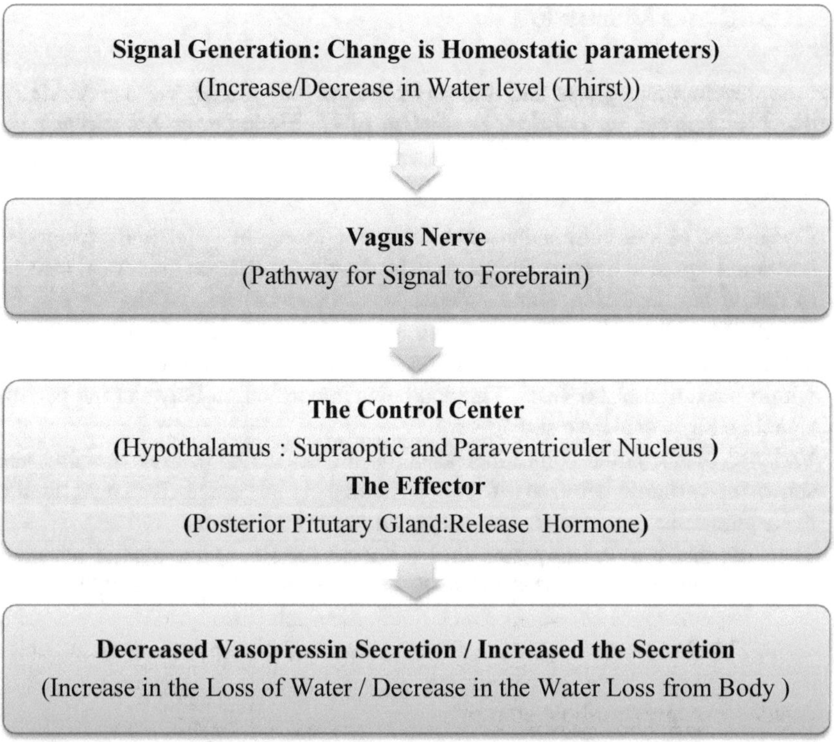

Fig. 2 Block diagram for the hypothalamic water balance process

to generate the hypothalamic signal. An entropy function **E (h)** is also used to measure uncertainty for the physiological data, i.e., hormonal signals. $Co(O(s), AD(s))$ is used to represent the covariance and correlation between the osmotic signal and hormonal signals. Since hormonal signal dependent on the osmotic signals, we use coefficient of correlation and determination to verify the utility of our model. The equation for the Hhpothalamic water balancing mathematical model is as follows:

$$HT = D4(O(s)) * frac(O(s)) + Sf(AD(s)) * E(h) \qquad (1)$$

$$\frac{dHT}{dt} = frac(h)D4'(O(s)) + E(h)Sf'(AD(s))$$

5 Result and Discussion

For the development phase and analysis phase of our model, we use MATLAB 2018a. For analysis, we consider coefficient of correlation, entropy, variance and skewness. These parameters depict different features of osmotic signals (O(s)) and hypothalamic signals (H(s)) which are as follows:

- Coefficient of correlation between the osmotic signal, O(s) and vasopressin hormonal signal at hypothalamus, ADH(s)), is the correlation numerical measure to define the dependence and independence between both of the signals. For HWBMM, if the value of N (total number of signals) is increased, then there is also an increase in coefficient of correlation. HWBMM the correlation calculated within the range of 60–80%. There is positive correlation between the osmotic signal and hypothalamic signal.
- Variance of hormonal signal increases the impact of the scaling function as it shows the variation between different hormonal signal values. Skewness gives us the asymmetric nature of the data (signals) toward the mean.
- The four different entropy measures are used for the hypothalamic hormonal signals, scalar entropy: $EM = -3.4900e + 03$, threshold entropy at 0.06: $EMt = 2677$, log entropy: $EMl = -3.4900e + 03$, Shannon entropy: $EMs = -2.2536e + 03$. Threshold entropy gives positive correlation, whereas scalar, Shannon and log entropy give negative correlation. So for the result, it is being analytically decided to apply threshold entropy.

For our randomly generated dataset, the threshold entropy is calculated at threshold value 0.06. Entropy measure provides more uncertainty and variation to our dataset (Figs. 3, 4 and 5).

- Since covariance and correlation can mathematically describe similar features and aspect of any dataset. For our simulated data, there is a positive covariance and correlation. Thus, our positive correlated hypothalamic and hormonal signals are interrelated and dependent (Fig. 6).

6 Conclusion

HWBMM describes the physiological process, osmoregulation (water balance), in basic scenario only. For an optimized and high utility model, we need to add more complex parameters which give more accuracy and efficient result. This version of HWBMM can be used as the basic study and implementation only as it gives the correlation between the osmotic signal and hypothalamic signals with an average range, i.e., 60–80%. Entropy is an important parameter which shows the repeated values and pattern. A simple scaling function can represent data but for better result, we need to introduce a scaling function with spatial parametric values. Hence, entropy is an impactful parameter for hormonal signals [18, 19].

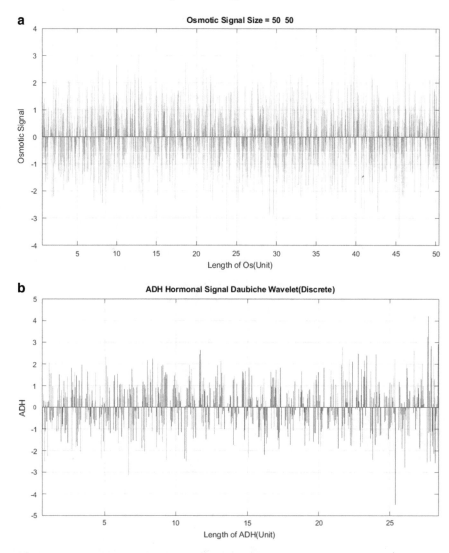

Fig. 3 a, b shows the osmotic signal and ADH hormonal signals respectively (2-D discrete wavelet transforms)

7 Future Work

Since it is a base model, there is a lot more scope for optimization in our model. We can analyze it with a different and multiparameter scaling function, multiscale entropy measure and fractal geometry. For more compatible model to real-life application, our model also needs to validate clinically with the help of an actual medical data, i.e., fMRI images.

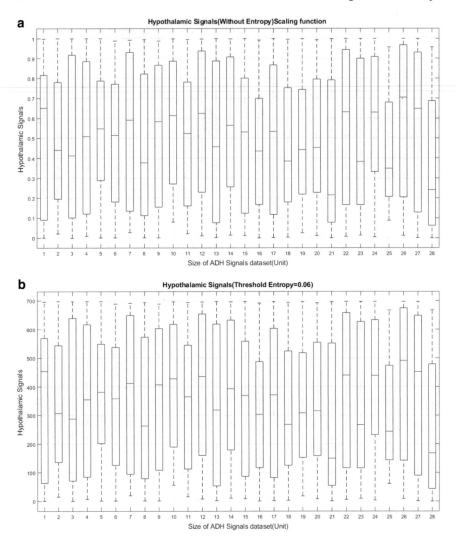

Fig. 4 Hypothalamic hormonal signal **a** without threshold ntropy **b** with threshold entropy at 0.06 (minimum)

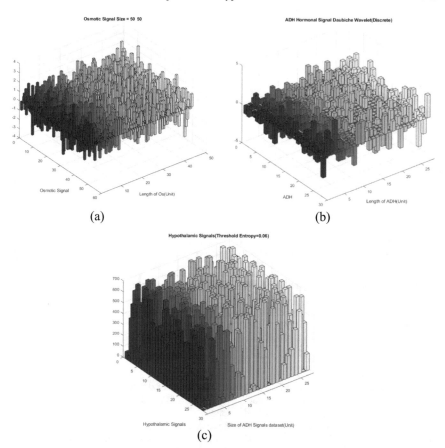

(a)

(b)

(c)

Fig. 5 Signals in 3-D representation, **a** osmotic signal which starts the process of osmolality **b** the ADH signal generated with the help of scaling function and wavelet transform (without entropy) **c** hypothalamic signal with threshold entropy

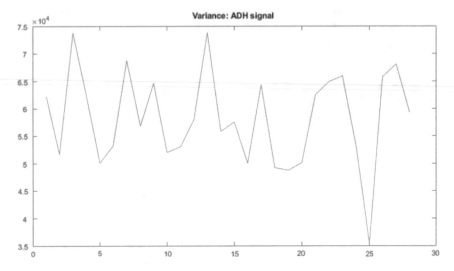

Fig. 6 Covariance in the hormonal signal and ADH signal at hypothalamus

References

1. Michael WS, Stephen SC, Porte D, Denis R (2000) Central nervous system control of food intake. Nature 404:661–671
2. Feher JJ (2016) Quantitative human physiology: an introduction. Edition 2, Chapter 9.2, "Hypothalamus and pituitary gland", 777–786 (2012), ISBN-10:0128008830, ISBN-13:9780128008836, Publisher: Elsevier Science
3. Koós I (2006) Mathematical models of endocrine systems. J Univ Comput Sci 12(9):1267–1277
4. Divya SM (2015) Mathematical model to represent the role of hypothalamus in hunger regulation. National seminar proceeding "information technology applications: strategies, issues and challenges". Published in conference preceding, ISBN: 978-93-85000-08-9
5. Divya SM (2015) Optimizing and validating the hypothalamic hunger regulation mathematical model. J Network Commun Emerg Technol (JNCET) 5(2):203–207. ISSN: 2395-5317, EverScience Publication
6. Conrad M, Hubold C, Fischer B, Peters A (2009) Modelling the hypothalamus–pituitary–adrenal system: homeostasis by interacting positive and negative feedback. J Biolog Phys 35(2):149–162
7. McKinley MJ, Johnson AK (2004) The physiological regulation of thirst and fluid intake. Physiology, 1–6
8. Bichet DG (2018) Vasopressin and the regulation of thirst. Ann Nutr Metab 72(2):3–7
9. Leib DE, Zimmerman CA, Knight ZA (2016) Thirst. Current Biol 26(24):R1260–R1265
10. Bichet DG (2019) Regulation of thirst and vasopressin release. Annu Rev Physiol 81(1):359–373
11. McKinley MJ, Denton DA, Ryan PJ, Yao ST, Stefanidis A, Oldfield BJ (2019) From sensory circumventricular organs to cerebral cortex: neural pathways controlling thirst and hunger. J Neuroendocrinology, e12689
12. McKiernan F, Houchins JA, Mattes RD (2008) Relationships between human thirsts, hunger, drinking, and feeding. Physiol Behav, 700–708
13. Zimmerman CA, Lin YC, Leib DE, Guo L, Huey EL, Daly GE, Knight ZA (2016) Thirst neurons anticipate the homeostatic consequences of eating and drinking. Nature 537(7622):680–684

14. Zimmerman CA, Leib DE, Knight ZA (2017) Neural circuits underlying thirst and fluid homeostasis. Nat Rev Neurosci 18(8):459–469
15. Gizowski C, Bourque CW (2018) Hypothalamic neurons controlling water homeostasis: it's about time. Curr Opin Physiol 5:45–50, ScienceDirect
16. Zimmerman CA, Huey EL, Ahn JS, Beutler LR, Tan CL, Kosar S, Bai L, Chen Y, Corpuz TV, Madisen L, Zeng H, Knight ZA (2019) A gut-to-brain signal of fluid osmolarity controls thirst satiation. Nature 568(7750):98–102
17. Heida JE, Minović I, van Faassen M, Kema IP, Boertien WE, Bakker SJL, van Beek AP, Gansevoort RT: Effect of vasopressin on the hypothalamic-pituitary-adrenal axis in ADPKD Patients during V2 Receptor Antagonism. Am J Nephrol 51:861-870. https://doi.org/10.1159/000511000(2020)
18. Thuraisingham RA, Gottwald GA (2006) On multi-scale entropy analysis for physiological data. Physica A. Stat Mech Appl 366:3233–32
19. Jänig W (2006) The integrative action of the autonomic nervous system: neurobiology of homeostasis. Cambridge University Press, Cambridge, UK

Importance of Deep Learning Models to Perform Segmentation on Medical Imaging Modalities

Preeti Sharma and Devershi Pallavi Bhatt

Abstract In the area of medical imaging, segmentation method is vital method to outline the area of interest in terms of pixels in 2D images or voxels in 3D images. Researchers are focusing on automated ways for accurate segmentation applied on images captured through different imaging modalities depending on the type of disease or infection in body. Deep neural network-based segmentation models are playing an important part for the accurate and fast segmentation. This paper presents the importance of deep learning models in the field of medical imaging focusing on the segmentation of different body parts from different imaging modalities.

Keywords Segmentation · Deep learning · Medical imaging

1 Introduction

Medical imaging technique provides health professionals crucial information about the internal organs and tissues of patient's body. There are numerous types of medical imaging technologies that give information about different body parts that help to identify diseases like pneumonia, brain injuries, cancer, internal bleeding, and other kind of complaints. Large amount of data is generated by medical imaging applications. High quality of imaging can advance medical decision making and can lessen the unnecessary procedures.

Segmentation is the process of dividing an image into different sections. A medical image is segmented so that the region of interest can be acquired for further processing. Threshold segmentation is the simplest segmentation method where each pixel in an image is set according to an intensity value. If pixel's intensity value is less than threshold vale (T), then it is switched to black pixel, otherwise pixel is switched to white pixel. Edge-based segmentation method is based on discontinuity detection when there is a rapid change occurs in intensity value. To construct object boundary, all edges are detected and then these edges are connected with each other, and region is segmented. In region-based segmentation method, image is segmented

P. Sharma · D. P. Bhatt (✉)
University, Manipal University, Jaipur, Rajasthan, India

© The Author(s), under exclusive license to Springer Nature Singapore Pte Ltd. 2022
P. Nanda et al. (eds.), *Data Engineering for Smart Systems*, Lecture Notes in Networks and Systems 238, https://doi.org/10.1007/978-981-16-2641-8_56

into regions and elements in a region having similar kind of characteristics. Two main techniques are region growing technique and region splitting and merging technique. In clustering-based segmentation technique, cluster of pixels having similar characteristics is formed in an image. Main concept behind this is that elements in same cluster are more similar compared to elements in other clusters. Two main types of clustering methods are as follows: One is the hierarchical method that is based on tree concept where root is whole database and nodes are clusters. Second is partition-based method to minimize objective function by using optimization methods. Artificial neural network (ANN)-based segmentation algorithms try to simulate human brain's learning process and make decisions like human does. In medical imaging, ANN-based models are currently most popular methods. Important information is extracted, and requisite objects are detached from the background. This process is achieved in two steps; first features are extracted and then neural network segments the image. Deep learning (DL) modes are based on ANN algorithms but have deeper networks than ANN. Researchers are applying deep neural networks for segmentation of images acquired by medical imaging techniques.

The incentive behind writing this paper is to present the importance of deep learning techniques to perform segmentation on images of different body parts, acquired by different kind of medical imaging modalities. For this purpose, studies that have used deep learning models for segmentation of various types of body parts or diseases have been reviewed. Likewise, some of the important deep learning models have been discussed that are basically used for segmentation of medical images. The rest of the paper is arranged as follows: Sect. 1 is introduction part, while in Sect. 2, a brief description of different kind of medical imaging modalities is given. Section 3 enlightens deep learning technique and popular deep learning models. Review of different studies is included in Sect. 4. Discussion of presented paper is included in Sect. 5, and conclusion and future scope are presented in Sect. 6.

2   General Description of MI Modalities

Medical imaging is a part of biological imaging that includes all the technologies that are used to generate pictorial depiction of human's internal body to get information or details to identify, control, or cure medical conditions. Medical imaging techniques started in twentieth century with the discovery of X-ray and from that there has been a lot of progress has been done by researchers in this field. Though X-ray is the oldest technique; CT scan, MRI, ultrasound and PET are also among the widely used imaging techniques.

2.1 Radiography/X-Ray

From all medical imaging techniques, X-rays are the eldest and widely used medical imaging technique to get pictures of internal parts of body. These images help to diagnose fractured bones, infection and any kind of injury, also finds presence of any foreign objects in soft tissue. X-ray images are also helpful for dentists. Bone tumor could be identified with the help of these images. Ionizing radiations is used by X-ray machine to generate images of internal parts of body.

2.2 Magnetic Resonance Imaging/MRI

MRI images give detailed information of internal tissues, and monitoring of patient's status during treatment is observed through MRI. Strong magnetic field is used to align rotating atomic nuclei generate pictures through MRI scanner. MRI-generated pictures are more thorough compared to other techniques. The non-aggressive quality of MRI scan makes more reliable, and this scan can be used for infants also. Drawback of this scan is that patient need to stay still during the process of scanning patients.

2.3 CT Scan

Computerized tomography scan is used to create cross-sectional images to give internal information of body parts like soft tissues, bones and blood vessels with help of rotating X-ray machines and computer. Body parts that can be visualized by CT scan are head, spine, heart, knee, chest, shoulders, and abdomen. CT scan images are mainly used for diagnosing diseases and assessing injuries. During CT scan, patient lies in a machine at looks as tunnel. A rotating X-ray machine keeps rotating to take pictures from different angles and transmitted to computer for further processing to give detail information.

2.4 Positron Emission Tomography (PET)

It is a nuclear imaging method that perceives gamma rays. Gamma rays are emitted by a radiotracer. Tracer is inserted into vein. For active neurons, glucose works as fuel. Gamma rays are captured by detector panels, and 3D images are generated. Generally, PET scans are used by radiologist to diagnose cancer, blood flow, and bone formation.

2.5　Ultrasound

Ultrasound is commonly known as sonography. Instead of using radiations, ultrasound uses sound waves of high frequency to take images of internal parts. These images are called sonograms. No use radiation makes it safe to scan fetal development. Other than this, it is also used to scan liver, kidney, abdomen, and heart.

3　Deep Learning Methods Overview

Machine learning (ML) methods are subpart of artificial intelligence (AI) technique where system perform automatic learning without any human interfere so that system could improve its decision making by observation and corrections. Deep learning is subcategory of machine learning that learns tasks and features from data directly. DL models use layered structure of algorithms known as artificial neural networks (ANN). Deep learning models learn by their own methods and tend to act like human brain that process data and generate patterns to take decisions; for this DL models require huge training process. Accuracy of deep learning algorithms is high compared to machine learning algorithms. The deep neural network has an input layer, multiple hidden layers, and an output layer. In healthcare industry, deep learning deliver answers to many problems. Deep learning uses effective processes for the diagnosis in state-of-the-art manner. DL networks can process data and extract information at a speed and scale which is not possible for humans. Deep learning can make task easy for radiologist by fast diagnosis. Though there are various kinds of deep neural networks, following are some deep neural models used for medical image segmentation that are presented in Table 1 (Fig. 1).

3.1　Convolutional Neural Network (CNN)

CNN or ConvNet [2] is multi-layer neural networks having learnable weights and biases. CNN performs objects detection, images recognition and classification, video

Table 1 Popular medical image segmentation deep neural network models

Model	Year	Authors	Proposed area
CNN	1980	Fukushima [2]	Visual pattern recognition
FCN	2015	Long [3]	Semantic segmentation through fully convolutional network
U-Net	2015	Ronneberger [4]	Biomedical image segmentation
V-Net	2016	Milletari [5]	3D image segmentation by taking the whole volume content into account at once

Fig. 1 General architecture
of deep neural network
(*Source* Ref. [1])

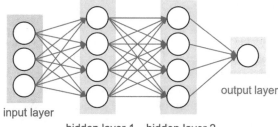

input layer

hidden layer 1   hidden layer 2

output layer

processing, etc. CNN study spatial hierarchies of features automatically by back-propagation. Its mathematical constructed architecture has three kinds of layers; these are convolution layers, next is pooling layer and finally a fully connected layer. These layers perform operations like 'convolution' and 'pooling'. Feature extraction is performed by convolution and pooling layers, whereas mapping of extracted features to final output is performed by fully connected layer.

Convolution layer: Convolution operation is a linear operation that is used for feature extraction. Input to the CNN called tensor is array of numbers, and kernel is also arrays of numbers of small size. Both array perform element-wise product and generate a feature map. Multiple kernels are applied to generate random number of feature maps that show diverse features of input tensors. Output of this linear operation is then conceded through activation function (nonlinear). One of the common activation functions is rectified linear unit (ReLU) function. Pooling layer: Pooling operation is a downsampling operation that decreases the image size by reducing the number of parameters though reserve important characteristics of the image. Max pooling and average pooling are two mostly used pooling operation. Fully connected layer/dense layer: After feature extraction from convolution operation and down-sampling by pooling operation features maps are transferred to fully connected layer that has the output nodes same as number of classes. A nonlinear activation function is performed that needs to be chosen according to task.

3.2 Fully Convolutional Network (FCN)

Long et al. [3] replaced the last fully connected layer of CNN with a fully convolutional layer so that the network can take random size input and can calculate the dense pixel-wise output. FCN perform pixel-wise prediction. Dense layer is replaced in FCN so that computation time and number of parameters can be reduced. This network has two paths, one is downsampling path that excerpt and infer contextual information and other is up sampling path that supports the detailed localization.

3.3   U-Net

U-Net is an improvement of FCN for fast and accurate segmentation of biomedical images developed by Ronneberger et al. [4]. This architecture is in U shape due to its two paths; one is contracting path and other one is expansive path. In between the contracting and expansive path is bottleneck. The contracting path is consisted of many contraction blocks that reduces the spatial information and increases the feature information. To learn difficult structures well, number of features maps or kernels gets double after every block. In expansive path, spatial information and features are combined by up-convolutions and from contracting path high-resolution features are concatenated. This halves feature channels numbers. U-Net model is effective and can be trained end-to-end even if image dataset is limited.

3.4   V-Net

Though being very popular model, U-Net process only 2D images, Milletari et al. [5] improved U-Net and proposed V-Net method for 3D image segmentation. V-Net was trained end-to-end on 3D MRI volumes. This architecture has four downsampling blocks and four upsampling blocks and a residual convolutional block. Compression and decompression are done by convolutional layers.

4   Review of Deep Learning-Based Segmentation Works on Different Medical Imaging Techniques

In year 2013, Prasoon et al. [6] presented a 2.5 D CNN-based segmentation model for voxel classification. Instead of using 3D CNN, authors combined three 2D CNN to classify voxels from 3D scans. Segmentation was performed on the low-field knee MRI scans segmenting the tibial articular cartilage. Proposed model was tested on 114 MRI scans.

In year 2015, Li et al. [7] presented CNN model to perform segmentation of liver tumor. Authors used 26 portal phase enhanced CT scans. In this study, five CNN models of different input patch size were designed for the segmentation of liver tumor. While authors Roth et al. [8] used abdomen CT scans for pancreas segmentation using CNN. Authors presented an effusively automated bottom-up model that was based on a hierarchical coarse-to-fine classification of super-pixels. CT scans of 82 patients were used in this study.

While in 2016, Pereira et al. [9] presented a novel CNN segmentation method for brain tumor in MRI images. CNN model was made over convolutional layers with kernel size of 3X3 that allows deeper architecture. Intensity normalization was investigated as a pre-processing step. Authors also found that data augmentation was effective for segmentation. Authors also compared deep architectures with shallow architectures and found that shallow architectures performed lower comparatively. For effective training of CNN, LReLU activation function was more effective than ReLU. Same year, Dalmıs et al. [10] proposed an automated U-Net architecture-based segmentation method that segment breast and fibro-glandular tissue (FGT) on breast MRI scans. Study was performed on 66 breast MRI. Authors used U-Net architecture in two ways, in first approach two 2C U-Nets were trained consecutively. First U-Net segment breast in whole MRI volume then second U-Net segment FGT from segmented breast. In second approach, only one 3C U-Net was used that segment volume into three regions that are FGT inside breast, fat inside breast, and non-breast.

In year 2017, Shryas et al. [11] modified the U-Net architecture and proposed a deep learning model that segments the brain region on MRI scans. Model focused on gliomas cancer and used MRI scans of 220 high grade gliomas (HGG) cases from BRATS 2015 dataset. In other paper, authors Lonney et al. [12] used deep learning on 3D ultrasounds. Authors used a previously proposed model DeepMedic[] based on CNN and presented a CNN model that can segment the first trimester placenta. For the training, testing and validation of CNN network 300 3D ultrasound scans were used.

Authors Zhu et al. [13] in year 2018 presented a segmentation model named AnatomyNet. The presented model segments the CT scan of head and neck (HaN) and detects organs-at-risk (OARs). AnatomyNet was built upon 3D U-net model with few extensions. For this study, 261 head and neck CT scans were used. Same year authors Blanc-Durand et al. [14] also used 3D U-net for segmentation of gliomas tumor from F-fluoro-ethyl-tyrosine (F-FET) PET brain images. PET brain images of 37 patients were included in this study. While AIT Skourt et al. [15] used U-net for the automated segmentation of lung parenchyma on lung CT scans. Authors presented that U-net network perform lung segmentation task accurately and in final images eliminate parts of trachea and bronchus area while keep nodules and blood vessel parts.

In 2019, a 3D U-net-based deeply supervised model was presented by Ye et al. [16] applied multi-depth fusion for the whole heart segmentation for the clinically analysis of cardiac status. Three main features of presented study were deeply supervised method, hybrid loss and multi-depth fusion. Sixty cardiac CT scans of whole heart substructures were included in this study. Same year, authors Chen et al. [17] presented a deep learning model named S-CNN that segment cervical tumor from PET images. Presented model was based on CNN. Authors added some pre-information like cervical tumor roundness and positioning information among bladder and cervix with CNN to improve segmentation accuracy. A total of 1176 3D PET images of 50 cervical tumor patients were used in this study. Whereas, CNN was applied on chest X-ray(CXR) images by Wessel et al. [18] to segment ribs. Authors used Mask R-CNN [19] model on 174 anterior–posterior CXR for the segmentation of individual ribs. Mask R-CNN segmentation method is an instance segmentation method that was originally used for real-time video processing. In another study by Panfilov et al. [20], deep learning was used to segmentation of knee cartilage on knee MRI images. Approach of presented study was based on U-Net model. Model segments knee MRI in five segment mask that are, no cartilage, patellar cartilage, femoral cartilage, menisci, tibial cartilage.

Automatic segmentation model of head and neck tumor and nodal metastates was proposed by Andrearczyk et al. [21] in year 2020. Authors focused on morphological and metabolic tissues of head and neck and used 2D and 3D V-Net model that was applied this model on PET-CT scans. A bimodal, i.e., a late fusion of PET and CT scans with 2D V-Net performed well for segmentation task.

5   Discussion

Segmentation is the method of dividing an image into different parts so that objects or classes can be differentiated from each other. Segmentation of medical imaginings is very crucial also risky part because region of interest (ROI) is to be selected by segmentation. Region or boundary lines of ROI must be clear and accurate. Deep neural models are making segmentation task easy and automatic for medical professionals. There has been a lot of work done by researchers to find automated deep neural networks to perform segmentation for different imaging modalities. Researchers have developed deep learning based models for segmentation of almost all body parts. Table 2 presents the research works done by researchers that shows that deep neural model has been used for segmentation of different organs or diseases using different medical imaging modalities. Table 2 presents the modalities used for diagnosis, diagnosing area, deep models used for segmentation and dice score achieved by these models. Figure 2 is a human body image showing almost all body parts with labeling. In Fig. 2, the labeled body parts that have been covered in different research works that are included in this study are marked by rectangle shaped box. From this figure, we can observe that deep neural-based models to perform segmentation have covered almost whole body.

6   Conclusion and Future Scope

Self-learning capability of deep neural networks has increased its popularity among the researchers. Medical imaging techniques like X-ray, CT scan, ultrasound, MRI, PET are used to look inside the human body and get information about any kind of disease or damage inside the body. Deep learning-based models are increasing the efficiency of segmentation of medical images. Deep neural networks are taking place of old segmentation methods.

In this paper, different research work has been reviewed that used deep neural networks for the segmentation of internal body parts using different imaging modalities. After reviewing these previous studies, we can perceive the importance of deep learning models in the field of medical imaging. Also we can observe that U-Net-based models have performed better than CNN-based models for the segmentation

Table 2 Different research works showing the importance of deep neural networks for almost whole body using different modalities

References/Year	Modality	Diagnosing area	Method	Dice Score
[6]/(2013)	MRI	Tibial articular cartilage from low field knee	2.5 D CNN (combined three 2D CNN)	79.52% (±9.10%)
[7]/(2015)	CT scans	Liver tumor	CNN	80.06% (±1.63%)
[8]/(2015)	CT scans	Pancreas	CNN	73.0%
[9]/(2016)	MRI	Brain tumor	CNN	78.0% (complete region), 65.0% (core region) 75.0% (enhancing region)
[10]/(2016)	MRI	Fibro-glandular tissue (FGT) on breast	U-Net	94.4% (breast segmentation using 2C U-Net), 85.0% (FGT segmentation using 3C U-Net)
[11]/(2017)	MRI	Brain region	U-Net	83.0% (whole tumor region), 75.0% (core tumor region), 72.0% (enhancing tumor region)
[12]/(2017)	3D Ultrasounds	First trimester placenta	CNN	73.0%
[13]/(2018)	CT scans	Organs-at-risk (OARs) on head and neck (HaN)	3D U-Net	79.25%
[14]/(2018)	PET	Gliomas tumor	3D U-Net	82.31% (+ 4.1%)
[15]/(2018)	CT scans	Lung parenchyma	U-Net	95.02%
[16]/(2019)	CT scans	Whole heart segmentation	3D U-net	90.73%
[17]/(2019)	PET images	Cervical tumor	CNN	84.0%
[18]/(2019)	Chest X-rays	Segmentation of individual ribs	Mask R-CNN	73.3%
[20]/(2019)	MRI images	Segmentation of knee cartilage on knee MRI images	U-Net model	90.7% (femoral cartilage)
[21]/(2019)	PET-CT scans	Head and neck tumor and nodal meta states	2D and 3D V-Net model	60.6%

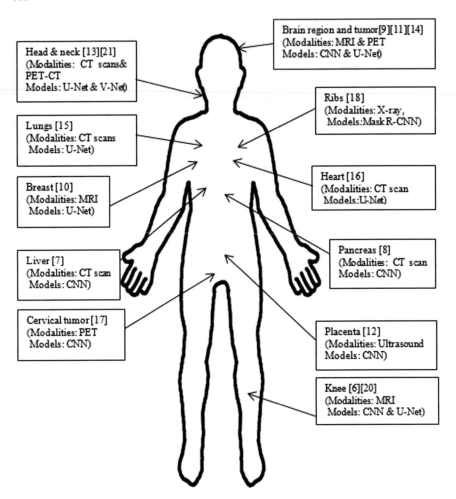

Fig. 2 Different deep learning models applied on different body parts

of knee cartilage on MRI scans of knee. Same applies for brain tumor segmentation where performance of U-Net-based models is better than CNN-based models. Almost all body parts or different disease has been covered in this review.

There are huge amount of research works that are focusing of segmentation of internal organs of human body so there is a lot of scope for future work. Presented study can extend by including research works that are focusing on body parts that have not been included in this study.

References

1. Musiol M (2016) Speeding up Deep Learning
2. Fukushima K, Miyake S (1982) Neocognitron: a self-organizing neural network model for a mechanism of visual pattern recognition. Competition and cooperation in neural nets. Springer, Berlin, Heidelberg, pp 267–285
3. Long J, Shelhamer E, Darrell T (2015) Fully convolutional networks for semantic segmentation. Proceedings of the IEEE conference on computer vision and pattern recognition
4. Ronneberger O, Fischer P, Brox T (2015) U-net: convolutional networks for biomedical image segmentation. International conference on medical image computing and computer-assisted intervention. Springer, Cham
5. Milletari F, Navab N, Ahmadi S-A (2106) V-net: fully convolutional neural networks for volumetric medical image segmentation. 2016 fourth international conference on 3D vision (3DV). IEEE
6. Prasoon A et al (2013) Deep feature learning for knee cartilage segmentation using a triplanar convolutional neural network. International conference on medical image computing and computer-assisted intervention. Springer, Berlin, Heidelberg
7. Li Wen (2015) Automatic segmentation of liver tumor in CT images with deep convolutional neural networks. J Comput Commun 3(11):146
8. Roth HR et al (2015) Deeporgan: multi-level deep convolutional networks for automated pancreas segmentation. International conference on medical image computing and computer-assisted intervention. Springer, Cham
9. Pereira S et al (2016) Brain tumor segmentation using convolutional neural networks in MRI images. IEEE Trans Med Imaging 35(5):1240–1251
10. Dalmış MU et al (2017) Using deep learning to segment breast and fibroglandular tissue in MRI volumes. Medical Phys 44(2):533–546
11. Shreyas V, Vinod P (2017) A deep learning architecture for brain tumor segmentation in MRI images. 2017 IEEE 19th International workshop on multimedia signal processing (MMSP). IEEE
12. Looney P et al (2017) Automatic 3D ultrasound segmentation of the first trimester placenta using deep learning. 2017 IEEE 14th international symposium on biomedical imaging (ISBI 2017). IEEE
13. Zhu W et al (2019) AnatomyNet: deep learning for fast and fully automated whole-volume segmentation of head and neck anatomy. Medical Phys 46(2):576–589
14. Blanc-Durand P et al (2018) Automatic lesion detection and segmentation of 18F-FET PET in gliomas: a full 3D U-Net convolutional neural network study. PLoS One 13(4):e0195798
15. Skourt BA, El Hassani A, Majda A (2018) Lung CT image segmentation using deep neural networks. Procedia Comput Sci 127:109–113
16. Ye C et al (2019) Multi-depth fusion network for whole-heart CT image segmentation. IEEE Access 7:23421–23429
17. Chen L et al (2019) Automatic PET cervical tumor segmentation by combining deep learning and anatomic prior. Phys Med Biol 64(8):085019
18. Wessel J et al (2109) Sequential rib labeling and segmentation in chest X-ray using Mask R-CNN. arXiv preprint arXiv:1908.08329
19. He K et al (2017) Mask r-cnn. Proceedings of the IEEE international conference on computer vision
20. Panfilov E et al (2019) Improving robustness of deep learning based knee mri segmentation: Mixup and adversarial domain adaptation. Proceedings of the IEEE international conference on computer vision workshops
21. Andrearczyk V et al (2020) Automatic segmentation of head and neck tumors and nodal metastases in PET-CT scans. Medical imaging with deep learning. PMLR

A Multi-component-Based Zero Trust Model to Mitigate the Threats in Internet of Medical Things

Y. Bevish Jinila, S. Prayla Shyry, and A. Christy

Abstract The advent of Internet of Medical Things (IoMT) has become a greater boom in the present scenario due to the pandemic COVID-19. Nowadays, remote monitoring of the patients is significant in elderly people as well as COVID-19-affected cases. It is known that the medical data is more sensitive and if not handled properly can result in adverse conditions. In this paper, a Zero Trust Model (ZTM) is introduced to handle the security of the IoMT data. This framework will definitely mitigate the occurrence if threats and further benefit the users.

Keywords IoMT · Trust model · Remote monitoring · Pandemic · IoT · Threat

1 Introduction

The Internet of Things (IoT) has evolved as an imperative approach for automation. It has been used in multiple areas like manufacturing sectors, traffic signals, homes, healthcare industries, and hospitals. In recent times, there has been a substantial change in the field of health care. The assistance of modernized equipment has led to the precise identification of diseases in humans. However, continuous monitoring of patients, especially the elderly is the need of the hour. IoT has been a promising technology since years, and it is believed to transform the universe at large based on trade and economy [1]. The advent of the Internet of Things (IoT) in the field of health care has embarked to remotely monitor the health of the patients by smart wearable devices and equipment. This has led to the evolution of the Internet of Medical Things (IoMT). According to a recent study [2], one-third of the IoT devices are being used in health care and are expected to be increased to a greater number, approximately $6.2 trillion.

Even though there is a sharp transition of usage of IoT in health care, the IoMT is more susceptible to security and privacy threats. There is a possibility of threat either in the physical layer, medium access control layer, network layer, or application layer. However, the vital attribute that is vulnerable to threats is the data. Over the years,

Y. Bevish Jinila (✉) · S. Prayla Shyry · A. Christy
Sathyabama Institute of Science and Technology, Chennai, India

millions of patient data is breached due to threats created by insiders, third-party vendors, and phishing attacks. Since the medical data should be highly confidential, it is definitely important to protect the medical data from unknown threats.

The rest of the paper is organized as follows. Section 2 briefly explains the need and security threats in IoMT. Section 3 describes the related work, Sect. 4 portrays the proposed Zero Trust Model (ZTM) on IoMT, and Sect. 5 briefly concludes the work on ZTM-IoMT and discusses the future work.

2   Internet of Medical Things

One of the prominent problems in Internet era is that the cause of threats is due to the increased trust on the network. The responsibility of security in IoMT relies upon multiple stakeholders. The administrators quite often claim that the security breaches are due to the inadequate security features of IoMT devices, and they often blame the manufactures. On the contrary, the manufactures claim that security breaches are due to ever-changing network threats and it is the responsibility of the administrators to frequently verify and update the intrusion detection system, so that the overall system can be secure and sage. So, there exists a wide gap which is not addressed and this requires an attention.

The Zero Trust architecture was first proposed in 2010 by John Kindervag to protect the confidential data of enterprise systems. This architecture was proposed in the sense that the enterprises should not trust and check the legality of the messages received. Instead, they should have zero trust on the messages and verify anything that is connecting to the enterprise network. So, by this approach the enterprises are not agreed to approve any messages without verifying its authenticity. By adopting the ZTM, the healthcare organizations will have the potential to withstand the security attacks by continuously monitoring the IoMT devices.

2.1   Need of IoMT

IoMT is a combination of data collected from the medical devices and software applications that help to connect both the patients and the healthcare professionals. The main objective of IoMT is to connect the medical devices deployed in the patient side and to remotely monitor the condition of the patients. The data acquired can also be analyzed, and the condition of the patient can be predicted.

Some examples of the IoMT devices to monitor the patients include fitness trackers like wrist bands, smart pills, and virtual consultations. Figure 1 shows the general architecture of IoMT.

The global pandemic COVID-19 has also accelerated the need of IoMT, and the data obtained is extremely sensitive and can be used for future predictions [3]. The data obtained includes the patient's condition, location, treatment details given and

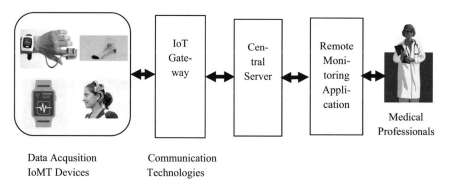

Fig. 1 IoMT general architecture

recovery details. Using the data collected, various studies can be done to improvise the healthcare facilities and taking care of the patients remotely.

2.2 Threats in IoMT

There are various types of threats associated with IoMT mainly with respect to user level and device level. The user-level threats include impersonation, Sybil, and duplicates [4–6].

3 Related Work

According to Kai et al. [7], trust assessment is essential for preserving data privacy. Medical Cloud Data Center, Health application market, and intelligent medical terminal are deployed for maintaining security and privacy.

According to Dwivedi et al. [8], healthcare devices can be classified as stationary medical devices, medical embedded devices, medical wearable devices, and wearable health monitoring devices. The role of intelligent medical terminals is enormous in present-day life. These medical terminals monitor the health condition of person and provides essential data on time. There remains lots of security challenges when data gets transmitted between the medical terminals.

The medical data of a person can be processed, encrypted, and stored on data cloud so that only the authorized users can access the data. By the increasing volume of cloud applications and cloud storage; there remains lots of challenges in storing sensitive healthcare data. Since healthcare data gets streamed online, handling of data in an incremental manner is also important.

When anonymized dataset is handled incrementally, two types of situations like handling of violation of privacy policy and over-anonymization happen.

The techniques adopted for privacy preservation are Generalization, Suppression, Pesudonymization, Bucketization, Slicing, Randomization, and Cryptography.

Anonymization is one of the techniques adopted for maintaining privacy where data is shared across cloud or other medical or IoT devices. Generalization, Anatomization, Anonymized data K-anonymity, and the L-diversity are some of the techniques adopted for privacy protection.

Consider a data holder has collected a series of records R1 at timestamp T1. K-anonymized version of R1 is denoted as K1. When the data holder collects a new series of records R2 at timestamp T2, the K-anonymized version of all datasets collected so far will be R1 U R2, representing the data collection of R2. Hence, the data collection during ith event at Time T_i is R_i. Publishing data release K2, for time periods T1 and T2, is the data recollected as R1 U R2. Data distortion which happens due to acceptance of live stream can be avoided by representing in the form of anonymization. Types of attacks that can happen with anonymization are: forward attack, cross attack, and backward attack.

Forward Attack: This is represented as F-attack (K1, K2). Consider a transaction has timestamp T_1 in which an intruder tries to attack his record in release K1 using the background knowledge K2. Since the transaction has similar record in K1 and K2, record k1 in K1 matches with another sensitive attribute k1 in K2. If k1 fails to track a similar record in K2, then k1 does not originate from the given transactions QID and k1 can be removed from the given transaction.

Let person A has 2 QIDs as [INDIA, TAMILNADU], where

qid1 = [INDIA, TAMILNADU] and the transaction which satisfies this are {a1, a2, a3, a4, a5}

qid2 = [INDIA, DISEASE] and the transaction matching are {b3, b4, b5, b6, b7}

qid1 has two groups g1 = {a1, a2, a3, a4} for identification of COVID19 and g2 = {b3, b4, b5, b6} for identification of FLU.

If all elements of g1 match with the QID of person A, K2 would have contained four entries of the form [INDIA, DISEASE, COVID19].

Cross Attack: This is represented by C-attack (K1, K2). Consider a transaction has timestamp T_1 in which an intruder tries to attack his record in release K1 using the background knowledge K2. Similar to F-attack, record k2 in K2 matches with another sensitive attribute k2 in K1. If k2 fails to track a similar record in K1, then k2 does not originate from the given transactions QID and k2 can be removed from the given transaction.

Let person A has 2 QIDs as [INDIA, TAMILNADU], where

qid1 = [INDIA, TAMILNADU] and the transaction which satisfies this are {a1, a2, a3, a4, a5}

qid2 = [INDIA, DISEASE] and the transaction matching are {b3, b4, b5, b6, b7}

qid2 has two groups g1 = {b3, b4} for identification of COVID19 and g2' = {b5, b6, b7} for identification of FLU.

If all elements of g2 do not match with the QID of person A, K1 would have contained two entries of the form [INDIA, DISEASE, FLU].

Backward Attack: This is represented by B-attack (K1, K2). Consider a transaction has timestamp T_2 in which an intruder tries to attack record in release K2 using the background knowledge K1. In this case, the transaction has a record in K2 but not in K1. If record k2 in K2 matches with another sensitive record K1, then k2 has timestamp T1 and k2 can be removed from the given transaction.

Let person A has 2 QIDs as [INDIA, TAMILNADU] where, qid1 = [INDIA, TAMILNADU] and the transaction which satisfies this are {a1, a2, a3, a4, a5}.

qid2 = [INDIA, DISEASE] and the transaction matching are {b3, b4, b5, b6, b7} qid2 has two groups g2 = {b3, b4} for identification of COVID19 and g2' = {b5, b6, b7} for identification of FLU. If all elements of g2' do not match with the QID of person A, K1 would have contained two entries of the form [INDIA, DISEASE, COVID19].

Preservation of privacy in the distributed data remains a challenge during updation. K-anonymity algorithm combines incremental data approach along with distributed data privacy considering time and space complexity. Xia and Tao [13] has introduced a metric named m-invariance for finding privacy preservation. Fung et al. (2015) proposed a model named BCF-anonymity for inserting and updating data. Numerous methods are available to maintain the privacy of Electronic Health Records (EHRs), namely k-anonymity, l-diversity, (a,k)-anonymity, t-closeness, etc. [9–14]. In order to maintain privacy, first the generalization method is deployed for classifying EHRs. Further preprocessing and anonymization are carried out as shown in Fig. 2.

4 Proposed Model

IoMT Zero Trust Model (IoMT-ZTM) is a model that helps to mitigate the availability of possible threats in IoMT. This model helps to analyze the possible threats and will generate an alert on the consumer side something is suspicious. Figure 3 shows the various areas where ZTM should be implemented.

In this proposed system, the user trust is taken into consideration. The user trust is the primary area where verification has to be done carefully to identify a valid user. The authenticity of a user is indispensable in ZTM. The basic method of identity verification is password-based authentication. But, if the password is weak or exposed somewhere, it will be ineffective and inefficient. To improvise the user authentication, multi-factor-based verification should be adopted. This multi-factor authentication adds additional layer of security. When it is implemented properly, it is obviously difficult for an adversary to steal the confidential data. Moreover, this will also mitigate the vulnerabilities present in the system. In this paper, a multi-component-based ZTM with multi-factor authentication is proposed.

ZTM adopts the methodology to trust no outsiders. Users behind the firewall are also not trusted. The main component of zero trust is data. Those organizations which have visibility of their data and the associated activities can detect the suspicious behavior; even other security solutions are compromised. The main aim of zero trust

Fig. 2 Schematic
representation of
anonymization of EHR

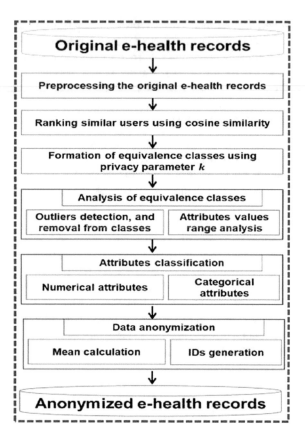

Fig. 3 Areas of Zero Trust
Model

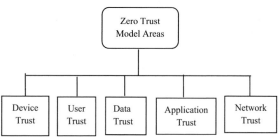

is to protect the devices from unusual threats. So, it will be beneficial if the trust of
the user is computed.

Figure 4 shows the working process of ZTM. The process includes identification
of sensitive data, limited access, detection of threats, using a baseline, and applying
analytics to segregate active internal and external attacks.

Fig. 4 Process flow of ZTM
based on user trust

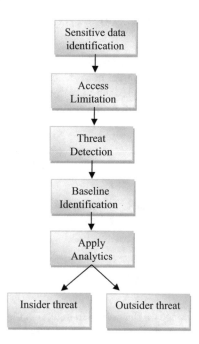

4.1   Sensitive Data Identification

The data retrieved from the sensors through remote monitoring of the patients is
tracked and analyzed for further predictions. It is always a general requirement,
where the sensitive data has to be identified and kept private from unauthorized
access. This first phase is meant for segregating the sensitive data from insensitive
data. And, user restrictions are applied to sensitive data.

4.2   Access Limitation

This phase will account for the limited access rights of the users. As various people
will be dealing with the medical data, it is always a main requirement to provide
access rights to the sensitive data. The basic method of access restriction is authentication. This is the first step in the verification process. There are various security
breaches possible with a single layer of security. So, an additional layer of security is
employed to secure the medical data. In this work, as a second layer of security, user
authentication, is verified through One-Time Password (OTP) using the registered
mobile number. Using this approach, the outside attacks can be avoided. Further,
to strengthen the authentication process, each patient's identity is verified through
options like face detection or fingerprint verification.

4.3 Threat Detection

Early detection of threats will definitely benefit the sustainability of the model. The proposed model holds a collection of pre-existing threat identities. All the data will be pre-verified with the threat database. If there is a match, then the threat alert will be generated. And, if there is no match, then it will be processed.

4.4 Baseline Identification

In this phase, based on the previous patterns, new patterns of threats and suspicious patterns are identified. The presence of anomalies is identified, and alert is generated.

4.5 Apply Analytics

Based on the collected data, it is important to analyze the data and predict the occurrence of threats. The Naïve Bayes classifier is used to analyze and classify the threats based on internal and external threats.

5 Conclusion

In this paper, a Zero Trust Model is proposed to secure the data in IoMT. The proposed model is a multi-component model which employs multi-factor authentication, which ensures more security to the medical data. In this model, the user trust is taken into account. The proposed model is a framework that can be adopted for more secure IoMT. In future, further investigations and analytics can be applied to the data captured and can be improvised.

References

1. Rodgers MM, Pai VM, Conroy RS (2015) Recent advances in wearable sensors for health monitoring. IEEE Sens J 15(6):3119–3126
2. A Guide to the Internet of Things Infographic, https://intel.com/content/www/us/en/internetof-things/infographics/guide-to-iot.html
3. Bevish Jinila Y, Thomas J, Shan P (2020) Internet of Things enabled approach for Hygiene Monitoring in Hospitals. International conference on computer, communication and signal processing, IEEE
4. Sathyabama B, Bevish JY (2019) Attacks in wireless sensor networks-a research. Int J Innov Technol Exploring Eng 8(2):775–783

5. Rubio JE, Roman R, Lopez J (2020) Integration of a threat traceability solution in the industrial internet of things. IEEE Trans Indust Inform
6. Fu K, Kohno T, Lopresti D, Mynatt E, Nahrstedt K, Patel S,... Zorn B (2020). Safety, security, and privacy threats posed by accelerating trends in the internet of things
7. Kai K, Pang Z-B, Wang C (2013) Security and privacy mechanism for health internet of things. J China Universities of Posts Telecommun 20(2):64–68
8. Dwivedi AD, Srivastava G, Dhar S, Singh R(2019) A decentralized privacy-preserving healthcare Blockchain for IOT. Sensors https://doi.org/10.3390/s19020326
9. Fung BCM, Wang K, Fu AW-C, Pei J (2015) Anonymity for continuous data publishing. In: Proceedings 11th international conference extending database technol. Adv. Database Technol. EDBT 08, p 264
10. Zhou B, Han Y, Pei J, Jiang B, Tao Y, Jia Y (2009) Continuous privacy preserving publishing of data streams. Proceedings 12th international conference extending database technol. Adv. Database Technol—EDBT'09, p 648
11. J. Pei, J. Xu, Z. Wang, W. Wang, K. Wang, Maintaining k-anonymity against incremental updates, Ssdbm, 2007
12. Doka K, Tsoumakos D, Koziris N (2011) Kanis: preserving k-anonymity over distributed data. Proceedings 5th International work. Pers. Access, Profile Manag. Context Aware. Databases
13. Xiao X, Tao Y (2007) M-invariance: towards privacy preserving re-publication of dynamic datasets. SIGMOD Conference, pp 689–700
14. He Y, Barman S, Naughton JF (2011) Preventing equivalence attacks in updated, anonymized data. Proceedings international conference data engineering, pp 529–540

A Modified Cuckoo Search for the *n*-Queens Problem

Ashish Jain, Manoj K. Bohra, Manoj K. Sharma, and Venkatesh G. Shankar

Abstract A modified cuckoo search algorithm is proposed in this paper to solve the *n*-queens problem. For solving the problem, the host nests are treated as permutations of a cluster of distinctive values. The main idea is that the cuckoos change their eggs (permutations) randomly using Lévy flight approach without affecting the distinctive characteristics of the solution. During evolution of the optimal solution, a cuckoo search parameter, namely fraction of worst nests F_w, is also fine-tuned. The results obtained show that the proposed cuckoo search method is favorable in solving the similar constraint satisfaction problem.

Keywords Permutation optimization · *n*-queens · Particle swarm optimization · Cuckoo search · Genetic algorithm

1 Introduction

The *n*-queens problem is perhaps the best instance of permutation problems. There are many areas in which the application of *n*-queens problem exists. To handle *n*-queens problem, many strategies have been proposed in the literature. A new strategy called modified cuckoo search to deal with the *n*-queens problem is proposed in this paper.

The considered problem is a traditional constraint satisfaction problem in the area of AI. The problem can be defined as placing *n*-queens on the $n \times n$ chessboard, so that no two queens attack to each other that is no two queens on the same diagonal,

A. Jain (✉) · M. K. Bohra · M. K. Sharma · V. G. Shankar
School of Computing and Information Technology, Manipal University Jaipur, Jaipur, India
e-mail: ashish.jain@jaipur.manipal.edu

M. K. Bohra
e-mail: manojkumar.bohra@jaipur.manipal.edu

M. K. Sharma
e-mail: manojkumar.sharma@jaipur.manipal.edu

V. G. Shankar
e-mail: venkateshg.shankar@jaipur.manipal.edu

© The Author(s), under exclusive license to Springer Nature Singapore Pte Ltd. 2022
P. Nanda et al. (eds.), *Data Engineering for Smart Systems*, Lecture Notes in Networks and Systems 238, https://doi.org/10.1007/978-981-16-2641-8_58

same row, and same column. During the past five-decades, the n-queens problem not only has served as the benchmark problem for the divide and conquer, backtracking, branch and bound, and soft computing algorithms. Moreover, it has several following uses: traffic routing, load balancing, very large-scale integration testing, optical parallel processing, computer task scheduling, and data compression [1]. The n-queens problem has three varieties: discovering all arrangements, discovering one arrangement, and finding a group of arrangements. In this paper, we center around finding the one arrangement inside a family.

In 2009, Yang and Deb [2] have proposed a population-based swarm intelligence strategy, namely cuckoo search. It is formed by motivation from the commit brood parasitic conduct of scarcely any cuckoo species in blend with Lévy flight conduct of certain flying creatures and organic product flies [2]. In the past years, cuckoo search has applied in different areas such as design optimization [3], structural optimization [4], reliability optimization [5], flexible manufacturing system optimization [6], automated cryptanalysis [7–9], and Twitter sentiment analysis [10], traveling salesman problem [11]. This method has also utilized to solve n-queens problem [12]. In this paper, we propose a modified cuckoo search technique to solve the n-queens problem. The changing permutations idea presented in this work is very different than those proposed in [12].

The remainder sections are arranged as follows: the basic cuckoo search strategy is reviewed in the nest section. Section 3 describes the modified cuckoo search method to deal with n-queens problem. Section 4 summarizes the experimental results, and Sect. 5 concludes the paper.

2 Standard Cuckoo Search

Cuckoo search is a population-based optimization algorithm, and similar to particle swarm optimization (PSO) and genetic algorithm (GA). In each of the algorithms, a population of random solutions are generated initially. Afterward, potential solutions are updated for searching optimum solution over generations. Unlike PSO, the cuckoo search has no velocity update concept for updating current solutions. And, unlike GA, the cuckoo search has no evolution operators such as mating and mutation. In the case of cuckoo search, the potential solutions called nests are updated in the problem space using current better available nests and Lévy flights.

Cuckoo search works dependent on the accompanying essential three principles: (a) each cuckoo fledgling regularly lays each egg in turn and arbitrarily dumps in any of the picked home; (b) among the best home's the place some excellent eggs have been now dumbed persisted to the following ages; (c) the quantity of accessible host homes is ordinarily fixed, and the egg laid by a separate cuckoo could be found uniquely by the host feathered creature with an of likelihood $p_a \in (0, 1)$.

Cuckoo search method starts with a random populace of individuals called "nests." A best nest (a nest with optimal value with respect to the objective function) is picked from the available nests. Afterward, from existing nests, one more nest is picked randomly, say, ith nest. Using ith nest and best nest, a new nest is generated via Lévy flights. In this process, a small fraction of worst nests is also abandoned by new nests. After numerous repetitions, the process stops, and we get a solution with good value regarding the objective function.

Algorithm 1 shows the standard template of the cuckoo search method, where from an existing nest $x_j(t)$, a new nest $x_j(t+1)$ is generated via Lévy flights as [2, 13, 14]:

$$xj(t+1) = xj(t) + \eta * l \tag{1}$$

For point-by-point depiction on the cuckoo search, the reader can refer [13]

Algorithm 1: Standard Cuckoo Search [2]

Generate initial population of m host nests x_i, where $i = 1, 2, \ldots, m$.

repeat

1. Select a cuckoo (i.e., candidate solution, say, x_j) randomly, and modify that solution to generate new solution by Levy flights
2. Compute the cost of new solution, let it be f_j
3. Randomly choose a nest among m host nests, say x_k
4. if $(f_i < f_k)$ comment: let, the problem has minimization objective
5.   Replace x_k by the new solution x_j
6. Abandon a fraction (P_a) of worst nests/solutions and construct new ones
7. Keep the best solutions
8. Rank the solutions and find the current best

until (Termination condition is satisfied)

Post-process results

---

3   Handling n-Queens Problem

Fitness Function

Without loss of generality, we can assume that the ith queen will be placed in the i row of the chessboard. Thus, each value in the permutation will represent the column number. In this way, the horizontal and vertical conflicts of placing queens are removed. The remaining challenge is to remove the diagonal conflicts. Thus, the fitness function can be defined as the number of collisions beside the diagonal of the chessboard [1]. The objective is changed to minimize the number of collisions, i.e., the fitness of an ideal solution must be zero.

Modified Cuckoo Search

In the standard cuckoo search strategy, the new nest is generated from the existing nest via Lévy flights (See Eq. (1). If this strategy is applied for generating permutation for the n-queens problem, then it is likely that during update two or more locations can get the identical value. That is the permutation rules may break. To overcome this problem, a modified nest update strategy is proposed which is similar to particle update position strategy of PSO which has been proposed in [1].

The proposed cuckoo search starts with a pool of solutions, where each solution will represent a permutation sequence. Call this pool of solutions as $Current_{solutions}$. Determine the cost of each of the solutions resided in the $Current_{solutions}$. Let one of the solutions exists in the $Current_{solutions}$ that has the lowest fitness value, call this solution as $Best_{solution}$ Select one more solution randomly from the current solutions, call this solution as $Current^i_{solutions}$. Using ith current solution, generate a new solution through Eq. (1), call this solution as $New_{solution}$. Since $New_{solution}$ will violate the rule of permutation sequence, therefore, update the $Current^i_{solution}$ using the $New_{solution}$ and the $Best_{solution}$. In order to understand the proposed update strategy, we take an example. For instance, we have given a 8*8 chessboard and 8 queens. Assume that initially we get the following $Best_{solution}$, $Current^i_{solution}$, and $New_{solution}$:

$$Best_{solution} = \{2, 4, 3, 1, 5, 6, 8, 7\}$$
$$Current^i_{solutions} = \{1, 3, 5, 2, 4, 8, 6, 7\}$$
$$New_{solution} = \{1.05, 3.07, 5.13, 2.03, 4.12, 8.21, 6.12, 7.07\}$$

Now, we update the $Current^i_{solutions}$ using the $New_{solution}$ and the $Best_{solution}$. Generate a random number between 1 and 8, say, 2. Now we scan the $New_{solution}$ from left to right and search for a value less than 2 which is 1.05 at position 1, call this position $P1$. Now, we find 1 in the $Best_{solution}$ which is situated at position 4, call this positon $P2$. Now interchange elements of $Current^i_{solutions}$ those are positioned at $P1$ and $P2$. Through this update strategy, we get the following updated current solution.

$$Current^i_{solutions} = \{2, 3, 5, 1, 4, 8, 6, 7\}$$

From the above-obtained updated solution, we can observe that the element 2 has come at position 1 which is actually the best position in the $Best_{solution}$ From the above idea, it is clear that we preserve the best positon of the element during updating ith current solution.

If we update ith current solution in this way, then a situation may arise that the ith current solution may become identical to the best solution. In that case the solution would stay in its current position forever. To resolve this issue, a swap mutation is used here that will interchange two elements of permutation sequence when it is identical to the best solution. Two positions of the elements are determined randomly.

After completion of above process, we abandon a fraction (0.02) of worst solutions from the pool and generate new solutions by again swapping two random positions elements of the best solution. The above-discussed strategy repeats for some number of iterations or the solution with zero fitness value has been found.

Table 1 Comparison of the results for different number of queens

Number of queens	Fitness evaluations needed to find a solution			
	The proposed approach	Hu et al. [1]	Kilic et al. [16]	Homaifar et al. [15]
20	2013.5	5669.7	6024	2043
50	11,534.3	14,991.4	19,879	59,227
100	31,683.7	36,799.4	44,578	244,208
200	86,592.6	93,439.9	86,747	340,991

4 Results

The modified cuckoo search strategy is implemented in Java and executed on an Intel Quad-Core processor i7 (@3.40Ghz). During execution, the cuckoo search parameter was set as follows: the population size was 20, $\eta = 0.02$, $\lambda = 1.5$, and abandon fraction was 0.02.

For a fair comparison of the proposed strategy with the existing strategy in the literature, we tested the cuckoo search strategy for 20 to 200 queens. The algorithm is tested 100 times for a single problem, and the average of function evaluation is then represented in the results (see Table 1) to reach a solution. It can be observed from Table 1 that the cuckoo search is able to find a solution in less number of fitness evaluations with respect to all the number of queens availability.

5 Conclusions

The motivation behind the investigation was to decide how well cuckoo search deals with permutation sequences. The *n*-queens issue was utilized to test the performance and legitimacy of the modified cuckoo search method. Based on the average number of fitness evaluations performance of the cuckoo search method has been compared with those of the available particle swarm optimization and genetic algorithms. The obtained results showed that cuckoo search is powerful to handle *n*-queens issue. Be that as it may, it despite everything should be checked whether this methodology can be reached out to other combinatorial permutation problems.

References

1. Hu X, Eberhart RC, Shi Y (2003) Swarm intelligence for permutation optimization: a case study of n-queens problem. In: Proceedings of the 2003 IEEE swarm intelligence symposium. SIS'03 (Cat. No. 03EX706) pp 243–246. IEEE
2. Yang XS, Deb S (2009) Cuckoo search via Lévy flights. In: 2009 World congress on nature & biologically inspired computing (NaBIC) pp 210–214. IEEE

3. Yang XS, Deb S (2013) Multiobjective cuckoo search for design optimization. Comput Oper Res 40(6):1616–1624
4. Gandomi AH, Yang XS, Alavi AH (2013) Cuckoo search algorithm: a metaheuristic approach to solve structural optimization problems. Eng Comput 29(1):17–35
5. Valian E, Tavakoli S, Mohanna S, Haghi A (2013) Improved cuckoo search for reliability optimization problems. Comput Ind Eng 64(1):459–468
6. Burnwal S, Deb S (2013) Scheduling optimization of flexible manufacturing system using cuckoo search-based approach. Int J Adv Manuf Technol 64(5–8):951–959
7. Jain A, Chaudhari NS (2018) A novel cuckoo search strategy for automated cryptanalysis: a case study on the reduced complex knapsack cryptosystem. Int J Syst Assur Eng Manag 9(4):942–961
8. Jain A, Chaudhari NS (2018) A novel cuckoo search technique for solving discrete optimization problems. Int J Syst Assur Eng Manag, 1–15
9. Jain A, Chaudhari NS (2019) An improved genetic algorithm and a new discrete cuckoo algorithm for solving the classical substitution cipher. Inte J Appl Metaheuristic Comput (IJAMC) 10(2):109–130
10. Pandey AC, Rajpoot DS, Saraswat M (2017) Twitter sentiment analysis using hybrid cuckoo search method. Inf Process Manage 53(4):764–779
11. Ouaarab A, Ahiod B, Yang XS (2014) Discrete cuckoo search algorithm for the travelling salesman problem. Neural Comput Appl 24(7–8):1659–1669
12. Sharma RG, Keswani B (2013) Implementation of N-Queens Puzzle using Meta- Heuristic Algorithm (Cuckoo Search). Int J Latest Trends Eng Technol 2(3):343–347
13. Yang XS (2014) Nature-inspired optimization algorithms. Elsevier, Amsterdam
14. Mantegna RN (1994) Fast, accurate algorithm for numerical simulation of Levy stable stochastic processes. Phys Rev E 49(5):46–77
15. Homaifar AA, Tumer I, Ali S (1992) The n-queens problem and genetic algo"1hmr. Proceedings of the IEEE Southeast sonference, pp 262–261
16. Kilic A, Kaya M (2001) A new local search algorithm based on genetic algorithms for the n-queen problem. Proceedings of the genetic and evolutionary computation conference (GECCO 2M)I) second won memetic algorithms (2nd WOMA). pp 158–161

A Survey on Diabetic Retinopathy Detection Using Deep Learning

Deepak Mane, Namrata Londhe, Namita Patil, Omkar Patil, and Prashant Vidhate

Abstract Diabetic retinopathy happens when there are high blood pressure and high sugar level in the body that damages the blood vessels and veins in retina. These arteries can become swollen and leaky, or they may close, block the flow of blood. Sometimes new, unusual blood arteries grow in the retina part. These unconditional changes can steal your eyesight. Manual examination and analysis of fundus images to detect morphological changes in the eyes are very sluggish and tedious. In the current scenario, deep learning has been set up as the most popular approach with superior performance in various areas and over traditional machine learning methods, especially in image analysis and treatment. In this paper, we adhere to traditional strategies mainly containing input Data acquisition, pre-data processing, segmentation and data preparation, feature measurement, feature extraction, model creation, model training, model testing on testing data, and outcome and analysis of the model. We have reviewed various algorithms and their challenges that help in the diagnosis of methods used in the detection of diabetic retinopathy.

Keywords Diabetic retinopathy · Retinal abnormalities · Deep learning · Image classification · Support vector machine · K-nearest neighbor · Deep convolutional neural network

1 Introduction

The disease, diabetic retinopathy is caused by chronic diabetes. People get symptoms of this disease; about 80% of people have diabetes for more than a decade or more years. It can affect various factors, such as diabetes, improper control, and early pregnancy in women. From the analysis, we can see many people suffering from diabetic retinopathy (DR). However, the tests for DR are still done manually from professionals in real life that take time and a long and boring process and due to poor communication and delayed results eventually leading to delayed and unconscious treatment of patients. DR can damage the retina's blood vessels due to

D. Mane (✉) · N. Londhe · N. Patil · O. Patil · P. Vidhate
JSPM'S Rajarshi Shahu College of Engineering, Pune, Maharashtra 411033, India

© The Author(s), under exclusive license to Springer Nature Singapore Pte Ltd. 2022
P. Nanda et al. (eds.), *Data Engineering for Smart Systems*, Lecture Notes in Networks and Systems 238, https://doi.org/10.1007/978-981-16-2641-8_59

diabetic complications, which later led to a loss of vision or loss of eyesight subsequently. Diabetic retinopathy is asymptotic in the initial stages, so most patients remain unaware of the condition or symptoms unless there are unusual limitations and restrictions in their eyesight. Therefore, this project aims to provide an automated, relevant, and sophisticated approach using image processing and pattern recognition, leading to the identification of DR in the early stages of a patient's symptoms.

A paper's overview is as follows: Sect. 2 describes the survey of related work in the field of diabetic retinopathy. Overall summary and observation are stated in Sect. 3. Section 4 describes the major techniques used for diabetic retinopathy. The challenges faced during the work are discussed in Sect. 5, and finally, conclusion is described in Sect. 6.

2   Related Work

In 2020, Thiagarajan et al. [1] proposed learning together in a machine learning model to ensure cross-cutting. Initially, they attempted to model databases using common machine learning techniques such as logistic regression, linear discriminant analysis, K-nearest neighbors algorithm, random forest algorithm, decision tree, SVM, and Naïve Bayes algorithm. High accuracy is achieved up to 80% from stacking 5 layers of CNN in the ML model, namely four layers with Relu and 1 softmax. This model is trained using an optimizer named Adam, having a learning rate of 0.001 to get quick and efficient results.

In 2020, Jadhav et al. [2] proposed a unique method to improving the automatic detection and classification of DR by abnormal retinal detection such as hard exudates, blood clotting of blood escaped from a ruptured blood vessel, etc. The classification of images is done in 4 categories of classes named class-1 (normal), class-2 (earlier), class-3 (moderate), and class-4 (Severe). The partition is done by using a DBN-based split algorithm. Best feature selection and weight updating on the DBN is made using an algorithm called MGS-ROA. The result shows that MGS_ROA_DBN has 30.1% more advanced accuracy than neural network(NN), 32.2% more advanced than KNN, and 17.1% more advanced than SVM.

In 2020, Gangwar and Ravi [3] proposed forming hybrid-ResNet-v2 deep learning structures for spotting the DR in the patient by considering the fundus-colored images. In this proposed model, they have used the previously trained Inception-ResNetV2, and they have appended a CNN block on that. They used the datasets named Messidor, and another one is ATPOS, which is available on Kaggle. They got a higher accuracy than Google Net by 6.3%. The next time they can try to use a generative network that is adversarial and used for data over-sampling.

In 2020, Heisler et al. [4] highlighted the role of ensemble learning techniques of deep learning as the application used for differentiating diabetic retinopathy (DR) from the colored fundus images present in the OCTA dataset and also from co-registered structural images. The neural network they built using only one type of data and optimized it using ResNet, DenseNet, and VGG19 architecture and pre-trained on ImageNet pre-defined weights. The Ensemble network was built using a calibrated VGG19 pre-trained model with 92% accuracy and 90% for major stacking methods. According to the paper, the Ensemble learning method increases CNN's specifications and performance for classifying the OCTA dataset for diabetic retinopathy use case.

In 2020, Shankar et al. [5] proposed an approach based on deep learning created for the automated colored fundus image classification and recognition. They used the SDL model to sort DR fundus images into different categories of classes. By model training on the MESIDOR database, they obtained the accuracy of 99.28%, which is highest, 98.54% for sensitivity, and 99.38% for specificity. According to their future measure and planning, they wanted to improve the current model by incorporating filtration methods to improve image quality. Simultaneously, the training is improved, and the accuracy of the testing is also improved.

In 2019, Wu and Hu [6] highlighted a migration learning technique or approach using the built-in Keras pre-trained model. They started with data enhancement, which included data enlargement, flicking images, wrapping, and adjusting the image's contrast. Next, the previously trained models used were VGG19, InceptionV3, Resnet50, and at last, classified all the images into 5 classes of DR. The accuracy of this method is reached up to 61% in the InceptionV3 model.

In 2017, Yu et al. [7] preferred a deep convolutional neural network(CNN) method based on detecting exudate at the pixel level. The first professional tagged exudates image mark in the CNN model, and it is retained as an off-line classifier. They found 91.96% accuracy pixel-wise for the convolutional neural network.

In 2017, Suryawanshi and Setpal [8] offered an approach based on stitching various lessons using the gray level co-occurrence matrix-(GLCM) and forward training on pattern separation problems using a 2-layer feed-forward transmitting network that outputs neural network neurons to achieve high accuracy. The proposed method achieves 90% accuracy.

In 2016, Roychowdhury et al. [9] proposed a unique approach that distinguishes neovascularisations in a wide range of optic disk (OD) (NVD) and other regions (NVE). This phase is performed separately to achieve a low false-positive neovascularization classification result. They have distinguished features of large vessels from the NVE OD region and smaller vessels in other parts of the NVE. They found 87% accuracy of Neovascularization on the disk (NVD) and 92% of Neovascularization (NVE).

In 2017, Dr. Ramani et al. [10] proposed a two-level division to differentiate DR. The first level classification is done by combining the best first trees. The second level division is done with a combination of J48Graft trees. These include various factors such as pre-processing, quality inspection, vessel removal, optic disk detection, red wound detection, and light wound detection. This method achieved the highest accuracy of 96.14% for J48Graft trees.

In 2018, Herliana et al. [11] proposed selecting features for DR detection and manipulating a combination of swarm optimization and neural network architecture. Their trained model obtained 71.76% accuracy for the neural network method, and the neural network (NN) + particular swarm optimization (PSO) method got an accuracy of 76.11%. The result of all the tests shows an increase in the accuracy of the NN model using a selection of particular swarm optimization feature method compared to the neural network model by 4.35%.

In 2018, Kumar and Kumar [12] proposed a model developed using SVM to divide fundus images into 6 classes. Simulation of this process was done in MatLab R2015a. The efficiency of the proposed diabetic retinopathy model is based on the sensitivity and specificity of the model. This paper represents an improved schema for the detection of DR with the precise determination of the microaneurysm area and number. In the future, the author wanted to propose a rapidly growing diabetes retinopathy diagnosis automated system that can detect the things considering cotton wool and abnormal blood vessels as features from the colored fundus images.

In 2017, Roy et al. [13] proposed a diabetic retinopathy classification using support vector machines (SVM)) and feature extraction depended on fuzzy C means algorithms. The proposed system is being tested on 1000 images, of which 600 are NO DR images, and the remaining are DR images. The proposed system got 96.23% accuracy as primary accuracy. For each of these models, they used the polynomial kernel of order 2 to train SVM models.

In 2020, Chetoui and Akhloufi [14] proposed a model based on the Area under Curve (AUC) algorithm. The experimental model is trained in two key datasets, namely EyePACS and APTOS 2019. According to each database, they have found 98.4% accuracy of Retinopathy images: referable diabetes and 99% for DR, causing threatening vision and eyesight. They developed a convolutional neural network CNN based on the previously created EfficientNETB7 convolutional neural network model. Their proposed procedure has a fascinating contribution to developing a diagnostic system that can assess the degree of magnitude of DR depending on the situation and the features which are extracted from the images. They developed an explanatory algorithm that is an application of Gradient weighted Class Activation and Mapping, which was used for visual explanation for the proposed convolutional NN model results.

In 2019, Alzami et al. [15] proposed a fractal analysis of the degree of diabetic retinopathy. The study concludes that fractal analysis can properly differentiate non-DR and DR patients based on patients' input data. Use of fractal dimensions with several dimensions, such as box size, size of information, and correlation dimension. In this paper, their model can only point to grade 1 and 3 images well, but they cannot find patterns in gentle DR images as per their analysis. So, in the future, they wanted to try to get and detect red wounds on the image to remove some of the features that are not required from the fundus images.

In 2019, Arora and Pandey [16] proposed a ConvNet-based algorithm for the separation of fundus-colored images as per the class of image. They also stated the feasibility of a deep convolutional neural network CNN approach for the problem of diabetic retinopathy detection. They have created their deep learning model to identify each image in 5 different categories by considering the features extracted from the colored fundus image illustrations. From a trained model based on fundus images, they have achieved 74% accuracy. Their model was tested based on the model's learning rate and epochs that happened for training the model. Depending on their future scope, they can be trying to use the pre-trained model that can give greater efficiency and accuracy to the model's results.

In 2017, Enrique V. et al. proposed an efficient algorithmic model to detect blood arteries, small aneurysms, optic disc, and hard exudates. Their proposed model can achieve almost 95% sensitivity and an average accuracy of 85%. According to their analysis, SVM shows improved results than other supervised machine learning algorithms. Depending on their future scope, they propose that they are working on obtaining and detecting soft exudates, other than solid exudates. I also wanted to try texture analysis to improve the accuracy and improve the sensitivity of the model.

3 Observation on Literature Survey

Mostly, from all observations extracted, we know that we always obtain low data accuracy using conventional machine learning algorithms in this type of classification. Still, somehow we get high accuracy from deep learning models. One can obtain a deep learning model by using different types of algorithms. Still, from all the survey we have done on the previous papers throughout our work, we can effortlessly say that the deep convolution neural network is one of the best methods which can be used for image classification problem statements in the medical sector. Deep learning is the best and effective approach to solve the image classification type of problems. Using a deep convolution neural network, we can get an efficient model by using various approaches, and one of the techniques we can use is CNN using hyperparameter optimization to get high accuracy with better efficiency. Summarization of several methods for diabetic retinopathy is represented in Table 1.

4 Techniques

Major Techniques used for diabetic retinopathy are.

4.1 Deep Learning

Deep learning is an application of ML, which is part of artificial intelligence that has its network which can learn unsupervised data from any data source. In such cases, data can be unstructured or unlabeled. Deep learning has vast applications; it has workings similar to the human brain for data processing, which is used to translate any language, detect objects, recognize speech, and take particular decisions from a specific set of patterns.

Table 1 Summarization of several methods for diabetic retinopathy

Year and method	Accuracy achieved	Output achieved and summary	Future scope	Remark
2020 [1]	Dataset: IRDiR Disease Grading Dataset	Maximum accuracy of 85% was attained in the architecture	The solution can be improvised to achieve	CNN is one of the latest methods which is used in image processing, detection, and classification. Using CNN, we get a high accuracy for the problem statement. Also, concerning that it is very easy to implement and all the processing on images is done automatically
Convolution neural network (CNN)	85%	containing a total of 5 CNN Layers (4 with ReLU activation, 1 with Softmax activation), trained to add Adam optimizer started using a model learning rate of $10-3$ to quickly convergence of the results	exceptional accuracy and minimize the categorization of false positives and false negatives, with the employing automatic hyper-parameter tuning. Additional reinforcements can be made with the confluence of highly performing deep learning and machine learning models	
2020 [2]		Diabetic retinopathy detection is successful. The extracted result revealed that the efficiency of MGS-ROA-DBN algorithm is 30.1% which is greater than NN, 32.2% which is greater than KNN, and 17.1% which is greater than SVM and DBN algorithms	The future extension of this project will consider the separation of unusual features include spots of wool of cotton, small yellow deposits of fats inside eyes, and so on. Also, the classification can be expanded with advanced deep learning algorithms with new variations of the optimization algorithm to obtain a higher accuracy and efficiency rate	A neural network contains a huge number of interconnected processing elements called a neuron. Neural network technique is generally used in use-cases where we have to deal with images Also, KNN, SVM, DBN, ROA are used for both classification and regression types of problems
Neural network	71%			
KNN	79%			
SVM	79%			
DBN (Deep belief network)	82%			
Modified gear and steering-based(ROA)	93%			

(continued)

Table 1 (continued)

Year and method	Accuracy achieved	Output achieved and summary	Future scope	Remark
2020 [3] Hybrid Inception ResNet-v2 model	72%	They suggested and developed the formation of a hybrid Inception-ResNet-v2 deep learning model which can efficiently solve the problem of diagnosing diabetic retinopathy in a patient	In the future scope, they proposed to use generative adversarial network instead of not just data over-sampling but also for the augmentation of data	Resnet-v2 has implemented using the artificial neural network method. Resnet-v2 has powerful representational capabilities so it improves the performance of computer vision applications
2020 [4] DenseNet VGG19 ResNet50	87% 77% 71%	The association of machine learning algorithms in a single model increases CNN's predictive accuracy of distinguishing DR transmissions in OCTA datasets images	Further work can use an association of both handmade properties and extracted learned features to improve the efficiency of the model for the given problem set	VGG19 is also one type of CNN model that is very powerful with 19 layers that can classify the 1000 object categories. ResNet and Densenet is a very powerful algorithm that enables a highly efficient implementation
2020 [5] M-AlexNet	96%	Test results have represented that the SDL model provides better detection accuracy than available models	The suggested method can be expanded by looking more at Inception techniques and AlexNet methods that can provide improved performance by tuning the model based on hyperparameters	Alexnet is a high-performance model that is implemented to detect the 1000 categories of images. Alexnet allows multi-GPU training which enables training of half neurons on one GPU and remaining on another GPU's

(continued)

Table 1 (continued)

Year and method	Accuracy achieved	Output achieved and summary	Future scope	Remark
2019 [6] InceptionV3 VGG19 RestNet50	 51% 49% 61%	This paper adopts a migration learning approach, using Keras's built-in trained model to fine-tune the model as per the new database to attain better classification accuracy based on the level or class of diabetic retinopathy images	There will be some large companies that have their big platforms that offer better algorithmic techniques, and supported with the alteration of artificial intelligence in Artificial intelligence industries	Inception v3 is usually used for computer vision applications where we have to work on high-scale images. Along with that VGG19, and Resnet50 is the pre-trained models that are highly efficient in prediction
2017 [7] Convolutional neural network	91%	This paper suggested a deep learning method at the pixel level for detecting exudate detection	In this paper, they suggested that a set of candidates come out first with the final morphological openings, and then nomination points are extended to CNN's deep networks for separation	CNN is one of the popular methods which are used for problems based on image classification and recognition. It is one of the most simple and efficient methods
2017 [8] Grey Level Co-occurrence Matrix (GLCM)	90%	This paper proposes a new approach using the MESSIDOR training database and the DRIVE test database, the proposed model uses to compress GLCM properties and converts points through Gaussian work into larger sizes and incorporates these points into a 2-layer training supply network	–	GLCM is one of the methods which works on co-occurring pixel values and uses the approach of texture analysis with various applications specifically in medical sciences

(continued)

Table 1 (continued)

Year and method	Accuracy achieved	Output achieved and summary	Future scope	Remark
2016 [9]		The sturdiness of the proposed technique is analyzed to separate the components of vessel segments within the classes as NVD and NVE, respectively, and to separate the common non-DR images from images having PDR	Future attempts will be focused on building an automated system for the detection of laser scars and nerve fibrosis which results in scaling model accuracy to enhance the separation and screening accuracy in the future	
Neovascularization at the disk (NVD)	87%			
Neovascularization (NVE)	92%			
2017 [10]		Diabetic retinopathy is detected successfully without feature extraction. All this classification is done by, an open-source data mining tool	In the future, feature extraction will increase the accuracy of improvisation in the efficiency and performance of the system	ADT tree is one of the machine learning methods generally used features to determine the label or class of given specific data
Alternating Decision Tree (ADTree)	93%			
J48Graft Trees	96%			
2018 [11] Particle swarm optimization(PSO) Dataset: Dataset of diabetic retinopathy Debrecen	76%	The accuracy obtained by only using a neural network method was 71.76% but after combining feature selection by Particle swarm optimization the accuracy can be increased up to 4.35%	The accuracy can be further improved by using methods such as C4.5 and Naïve Bayes	Particle swarm optimization is a computational method that optimizes a problem by repeatedly trying to improve a solution about a given measure of quality

(continued)

Table 1 (continued)

Year and method	Accuracy achieved	Output achieved and summary	Future scope	Remark
2018 [12] Linear support vector machine (SVM) Dataset: DIABETDB1 dataset	92%	For the detection of microaneurysms, average filtering, morphological process, Principal Component Analysis that is PCA, contrast limited adaptive histogram equalization which is CLAHE, has been implemented and classification of diabetes retinopathy has done using SVM	The future implementation idea will be to implement the most advanced diabetes diagnostic system using the radial basis function neural network that is RBFNN, feed-forward neural network which is FFNN that look for cotton wool and abnormally sized blood arteries as features	SVM is one of the machine learning algorithms which is capable of classification and regression type of problem. In the case of SVM, we have to manually find the features
2017 [13] Support vector machine (SVM)	92%	It mainly uses fuzzy C methods to detect exudates, detect and remove optical disk is done using Convex Hull and to sort the fundus images into Normal and Non-proliferative diabetic retinopathy or proliferative diabetes retinopathy they have used support vector machines (SVM) as primary classification algorithm	The scope of the future is to enhance the accuracy and efficiency of the automated classification system being used	SVM or support vector machine is a machine learning algorithm used for classification and regression problems. It can solve linear and nonlinear problems and work efficiently for many practical problems

(continued)

Table 1 (continued)

Year and method	Accuracy achieved	Output achieved and summary	Future scope	Remark
2020 [14] Deep convolutional neural network (CNN) Dataset: EyePACS and APTOS 2019	96%	The algorithm used was a convolutional neural network_(CNN), i.e., based on the EfficientNETB7 CNN model. The model is trained in the diagnosis of vision-threatening diabetic retinopathy (VTDR) and referable diabetic retinopathy (RDR)	Future work will include multi-image testing on more images, improved image quality for improving accuracy, and creating a differentiated model for other types of medical imaging and other similar diseases classification and detection	Deep convolutional neural networks have proven to work very effectively in areas such as image recognition and classification
2019 [15] Random Forest Classifier Dataset: MESSIDOR dataset	80%	Here to distinguish between healthy patient data with diabetic patients and the identification of the severity of diabetes they have used the Fractal and Random Forest Classifier as primary algorithms	Analysis of features, such as multivariate and univariate or statistics of that model can be done. Apart from this red wound detection will be done to efficiently remove the unnecessary part from the features so that based on other features remaining accuracy of the model are improved	Random forest algorithms work efficiently in areas of image processing and image classification. Also, it handles the missing values and maintains the accuracy for a large proportion of data
2019 [16] Deep Neural Network	93%	The algorithm used was an application of deep learning that automatically rectifies the pattern and differentiates retina images into five categories or classes based on the severity of diabetic retinopathy which is determined from the features extracted	The scope of the future will include the use of a variety of alternative algorithms to enhance the efficiency and correctness of diabetic retinopathy diagnosis and detection systems	The deep neural network is used to extract useful patterns in the digital representation of data, i.e., images. It is an efficient way to work on image processing problems

(continued)

Table 1 (continued)

Year and method	Accuracy achieved	Output achieved and summary	Future scope	Remark
2017 [17] Support vector machine (SVM)	94%	The algorithm proposed was an SVM based to determine the degree or class of diabetic retinopathy for each retinal image by taking into consideration features that are manually extracted from the images	Future works will include the identification of cotton-wool spots, excluding solid exudates, and to improve the accuracy of model texture analysis is used so that the sensitivity of the diabetic retinopathy classification is also improved	SVM is memory efficient and used when the quantity of dimensions is greater than the quantity of samples

4.2 *Convolutional Neural Network (CNN)*

A convolutional neural network (CNN) is one of the sub-branches and application of deep learning that is, and it can be called one of the techniques in deep learning. In the simple word is one of the deep learning algorithms which can take and process the input, provide importance to various aspects and properties, and will be able to differentiate or recognize the images by extracting the patterns from the images. The basic component in CNN is neurons. In CNN, the main and important operation done is convolution. It means that for extracting information from pixels, a matrix is moved along the pixels, and for each window, it calculates some value as a feature. The architecture of a ConvNet can efficiently categories data like Neural Networks, which is proven powerful in areas like image recognition and classification. Each level of training process is represented in Fig. 1.

In CNN, the base operation that generally happens is convolution. The convolution function draws out unlike features from the given image input. The first layer of convolution extracts small features like lines, edges, and corners. Other deep layers pull out big and visible features from the given image. Operation of convolution starts from the top-left corner of the matrix and ends at the bottom-right corner of the matrix. Different variations of CNN algorithms:

- **LeNet**: It is a tiny and straightforward algorithm used for beginner's level. LeNet generally used for teaching the basics of CNN. It is the only one-layered convolutional neural network that consists of two blocks of convolutional layers and average pooling layers followed by a flattening convolutional layer, then two fully connected layers, and at last a softmax classifier used in the dense layer, which is at last.
- **AlexNet**: AlexNet is applying a convolutional neural network CNN that has made a significant contribution to the area of machine learning and deep learning, mainly in the application of ML and deep learning. It consists of 8 layers, i.e., the first 5 were layers of convolutional, some of which were followed by layers of max pooling, and the last 3 were fully connected layers. It has the activation function named ReLU, which has shown increased training performance over tan and sigmoid.
- **VGGNet**: VGG is also called very deep convolutional network used for large scale of images classification and image recognition. This network has its directness,

Fig. 1 Training process at each level

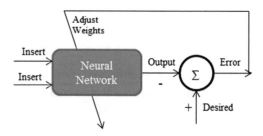

using only 3*3 convolutional layers placed upon each other to make the model deep and reducing volume as per layers increase is covered by max-pooling layers. Finally, two highly connected layers have come after a softmax classifier, which is used in dense layers of the model.

- **ResNet**: A residual neural network (ResNet) is also called an artificial neural network (ANN), which can build its structure known as pyramidal cells. The fundamental advantage of ResNet is that it allows the operation of training of the model, i.e., an extremely deep neural network with 150 plus layers can be trained successfully.
- **GoogLeNet**: GoogLeNet is one of the CNN applications that is 22 layers deep and very huge in performance.

4.3 K-Nearest Neighbor (KNN)

KNN is a pattern division algorithm that helps us determine which class of new input (test value) is in which neighbors and distance range is selected. This algorithm finds the new class of particular input based on the distance calculated between the input and nearest points, so the region for which the input has less value that region is stated as the class of that input. KNN Algorithm first recognizes the k points from the training data nearest to the test values and calculates the distance between them. The test value will belong to the category whose distance is lesser than the input. The KNN Algorithm starts scoring k in training details closer to the test value and calculates the distance between them. The test value will be for the lowest range.

For calculating distance, there are various methods:

- Euclidean Method:

$$d(p, q) = d(q, p) = \sqrt{(q_1 - p_1)^2 + (q_2 - p_2)^2 + \cdots + (q_n - p_n)^2} \quad (1)$$

The Euclidian Distance formula can be given as

$$ED = \sqrt{\sum_{i-1}^{n} (q_i - p_i)^2} \quad (2)$$

- Manhattan Method:

$$\sum_{i=1}^{n} |x_i - y_i| \quad (3)$$

Here: n = number of the dimension.
x = data point available from dataset.
y = new data point which is to be predicted.

– Minkowski Method:

$$\left(\sum_{i=1}^{n} (|x_i - y_i|)^{1/q} \right) \tag{4}$$

The case where q = 1 is similar to the Manhattan distance and the case where q = 2 is similar to the Euclidean distance.

4.4 Support Vector Machine (SVM)

SVM is a binary separator based on supervised learning that offers better performance than other supervised machine learning-based classifiers. SVM distinguishes between the two classes by building a hyperplane at the top of the feature to split. It is used to classify or for regression problems. It uses a kernel strategy to modify your data and, based on these changes, finds the perfect boundary between the possible results. An support vector machine (SVM) is an ML classifier algorithm that enlarges a line of margin. The SVM classifier aims to find a dividing line or (n-1) plane that separates the two existing classes in N-dimensional space which is represented in Fig. 2.

Let us take a glance at the equation for a straight line whose slope is m and intercept c. The equation now becomes mx + c = 0. In SVM, the hyperplane equation dividing the points (for classifying) can now easily be written as:

$$wT (x) + b = 0 \tag{5}$$

where b = Intercept and bias term of the hyperplane equation.

Fig. 2 Equation of straight line

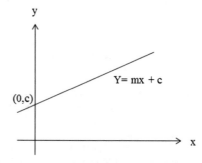

5 Challenges

All the clarity of fundus dataset images is unsatisfactory. We cannot predict the accuracy more precisely. A large amount of data causes overfitting, which leads to the detection of images as a single class, which is problematic while performing operations on the fundus images. The model used in diabetic retinopathy detection is quite large; this is not possible to run the models on a simple machine, so high-graphic card systems are required. The fundus dataset used for diabetic retinopathy detection has both diabetic retinopathy (DR) and non-diabetic retinopathy (NDR) images. Still, the amount of NDR images is more as compared to DR images leading to data inequality.

6 Conclusion

In this paper, we put forward the method of automated systems to detect the case of diabetic retinopathy present in the patients having diabetes and help them detect and classify the cases of DR before early stages symptoms to heal it with ease and with no difficulties. This paper also represents the various techniques used for the detection and diagnosis of early stage diabetic retinopathy. In the future extension, the classification can be enlarged by reliable machine learning algorithms or deep learning algorithms with new varieties of optimization algorithms for achieving increased detection rate and high classification accuracy. It is also essential to provide clinical history and other clinical data resources like fundus-colored images for cross-modality early stage DR detection.

References

1. Thiagarajan AS et al (2020) Diabetic retinopathy detection using deep learning techniques. J Comput Sci 16:305–313
2. Jadhav A et al (2020) Optimal feature selection-based diabetic retinopathy detection using improved rider optimization algorithm enabled with deep learning. Evol Intell pp1–18
3. Gangwar AK, Ravi V (2020) Diabetic retinopathy detection using transfer learning and deep learning. In: Evolution in computational intelligence, advances in intelligent systems and computing, vol 1176
4. Heisler M et al (2020) Ensemble deep learning for diabetic retinopathy detection using optical coherence tomography angiography. Transl Vis Sci Technol 9(2)
5. Shankar K et al (2020) Automated detection and classification of fundus diabetic retinopathy images using synergic deep learning model. Pattern Recogn Lett 133:210–216
6. Wu, Y-C, Hu Z (2019) Recognition of diabetic retinopathy basedon transfer learning. In: IEEE 4th international conference on cloud computing and big data analysis (ICCCBDA) pp 398–401
7. Yu S et al (2017) Exudate detection for diabetic retinopathy with convolutional neural networks. In: 39th annual international conference of the IEEE engineering in medicine and biology society (EMBC), pp 1744–1747

8. Suryawanshi V, Setpal S (2017) Guassian transformed GLCM features for classifying diabetic retinopathy. In: International conference on energy, communication, data analytics and soft computing (ICECDS) pp 1108–1111
9. Roychowdhury S et al (2016) Automated detection of neovascularization for proliferative diabetic retinopathy screening. In: 38th annual international conference of the IEEE engineering in medicine and biology society (EMBC), pp 1300–1303
10. Ramani RG et al (2017) Automatic diabetic retinopathy detection through ensemble classification techniques automated diabetic retionapthy classification. In: IEEE international conference on computational intelligence and computing research (ICCIC), pp 1–4.
11. Herliana A et al (2018) Feature selection of diabetic retinopathy disease using particle swarm optimization and neural network. In: 6th international conference on cyber and IT service management (CITSM), pp 1–4
12. Kumar S, Kumar B (2018) Diabetic retinopathy detection by extracting area and number of microaneurysm from colour fundus image. In: 5th international conference on signal processing and integrated networks (SPIN), pp 359–364
13. Roy A et al (2016) Filter and fuzzy c means based feature extraction and classification of diabetic retinopathy using support vector machines. In: International conference on communication and signal processing (ICCSP), pp 1844–1848
14. Chetoui M, Akhloufi M (2020) Explainable diabetic retinopathy using EfficientNET*. In: 42nd Annual international conference of the IEEE engineering in medicine & biology society (EMBC), pp 1966–1969
15. Alzami F et al (2019) Diabetic retinopathy grade classification based on fractal analysis and random forest. In: International seminar on application for technology of information and communication (iSemantic) pp 272–276
16. Arora M, Pandey M (2019) Deep neural network for diabetic retinopathy detection. In: International conference on machine learning, big data, cloud and parallel computing (COMITCon). pp 189–193
17. Carrera E et al (2017) Automated detection of diabetic retinopathy using SVM. In: IEEE 14th international conference on electronics, electrical engineering and computing (INTERCON), pp 1–4

A Survey on Alzheimer's Disease Detection and Classification

D. T. Mane, Mehul Patel, Madhavi Sawant, Karina Maiyani, and Divya Patil

Abstract Early Alzheimer's Disease (AD) is essential so that may take a preventive step. Current AD detection techniques depend on checks for cognitive impairment that regrettably now no longer offer an accurate prognosis until the affected person has superior beyond a moderate AD. An incorporated and correct gadget for the prognosis and class of mind issues has consequently been suggested. A smartphone is used as a tool to gather sensor datasets. Another hardware tool used is computer and software tools such as python IDE, R studio, and Anaconda. This paper depicts a detailed literature survey on various techniques for identifying AD from the brain image dataset. This survey will help researchers to give direction to their work in this domain. This paper gives a brief idea about various machine learning algorithms and challenges faced during the work. The AD is successfully detected and classified as a result.

Keywords Pattern classification · Convolutional Neural Network · Alzheimer's disease detection · Support vector machine

1 Introduction

Alzheimer's disorder has emerged as a big neurodegenerative mind disorder in aged people in latest years. There are approximately 44 million dementia sufferers globally, consistent with a examine launched via way of means of Alzheimer's Disease International, and the parent will attain 76 million via way of means of 2030 and one hundred thirty-five million via way of means of 2050. Patients with Alzheimer's disorder (AD) account for among 50% and 75% of those sufferers [1], characterized via way of insidious means with revolutionary deterioration of episodic reminiscence [2]. Mild cognitive impairment (MCI) is a disease wherein someone has variations in concept talents which are slight but noticeable. MCI with each is much additional liable than humans exclusive of it to develop AD [3]. Alzheimer's disorder has emerged as a not unusual place neurodegenerative mind disease in aged humans

---

D. T. Mane (✉) · M. Patel · M. Sawant · K. Maiyani · D. Patil
JSPM's Rajarshi Shahu College of Engineering, Pune, Maharashtra 411033, India

© The Author(s), under exclusive license to Springer Nature Singapore Pte Ltd. 2022
P. Nanda et al. (eds.), *Data Engineering for Smart Systems*, Lecture Notes in Networks and Systems 238, https://doi.org/10.1007/978-981-16-2641-8_60

in latest years. There are approximately 44 million dementia sufferers worldwide, consistent with a examine launched via way of means of Alzheimer's disease international, and the parent will attain 76 million via way of means of 2030 and one hundred thirty-five million via way of means of 2050. MCI is a disease wherein someone has modifications in questioning capacity which are slight but noticeable. Individuals with MC are much more likely than humans without it to develop AD [3]. While no drug treatments are to be had to deal with AD, sure drug treatments had been used to delay the onset of such signs and to lower the mental impact on sufferers, including reminiscence loss [4]. Accurate analysis withinside the early level of AD sufferers or MCI is consequently very critical. Application to the processing of images, particularly withinside the subject of scientific imaging. X-Ray, CT experiment, and MRI are the numerous imaging modalities, to call some images of the frame components to encompass the general view of the inner organs. For higher comprehension and examination, the diseased or affected organs of subject want to be segmented and remoted from the rest. The segmentation strategies will spotlight the smallest vicinity of hobby imparting the entire data. The strategies of segmentation are regularly automated or semiautomatic. Segmentation of the mind is a subject that has immensely involved researchers. In reading the volumetric modifications in the mind, segmentation of brain images is a critical and prerequisite level. A series of 2D images that require excessive computation, MRI consists of big data. When the images constitute a disease withinside the mind, the computational complexities emerge as amplified. The extent of numerous cranial tissues is stimulated via way of means of numerous neural issues. In diagnosing neural issues including a couple of sclerosis, stroke, and Alzheimer's disorder, segmentation, and volumetric evaluation are applied. A sturdy set of rules is mentioned right here that looks after the depth versions of pixels. This is encouraged via way of means of the purpose to offer correct and value powerful approach for our clinicians/neurologists withinside the analysis and remedy of AD. In extra than forty percentage of the older population, AD money owed for dementia, inflicting notable social subject. Eleven of them have been detected for early AD in a survey [1] on 103 topics of certified AD aged candidates (elderly 60–80). The scientific presentation of the disorder can be subtle, however, that allows you to facilitate early drug intervention and stronger prognosis, early analysis is critical. Early remedy will degenerate the development of the disorder, and therefore, it becomes very critical to apprehend the onset of AD early in a patient. In the early tiers of the disorder, the deterioration of reminiscence and cognitive loss may be very slow and is going unnoticed. Only at a completely later level does the disorder have a tendency to want care. Outline of this paper is as follows: Various papers are discussed in related work which is in Sect. 2. In Sect. 3, techniques which are being used are explained. In Sect. 4, challenges faced during work are described. Conclusion is given in Sect. 5.

2 Related Work

An easy alternative approach that costs less and much efficient is represented by [5]. Deep Learning shows Machine Intelligence's truly cutting edge. Biologically induced Multilayer perceptron Convolution Neural Networks (CNN) are particularly good at producing good image processing results. It presents deep CNN to detect Alzheimer's Disease. A method of learning is proposed by [6] to improve the data set technique used for little numbers of samples and develops a complete 3D convolutional method Model of Dense Net classification, which gives information about the image function, but also boost the model's generalization ability. In the current research [7], we checked the effectiveness of classifying well subject and AD patients using brain shape information. A P-type Fourier descriptor was used as shape details, and the lateral ventricle except the lucidum septum was analyzed. We conducted classification using a support vector machine using a combination of several descriptors as features. The findings showed 87.5% accuracy of classification, which was superior to the accuracy obtained using volume ratio to intracranial volume (81.5%), commonly used for traditional morphological modification assessment. Aziz and others talked about that the algorithms used for information mining may be implemented primarily based totally on the similarity of their attributes to corporations of various subjects in [8]. The use of information mining in medicinal drug enables to expect diverse sicknesses, apprehend the type of sicknesses in particular in neuroscience, and apprehend biomedicine. This evaluation's major goal is to use device studying and information mining strategies to the dataset of the ADNI to pick out the diverse tiers of AD. In this paper, six exclusive device studying and information mining algorithms are implemented to the ADNI dataset, such as k-nearest neighbors (k-NN), selection tree, rule induction, generalized linear version (GLM), and deep studying algorithm, to be able to outline the five exclusive tiers of the AD and to apprehend the maximum exclusive attributes through the range of times for EMCI and SMC instructions, the accuracy of the AD degree type can be besides more advantageous so that may educate the version for all instructions with ok and balanced information. Clinical research withinside the beyond has proven that AD pathology starts 10 to fifteen years earlier than obvious medical symptoms, and symptoms of cognitive disorder start to arise in sufferers identified with AD. Early prognosis of AD the usage of feasible biomarkers of early degree cerebrospinal fluid (CSF) might also be beneficial within a medical trial layout. The right remedy for sufferers with AD is explained in [9]. Three best-regarded ML algorithms had been utilized on this paper to pick out topics for impaired and secure manipulate primarily based totally on decided on attributes. The medical dataset for Alzheimer's sickness turned into acquired from the Kaggle dataset. The medical dataset represents a medical pattern of 333 topics containing MCI sufferers (n = 91) and secure manipulate topics (n = 242). A web predictor gadget can be advanced, withinside the destiny to check topics with preliminary tiers of cognitive impairment. The study explored in [10], processes to gadget getting to know to use scientific information to forecast the improvement of AD in destiny years. The All pairs technique, a unique method advanced for this evaluation related to assessing all viable pairs of temporal information factors for every patient, turned into used to technique information from 1737 sufferers. On the processed information, gadgets getting to know fashions had been educated and examined using a separate checking out information set (one hundred ten sufferers). Future paintings may want to contain checking out the version's output on different information units similar to the ADNI information set and the version's overall performance the usage of specific biomarkers than the ones decided on for this paper also to examine the potential of the gadget getting to know version noted on this paper to render correct critiques of AD sufferers. The technique is generated by integrating three feature extraction methods: histogram of directed gradients (HOG), local binary patterns, and gray-level-cooccurrence matrix is explained in [11]. The brain images were obtained from the dataset first. Then through histogram equalization, removed the unnecessary noises in recorded brain images. Then using the otsu threshold technique used the preprocessed images for skull removal. After that, use FCM (Fuzzy C-Means) to segmentation white matter, cerebro-spinal fluid (CSF), and gray matter. After hybrid feature extraction was applied, the relief feature choice was used to select the optimal feature subsets after obtaining feature information. These characteristic values were then given to identify the three Alzheimer's groups as input for SVM: normal, Alzheimer's disease, and. The ADNI dataset was used and provided 85% accuracy. This study's future scope involves introducing a new unsupervised classification technique with high-level functionality to enhance the classification of AD further. Machine learning techniques were used to identify MRI images with Alzheimer's disease, represented by Heba Elshatoury and EgilsAvots [12]. The data used for the study are 3-D brain MRI scans. To train classifiers, in which the accuracies were compared, used supervised learning algorithms. For all slices of all images, histograms are used. Then, selected unique slices for further analysis based on the highest results. The ADNI dataset was used and gave 69.5% accuracy. This research's future scope includes further study that is still being carried out to enhance the accuracy of classifier and incorporate more modalities by using positron emission tomography images.

The novel method for classifying a given MR picture as normal or abnormal was Zhang and Wu [13]. To boost the generalization of KSVM, the technique of k fold stratified used cross-validation. In this paper, they first used wavelet transformation and feature reduction method to extract features from images. The decreased features were submitted to the vector machine support kernel (KSVM). Cross validation is used to boost the KSVM. The ADNI dataset was used and provided 95% accuracy. The paper [14] explains factors line noise, post-processing strategies discussed which are useful clinical applications. This research uses MRI images and their measurements to investigate the efficacy of the Random Forest classifier. Here, 225 HC, 185 AD are split randomly into datasets for training and research, and the ADNI database is used. In the paper [15], the model method of the SVM is used to model and identify and forecast different processes of AD to make an auxiliary diagnosis of the disease based on MRI imaging data. Here, SVM model uses MRI data to produce results of classification. This demonstrates that there is good potential for the SVM methodology to adapt to AD disease and can be used to test uncertain samples.

A positive discipline of studies is represented in [16] the utility of system gaining knowledge of predicting Alzheimer's disorder from neuroimaging evidence. Using diverse dimension types, together with demographic, bodily, and cognitive statistics, the TADPOLE undertaking goals to expect the onset of Alzheimer's disorder. As monitoring has progressed, the individuals in ADNI every advanced a time collection of measurements. Three categorized units of those time collection shape the TADPOLE leaderboard dataset, LB1, LB2, and LB4, respectively. We have proven how may be found out the connection among pairs of statistical factors for predicting Alzheimer's disorder to use a system gaining knowledge of gadgets at great time separations. Different quantitative clinical rules for listing demented and non-demented patients were explored in the recent AD-based study [17] based on MRI numerical metrics provided by Washington ADRC. Various related features were also analyzed that were highly dependent on defining AD patients' categories and performing Exploratory Data Analysis (EDA) among different elements in a given dataset. The data collection used in this analysis was derived from elderly adults MRI information. Each record includes 15 characteristics, including age, sex, and other MRI measurements. In this one, the process offers 87% precision compared to 87% accuracy. In the paper [18], Consensus-Based combining is the new method used for fusing a troupe of classifiers. After comparing all the outputs of the classifiers, CCM adapts the weight iteratively an ultimately, all the weights meet toward ending weight place. In the paper [19], the genetic programming algorithm was tailored to solve attribute reduction and demonstrated the use of an intelligent genetic programming mechanism to create new programs to increase the chance of seeking high-quality reductions. The summary of the various methods is represented in Table 1.

2.1 The Summary of the Literature Survey

Summarization of several methods for Alzheimer's Disease detection and classification is represented in Table 1.

3 Techniques

3.1 Support Vector Machine

SVM is a supervised learning method that looks at knowledge and classifies it into one of two groups. In addition to carrying out a linear classification, SVMs can also effectively carry out a nonlinear classification. SVM aims to find the best dividing boundary between the data. We deal with the vector space in the help vector machine so that the separating line is actually a separating hyperplane. Nothing but a decision boundary is a hyperplane. As the best hyperplane, the hyperplane which has the

Table 1 The summary of the various methods

Method/dataset	Accuracy achieved	Output achieved	Summary	Future scope
(2019), [15] Method: CNN (Convolutional Neural Network Dataset: Smartphone-based single sensor dataset, Multi-sensor based dataset	85%	Alzheimer's Disease-based classification successful	Classifier categorizes every day activities that are used to track mind tissue degradation	This work can be extended in future research where prediction can be improved
(2019), [20] Method: KNN, Decision tree Dataset: ADNI	88.24%	Five different stages of Alzheimer's disease were classified	various algorithms were tested like KNN, DT, Naïve Bayes, CNN, deep learning algorithm to check the accuracy of each	The classification accuracy can be improved
(2017), [9] Method: Naïve Bayes, Sequential minimal optimization Dataset: Kaggle	82.19%	Classification of impaired and healthy control is implemented successfully	Three best known ML algorithms NB, SMO, J48 were used for classification	In future, an online prediction system can be build
(2019), [13] Method: All pairs technique Dataset: ADNI	86.66%	Prediction of AD is implemented successfully	Using separate testing datasets, ML models were trained and evaluated	Evaluation of the models performance on other datasets
(2019), [11] Method: Support vector machine (SVM) Dataset: ADNI	85%	Three stages of AD are classified successfully. (normal, AD, mild cognitive impairment)	Segmentation was carried out using Fuzzy-C-Means and SVM is used for classification	In the future work, A new unsupervised classification methodology can be implemented
(2019), [12] Method: SVM Dataset: ADNI	69.5%	Classification of MRI images between AD and cognitively normal is implemented	MRI brain scans with axial, sagittal, coronal view were used	Future study will be carried out to improve accuracy of classifier

(continued)

Table 1 (continued)

Method/dataset	Accuracy achieved	Output achieved	Summary	Future scope
(2019), [15] Method: SVM Dataset: ADNI	94%	Classification of impaired and healthy control is implemented successfully	Classification is done through SVM	Technique will be evaluated within mild cognitive impairment and other diseases
(2012), [21] Method: NMF-SVM Dataset: SPECT	91%	Prediction of AD is implemented successfully	They have proposed a tool based on non-negative matrix factorization	Model will be evaluated for the classification of MCI and non-cognitive
(2019), [22] Method: Random Forest Dataset: unitprot	85.5%	Classification of AD is carried out successfully	Random Forest algorithm was used for the classification	Accuracy can be improved
(2019), [17] Method: Random Forest Dataset: ADNI	82%	Classification of images between AD and cognitively normal is implemented	With and without neuroanatomical constraints, Various structural MRI measures used to training	Future study will be carried out to provide better accuracy
(2019), [16] Method: Random Forest Dataset: ADNI	84%	Classification of impaired and healthy control is implemented successfully	They have concentrated particularly on the major risk factors which are responsible for AD using Random Forest classifier	In future will try to use efficient techniques and accuracy
(2020), [23] Method: EEG microstate complexity Dataset: IWG-2 and DSM-IV	85%	Prediction of AD is implemented successfully	The idea is that molecular and neuroimaging markers do not measure brain function, so it is likely that combining them with functional markers will further enhance early diagnosis of AD specificity	In future will try to increase the efficiency and accuracy

(continued)

Table 1 (continued)

Method/dataset	Accuracy achieved	Output achieved	Summary	Future scope
(2017), [24] Method: EEG and SVM Dataset: ADNI	94%	Classification of AD is carried out successfully	Wavelet, spectral, and difficulty based pattern were calculated and classified using SVM	It will include automated EEG data diagnosis and classification using different classifiers such as NNs, L DA, etc.
(2014), [25] Method: SVM, Decision tree Dataset: ADNI	80%	Classification of MRI images is implemented	Eitj atlas registered normalization was preprocessed with MRI images and the decision tree for classification kernel support vector machine was used	Future study will be carried out to provide better accuracy
(2012), [26] Method: KSVM Dataset: ADNI	95%	MRI images are successfully classified	They have extracted the features by applying principal component analysis (PCA)and were classified using kernel SVM	In future wavelet transform used to speed up computations
(2020), [27] Method: logistic regression, decision tree Dataset: OASIS Longitudinal MRI dataset	84%	Prediction AD and cognitively normal is implemented successfully	ML models were trained for identifying the patients on the basis of fourteen features available in MRI dataset	In future, this is utilized in applying learning models in different areas like health, agriculture, etc.

widest margin between the support vectors is considered. It is essentially a line between the data classes that divides them. There are two other lines in SVM other than the hyperplane that produce a margin. In the SVM Classifier, the hypothesis function h is defined as:

$$h(xi) = \begin{cases} +1 \ if \ w \cdot x + b \geq 0 \\ -1 \ if \ w \cdot x + b \ < 0 \end{cases} \tag{1}$$

3.2 Convolutional Neural Network

A deep learning network used for image recognition is the Convolutional Neural Network. They can be used for analysis and arrangement of other data as well. In Neural Network, there are three layers, i.e., input layers, hidden layers, and output. Input layer admits the input inside dissimilar ways, and the hidden layers measure the input, and the output layer gives the result. Each of these layers contains neurons, and the weight of each neuron is included. They are only connected to their neighboring neurons, and all have the same weight rather than being connected to any neuron of the previous layer, as they perceive the data as unique. The Convolution Neural Network works differently. There are several layers of Convolutional Neural Networks. There are a few layers, the convolutional layer and the pooling layer, that make it unique. However, it also has a multilayer and RELU layer like other Neural Networks. RELU layer acts as an activation mechanism, and the completely connected layer helps us perform classification on the dataset.

$$(f * g)(t) = \int_{-\infty}^{\infty} f(\tau)g(t - \tau)d\tau \qquad (2)$$

3.3 Decision Tree

One of the most widely used algorithms in the outside world is the decision tree. It can be used for both regression and classification. It is used for classification most of the time. The decision tree algorithm identifies and expresses the connection between the target column and the independent variables as a tree structure. A tree of decisions consists of nodes and arcs. Every node the decision tree algorithm represents an attribute, and each terminal node represents a decision. We normally break down our data in the decision tree by making a decision based on a set of yes and no questions. A classification rule is called the path from the root to the leaf. Each internal attribute represents the characteristics on which a specific decision needs to be taken. Decision nodes are like tests in which any particular condition is checked, and the outcome of this test is in the form of yes and no or high and low. The decision tree aims at building a training model to predict the class of the objective variable by the decision rules, using previous data. We start from the tree's root to predict a class, then compare the values of parent attribute by record attribute, and then follow the stem based on comparison and jump to the net node. We may use a simple squared equation as a cost function for a regression tree:

$$E = (Y - \hat{Y})^2 \qquad (3)$$

3.4 Logistic Regression

A supervised learning algorithm is a logistic regression. It can be used for binary classification or classification of multiple classes. Binary classification if we have two possible outcomes, such as whether or not a person is infected with the illness. And multi-class classification means that we have multiple results, such as influenza or an allergy or cold. To predict the outcome variable, which is categorical, logistic regression is used. The categorical variable is like a variable that can only have a limited and specific value. The threshold value in logistic regression indicates the likelihood of winning or losing. Logistic regression contains one sigmoid function; the sigmoid curve's S shape curve is called the sigmoid curve. We can get the output in the classification type through the use of this sigmoid function. The cost function for logistic regression is:

$$J(\emptyset_0, \emptyset_1) = \frac{1}{2m} \sum_{i=1}^{m} (h\emptyset(x^{(i)}) - y^{(i)})^2 \tag{4}$$

4 Challenges

As dementia is a progressive disease, there are various techniques used to study patients of dementia. Some of the techniques need to collect daily results, and some need to analyze other health issues rising because of it. Collecting data daily, keeping track of patients is quite a struggle as the data needs to be complete and updated to achieve the desired prediction and accuracy. This is the reason that developers find it difficult to collect the required data. Another challenge faced related to data is data engineering. Data engineering means to work with data that includes data cleaning, feature extraction, modeling, etc. Pre-processing data is another challenge; as the data comes from different sources, it needs to be processed. If the data is complete and accurate, then it becomes easy to work with it. But if the data is incomplete and ambiguous, it is time-consuming and flinty to achieve the desired prediction. Although several techniques solve the problem, selecting an algorithm is one of the biggest challenges faced while working on Alzheimer's disease classification. Applying the right machine learning methods, choosing proper parameters to tune the algorithms, adopting the right strategies are some other challenges related to methods and techniques.

- **Data quality**: I. Getting enough data II. Completeness of the features III. The ability to aggregate from different sites (differences in data collection methods, privacy-sensitive patient data) IV. Collecting regularly tracked and monitored data.

- **Model bias**: We need to cross-verify that the model we have learned is appropriate to use on a particular patient, e.g., there could be bias in the population that was learned. Hence to classify the patient, we should be aware of model bias.
- **Model validation**: The model needs to be tested on clinical samples or test set for validation purposes as the prototype may vary from the real application.

5 Conclusion

After referring to various papers, we can conclude that there are several challenges that we need to overcome. For that, we can perform fair experiments. This is where many methods fail and do not lead to any significant impact. Every algorithm has its own advantages and accuracy. The AD classification is done successfully using Random Forest and various datasets. The UniProt dataset gave 85.5% accuracy and needed to work on it to achieve more accuracy. The dataset named SPECT is used for implementing the SVM technique, which gave 91% accuracy for the prediction of AD. The ADNI dataset gave 94% accuracy where the classification of AD is carried out successfully. The SVM technique which is used for classification, this technique is used with various datasets such as with ADNI it provides 94% accuracy where the classification of impaired and healthy control is implemented successfully; the ADNI dataset provided the highest accuracy of 95% where the classification of MRI images as normal and abnormal is implemented successfully. Using the EEG and SVM method, AD classification is carried out successfully with 94% accuracy, and the ADNI dataset is used. CNN provides 85% accuracy and is an effective method to implement. At least we can conclude that researchers can use this paper as a reference to get an overall idea about previous work in the domain of AD detection and classification to get ahead with their research.

References

1. Prince M, Albanese E, Guerchet M et al (2014) World Alzheimer Report 2014: dementia and risk reduction an analysis of protective and modifiable factors
2. Blennow K et al (2006) Alzheimer's disease. The Lancet 368:387–403
3. Fargo K, Bleiler L (2014) 2014 Alzheimer's disease facts and figures. Alzheimers Dement 10:47–92
4. Yiannopoulou KG, Papageorgiou SG (2013) Current and future treatments for Alzheimer's disease. Ther Adv Neurol Disord 6(1):19–33
5. Ullah HMT (2018) Alzheimer's disease and dementia detection from 3D brain MRI data using deep convolutional. In: 3rd international conference for convergence in technology (I2CT)
6. He G et al (2019) Alzheimer's disease diagnosis model based on three-dimensional full convolutional DenseNet. In: 10th international conference on information technology in medicine and education (ITME), pp 13–17
7. Fuse H et al (2018) Detection of Alzheimer's disease with shape analysis of MRI images. In: Joint 10th international conference on soft computing and intelligent systems (SCIS) and 19th international symposium on advanced intelligent systems (ISIS), pp 1031–1034

8. Shahbaz M, Ali S, Guergachi A, Niazi A, Umer A (2019) Classification of Alzheimer's disease using machine learning techniques . In: International conference on data science, technology and applications (DATA 2019), pp 296–303

9. Hassan SA, Khan T (2017) A machine learning model to predict the onset of Alzheimer disease using potential cerebrospinal fluid (CSF) biomarkers. Int J Adv Comput Sci Appl (IJACSA) 8(12)

10. Albright J (2019) Forecasting the progression of Alzheimer's disease using neural networks and a novel preprocessing algorithm. Alzheimer's Dement Transl Res Clin Interv 5:483–491

11. Suresha HS, Parthasarathi S (2019) Relieff feature selection based Alzheimer disease classification using hybrid features and support vector machine in magnetic resonance imaging. Int J Comput Eng Technol 10:124–137

12. Elshatoury H, Avots E (2019) Volumetric histogram-based alzheimer's disease using support vector machine. J Alzheimer's Dis

13. Zhang Y, Wu L (2012) An MR brain image classifier via principal component analysis and kernel support vector machine. School of Information Science and Engineering, vol 130

14. Lebedev AV, Westman E, Westen GV, Kramberger MG, Lundervold A, Aarsland D, Soininen H, Kłoszewska I, Mecocci P, Tsolaki M, Vellas B, Lovestone S, Simmons A (2014) Random Forest ensembles for detection and prediction of Alzheimer's disease with a good between-cohort robustness. NeuroImage: Clin 6:115–125

15. Fan Z, Xu F, Qi X, Li C, Yao L (2019) Classification of Alzheimer's disease based on brain MRI and machine learning. Neural Comput Appl 32:1927–1936

16. Moore PJ, Lyons T, Gallacher J (2019) Random forest prediction of Alzheimer's disease using pairwise selection from time series data. PLoS ONE, vol 14

17. Khan A, Zubair S (2019) Usage of random forest ensemble classifier based imputation and its potential in the diagnosis of Alzheimer's disease. Int J Sci Technol Res 8:271–275

18. Alzubi OA, Alzubi JA, Tedmori S, Rashaideh H, Almomani O (2018) Consensus-based combining method for classifier ensembles. Int Arab J Inf Technol 15:76–86

19. Alweshah M, Alzubi OA, Alzubi JA, Alaqeel S (2016) Solving attribute reduction problem using wrapper genetic programming. Int J Comput Sci Netw Secur

20. Raza M, Awais M, Ellahi W, Aslam N, Nguyen HX, Le HM (2019) Diagnosis and monitoring of Alzheimer's patients using classical and deep learning techniques. Expert Syst Appl 136:353–364

21. Padilla P, Lopez M, Ramirez J, Salas-Gonzalez D, Alvarez I (2012) NMF-SVN based CAD tool applied to functional brain images for the diagnosis of Alzheimer's disease. IEEE Trans Med Imaging 31

22. Xu L, Liang G, Liao C, Chen G, Chang C (2019) K-Skip-n-gram-RF: a random forest based method for Alzheimer's disease protein identification. Front Genet 10

23. Tait L, Tamagnini F, Stothart G, Barvas E, Monaldini C, Frusciante R, Volpini M, Guttmann S, Coulthard E, Brown JT, Kazanina N, Goodfellow M (2020) EEG microstate complexity for aiding early diagnosis of Alzheimer's disease. Sci Rep 10

24. Kulkarni N, Bairagi V (2017) Extracting salient features for EEG-based diagnosis of Alzheimer's disease using support vector machine classifier. IETE J Res 63:11–22

25. Zhang Y, Wang S, Dong Z (2014) Classification of Alzheimer disease based on structural magnetic resonance imaging by Kernel support vector machine decision tree. Prog Electromagn Res 144:171–184

26. Zhang Y, Wu L (2012) An MR brain images classifier via principal component analysis and kernel support vector machine. Prog Electromagn Res 130:369–388

27. Vidushi AR, Shrivastava AK (2020) Diagnosis of Alzheimer disease using machine learning approaches. Int J Adv Sci Technol 2(4)

An Approach for Graph Coloring Problem Using Grouping of Vertices

Prakash C. Sharma, Santosh Kumar Vishwakarma, Nirmal K. Gupta, and Ashish Jain

Abstract The algorithm works by dividing the nodes of a graph G into two groups; one is non-visited type of groups including the nodes that are not colored and visited type of groups including the nodes that are already colored and hence finds minimum number of colors that have been filled into visited nodes. An assumption is taken that k number of colors is already given, and the colors are selected from the same k colors. The proposed algorithm is implemented on random graphs along with some well-known graph coloring DIMACS benchmarks. In this research paper, an efficient graph color algorithm is proposed that uses a reduced number of colors for the well-known graph coloring problem. This projected algorithm can be applied to all types of graphs.

Keywords Graph coloring · Neighbors of a vertex · Grouping of vertices · Combinational optimization · DIMACS

1 Outline

There are many NP-classes combinatorial issues, and graph coloring problem (GCP) is one of them [1–3]. Thus, researchers strained to discover the solution of the GCP by implementing some novel methods compared to the former standing methods [4–8] for their projected techniques so that its running complexity could be additionally reduced. GCP has been explored in various literature due to extensive types of its practices in numerous real-life technical applications. Assume a directionless graph

P. C. Sharma · S. K. Vishwakarma (✉) · N. K. Gupta · A. Jain
School of Computing and Information Technology, Manipal University Jaipur, Jaipur, India
e-mail: santosh.kumar@jaipur.manipal.edu

P. C. Sharma
e-mail: prakashsharma12@gmail.com

N. K. Gupta
e-mail: nirmalkumar.gupta@jaipur.manipal.edu

A. Jain
e-mail: ashish.jain@jaipur.manipal.edu

G which has mainly two components; one is N which is the set of nodes (vertices) and second one is L which is collection of links between nodes in the graph G.

The problem of coloring is to get the minimum colors which can apply in allocating the colors to the nodes in the graph with the limitations as no neighboring nodes in graph have allocated similar color. The graph is painted by least numeral of different colors, which is known by "chromatic number" of the graph.

There are many applications where GCP has applied; few of them are satellite scheduling problem [9], timetable scheduling [1, 10], register allocation in CPU [11], air traffic management [1], channel assignment problem in cellular network [10, 12, 13] and all other resource scheduling issues. Graph coloring concept can be reduced to SAT problems and vice versa [12–14] to understand issues related to NP-complete problems.

The planned technique is constantly completed by allocating suitable color to all node in the graph. As it certainly not produced other implicit limitations throughout coloring process, so, the running period is effective and minimum. The method established in this manuscript is a very simple algorithm. As per definition of graph, we know that each node is related (linked) to other nodes straight or indirectly, the straight linked nodes are adjacent node of each other. This can be two types of neighbors: 1st one is traversed neighbors (T-neighbor) and 2nd one is not-traversed neighbors (NT-neighbors). In such situation, we marked traversed neighbors, T-neighbor and not-traversed neighbors, NT-neighbors. The algorithm will work on pool of k colors (C1, C2, …, Ck). This algorithm takes colors from the pool of k color and thus efficiently assigns the distinct colors to graph as per graph coloring constraints using a smaller number of colors. We will explain our proposed method with the help of an instance of graph and do experiments to see how our results will be compared with previous ones.

This manuscript is organized as follows: In Segment 2, we reviewed our proposed graph coloring strategies through pseudocode. In Segment 3, we described our approach of graph coloring by an example. In Segment 4, we showed experimental results and comparisons. Last Segment 5 summarized this research manuscript.

2   Proposed Methodology

Under this segment, there is discussion about our approach for effective coloring of graph with smaller count of colors. Our algorithm will take input in the form of a graph G = (N, L), along with a pool of k colors (C1, C2, …, Ck). Outcome of our approach will be the graph G with colored nodes by proper coloring scheme. Our approach manages a list of nodes and then designates it by numbers starting from 1 to N where N is the overall count of nodes for the given graph. This assigns a color to the node 1 with the first color from the given pool of k colors. Then, it starts traversing every node in a sequential order. It picks a node and checks whether it is colored or not. If the node is not assigned any color, then it calculates all its N neighbors and traverses one by one. Once a node is traversed, it is colored in such a way that no two contiguous nodes have the same color.

Our approach uses a pool of k colors (C1, C2, ..., Ck). Each time when it picks a node j for assigning color, it searches in the pool of k colors (C1, C2, ..., Ck) to calculate the next color to be assigned for j. After traversing all nodes, this method gives outcomes as a colored graph along with smaller count of colors then finally stops procedure. Further, we explained our procedure in Fig. 1 as flowchart.

Our research paper is proposing a novel method to color nodes (vertices) of a graph. Two groups are formed by the algorithm that is visited node group and non-visited node group. The flow chart of the proposed procedure is shown in Fig. 1.

Fig. 1 Flow chart of proposed algorithm

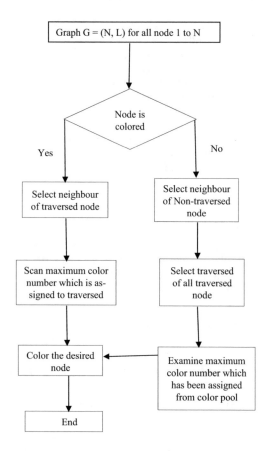

Proposed Approach (Algorithm):

Input: A Graph $G = (N, A)$ with number of node N; Number of colors $= k$.

Output: The input graph $G = (N, A)$ with number of color k which efficiently colored node.

Start
Assume the nodes of G be designated by numbers from 1 to N
Color node 1 by the 1st color from pool of k colors (C1, C2, ..., Ck).
For each nodes i $(1 \leq i \leq N)$ of G
   If node i is painted by color, then.
      For every N neighbours j $(1 \leq j \leq N)$ of the node i
         Find each of l its V-Type neighbours
            Scan the pool of k colors (C1, C2, ..., Ck) starting to find the 1st color C_k
               which is not assigned in any NT-neighbour
         Color the node j by color C_k
      Close For
   Close If
Else
Determine every NT-adjacent (neighbour) of node i
Examine the pool of colors (1, 2, ..., k) by scanning from starting to find a 1st color
from pool of k colors (C1, C2, ..., Ck) that is not assigned in any NT-neighbour
Color the node i from pool of k colors $(C_1, C_2, ..., C_k)$.
Close Else
Close For
Terminate

In the form of pseudocode, we can write above procedure as below:
Initialize $i = 1$ and $n = N$
While $(i \leq n)$
{
if colored $(Ni) =$ true
then
{
N_Visited _Neighbor $i =$ select_non_visited (Ni)
$j = 1$
While $(j \leq$ N_visited_Neighbor $i)$
{
visited_Node j $=$ select_visited_Node (Vj)
Ck $=$ Scan_visited Node_color (visited_Node j)
Color (Vj, Ck $+ 1)$
j++
}
else
{

```
V_Neighbor i = Select_visited Node (Vi)
Ck = Scan_visited Node (visited_Neighbor i)
Color (Vi, Ck + 1)
}
i++
}}
```

3   Illustration by an Example

We try to understand the working of our approach through considering a graph as input which is displayed in Fig. 2 along with a pool of k colors (C1, C2, ..., Ck). Assume the pool of colors having blue, red, green, yellow, pink, and gray color correspondingly. Firstly, designate numbers each node of graph from 1 to 10. Our approach paints the node 1 with blue. For i = 1, it picks node 1, as node 1 is previously colored, so it starts assigning colors its NT-neighbor j. For j = 1, it picks node 2 and determines its every NT-neighbors, node 2 has only one NT-neighbor node 1, so our method assigns a color red to node 2. Now for i = 2, it picks node 2, that is already assigned a color, so it begins assigning colors its NT-neighbors. Node 2 has 5 NT-neighbors. So, our procedure runs from j = 1 to 5. For j = 1, method assigns colors to nodes 10 with blue, for j = 2 it colors node 3 with blue, for j = 3 it colors node 9 with blue, for j = 4 it colors node 4 with green, and for j = 5, it colors node 7 by green. Subsequently, nodes 4 and 7 are already painted by blue and red in its NT-neighbors that is why now select green. On behalf of i = 3 to 10, methodology trails the similar iteration so that outcome will be a proper colored graph. The flow of our procedure is displayed on Table 1 which is outcome for input graph displayed in Fig. 2.

Fig. 2 Random graph

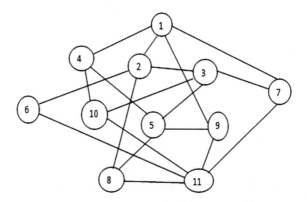

Table 1 Steps of implementation of proposed algorithm to Graph in Fig. 2

Steps	Current vertex	Color	Non-traversed neighbor	Traversed neighbor
1	1	Blue	2, 4, 7, 9	-
2	2	Red	–	1
3	4	Red	–	1
4	7	Red	–	1
5	9	Red	–	1
6	3	Blue	5, 10	2, 7
7	5	Green	–	3, 4, 9
8	10	Green	–	3, 4
9	6	Blue	11	2
10	11	Yellow	–	6, 7, 9, 10
11	8	Blue	–	2, 5, 11

Implementation with Example: The proposed algorithm can be implemented globally for all graphs. The graph shown in Fig. 2 is used to evaluate the performance of the proposed algorithm.

No. of vertices in graph (n) = 11.

Let the color vector C = [blue, red, green, yellow, pink, gray…]. On implementing the proposed algorithm for above graph in a very first step the vertex 1 is colored using the color blue, and all the next steps are shown in Table 1.

The algorithm uses 4 colors that are blue, red, green, and yellow. Initially, the vertices of the graph marked by 1 to 11. Node 1 is colored by color blue, and now in step 2, the node 1 is previously colored thus the situation begins coloring its non-visited neighbor. Vertex 2 has only 1 visited neighbor, i.e., vertex 1, and on scanning the color of vertex 1 is blue color; therefore, the algorithm colors the vertex 2 by red color. Currently, this considers node 2 that is previously colored therefore scanning is processed as a result vertex 6,4,7, and 9 are vertices found to be the non-visited neighbor of vertex 3, and the algorithm will continue until all the vertices of the graph are colored. As a result, the graph is colored using 4 colors.

4   Result and Discussion

The proposed algorithm is implemented for some random graphs that have different number of vertices and edge densities. The comparison among graph coloring using genetic algorithm [7] KGA and result of our approach Knew is shown in Table 2. As it is clear that most of the cases, the proposed algorithm of graph coloring provides better results than our new algorithm.

Table 2 Experimental results and comparisons of ECG and our proposed algorithm

Vertices	Edges	K_{GA} [7]	K_{new}
450	16,680	18	15
138	493	6	6
250	3218	10	8
64	728	12	9
95	755	9	7
191	2360	12	11
81	1056	12	10

5 Conclusion

There has been presented an effective approach to resolve the graph coloring problem. Results and experiments demonstrate that this approach can be applied for any type of graph. Though neighbors are classified into two types, therefore we have mainly focused NT-neighbors only, it reduces the running time. Our proposed approach is user friendly and easily understandable.

References

1. Brelaz D (1979) New methods to color vertices of a graph. Commun ACM 22:251–256
2. Garey MR, Johnson DS (1979) Computers and intractability: a guide to the theory of NPcompleteness. W.H. Freeman and Company, San Francisco
3. Brelez D (1979) New methods to color the vertices of a graph. Commun ACM 22(4):251–256
4. Turner JS (1988) Almost all k-colorable graphs are easy to color. J Algorithms 9(1):63–82
5. Sharma PC, Chaudhari NS (2019) A tree based novel approach for graph coloring problem using maximal independent set. Springer J Wirel Pers Commun 110(3):1143–1155. (Sept 2019)
6. Sharma PC, Chaudhari NS (2015) Maximal independent set based approach for graph coloring problem. Int J Comput Eng Appl (IJCEA), IX(II):205–214. (February 2015)
7. Marino A, Damper RI (2000) Breaking the symmetry of the graph colouring problem with genetic algorithms. In: Genetic and evolutionary computation conference (GECCO-2000)
8. Al-Omari H, Sabri KE (2006) New graph colouring algorithms. Am J Math Stat 2(4):739–741
9. Zufferey N, Amstutz P, Giaccari P (2008) Graph colouring approaches for a satellite range scheduling problem. J Sched 11(4):263–277
10. Smith DH, Hurley S, Thiel SU (1998) Improving heuristics for the frequency assignment problem. Eur J Oper Res 107(1):76–86
11. de Werra D, Eisenbeis C, Lelait S, Marmol B (1999) On a graph-theoretical model for cyclic register allocation. Discret Appl Math 93(2–3):191–203
12. Sharma PC, Chaudhari NS (2016) Investigation of satisfiability based solution approach for graph coloring problem. Int J Eng Adv Technol (IJEAT) 6(1):106–112
13. Sharma PC, Chaudhari NS (2012) A new reduction from 3-SAT to graph K-colorability for frequency assignment problem. Int J Comput Appl (IJCA) 1:23–27. (Special Issue on Optimization and On-chip Communication ooc, Feb2012)

P. C. Sharma et al.

14. Sharma PC, Chaudhari NS (2011) Polynomial 3-SAT encoding for k-colorability of graph. Int J Comput Appl (IJCA) 1:19–24. (Special Issue on Evolution in Networks and Computer Communications)

Role of PID Control Techniques in Process Control System: A Review

Vandana Dubey, Harsh Goud, and Prakash C. Sharma

Abstract Process control system (PCS) is the mixture of chemical engineering and control engineering. Process control is the skill to supervise and alter a process to offer a preferred output. It is used in industry to sustain worth and improve presentation. Preferred output can be achieved with the use of proportional–integral–derivative (PID) control in process control system. The majority of the process control systems used PID controller, for the reason of its easy configuration, ease of realization, and energetic investigation in tuning the PID. The methods discussed in the paper are classified from conventional to artificial intelligence (AI) employed for the PID controller. This paper aim is to concentrate on the journalism evaluation of PID controller in a period of process control system. The most important reason of this review paper is to present in comprehensive for the group of people to know the control of PID controller in industrial control systems.

Keywords PID controller · Process control · Artificial intelligence · Tuning techniques

1 Introduction

Process control system (PCS) at times called industrial control systems (ICS). A PCS is completed with a set of electronic devices that offer constancy, correctness, and abolish injurious shift statuses in construction processes. Feedback loops have been calculating nonstop processes in PCS since 1700s [1]. In 1769, James Watt's urbanized steam engine and governor and that was received as an earliest negative feedback tool [2]. J. C. Maxwell formulates an arithmetic model of vapor engine for governor control in 1868 [3]. Two categories of the governors are listed by Maxwell: genuine and moderator governors. In keeping with current technologies, established

V. Dubey · P. C. Sharma (✉)
School of Computing & Information Technology, Manipal University Jaipur, Jaipur, India

H. Goud
Department of Electronics and Communication Engineering, Indian Institute of Information Technology Nagpur, Nagpur, India

© The Author(s), under exclusive license to Springer Nature Singapore Pte Ltd. 2022
P. Nanda et al. (eds.), *Data Engineering for Smart Systems*, Lecture Notes in Networks and Systems 238, https://doi.org/10.1007/978-981-16-2641-8_62

the moderator as the only P controller action, while certain governors as PI controllers action [4].

PID controller is an initial control technique for process manufacturing, which was invented in the early 1900s. Minorsky's contribution was primarily discarded by naval operators, while it was a big support for the succeeding appearance of present PID controllers. In 1911, first PID controller was developed by Elmer Sperry, which was used by the US Navy [5]. Full-scope pneumatic controller, "Pre-act/Hyper-reset" and "reset" (floating) action were invented by the Taylor Instrument Companies.

Due to of these two actions, a new controller was developed that is called PID. The PID control is broadly used technology in production for the control of business-critical manufacturing processes. Proportional controller had the problem of steady-state error which be reduced via regulating the point in a specified range until error becomes zero. Proportional integral controller was developed after resetting the integrated error. First pneumatic control instrument was invented in 1940 by Taylor instrument companies with a derivative action for overcoming the problem of over-shooting [6].

For tuning, the PID control's parameters "Zieglers and Nichols alteration method" were implemented in 1942 by Zieglers and Nichols [7]. In mid-1950s, automatic PID controller was developed for industrial use. Tuning of PID controllers is able to appear a mystery. Therefore, investigators have extra ears on PID alteration such as auto and self-alteration [8, 9], genetic alteration [10], optimal and robust alteration [11], and many more. As well, AI-based PID controlling strategies presented in [12], PID fuzzy controlling in [13], optimal PID controlling technique propose in [14], adaptive PID controller [15], and fractional order PID [16] are proposed.

With the use of artificial intelligent-based PID tuning techniques, PCS may get so many benefits like enhance system throughput without extra expenditures, enhance automation and reduce human involvement, increase system ability to take on extra work, improve power efficiency. PID is well-known controllers within all industrial purposes due to their straightforward configuration, acceptable control outcome, and satisfactory robustness. Above 95% of the instruments in PCS are controlled by the PID controller, in 98% cases, pulp and paper factories also used PID controlling strategy [17]. Even though this is a long-establish kind of control technique in general. As in mid of twentieth century, one of the conclusions reached at the conference of IFAC on "Advances in PID control" Brescia, Italy-2012, was that at the foundation level the PID controllers will continue as the favorite control techniques in malice of other hopeful proposals, for example, predictive model control techniques [18].

The paper organization is as follows: Sect. 2 focused on the basics of the PID controller followed by the PID tuning techniques like conventional and artificial intelligence, Sect. 3 describes the fundamental of Process control system followed by different reactors used in it. The last section presents the conclusion of the work.

2 PID Controller

2.1 Basics of PID Controller

The PID controller is defined as stated into their characteristics of controller gains as shown in Fig. 1. The response of the controller is the function of controller gains. Controller parameters are managed both steady state and transient responses, and PID controller provides a proficient result in genuine world control troubles [5].

Where e (t) = r (t) − y (t) is the error occurred between output and reference input.

The weighted amount of these three (K_p: proportional, K_i: integral, and K_d: derivative) events is used to regulate the method using a control aspect such as manipulation of the process variable. The usual PID controller is in the subsequent form:

$$u(t) = K_p e(t) + K_i \int_0^t e(t)dt + K_d \frac{d}{dt} e(t)$$

By suitable tuning the controller parameters in the PID techniques, the controller parameters can offer a very effective control action planned for specific process necessities. The proportional word examines suitable proportional regulate for error (which is the variation between the set position and response) to the control output. The integral word incorporates the measured variable above time, and it adjusts the response by dipping the offset from process variable. Derivative word examines the speed of the measured variable and consequently adjusts the output when there are abnormal changes. There searchers alter every constraint of the control parameters functions to attain the required concert from the system. Table 1 illustrates the outcome of growing parameter separately [19].

PID controllers are usually in simple execution and simple structure. Hence, PID controllers are mostly employed in the process and manufacturing industries,

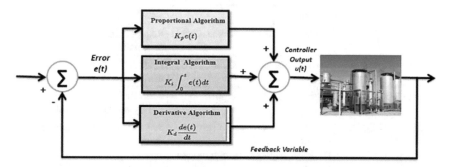

Fig. 1 PID controller

Table 1 Outcome of rising PID's parameter separately

Parameter	Over-shoot	Steady-state error	Rise time	Settling time	Stability
K_p	Enlarge	Reduce	Reduce	Slight Vary	Debase
K_i	Enlarge	Eradicate	Reduce	Enlarge	Debase
K_d	Reduce	No outcome	Slight Vary	Reduce	Get better if K_d small

robotics, power electronics, and biomedical engineering. PID controllers send excellent concert through price/profit relation which is complicated for other controllers. They are also widespread in contemporary uses, similar to auto-driving cars, unmanned airliner means of transportation, and self-governing robots for the alike reason [20]. In the majority of the control request, 90 to 95% of the control fields are of PID controller form. The broad utilization of PID controller has encouraged the continuous research and progress to "acquire the greatest output of PID," and "the investigation is on to discover the subsequent key tools or method for PID tuning."

2.2 PID Tuning

Conventional Tuning: There are two ways to analyze PID controller: frequency domain analysis and time domain analysis [21]. In PID tuning techniques, most of the PID tuning techniques are not accurately tuned, some of them work in manual and rest of all closed-loop tunings use default hard settings [22]. For that reason, tuning of PID controller is an important task in education as well as in industrial system. Table 2 [23] shows some advantages and disadvantages of tuning methods.

More than thousands of the literature reviews are available related to PID controller tuning technique and regulations. First PID controller tuning rule was presented by Ziegler and Nichols [7]. Model-based tuning techniques are also available for

Table 2 Comparison between conventional and AI techniques

Techniques	Benefits	Drawbacks
Manual tuning	Mathematics is not essential, Online	Required knowledgeable personality
Ziegler-Nichols	Verified technique, Online	Process disturbance, hit and trial method, extremely hostile tuning
Cohen-Coon	High-quality Process replica	Some mathematics; Offline; merely good for first-order system
Artificial intelligence	Reliable tuning; Online otherwise Offline-can utilize computer-automated power system design method	Some expenditure or guidance required

decreasing remarkable concert criteria which are based on an integrated error principle, achieving some toughness specification. Additional tuning techniques offer a transaction among toughness and performance, or between adjustment and servo modes. Another technique is based on the frequency response [24]. These methods employ toughness situation in the frequency domain, such as maximum sensitivity, phase margin, and gain margin.

Artificial Intelligence Tuning: Classical techniques, for example, Ziegler-Nichols rule do not offer best PID tuning parameters and generally outcomes in closed-loop response described by swinging and a big over shoot. Artificial intelligence approaches involve nonstop PID tuning, based on the dynamic design to get the finest outcome by the choice of the performance key [12]. Nowadays, a lot of AI techniques have been planned to automatic tune the controller parameters, for example, neural networks (NNs), particle swarm optimization (PSO) algorithms [25], genetic algorithms [10], the bat algorithm [26], fuzzy logic [13], the whale optimization algorithm (WOA) [27], ant colony optimization (ACO) [28], etc. There are so many novel AI algorithms are proposed for getting superior result in process control system.

3   Process Control System (PCS)

Process control systems (PCS), occasionally called industrial control systems (ICS), function as pieces of apparatus along the manufacturing line during industrial work that analyze the process in a diversity of ways and return data for observing and trouble shooting. A process control system comprises a data set and sharing structure to collect and store information from a variety of information set, while for information gaining or generating PCS use own accurate method.

To guesstimate the vital variables of PCS, three idealized models are used.

- Batch Reactor Model (BR-M)
- Continuous Stirred Tank Reactor Model (CSTR-M)
- Plug Flow Reactor Model (PFR-M).

PCS's response is either sluggish or oscillatory, so, these problems can be overcome by using AI-based PID controller.

3.1   BR-M

BR lies in the easy reactor's category. The reaction should proceed within time when BR is overloaded with material. BR controlling parameters are volume, pressure, and temperature. BR never reaches in steady-state condition. There are input–output ports as shown in Fig. 2, for sensors and materials and are usually employed in low-scale

Fig. 2 BR

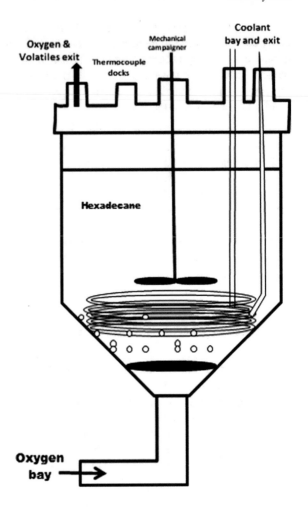

manufacturing and reaction with biological resources, such as in pulp fermentation and enzyme making. A pressure reactor (PR) is the one example of BR [29].

1. It is quite complicated in BR to control temperature due to of manufacturing necessities. Temperature controlling problem of BR can be overcome by using a robust PID control technique. Generally, the cascade arrangement of PID control technique is used inside the industrial applications for temperature controlling of the reactor [21]. A split-range controller is one of the favorites in the cascade control's slave loop to work simultaneously with diverse effects of 2 actuators, in industrial control engineering. In the case of two modes of action, proportional splitting is used. High control performance of jacketed batch reactors can be getting by using Split-range algorithm with the combination of classical PID-based cascade temperature controller. For getting better performance, this algorithm can also be used with model-based temperature controllers. The

cascade structure of PID control with Split-range algorithm becomes simpler in design and can be put into action in presented systems with no somber changes. The steady-state temperature of jacket inlet system can also be controlled by using PID control technique [30].

2. There are so many AI-based PID controllers with many advantages used in different industrial plants effectively. For getting outstanding control performance of batch reactor, the parameters of PID controller must be correctly tuned [31].

Batch process is a time-varying system due to different production conditions for each cycle, so it is hard to obtain precise mathematical model for batch reactor. The problem of batch reactor can be overcome by using PID controller, but occasionally, it may be unsuccessful. This problem can also be overcome by using proper tuning algorithm in PID controller to get the production requirements during ecological variations.

A dynamic matrix control (DMC)-based PID controller [32] is presented in a coke furnace for controlling the residual oil outlet temperature. BR can also be controlled by using a novel control approach, i.e., is DMC optimizing PID control technique (DOP). The parameters of closed-loop system such as stability and speed can be controlled with the help of this novel (i.e., is DOP) technique. DOP has the trouble-free structure as classical PID controller and high-quality power concert as DMC which plays a very important role in practical batch reactor system. In [33], researchers show that predictive functional control (PFC)-PID controller provides a better result in batch reactor temperature controlling as well as in a number of multifaceted plants with uncertainties, large time delay, and time-varying parameters. PID control techniques, fuzzy control approach, and online attuned PID constraint are combined in [34] and give a relationship between conventional PID control technique, PID fuzzy control approach, and piecewise fuzzy PID control, also presented a Smith fuzzy PID control technique for chemical industries.

In [35], researcher studied a neural network online learning-based adaptive control strategy for super-heated steam temperature controlling system, which is an equalizer of PID control, and a new adaptive PID algorithm with the combination of expert system and neural networks for achieving the necessary temperature performance index for unlike operating modes. The researcher used expert generalized PID (GPID) to design the primary weighting standards setter, with the help of traditional value for the assurance of little temperature over-shoot and system security.

3.2   CSTR-M

This reactor lies in the category of unremitting (continuous) reactor due to of its continuous input–output material flow. Continuous flow stirred tank reactor, mixed-flow reactor or ideal mixers are the other name of CSTR. Reactor's concentration at any point is assumed to be same as outlet concentration in the same reactor [29] as shown in Fig. 3.

Fig. 3 CSTR

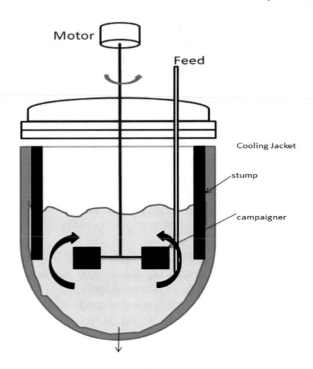

Generally, the continuous ideally stirred tank reactor (CISTR) is used to model the behavior of a CSTR. To simplify engineering calculations, the CISTR model is generally preferred which is also used to depict research reactors. For continuous hydrogen production, CSTR regularly gives better response [36]. Availability of hydrogen-producing bacteria is more in CSTRs as compare to batch reactor systems and is better settled in diverse liquor, and hence, mass transfer resistance becomes less. High levels of biomass inventory cannot be preserved by using CSTR due to of its intrinsic structure and operating pattern. This problem of CSTR can be controlled by using intelligent controller such as PID controller, neural network, fuzzy controller, optimal controller, internal control principle every controller has several pros and cons. Still that classical PID controller plays an important role in CSTR controlling because of its ease, stability, and low price.

The PID controllers are greatly utilized to improve the performance of various processes in industries. By fine tuning of PID parameters, maximum control can be achieved in process control system. internal model control (IMC) lies in close loop as well as in open-loop system. The fundamental plan behind internal model method is to calculate the values of the controlling parameter and to put it's to a prearranged result which is prepared for prearranged (internal) model. IMC strategy uses model-based controlling strategy for PID alteration [37].

Conventional techniques like Z-N (Ziegler-Nichols) produce slow response for nonlinear type of systems generally they generate large over-shoot in such system, for that reason AI-based techniques like IMC and particle swarm optimization (PSO) are in use to enhance the ability of conventional approach [38]. The limits of classical PID techniques can also be overcome by using AI-based auto-tuning methods and their hybrids. The problem of optimal PID controller can be enhanced by using PSO method [25]. Numerous PID tuning methods have been proposed in current years for nonlinear system [39]. Realization of AI techniques can be used to improve PID controller's gain. Swarm intelligence (SI) and evolutionary algorithm (EI) lie under the category of population-based AI algorithm [40]. The SI works on the principle of the joint performance of self-organized systems which is a narrative AI method. The focal point of SI-based method is artificial bee colony (ABC) which is best for adaptive controller design. The required value of PID control parameter like K_p, K_i, and K_d can be obtained by using this new population-based approach.

The Massachusetts Institute of Technology (MIT) law is best to design model reference adaptive control (MRAC)-based controller. The unidentified parameter variation and the ecological changes can be controlled very well by using this adaptive controller than the renowned and permanent gain PID controllers [41]. For CSTR temperature controlling [42], a novel controller is invented for steady-state as well as for transient response and outcomes is achieved in stipulations of time-response condition of CSTR-M and shows a comparison between MIT, ABC-MIT, traditional PID, and ABC PID. High value of performance can be observed by using ABC-MIT-PID controller. ABC Algorithm can be used to achieve the excellent regulations of PID parameter. ABC-MIT plays an important role in CSTR's temperature controlling than other PID controllers.

By using appropriate PID tuning techniques, system response can be enhanced, and error can be minimized to zero value. Environmental changes and unidentified parameter variant can be well controlled by the adaptive controller in comparison with the basic and flat gain PID controller.

3.3   Plug Flow reactor model (PFR-M)

Piston flow or perfect flow reactor is name of Plug flow reactor. Flows of input/output material are continuous and turbulent in case of Plug flow reactor and move during the reactor as a "plug" [29], as shown in Fig. 4.

The writer of [43] deals with the trouble of gain schedule control plan which guaranteed the stability of closed loop and ensured outlay for every alteration of gain schedule control. The Lyapunov theory of stability was used for the planned course of action. For condensing boiler's temperature controlling, a gain-scheduled PI controller and a lusty PID control technique were presented in [44]. According to researcher, system's close-loop performance can be improved, with the use of gain-scheduled policy, while the controller's gain is polynomial functions of water flow. This strategy was experimentally demonstrated in magnetic-levitation-system (MLS)

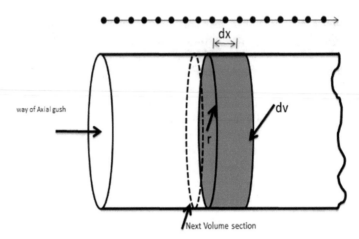

Fig. 4 PFR

which defeats the problem of state-feedback controller and classical controller. The tubular reactors generally used PID controller to become stable and get appropriate positions of transducer and actuator to guaranteed outcome [45]. The designs of PID controllers also depend on native linear models so that various operations can be controlled by input–output linearization [46].

4 Conclusion

PID is generally very simple and relevant control technique basically for industrial applications. In case of batch reactor, the reaction should proceed within time when materials are loaded into it; due to of CSTR's intrinsic structure and operating pattern, high levels of biomass inventory cannot be preserved in it; Plug flow reactor has continuous flow of input/output material into it. Therefore, all these problems require proper control of the reactor. The performance of various reactors can be enhanced by using intelligent PID controlling techniques. PID control plays a very vital role in process control system such as in reactor's temperature, Input–Output flow of the material, system's stability.

References

1. Franklin GF, Powell JD, Emami-Naeini A (2010) Feedback control of dynamic systems. 6th edn, pp 19–27
2. Kang C-G (2016) Origin of stability analysis: "on governors" [historical perspectives]. IEEE Control Syst Mag 36(5):77–88
3. Åström KJ, Kumar PR (2014) Control: a perspective. Automatica 50(1):3–43

4. Medaglia JD (2019) Clarifying cognitive control and the controllable connectome. Wiley Interdiscip Rev Cognit Sci 10(1):1471
5. Bennett S (2000) The past of PID controllers. Ann Rev Control 25:43–53
6. Bennett S (1993) Development of the PID controller. IEEE Control Syst Mag 13(6):58–62
7. Ziegler JG, Nichols NB (1942) Optimum settings for automatic controllers. Trans ASME 64:759–768
8. Gawthrop P (1986) Self-tuning PID controllers: algorithms and implementation. IEEE Trans Autom Contr 31(3):201–209
9. Zhuang M, Atherton DP (1993) Automatic tuning of optimum PID controllers. In IEEE Proceedings D (Control Theory and Applications), vol 140. IET, pp 216–224
10. Porter B, Jones AH (1992) Genetic tuning of digital PID controllers. Electr Lett 28(9):843–844
11. Kristiansson B, Lennartson B (2002) Robust and optimal tuning of PI and PID controllers. IEEE Proc Contr Theory Appl 149(1):17–25
12. Chan YF, Moallem M, Wang W (2007) Design and implementation of modular FPGA-based PID controllers. IEEE Trans Ind Electro 54(4):1898–1906
13. Tzafestas S, Papanikolopoulos NP (1990) Incremental fuzzy expert PID control. IEEE Trans Ind Electr 37(5):365–371
14. Verma B, Padhy PK (2018) Optimal PID controller design with adjustable maximum sensitivity. IET Contr Theory Appl 12(8):1156–1165
15. Kaya Y, Yamamura S (1962) A self-adaptive system with a variable-parameter PID controller. Trans Am Instit Electr Eng Part II Appl Ind 80(6):378–386
16. Ranjbaran K, Tabatabaei M (2018) Fractional order [PI], [PD] and [PI] [PD] controller design using Bode's integrals. Int J Dyn Contr 6(1):200–212
17. Srivastava S, Pandit VS (2016) A PI/PID controller for time delay systems with desired closed loop time response and guaranteed gain and phase margins. J Process Control 37:70–77
18. Alcántara S, Vilanova R, Pedret C (2013) PID control in terms of robustness/performance and servo/regulator trade-offs: A unifying approach to balanced autotuning. J Process Control 23:527–542
19. Ang KH, Chong G, Li Y (2005) PID control system analysis, design, and technology. IEEE Trans Control Syst Technol 13(4):559–576
20. Díaz-Rodríguez ID, Han S, Bhattacharyya SP (2019) Analytical design of PID controllers. Springer, Berlin. ISBN:978-3-030-18227-4
21. Aguilar R, Poznyak A, Martínez-Guerra R, Maya-Yescas R (2002) Temperature control in catalytic cracking reactors via a robust PID controller. J Process Control 12(6):695–705
22. Van Overschee P, Moons C, Van Brempt W, Vanvuchelen P, De Moor B (2000) RaPID: the end of heuristic PID tuning. In: Proceedings of the PID'00: IFAC workshop on digital control, Terrasa, Spain, 4–7 Apr 2000, pp 687–692
23. Kumar A, Morya R, Vashishath M (2013) Performance comparison between various tuning strategies: ciancone, cohen coon & ziegler-nicholas tuning methods. Int J Comput Technol 5(1):60–68
24. Morilla FSD (2000) Methodologies for the tuning of PID controllers in the frequency domain. IFAC Proc Vol 147–152
25. Ye H-T, Li Z-Q (2015) PID neural network decoupling control based on hybrid particle swarm optimization and differential evolution. Int J Autom Comput pp 1–6. (Nov 2015)
26. Nor'Azlan, NA, Selamat NA, Yahya NM (2018) Multivariable PID controller design tuning using bat algorithm for activated sludge process. In: IOP Conference Series: Materials Science and Engineering, vol 342, p 12030
27. Mosaad AM, Attia MA, Abdelaziz AY (2019) Whale optimization algorithm to tune PID and PIDA controllers on AVR system. Ain Shams Eng J 10(4):755–767
28. Kaliannan J, Baskaran A, Dey N, Ashour SA (2016) 'Ant colony optimization algorithm based PID controller for LFC of single area power system with non-linearity and boiler dynamics.' World J Model Simul 12(1):3–14
29. Foutch GL, Johannes AH (2003) Reactors in process engineering. In: Encyclopedia of physical science and technology, 00654-2

30. Balaton MG, Nagy L, Szeifert F (2012) Model-based split-range algorithm for the temperature control of a batch reactor
31. Yamamoto T, Yamada T, Fujii T, Hosokawa H (2006) Design and implementation of a GPC-based auto-tuning PID controller. In: IEEE international conference on industrial technology, pp 1920–1924
32. Xiao J, Shao C, Zhu L, Song T (2015). Batch reactor temperature control based on DMC-optimization PID. In: 2015 IEEE international conference on cyber technology in automation, control, and intelligent systems (CYBER). IEEE, pp 371–376. (June 2015)
33. Ridong Z, Sheng Wu, Furong G (2014) Improved PI controller based on predictive functional control for liquid level regulation in a coke fractionation tower. J Process Control 139(15):70–75
34. Zhang WL, Jiang HP (2005) The design and simulation of fuzzy PID controller with parameter adjustment for polymerizing-kettle. J Qingdao Univ Sci Technol 26(2):132–135
35. Zhao X, Zheng W (2009) Generalized pid algorithm for batch reactor's temperature control. In: 2009 WRI global congress on intelligent systems, vol 1. IEEE, pp 426–430. (May 2009)
36. Show KY, Lee DJ (2013) Bioreactor and bioprocess design for biohydrogen production. In: Biohydrogen. Elsevier, pp 317–337
37. Wayne Bequette B (2003) Process control, modeling, design and simulation. Prentice-Hall of India Private Limited, India
38. Khanduja N, Bhushan B (2016) Intelligent control of CSTR using IMC-PID and PSO-PID controller. In: 2016 IEEE 1st international conference on power electronics, intelligent control and energy systems (ICPEICES). IEEE, pp 1–6. (July 2016)
39. Chen C-T, Dai C-S (2001) Robust controller design for a class of nonlinear uncertain chemical processes. J Process Control 11:469–482
40. Kashan MH, Nahavandi N, Kashan AH (2012) DisABC: a new artificial bee colony algorithm for binary optimization. Appl Soft Comput 12:342–352
41. Neogi B, Islam SS, Chakraborty P, Barui S, Das A (2018) Introducing MIT rule toward improvement of adaptive mechanical prosthetic arm control model. In: Progress in intelligent computing techniques: theory, practice, and applications. Springer, Singapore, pp 379–388
42. Goud H, Swarnkar P (2018) Signal synthesis model reference adaptive controller with artificial intelligent technique for a control of continuous stirred tank reactor. Int J Chem React Eng 17(2)
43. Veselý V, Ilka A (2013) Gain-scheduled PID controller design. J Process Control 23:1141–1148
44. de Oliveira V, Karimi A (2012) Robust and gain-scheduled PID controller design for condensing boilers by linear programming. In: IFAC proceedings, vol 45, pp 335–340
45. Aglar R, Poznyak A, Martínez Guerra R, Maya Yescas R (2002) Temperature control in catalytic cracking reactors via robust PID controller. J Process Control 12(6):695–705
46. Mikhalevich S, Rossi F, Manenti F, Baydali S (2015) Robust PI/PID controller design for the reliable control of plug flow reactor. Chem Eng Trans 43:1525–1530

Investigating Cancer Survivability of Juvenile Lymphoma Cancer Patients Using Hard Voting Ensemble Technique

Amit Pandey, Rabira Galeta, and Tucha Kedir

Abstract Lymphoma is a type of cancer occurring in the lymphatic system of body. The lymphatic system is part of the immune system of our body. According to an estimate, around 8,480 people, which included 4,690 men and 3,790 women, were diagnosed with Hodgkin lymphoma in the year 2020 in the USA. According to another estimate, about 970 deaths, which included 570 men and 400 women, took place in the year 2020 because of lymphoma. This study investigates a model, based on hard voting classifier for predicting the survivability of juvenile lymphomas cancer patients. Initially, we have applied multinomial naive Bayes, logistic regression and support vector machine classifiers for making independent predictions for next five years and next ten years survivability, respectively. Further, we have used one of the ensemble methods, the hard voting classifier, for making the final predictions. This study also compares the accuracy of various considered independent classifiers. For evaluating the validity of the classifiers, we have calculated the accuracy scores for each approach using the obtained confusion matrices. The hard voting classifier has predicted the five year survivability with 99.12% accuracy and the ten year survivability with the accuracy of 99.56%.

Keywords Lymphoma · Hodgkin lymphoma · Non-Hodgkin lymphoma · Cancer survivability · Juvenile cancer

1 Introduction

The lymphoma is a type of cancer occurring in the lymphatic system, which is an important part of germ fighting mechanism of our body. The lymphoma can be further classified in to two main categories Hodgkin lymphoma and Non-Hodgkin lymphoma[1–4]. According to an estimate, about 8,480 people were infected with

This research is funded by the Bule Hora University, Ethiopia. (Under the Ministry of Science and Higher Education, Ethiopia)

A. Pandey (✉) · R. Galeta · T. Kedir
College of Informatics, Bule Hora University, Bule Hora, Ethiopia
e-mail: amit.pandey@live.com

© The Author(s), under exclusive license to Springer Nature Singapore Pte Ltd. 2022
P. Nanda et al. (eds.), *Data Engineering for Smart Systems*, Lecture Notes in Networks and Systems 238, https://doi.org/10.1007/978-981-16-2641-8_63

Hodgkin lymphoma in the year 2020 in the USA and about 970 deaths were recorded in the same year because of it [5]. On the other hand, according to another estimate around 77,240 people, which included 42,380 men and 34,860 women, were diagnosed with non-Hodgkin lymphoma in the year 2020 in the USA and about 19,940 deaths, which included 11,460 men and 8,480 women, took place in the year 2020 because of non-Hodgkin lymphoma [6].

The Hodgkin lymphoma affects adults and children both. It is mainly found in two age groups. First one is group of people in their early adulthood or who are in their twenties, and second group consists of elderly people aged more than fifty five. Out of total cancer cases diagnosed in the age group of 15–19 years, Hodgkin lymphoma accounts for 13% of them alone. In the year 2020, about 800 new cases of this type were diagnosed in this age group [7, 8]. Further, about 1,050 children below the age of 20 years are diagnosed with the non-Hodgkin lymphoma every year in the USA, account for the 7 % of the total cancer cases diagnosed in this age group. Study shows that boys are more prone to get non-Hodgkin lymphoma when compared to girls [9].

This study investigates the survivability of juvenile lymphoma cancer patients in two cases, the five year survivability and ten year survivability after being diagnosed with lymphoma. Initially, we have applied multinomial naive Bayes, logistic regression and support vector machine classifiers (SVM) for making independent predictions. Then, we have used, one of the ensemble methods, the hard voting classifier for making the final predictions. Further, the confusion matrices were used to evaluate the accuracy score to make the comparison among the applied classifiers.

2   Related Work

Research on cancer is prevailing topic. Earlier many researchers have worked on prediction and survivability estimation of various cancers, such as lung cancer and breast cancer. In 2020, Abdar, M. et. al. proposed a nested ensemble machine learning model using the voting and stacking techniques for the automated diagnosis of breast cancer. The proposed model was a two-layered classifier model that identifies benign tumors from the malignant ones [10]. Also, Kate, R. J. et. al. in 2017 proposed a machine learning based model to predict breast cancer survivability in various stages of the disease. In this approach, basically the model was trained on stage specific data to improve the prediction accuracy [11]. Further in 2017, Lynch Chip, M. et. al. in his study applied support vector machine, linear regression and decision tree classifiers over the attributes like age, tumor size, tumor grade and gender for estimating the survivability of a lung cancer patient [12].

Shipp, M. A. et. al. in his study used gene expression profiling using machine learning techniques for predicting the diffuse large B-cell lymphoma. The study progresses with the identification of marker genes using the correlation analysis, followed by application of techniques like weighted K-nearest neighbors and support vector machine for making the predictions [13].

Kari, L. et.al. in 2003 used cDNA arrays for investigating the gene expression patterns for making the survivability predictions for the patients suffering from leukemic phase of cutaneous T cell lymphoma. The model used penalized discriminant analysis for the analysis [14].

In 2011, Alizadeh, A. A. et. al. in his study used Cox regression analysis for predicting the survivability of patients suffering from diffuse large B-cell lymphoma. The study investigated the gene expression patterns of two genes associated with tumor and microenvironment [15].

Further in 2019, Mettler, J. et. al. used the size of metabolic tumor for predicting the advanced stage Hodgkin lymphoma. The study analyzed the metabolic tumor volume using the PET scans. Then, Pearson product moment correlation coefficients were used to compare thresholds. Further, a baseline metabolic tumor volume was set for predictor response. The study used the Cox regression for analyzing the prognostic impact [16].

In addition to the above researches, the current study is focused on a particular age group of lymphoma patients. In this study, we have focused our investigation to evaluate the only the survivability of juvenile lymphoma patients.

3 Methodology

The obtained dataset used in the study is comprised of 13 attributes. Gender, number of malignant tumors, duration of survival and age at diagnosis are the major attributes contributing toward prediction. Data is initially treated with multinomial naive Bayes, logistic regression and support vector machine classifiers independently for making predictions. Further, hard voting classifier is used for making the final prediction.

3.1 Naive Bayes

The naïve Bayes classifier incorporates the Bayes theorem that is used to evaluate the probability of occurring event based on some previous knowledge of any associated condition. The Bayes theorem can be expressed as shown below in Eq. 1:

$$Pr(A/B) = \frac{Pr(B/A).Pr(A)}{Pr(B)} \tag{1}$$

Here,

$Pr(A/B)$ is probability of A on a given event B. It is also called posterior probability.

$Pr(B/A)$ is probability of B provided that the hypothesis A is True.

$Pr(A)$ is the probability of the hypothesis A to be True.

$Pr(B)$ is the probability of the event B.

3.2 Logistic Regression

Logistic regression is evaluated using the Sigmoid function, expressed as *Sgm(n)* in Eq. 2. The value of this function varies from 0 to 1. When $n \rightarrow +\infty$ the value of *Sgm(n)* becomes $+1$ and when $n \rightarrow -\infty$, the value of *Sgm(n)* becomes 0.

$$Sgm(n) = \frac{1}{(1 + e^{-n})} \qquad (2)$$

Further, as input to this sigmoid function we will use regression function *β(x)* to define a clear hyper-plane that will classify the given inputs into two classes. See Eq. 3.

$$h(x) = Sgm(\beta(x)) \qquad (3)$$

3.3 Support Vector Machine (SVM)

The SVM classifier figures out set of few data points as support vectors, which are capable of setting up a clear hyper-plane for classifying the given dataset into two classes. Initially a partition is set. Then width along the partitioning plane is maximized until any data point is reached. These points are the support vectors. The partitioning hyper-plane with maximum width will be considered as the final classifier. Mathematically, a linear SVM classifier can be expressed as mentioned in Eq. 4

$$W^T x + b = \left\{ {1 \atop -1} \right. \qquad (4)$$

3.4 Voting Classifier

Conceptually voting classifier is assembly of various independent machine learning classifiers. It uses the majority vote or the average predicted probabilities for predicting the classes. This approach overcomes the weakness of any individual contributing classifier in making the final more precise predictions. In this study, we have used the hard voting classifier that uses the majority voted class by the contributing classifiers as the final prediction (see Fig. 1).

Fig. 1 Hard voting classifier

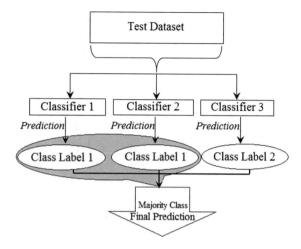

4 Results and Discussion

This section discusses the prediction performance results of multinomial naive Bayes, logistic regression, support vector machine and hard voting classifiers. For validation purpose the *accuracy scores, precision and recall* are computed, from the recorded confusion metrics, individually for each classifier. Further, these *precision, recall and accuracy scores* are compared to establish the prominent performance of hard voting classifier.

4.1 Results from Support Vector Machine Classifier with Gaussian Kernel

This section investigates the prediction performance of support vector machine classifier. We have used the nonlinear support vector machine with the Gaussian kernel for making the predictions. The experimental results obtained for the used dataset are shown in Table 1. The highest accuracy recorded for the support vector machine classifier with the Gaussian kernel for the obtained dataset is 98.23% in predicting the 5 years survivability and 99.56% in predicting the 10 years survivability.

Table 1 Results for support vector machine classifier with the Gaussian kernel

	5 years survivability (%)	10 years survivability (%)
Precision	99.18	100
Recall	97.58	97.73

Table 2 Results for multinomial naive Bayes classifier

	5 years survivability (%)	10 years survivability (%)
Precision	85.71	46.25
Recall	91.94	84.09

Table 3 Results for logistic regression classifier

	5 years survivability (%)	10 years survivability (%)
Precision	100	97.73
Recall	98.39	97.73

4.2 Results from Naive Bayes Classifier

This section investigates the prediction performance of naïve Bayes classifier. The experimental results obtained for the used dataset are shown in Table 2. The highest accuracy recorded for multinomial naive Bayes classifier for obtained dataset is 87.17% in predicting the 5 years survivability and 77.88% in predicting the 10 years survivability.

4.3 Results from Logistic Regression Classifier

In this section, we will discuss the prediction performance of logistic regression classifier. The acquired experimental results for the used dataset are shown in Table 3. The highest accuracy registered for the logistic regression classifier for the obtained dataset is 99.12% in predicting the 5 years survivability and 99.12% in predicting the 10 years survivability.

4.4 Results from Hard Voting Classifier

This section discusses about the prediction performance of the hard voting classifier. Here, we have used an ensemble technique for making the final predictions. The acquired experimental outcomes for the obtained dataset are shown in Table 4.

Table 4 Results for hard voting classifier

	5 years survivability (%)	10 years survivability (%)
Precision	100	100
Recall	98.39	97.73

Table 5 Comparing the various accuracy scores

	SVM (%)	Naïve Bayes (%)	Logistic Regression (%)	Hard Voting (%)
5 Years Survivability	98.23	87.17	99.12	99.12
10 Years Survivability	99.56	77.88	99.12	99.56

The highest accuracy registered for the hard voting classifier for the used dataset is 99.12% in predicting the 5 years survivability and 99.56% in predicting the 10 years survivability.

4.5 Comparing the Accuracy Scores of Various Used Classifiers

This section compares the accuracy scores of various independent classifiers with that of hard voting classifier. As shown in Table 5, the hard voting classifier chooses the majority class predicted by neglecting the weak predictions for making a more precise final prediction. It is deducible from Table 5 that hard voting classifier picks the best predictions from the outcome of the contributing classifiers and use them to make the final predictions with highest accuracy.

5 Conclusion

The study performed investigates the survivability of juveniles lymphoma patients. The precision, recall and accuracy scores for the hard voting classifier for 5 year survivability are 100, 98.39 and 99.12%, respectively. Further, the precision, recall and accuracy scores for the hard voting classifier for 10 year survivability are 100, 97.73 and 99.56%, respectively. It is clearly deducible from the outcomes of the current study that among the employed classifiers the hard voting classifier has recorded the highest accuracy in making the survivability prediction for the juvenile lymphoma patients.

References

1. Lymphoma. https://en.wikipedia.org/wiki/Lymphoma. Accessed 01 Jan 2021
2. What Is Lymphoma? https://www.webmd.com/cancer/lymphoma/lymphoma-cancer#1. Accessed 01 Jan 2021
3. What to know about lymphoma. https://www.medicalnewstoday.com/articles/146136. Accessed 01 Jan 2021

4. Lymphoma. https://www.mayoclinic.org/diseases-conditions/lymphoma/symptoms-causes/syc-20352638. Accessed 01 Jan 2021

5. Lymphoma-Hodgkin: Statistics. https://www.cancer.net/cancer-types/lymphoma-hodgkin. Accessed 01 Jan 2021

6. Lymphoma-Non-Hodgkin: Statistics. https://www.cancer.net/cancer-types/lymphoma-non-hodgkin/statistics. Accessed 01 Jan 2021

7. Lymphoma-Hodgkin-Childhood: Statistics. https://www.cancer.net/cancer-types/lymphoma-hodgkin-childhood/statistics. Accessed 01 Jan 2021

8. Childhood Hodgkin Lymphoma Disease–Stages and Prognosis. https://www.acco.org/blog/childhood-hodgkin-lymphoma-disease-stages-and-prognosis/. Accessed 01 Jan 2021

9. Lymphoma-Non-Hodgkin-Childhood: Statistics. https://www.cancer.net/cancer-types/lymphoma-non-hodgkin-childhood/statistics. Accessed 01 Jan 2021

10. Abdar M, Zomorodi-Moghadam M, Zhou X, Gururajan R, Tao X, Barua PD, Gururajan R (2020) A new nested ensemble technique for automated diagnosis of breast cancer. Pattern Recogn Lett 132:123–131

11. Kate RJ, Nadig R (2017) Stage-specific predictive models for breast cancer survivability. Int J Med Inform 97:304–311

12. Lynch CM, Behnaz A, Fuqua JD, De Carlo Alexandra R, Bartholomai JA, Balgemann RN, van Berkel VH, Frieboes HB (2017) Prediction of lung cancer patient survival via supervised machine learning classification techniques. Int J Med Inform 108:1–8

13. Shipp MA, Ross KN, Tamayo P, Weng AP, Kutok JL, Aguiar RC, Ray TS (2002) Diffuse large B-cell lymphoma outcome prediction by gene-expression profiling and supervised machine learning. Nat Med 8(1):68–74

14. Kari L, Loboda A, Nebozhyn M, Rook AH, Vonderheid EC, Nichols C, Wysocka M (2003) Classification and prediction of survival in patients with the leukemic phase of cutaneous T cell lymphoma. J Exp Med 197(11):1477–1488

15. Alizadeh AA, Gentles AJ, Alencar AJ, Liu CL, Kohrt HE, Houot R, Gascoyne RD (2011) Prediction of survival in diffuse large B-cell lymphoma based on the expression of 2 genes reflecting tumor and microenvironment. Blood J Am Soc Hematol 118(5):1350–1358

16. Mettler J, Müller H, Voltin CA, Baues C, Klaeser B, Moccia A, Dietlein M (2019) Metabolic tumor volume for response prediction in advanced-stage hodgkin lymphoma. J Nucl Med 60(2):207–211

Author Index

A
Abdulhussain, Wadhah A., 431
Abid, Ahsin, 531
Abu Ishak Mahy, Md., 543
Agarwal, Garima, 333
Agarwal, Mahek, 361
Agarwal, Pinky, 77
Aggarwal, Aditya, 205
Ahmmed, Asif, 557
Aishwarya, B. S., 521
Amulya, R., 521
Anh, Ho Nhat, 497
Ankur, Adarsh, 353
Aziz, Azizi Ab, 431

B
Bairwa, Amit Kumar, 183
Barekar, Praful, 109
Basak, Joyanta, 543
Basu, Joyanta, 511
Basu, Tapan Kumar, 511
Bevish Jinila, Y., 605
Bhagit, Darshana, 37
Bhardwaj, Sumit, 353
Bhatt, Devershi Pallavi, 593
Biswas, Al Amin, 531
Bohra, Manoj K., 615

C
Chakraborty, Pinaki, 205
Chakraverty, Shampa, 205
Chamoli, Tanishq, 271
Chaurasia, Sandeep, 341
Cherukuru, Rajkishan, 377

Christy, A., 605

D
Dahiya, Parv, 255, 263
Damahe, Lalit, 109
Dhaka, Ramgopal, 57
Dihingia, Leena, 313
Diwe, Saurabh, 109
Doshi, Adit, 123
Dubey, Vandana, 659

G
Galeta, Rabira, 671
Garg, Sahil, 255, 263
Garg, Sonam, 271
Ghosh, Aninda, 27
Goud, Harsh, 659
Goyal, Palash, 1
Goyal, Somya, 479
Gupta, Ankit, 271
Gupta, Gaurav, 205
Gupta, Nirmal Kumar, 323, 651
Gupta, Punit, 353
Gupta, Rohit Kumar, 401
Gupta, Shivani, 299

H
Hasan, Iqbal, 191
Hasibul Islam Shad, Sk., 557

I
Imdadul Islam, M., 531

J

Jain, Anamika, 57
Jain, Ashish, 323, 615, 651
Jain, Kusum Lata, 299
Jain, Tarun, 361, 377
Jangid, Divya, 581
Jangid, Mahesh, 245
Joshi, Sandeep, 183
Joy, Deborah T., 225
Jueal Mia, Md., 543, 557

K

Kakde, Sandeep, 109
Kamble, Shailesh, 109
Kanika, 205
Kant, Moksh, 341
Kapadia, Muskan, 123
Kapoor, Rohan, 401
Kasliwal, Yash, 461
Katarya, Rahul, 461
Kaur, Mandeep, 71
Kavya, A. P., 521
Kavyashree, B., 521
Kedir, Tucha, 469, 671
Khoa, Bui Thanh, 147, 497
Kirthiga, M., 569
Kothambawala, Taikhum, 461
Kotriwal, Sparsh, 489
Kulkarni, Soummya, 37
Kumar, Ashish, 361, 377
Kumar, Lalit, 1
Kumar, Rajesh, 235, 469
Kumar, Rishav, 1
Kumar, Saurabh, 411
Kumar, Shishir, 13
Kumar, Sushil, 173

L

Londhe, Namrata, 621
Ly, Nguyen Minh, 497

M

Mache, Sidhharth, 423
Mache, Suhas, 423
Madan, Manan, 205
Maiyani, Karina, 639
Majumder, Swanirbhar, 511
Malik, Ashish, 391
Malik, Karan, 1
Malini, A., 45, 89
Mane, Deepak, 621

Mane, D. T., 639
Mathur, Pratistha, 77
Mazumder, Sourov, 543, 557
Mishra, Shekhar, 401
Modak, Masooda, 37
Mohammed, Abdella K., 235
Mukherjee, Saurabh, 581
Mundra, Ankit, 401

N

Nagpal, Kabir, 245
Nanavati, Nirali, 123
Nand, Parma, 71
Nawal, Meenakshi, 299
Niharika, S., 521
Nimbhore, Sunil, 423
Nirmal Raja, K. L., 569

P

Pachauri, Bhoopendra, 57
Pandey, Amit, 235, 469, 671
Pandit, Hardik B., 159
Parveen, 173
Patel, Dipjayaben, 133
Patel, Helly, 123
Patel, Mehul, 639
Patel, Rikin, 123
Patil, Divya, 639
Patil, Namita, 621
Patil, Omkar, 621
Paul, Tuhin Utsab, 27
Prasad Sharma, Devi, 489
Prathibha, Soma, 569
Prayla Shyry, S., 605

R

Raguru, Jaya Krishna, 489
Rakesh, Nitin, 71
Rishi, Priyanka, 411
Ritu, Nadia Afrin, 531
Rizvi, Sam, 191

S

Sanjith, Sivapuram Sai, 377
Sarmah, Priyankoo, 313
Satasiya, Brijesh, 123
Sattar, Abdus, 451
Sawant, Madhavi, 639
Saxena, Siddhartha, 489
Shah, Priyanka, 159
Shankar, Venkatesh G., 615

Sharma, Ankush, 411
Sharma, Chitra, 333
Sharma, Harish, 341
Sharma, Manoj Kumar, 391, 615
Sharma, Prakash Chandra, 323, 651, 659
Sharma, Preeti, 593
Sharma, Utkarsh, 13
Shrestha, Rashmee, 71
Shrivastav, Dhruv, 1
Shyamkumar, M., 569
Singh, Lovedeep, 101
Singh, Neha, 443
Singh, Rishipal, 173
Sinha, Shweta, 291
Sinwar, Deepak, 235, 469
Sohel, Salowa Binte, 543
Sridevi, P. V., 281
Srivastava, Utpal, 225
Sultan, Tipu, 557
Suneetha, Regidi, 281

T
Tariqul Islam, Md., 213

Thada, Vikas, 225
Thakur, Shreyas, 461
Tiwari, Pradeep Kumar, 333
Tomar, Bhuvidha Singh, 353
Truong, Nguyen Xuan, 497
Tusher, Abdur Nur, 213

U
Usrika, Sumaya Akter, 451

V
Varkhedi, Mahima Manohar, 521
Verma, Shriya, 271
Verma, Vivek Kumar, 361, 377, 401
Vidhate, Prashant, 621
Virmani, Deepali, 443
Vishwakarma, Santosh Kumar, 323, 651

Y
Yadav, Anju, 77, 361, 377
Yugakiruthika, A. B., 45, 89

Printed in the United States
by Baker & Taylor Publisher Services